T0323650

Logic, Language, and Mathematics

Logic, Language, and Mathematics

Themes from the Philosophy of Crispin Wright

EDITED BY
Alexander Miller

OXFORD
UNIVERSITY PRESS

OXFORD
UNIVERSITY PRESS

Great Clarendon Street, Oxford, OX2 6DP,
United Kingdom

Oxford University Press is a department of the University of Oxford.
It furthers the University's objective of excellence in research, scholarship,
and education by publishing worldwide. Oxford is a registered trade mark of
Oxford University Press in the UK and in certain other countries

Published in the United States of America by Oxford University Press
198 Madison Avenue, New York, NY 10016, United States of America

British Library Cataloguing in Publication Data
Data available

Library of Congress Control Number: 2019951166

ISBN 978–0–19–927834–3

Printed and bound in Great Britain by
Clays Ltd, Elcograf S.p.A.

To the memory of William Demopoulos and Bob Hale

Contents

Replies by Crispin Wright

Preface

This is the much anticipated companion to *Mind, Meaning and Knowledge: Themes from the Philosophy of Crispin Wright*, edited by Annalisa Coliva, and published by Oxford University Press in 2012. It consists of nine specially commissioned chapters, together with a previously unpublished paper by George Boolos (edited by Riki Heck) and a set of replies by Crispin Wright. As in the 2012 volume, Wright has eschewed the standard approach of writing a brief response to each individual chapter, and has instead written four substantial chapters offering a thematic overview of the relevant topic while incorporating responses to the specific contributions in the corresponding four sections of the book. It's no exaggeration to say that the four sets of replies (amounting to over 80,000 words!), themselves collectively constitute the equivalent of a small monograph in their own right. The breadth and depth they exhibit is as clear an indication one could have of the breadth and depth of Wright's vast contribution to analytic philosophy in a career spanning over half a century and still very much going strong.

It is a testament to the influence Wright's work has had in twentieth century and contemporary analytic philosophy that three of the sections elaborate on positions first broached in relatively early publications that have become essential points of reference for anyone seriously approaching the relevant topics. The five papers in Part I deal with issues relating to the neo-Fregean platonism and logicism explored in Wright's seminal 1983 book *Frege's Conception of Numbers as Objects*; Part II deals with vagueness and the Sorites paradox, the subject of Wright's influential papers "On the Coherence of Vague Predicates" (1975) and "Language Mastery and the Sorites Paradox" (1976); while Part III looks in depth at issues relating to the fall out for the philosophy of logic of the Dummettian agenda that was the focus of Wright's *Realism, Meaning and Truth* (first edition 1986, second edition 1993). Part IV consists of a discussion by Wright's long-time collaborator and friend, Bob Hale, of Wright's critique of the anti-physicalist argument to be found in the concluding pages of Saul Kripke's *Naming and Necessity*. This critique appears in a hitherto insufficiently celebrated 2002 paper by Wright, "The Conceivability of Naturalism," and it is to be hoped that the exchange between Hale and Wright in the present volume brings Wright's 2002 paper to the attention of the wider audience that it deserves.

I'm grateful to the University of Otago for financial assistance in preparing the volume for publication, and to Hannah Clark-Younger for help with converting LaTeX files into Word. Thanks, too, to Joy Mellor for copy-editing the manuscript, to Finn Butler for compiling the Index, and to the production team led by Ganesan Kayalvizhi. I'm very grateful indeed to all of the contributors to the volume, not only for their excellent chapters, but for the patience and good humour they have exhibited in the long process of bringing the volume into print. In this regard, special thanks are due to Peter Momtchiloff, who has displayed the patience of a veritable saint since we started planning the volume some fifteen years ago. Most of all, though, I'm grateful to Crispin Wright for the vast effort he has put into the replies. All serious philosophers should be immensely thankful to him for the body of work he has contributed to our discipline over his career and for the generosity he has exhibited in enabling others to make their own contributions to the subject. It would be difficult to express such thanks in an adequate manner: hopefully this volume goes a little way in the right direction.

Alex Miller

Dunedin
27 September 2018

Bio-bibliographic Note

Crispin J. G. Wright, born in 1942 in Surrey, was educated at Birkenhead School and at Trinity College, Cambridge, graduating in Moral Sciences in 1964 and taking a PhD in 1968. He took an Oxford BPhil in 1969 and was elected Prize Fellow and then Research Fellow at All Souls College, Oxford, where he worked until 1978. He then moved to the University of St Andrews, where he became the first Bishop Wardlaw University Professor in 1997. From fall 2008, he has been Professor at New York University, and, since 2015, Professor of Philosophical Research at the University of Stirling and Regius Professor of Logic Emeritus at the University of Aberdeen. He has also taught at the University of Michigan, Oxford University, Columbia University, and Princeton University. He is a Fellow of the British Academy, of the Royal Society of Edinburgh, and of the American Academy of Arts and Sciences, and a Member of the Academia Europaea. He is the founder of Arché—Philosophical Research Centre for Logic, Language, Metaphysics and Epistemology, at St Andrews, which he left in September 2009 to take up leadership of the Northern Institute of Philosophy at the University of Aberdeen.

BOOKS

In preparation	*The Riddle of Vagueness* (Collected Papers Vol. 5), Oxford: Oxford University Press.
In preparation	*Imploding the Demon* (Collected Papers Vol. 4), Oxford: Oxford University Press.
In preparation	second edition of *Frege's Conception of Numbers as Objects*, original pagination but with a new extended postscript and bibliography, Oxford: Oxford University Press.
In preparation	*Expressivism and Self Knowledge*, co-authored with Dorit Bar-On, Oxford: Wiley Great Debates Series.
2003	*Saving the Differences: Essays on Themes from* Truth and Objectivity (Collected Papers Vol. 3), Cambridge, MA: Harvard University Press.
2001	With Bob Hale, *The Reason's Proper Study*, Oxford: Oxford University Press.
2001	German translation of *Truth and Objectivity* published as *Wahrheit und Objectivität*, Frankfurt: Suhrkamp.
2001	*Rails to Infinity: Essays on Themes from Wittgenstein's* Philosophical Investigations (Collected Papers Vol. 2), Cambridge, MA: Harvard University Press.
1992	*Truth and Objectivity*, Cambridge, MA: Harvard University Press.
1993	Expanded paperback second edition of *Realism: Meaning and Truth* (Collected Papers Vol. 1), Oxford: Blackwell.
1993	Limited edition re-issue of *Wittgenstein on the Foundations of Mathematics* in hardcover by Gregg Revivals.
1986	*Realism: Meaning and Truth* (Collected Papers Vol. 1), Oxford: Blackwell.
1983	*Frege's Conception of Numbers as Objects*, Scots Philosophical Monographs Series, Aberdeen: Aberdeen University Press.
1980	*Wittgenstein on the Foundations of Mathematics*, London: Duckworth.

EDITED VOLUMES

2017   Revised and expanded second edition of *A Companion to the Philosophy of Language* in 2 volumes, co-edited with Bob Hale and Alexander Miller, Oxford: Wiley-Blackwell.

2002   *Rule-Following and Meaning*, co-edited with Alexander Miller, Chesham: Acumen.

1998   *Knowing Our Own Minds*, co-edited with Cynthia MacDonald and Barry C. Smith, Oxford: Oxford University Press.

1997   *A Companion to the Philosophy of Language*, co-edited with Bob Hale, Oxford: Blackwell.

1993   *Reality, Representation and Projection*, co-edited with John Haldane, Oxford: Oxford University Press.

1988   *Mind, Psychoanalysis and Science*, co-edited with Peter Clark, Oxford: Blackwell.

1987   *Fact, Science and Morality: Essays on A. J. Ayer's* Language, Truth and Logic, co-edited with Graham MacDonald, Oxford: Blackwell.

1984   *Synthese* 58 (3), special number entitled *Essays on Wittgenstein's Later Philosophy*.

1984   *Philosophical Quarterly* 34 (136), special number on Frege, paperbacked as *Frege: Tradition and Influence*, Oxford: Blackwell.

TRANSLATION

2013   *Gottlob Frege: Basic Laws of Arithmetic*, with Philip Ebert and Marcus Rossberg (principal translators), Oxford: Oxford University Press.

FESTSCHRIFTEN

2012   Annalisa Coliva (ed.) *Mind, Meaning and Knowledge: Themes from the Philosophy of Crispin Wright*, Oxford: Oxford University Press (Wright's *Replies* are at pages 377–486).

2009   Jesper Kallestrup and Duncan Pritchard (eds.) *The Philosophy of Crispin Wright, Synthese* special number 171 (3).

ARTICLES AND CRITICAL STUDIES

General Epistemology

Forthcoming   With Giacomo Melis, "Oxonian Scepticism about the A Priori," in Dylan Dodd and Elia Zardini (eds.), *Beyond Sense? New Essays on the Significance, Grounds, and Extent of the A Priori*, Oxford: Oxford University Press.

2018   Reprint of "Facts and Certainty" with a new retrospect in G. Anthony Bruno and Abby Rutherford (eds.), *Skepticism: Historical and Contemporary Inquiries*, London: Routledge, 61–113.

2014   "On Epistemic Entitlement (II): Welfare State Epistemology," in Dylan Dodd and Elia Zardini (eds.), *Scepticism and Perceptual Justification*, Oxford: Oxford University Press, 213–47.

2012   "Warrant Transmission and Entitlement," Part IV of "Replies," in Annalisa Coliva (ed.), *Mind, Meaning and Knowledge: Themes from the Philosophy of Crispin Wright*, Oxford: Oxford University Press, 451–86.

2011   "McKinsey One More Time," in Anthony Hatzimoysis (ed.), *Self-Knowledge*, Oxford: Oxford University Press, 80–104.

2011   "Frictional Coherentism? A Comment on Chapter 10 of Ernest Sosa's
          Reflective Knowledge," *Philosophical Studies* 153 (1), 29–41.

2008   "Internal–External: Doxastic Norms and the Defusing of Sceptical Paradox,"
          in the *Journal of Philosophy* CV (9), a special number on Epistemic Norms,
          edited by John Collins and Christopher Peacocke, 501–17.

2008   Comment on John McDowell's "The Disjunctive Conception of Experience
          as Material for a Transcendental Argument," in A. Haddock and
          F. MacPherson (eds.), *Disjunctivism: Perception, Action, Knowledge,* Oxford:
          Oxford University Press, 390–404.

2007   "The Perils of Dogmatism," in Susana Nuccetelli and Gary Seay (eds.),
          Themes from G. E. Moore: New Essays in Epistemology and Ethics, Oxford:
          Oxford University Press, 25–48.

2004   "Hinge Propositions and the Serenity Prayer," in W. Loffler and
          P. Weingartner (eds.), *Knowledge and Belief,* Proceedings of the 26th
          International Wittgenstein Symposium, Vienna: Holder–Pickler–Tempsky,
          287–306.

2004   "Warrant for Nothing: On Epistemic Entitlement," *Aristotelian Society
          Supplementary Volume* LXXVIII, 167–212.

2004   "Scepticism, Certainty, Moore and Wittgenstein," in M. Kolbel and B. Weiss
          (eds.), *Wittgenstein's Lasting Significance,* London: Routledge.

2004   "Wittgensteinian Certainties," in D. McManus, ed. *Wittgenstein and
          Scepticism,* London: Routledge, 22–55.

2003   "Some Reflections on the Acquisition of Warrant by Inference," in
          S. Nuccetelli (ed.), *New Essays On Semantic Externalism and Self-Knowledge,*
          Cambridge, MA: MIT Bradford, 57–77.

2002   "(Anti)-Sceptics, Simple and Subtle: G.E. Moore and John McDowell," in
          Philosophy and Phenomenological Research LXV, 330–48.

2000   "Replies," in *Philosophical Issues* 10, 201–19.

2000   "Cogency and Question-Begging: Some Reflections on McKinsey's Paradox
          and Putnam's Proof," in *Philosophical Issues* 10, 140–63.

1998   "McDowell's Oscillation," contribution to a Book Symposium on John
          McDowell's *Mind and World* in *Philosophy and Phenomenological Research*
          LVIII (2), 395–402.

1996   "Human Nature?," critical study of John McDowell, *Mind and World,* in the
          European Journal of Philosophy 4, 235–54. Reprinted with a new postscript in
          Nicholas Smith (ed.), *Reading McDowell: On Mind and World,* London:
          Routledge, 2002, 140–73.

1994   Extended version of "On Putnam's Proof that we are not Brains-in-a-Vat," in
          P. Clark and R. Hale (eds.), *Reading Putnam,* Oxford: Blackwell, 216–41.

1991   "On Putnam's Proof that we are not Brains-in-a-Vat," in *Proceedings of the
          Aristotelian Society* XCII, 67–94.

1991   "Scepticism and Dreaming: Imploding the Demon," *Mind* C, 87–116.

1985   "Facts and Certainty," the Henriette Hertz Philosophical Lecture for the
          British Academy, December 1985, in *Proceedings of the British Academy*
          LXXI, 429–72. Reprinted in T. Baldwin and T. Smiley (eds.), *Studies in the
          Philosophy of Logic and Knowledge,* Oxford: Oxford University Press, 2004),
          51–94, and in Bruno and Rutherford (eds.) (see above.)

1984   "Comment on Lowe," *Analysis* 44, 183–5.

1983   "Keeping Track of Nozick," *Analysis* 43, 134–40.

Ethics

1995	"Truth in Ethics," *Ratio* (New Series) VIII, 209–26.
1988	"Moral Values, Projection and Secondary Qualities," Inaugural Address to the 1988 Joint Session of the Mind Association and Aristotelian Society, *Proceedings of the Aristotelian Society Supplementary Volume* LXII, 1–26.
1984	"The Moral Organism," in A. Manser and G. Stock (eds.), *The Philosophy of F.H. Bradley*, Oxford University Press, 77–97.

Relativism and Contextualism

Forthcoming	"Relativism and Representation: Comment on Daan Evers, 'Relativism and the Metaphysics of Value,'" in Daan Evers and Louise Hanson (eds.), *Meta-Aesthetics*, Oxford: Oxford University Press.
2017	"The Variability of 'Knows': An Opinionated Overview," in Jonathan Jenkins Ichikawa (ed.), *The Routledge Handbook of Epistemic Contextualism*, London: Routledge, 13–31.
2017	With Filippo Ferrari, "Talking with Vultures,"*Critical Study of Herman Cappelen and John Hawthorne: Relativism and Monadic Truth*, Oxford: Oxford University Press 2009; *Mind* 126 (503), 911–36.
2016	"Assessment-Sensitivity and the Manifestation Challenge" contribution to a Book Symposium on John MacFarlane *Assessment Sensitivity: Relative Truth and its Applications*, Oxford: Oxford University Press, 2014, *Philosophy and Phenomenological Research* XCII (1), 189–96.
2012	"Alethic Relativism and Faultless Disagreement," in Part III of "Replies," in Annalisa Coliva (ed.), *Mind, Meaning and Knowledge: Themes from the Philosophy of Crispin Wright*, Oxford: Oxford University Press, 435–50.
2009	With Sebastiano Moruzzi, "Trumping Assessments and the Aristotelian Future," in Berit Brogaard (ed.), *Synthese* 166 (2), 309–31.
2008	"Fear of Relativism,"*Philosophical Studies* 141 (3), 379–90, contribution to a book symposium on Paul Boghossian's *Fear of Knowledge*, Oxford: Oxford University Press, 2006.
2008	"Relativism about Truth Itself: Haphazard Thoughts about the Very Idea," in Manuel Garcia-Carpintero and Max Kolbel (eds.), *Relative Truth*, Oxford: Oxford University Press, 157–85.
2007	"New Age Relativism and Epistemic Possibility: The Question of Evidence," *Philosophical Issues*, volume on *The Metaphysics of Epistemology*, 17, 262–83.
2006	"Intuitionism, Realism, Relativism and Rhubarb," in P. Greenough and M. Lynch (eds.), *Truth and Realism*, Oxford: Oxford University Press, 38–60.
2002	"Relativism and Classical Logic," in *Logic, Language and Thought*, Anthony O'Hear (ed.), Cambridge: Cambridge University Press, 95–118.

Realism and Truth

2013	"A Plurality of Pluralisms," in Nikolaj Pedersen and Cory Wright (eds.), *Truth Pluralism: Current Debates*, Oxford: Oxford University Press, 123–53.
2012	"Meaning and Assertibility: Some Reflections on Paolo Casalegno's 'The Problem of Non-Conclusivenes'," in *Dialectica* 66 (2), special number on the Philosophy of Paolo Casalegno, E. Paganini (ed.), 249–66.
2012	Part III of "Replies," in Annalisa Coliva (ed.), *Mind, Meaning and Knowledge: Themes from the Philosophy of Crispin Wright*, Oxford: Oxford University Press, 418–35.

2010   With Bob Hale, "Assertibilist Truth and Objective Content: Still Inexplicit?," in Bernhard Weiss and Jeremy Wanderer (eds.), *Reading Brandom: Making It Explicit*, London: Routledge, 276–93.

2002   "What Could Anti-Realism about Ordinary Psychology Possibly Be?," in *The Philosophical Review* 111 (2) 205–33.

2001   "Minimalism, Deflationism, Pragmatism, Pluralism," in Michael P. Lynch (ed.), *The Nature of Truth: Classic and Contemporary Perspectives*, Cambridge, MA: MIT Press, 751–87.

2000   "Truth as Sort of Epistemic: Putnam's Peregrinations," *The Journal of Philosophy* XCVII (6), 335–64.

1998   "Truth: A Traditional Debate Reviewed," in supplementary volume 24 of the *Canadian Journal of Philosophy* on Pragmatism, guest edited by Cheryl Misak, 31–74; reprinted in *Truth*, ed. Simon Blackburn and Keith Simmons Oxford: Oxford University Press, 1999, 203–38; German version in Matthias Vogel and Lutz Wingert (eds.), *Unsere Welt gegeben oder gemacht? Menschliches Erkennen zwischen Entdeckung und Konstruktion*, Frankfurt/M., Suhrkamp, 1999.

1998   "Euthyphronism and the Physicality of Colour," in R. Casati and C. Tappolet, guest editors, *European Review of Philosophy*, vol. 3, special number on Response-Dependence, 15–30.

1998   "Comrades against Quietism: Reply to Simon Blackburn on *Truth and Objectivity*," *Mind* CVII, 183–203.

1997   With Bob Hale, "Putnam's Model-Theoretic Argument against Metaphysical Realism," in Bob Hale and Crispin Wright (eds.), *A Companion to the Philosophy of Language*, Oxford: Blackwell, 427–57.

1996   Book Précis and Response to Commentators (Horgan, Horwich, Pettit, Sainsbury, van Cleve, Williamson) for a Book Symposium on *Truth and Objectivity* in *Philosophy and Phenomenological Research* LVI (4), 863–8 (précis) and 911–41 (responses).

1995   "Can there be a Rationally Compelling Argument for Anti-realism about Ordinary ('Folk') Psychology?," in *Philosophical Issues* 6 (number containing the proceedings of the SOFIA conference on "Content" held at Lisbon in May 1994), 197–221.

1995   "Truth and Coherence," critical study of Ralph C. S. Walker, *The Coherence Theory of Truth*, in *Synthese* 103, 279–302.

1994   "Realism, Pure and Simple?": Reply to Timothy Williamson's Critical Study of *Truth and Objectivity*, in *International Journal of Philosophical Studies* 2, 147–61.

1994   "Response to Jackson": Reply to Frank Jackson's Critical Study of *Truth and Objectivity*, in *Philosophical Books* 35, 169–75.

1993   "Eliminative Materialism: Going Concern or Passing Fancy?," in *Mind and Language* (Forum on Eliminativism) 8, 316–26.

1993   "Scientific Realism and Observation Statements," in D. Bell and W. Vossenkuhl (eds.), *Science and Subjectivity: The Vienna Circle and Twentieth Century Philosophy*, Berlin: Akademie Verlag 1993, 21–46. Reprinted in the *International Journal of Philosophical Studies* 1, 231–54. Slightly shortened version reprinted in Italian as "Il Realismo Scientifico e gli Asserti Osservativi," in Alessando Pagnini (ed.), *Realismo/Antirealismo: Aspetti del Dibattito Epistemologico Contemporaneo*, Rome: La Nuova Italia, 1995, 205–35.

1993	"Anti-realism: The Contemporary Debate—Whither Now?," in J. Haldane, and C. Wright (eds.), Reality: *Representation and Projection*, Oxford: Oxford University Press, 63–84.
1993	"On an Argument Concerning Classical Negation," *Mind* CII, 123–31.
1992	"Wohin fuhrt die aktuelle Realismusdebatte?," in Wolfgang R. Kohler (ed.), *Realismus und Antirealismus*, suhrkamp taschenbuch wissenschaft 976, Berlin: Suhrkamp, 300–34.
1989	"The Verification Principle—Another Puncture, Another Patch," *Mind* XCVIII, 611–22.
1989	"Misunderstandings Made Manifest, a Response to Simon Blackburn," in Wettstein, Uehling, and French (eds.), *Contemporary Perspectives in the Philosophy of Language II, Midwest Studies in Philosophy*, vol. XIV, Paris: University of Notre Dame Press, 48–67.
1988	"Realism, Anti-Realism, Irrealism, Quasi-Realism," 1987 Gareth Evans Memorial Lecture, in Wettstein, Uehling, and French (eds.), *Realism and Anti-Realism, Midwest Studies in Philosophy*, vol. XII, University of Minnesota Press, 25–49. Reprinted in J. Kim and E. Sosa (eds.), *Metaphysics: An Anthology* Oxford: Blackwell 1999.
1986	"Scientific Realism, Observation and the Verification Principle," in G. MacDonald and C. Wright (eds.), *Fact, Science and Morality*, Oxford: Blackwell, 1986/US edition 1987, 247–74.
1985	Review of Simon Blackburn's *Spreading the Word*, in *Mind* XCIV, 310–19.
1984	"Second Thoughts about Criteria," *Synthese* 58, 383–405.
1982	"Anti-realist Semantics: The Role of *Criteria*," Royal Institute of Philosophy lecture, October 1978, in *Idealism: Past and Present*, Cambridge: Cambridge University Press, 225–48.
1980	"Realism, Truth-value Links, Other Minds and the Past," in J. Dancy (ed.), *Papers on Language and Logic*, Keele University Library, 192–226; reprinted in *Ratio*, XXII (December), 112–32.
1979	"Strawson on Anti-Realism," *Synthese* 40, 283–99.
1978	Review article on S. Guttenplan (ed.), *Mind and Language*, Oxford: Oxford University Press, 1975, in the *Journal of Literary Semantics*, VII (2), 99–107.
1976	"Truth-Conditions and Criteria," reply to Roger Scruton, *Proceedings of the Aristotelian Society Supplementary Volume*, 217–45.

Metasemantics

2012	With Bob Hale, "Horse Sense," in Robert May and Charles Parsons, eds, special number of the *Journal of Philosophy* 109, 85–131.
1998	"Why Frege Does Not Deserve His Grain of Salt: A Note on the Paradox of 'The Concept Horse' and the Ascription of *Bedeutungen* to Predicates," in J. Brandl and P. Sullivan (eds.), *Grazer Philosophische Studien* 55, *New Essays on the Philosophy of Michael Dummett*, Vienna: Rodopi, 239–63.
1997	"The Indeterminacy of Translation," in Bob Hale and Crispin Wright (eds.), *A Companion to the Philosophy of Language*, Oxford: Blackwell, 397–426.
1986	"Theories of Meaning and Speakers' Knowledge," in S. Shanker (ed.), *Philosophy in Britain Today*, London: Croom Helm, 267–307.
1986	"How Can the Theory of Meaning Be a Philosophical Project?," *Mind and Language* 1 (1), 31–44.

Philosophy of Logic and Modality

2019   "Logical Non-Cognitivism", *Philosophical Issues* 28, 425–50.

2018   "How High the Sky? Rumfitt on the (Putative) Indeterminacy of the Set-Theoretic Universe," contribution to a Book Symposium on Ian Rumfitt, *The Boundary Stones of Thought*, Oxford: Oxford University Press, 2015, *Philosophical Studies* 175, 2067–78.

2018   "Counter-Conceivability Again," in Ivette Fred and Jessica Leech (eds.), *Being Necessary: Themes of Ontology and Modality from the Work of Bob Hale*, Oxford: Oxford University Press, 266–82.

2015   With Bob Hale, "Bolzano's Definition of Analytic Propositions," *Grazer Philosophische Studien* 91 (1), 323–64.

2014   Comment on Paul Boghossian, "The Nature of Inference," *Philosophical Studies* 169, 1–11.

2012   "The Pain of Rejection, the Sweetness of Revenge," contribution to a book symposium on Mark Richard's *When Truth Gives Out*, Oxford: Oxford University Press, 2008, *Philosophical Studies* 160 (3), 465–76.

2007   "On Quantifying into Predicate Position: Steps towards a New(tralist) Perspective," in Michael Potter (ed.), *Mathematical Knowledge*, Oxford: Oxford University Press, 150–74.

2004   "Intuition, Entitlement and the Epistemology of Logical Laws," *Dialectica* 58, 155–75.

2002   "The Conceivability of Naturalism," in T. Gendler and J. O'Leary-Hawthorne (eds.), *Imagination, Conceivability and Possibility*, Oxford: The Clarendon Press, 401–39.

2001   "On Knowing What is Necessary: Three Limitations of Peacocke's Account," contribution to a Book Symposium on Christopher Peacocke's *Being Known*, in *Philosophy and Phenomenological Research* LXIV, 656–63.

2001   "On Basic Logical Knowledge," response to Paul Boghossian's "How Are Objective Epistemic Reasons Possible?," in *Philosophical Studies* 106, 41–85; reprinted in J. Bermudez and A. Millar (eds.), *Reason and Nature*, Oxford: The Clarendon Press, 49–84.

2000   With Bob Hale, "Implicit Definition and the A Priori," in Paul Boghossian and Christopher Peacocke (eds.), *New Essays on the A Priori*, Oxford: The Clarendon Press, 286–319.

1989   "Necessity, Caution and Scepticism," *Proceedings of the Aristotelian Society Supplementary Volume* LXIII, 203–38.

1986   "Inventing Logical Necessity," in J. Butterfield (ed.), *Language, Mind and Logic*, Cambridge: Cambridge University Press, 187–209.

1985   "In Defence of the Conventional Wisdom," in I. Hacking (ed.), *Exercises in Analysis*, essays for Casimir Lewy by his pupils, Cambridge: Cambridge University Press, 171–97.

1983   "What the Liar Really Says," critical study of James Cargile, *Paradoxes, A Study in Form and Predication*, Cambridge: Cambridge University Press 1979, in *British Journal for the Philosophy of Science* 34, 277–88.

1981   "Dummett and Revisionism," Critical Notice of M. Dummett's *Truth and Other Enigmas*, in *Philosophical Quarterly* 31, 47–67; reprinted in B. Taylor (ed.), *Michael Dummett, Contributions to Philosophy*, Leiden: Martinus Nijhoff, 1987.

Philosophy of Mathematics (i)—General Issues

2008   Contribution to *Philosophy of Mathematics: 5 Questions*, Vincent Hendriks
               and Hannes Leitgeb, eds., London: Automatic Press/VIP, 301–11.

2006   With Stewart Shapiro, "All Things Indefinitely Extensible," in A. Rayo and
               G. Uzquiano (eds.), *Absolute Generality*, Oxford: Oxford University Press,
               255–304.

1995   "Intuitionists Are Not (Turing) Machines," in *Philosophia Mathematica* 3
               (3), 86–102.

1994   "About 'The Philosophical Significance of Godel's Theorem': Some Issues,"
               in B. F. McGuinness and G. Oliveri (eds.), *The Philosophy of Michael
               Dummett*, Synthese Library vol. 239, Alphen aan den Rijn: Kluwer, 167–202.

1994   With Bob Hale, "A Reductio ad Surdum? Field on the Contingency of
               Mathematical Objects," *Mind* CIII, 169–84.

1992   With Bob Hale, "Nominalism and the Contingency of Abstract Objects,"
               Journal of Philosophy LXXXIX, 111–35.

1991   "Wittgenstein on Mathematical Proof," in A. Phillips Griffiths (ed.),
               Wittgenstein Centenary Essays, Royal Institute of Philosophy Supplement 28,
               Cambridge: Cambridge University Press, 79–99.

1988   "Why Numbers Can Believably Be: A Reply to Hartry Field," in Philip
               Kitcher (ed.), *Revue Internationale de Philosophie*, special number on
               Philosophy of Mathematics, 42, 425–73.

1985   "Skolem and the Sceptic," symposium with Paul Benacerraf, *Proceedings of
               the Aristotelian Society Supplementary Volume* LIX, 117–37.

1982   "Strict Finitism" *Synthese* 51, 203–82.

Philosophy of Mathematics (ii)—Frege and Abstractionism

2019   "How Did the Serpent of Inconsistency Enter Frege's Paradise?," in Philip
               Ebert and Marcus Rossberg (eds.), *Essays on Frege's Basic Laws of Arithmetic*,
               Oxford: Oxford University Press, 411–36.

2016   "An Entitlement to Hume's Principle?," in Philip Ebert and Marcus Rossberg
               (eds.), *Abstractionism*, Oxford: Oxford University Press, 154–78.

2016   "Whence the Paradox? Axiom V and Indefinite Extensibility," in Michael
               Frauchiger (ed.), *Truth, Meaning, Justification, and Reality: Themes from
               Dummett*, The Lauener Library of Analytical philosophy vol. 4, Berlin: De
               Gruyter.

2012   "Frege and Benacerraf's Problem," in Robert DiSalle, Mélanie Frappier, and
               Derek Brown (eds.), *Analysis and Interpretation in the Exact Sciences: Essays
               in Honour of William Demopoulos*, Dordrecht: Springer, 117–33.

2009   With Bob Hale, "The Metaontology of Abstraction," in David Chalmers et al.
               (eds.), *Metametaphysics*, Oxford: Oxford University Press, 178–212.

2009   With Bob Hale, "Focus Restored: Comment on John Macfarlane's 'Double
               Vision,'" in *Synthese* 170 (3) 457–82, special number on Bad Company
               edited by Oystein Linnebo.

2009   "The Metaphysics and Epistemology of Abstraction," in Alexander Hieke
               and Hannes Leitgeb (eds.), *Reduction-Abstraction-Analysis*, 195–216.
               Publications of the Austrian Ludwig Wittgenstein Society: New Series, vol.
               11, Frankfurt: Heusenstamm bei Frankfurt, Ontos Verlag.

2008   With Bob Hale, "Abstraction and Additional Nature," in *Philosophia
               Mathematica* 16 (2), 182–208.

2005   With Bob Hale, "Logicism in the 21st Century," in Stewart Shapiro (ed.), *The Oxford Handbook in the Philosophy of Mathematics and Logic*, Oxford: Oxford University Press.

2003   With Bob Hale, "Responses to Commentators"—book symposium on *The Reason's Proper Study*, *Philosophical Books* 44 (3), 245–63.

2002   With Bob Hale, "Benacerraf's Dilemma Revisited," *European Journal of Philosophy* 10 (1), 101–29.

2001   With Bob Hale, "To Bury Caesar . . . ," in Bob Hale and Crispin Wright (eds.), *The Reason's Proper Study*, Oxford: The Clarendon Press, 335–96.

2000   "Neo-Fregean Foundations for Real Analysis: Some Reflections on Frege's Constraint," *Notre Dame Journal of Formal Logic* 41, 317–34.

1999   "Is Hume's Principle Analytic?," *Notre Dame Journal of Formal Logic* 40 (1), comprising a Festschrift for the late George Boolos, 6–30.

1998   "Response to Dummett," in M. Schirn (ed.), *Philosophy of Mathematics Today*, Oxford: Oxford University Press, 389–405.

1998   "On the (Harmless) Impredicativity of Hume's Principle," in M. Schirn (ed.), *Philosophy of Mathematics Today*, Oxford: Oxford University Press, 339–68.

1997   "On the Philosophical Significance of Frege's Theorem," in R. Heck (ed.), *Language, Thought and Logic: Essays in Honour of Michael Dummett*, Oxford: Oxford University Press, 201–44.

1995   Critical Study of Michael Dummett's *Frege: Philosophy of Mathematics* in *Philosophical Books* 36, 89–102.

1990   "Field and Fregean Platonism," in *Physicalism in Mathematics*, Andrew Irvine (ed.), Kluwer Academic Publishers, 73–93. Reprinted in French translation in M. Marion and A. Voizard (eds.), *Frege: Logique et Philosophie*, Montreal, Harmattan 1998.

1983   "Recent Work on Frege," critical study of Gregory Currie, *Frege: An Introduction to his Philosophy*, London: Harvester, 1982; Hans Sluga, *Gottlob Frege*, London: Routledge, 1980; Michael Dummett, *The Interpretation of Frege's Philosophy*, London: Duckworth, 1981, *Inquiry* 26, 363–81.

Vagueness and the Sorites

2019   "Intuitionism and the Sorites Paradox," in Elia Zardini and Sergi Oms (eds.), *The Sorites Paradox*, Cambridge: Cambridge University Press, 95–117.

2016   "On the Characterisation of Borderline Cases," in Gary Ostertag (ed.), *Meanings and Other Things: Essays on Stephen Schiffer*, Cambridge: MA: MIT Press.

2010   "The Illusion of Higher-order Vagueness," in Richard Dietz and Sebastiano Morruzzi (eds.), *Cuts and Clouds*, Oxford: Oxford University Press, 523–49.

2007   "'Wang's Paradox,'" in the *Library of Living Philosophers* volume XXI, *The Philosophy of Michael Dummett*, ed. R. Auxier and L. Hahn, Chicago, IL: Open Court, 415–44.

2006   "Vagueness-related Partial Belief and the Constitution of Borderline Cases," in *Philosophy and Phenomenological Research* Book Symposium on Stephen Schiffer's *The Things We Mean*, Oxford: Oxford University Press, 2003.

2005   "Vagueness, Response Dependence and Rule-Following—Some Reflections," in S. Moruzzi and A. Sereni (eds.), *Issues on Vagueness*. Second Workshop on Vagueness, Padova: Il Poligrafo, 2005.

2003   "Vagueness: A Fifth Column Approach," in J. C. Beall and Michael
            Glanzberg (eds.), *Liars & Heaps: New Essays on Paradox*, Oxford: Oxford
            University Press, 84–105.
2003   "Rosenkranz on Quandary, Vagueness and Intuitionism," *Mind* 112, 465–7.
2001   "On Being in a Quandary: Relativism, Vagueness, Logical Revisionism,"
            Mind CX, 45–98; reprinted in *The Philosopher's Annual*, vol. 24.
1995   "The Epistemic Conception of Vagueness," in a supplement to *The Southern
            Journal of Philosophy*, XXXIII, special number on Vagueness, 133–59.
            Proceedings of the 1994 Spindel Conference held at Memphis, October 1994.
1992   "Is Higher Order Vagueness Coherent?," *Analysis* 52, 129–39.
1989   "The Sorites Paradox and Its Significance for the Interpretation of Semantic
            Theory," in *Recherches sur la Philosophie et le Langage*, Grenoble, Imprimerie
            de la Bibliotheque Interuniversitaire et des Sciences Sociales de Grenoble,
            119–36. Reprinted with a new appendix on higher-order vagueness in
            N. Cooper and P. Engel (eds.), *New Inquiries into Meaning and Truth*,
            London: Harvester Wheatsheaf, 1991, 135–62.
1987   "Further Reflections on the Sorites Paradox," *Philosophical Topics* XV (1),
            227–90. Reprinted in R. Keefe and P. Smith (eds.), *Vagueness: A Reader*,
            Cambridge, MA: Bradford/MIT, 1996.
1986   "The Sorites Paradox," *Quaderni di Semantica* VII (2), 277–91.
1985   With Stephen Read, "Hairier than Putnam Thought," *Analysis* 45, 56–8.
1976   "Language-Mastery and the Sorites Paradox," in G. Evans and J. McDowell
            (eds.), *Truth and Meaning*, Oxford: Oxford University Press, 223–47.
            Reprinted in R. Keefe and P. Smith (eds.), *Vagueness: A Reader*, Cambridge,
            MA: Bradford/MIT 1996.
1975   "On the Coherence of Vague Predicates," *Synthese* 30, 325–65.

The Self and Self-knowledge
2015   "Self-knowledge: The Reality of Privileged Access," in Sanford Goldberg
            (ed.), *Externalism, Self-Knowledge, and Skepticism: New Essays*, Cambridge:
            Cambridge University Press, 49–74.
2012   "Reflections on François Recanati's, 'Immunity to Error through
            Misidentification: What It Is and Where It Comes From,'" in Simon Prosser
            and François Recanati (eds.), *Immunity to Error Through Misidentification:
            New Essays*, Cambridge: Cambridge University Press, 247–79.
2012   "Knowledge of Our Own Minds and Meanings," Part II of "Replies," in
            Annalisa Coliva (ed.), *Mind, Meaning and Knowledge: Themes from the
            Philosophy of Crispin Wright*, Oxford: Oxford University Press, 402–17.
1998   "Self-knowledge: The Wittgensteinian Legacy," in C. Wright, B. Smith, and
            C. Macdonald (eds.), *Knowing Our Own Minds*, Oxford: Oxford University
            Press, 15–45.
1998   Introduction, in C. Wright, B. Smith, and C. Macdonald (eds.), *Knowing Our
            Own Minds*, Oxford: Oxford University Press, 15–45.
1998   "Self-knowledge: The Wittgensteinian Legacy" (shortened version) in
            Anthony O'Hear (ed.), *Current Issues in Philosophy of Mind*, Royal
            Institute of Philosophy Supplement 43, Cambridge: Cambridge University
            Press, 101–22.

Wittgenstein, Privacy, Rule-following

2012   "The Rule-Following Considerations and the Normativity of Meaning," Part I
            of "Replies," in Annalisa Coliva (ed.), *Mind, Meaning and Knowledge:
            Themes from the Philosophy of Crispin Wright*, Oxford: Oxford University
            Press, 379–401.

2007   "Rule-following without Reasons: Wittgenstein's Quietism and the
            Constitutive Question," in *Wittgenstein and Reason*, ed. John Preston, *Ratio*
            XX (4).

1989   "Wittgenstein's Later Philosophy of Mind: Sensation, Privacy and Intention,"
            Journal of Philosophy LXXVI (11) 622–34. Extended version in Klaus Puhl
            (ed.), *Meaning Scepticism*, Berlin: de Gruyter, 1991, 126–47.

1989   Critical Study of Colin McGinn's *Wittgenstein on Meaning* in *Mind* XCVIII,
            289–305.

1989   "Wittgenstein's Rule-Following Considerations and the Central Project of
            Theoretical Linguistics," in Alexander George (ed.), *Reflections on Chomsky*,
            Oxford: Blackwell, 233–64.

1987   "On Making Up One's Mind: Wittgenstein on Intention," in P. Weingartner
            and Gerhard Schurz (eds.), *Logic, Philosophy of Science and Epistemology*, the
            Proceedings of the XIth International Wittgenstein Symposium, Vienna:
            Holder–Pickler–Tempsky, 391–404.

1986   "Does *Philosophical Investigations* I, 258–60 Suggest a Cogent Argument
            Against Private Language?," in P. Pettit and J. McDowell (eds.), *Subject,
            Thought and Context*, Oxford: Oxford University Press, 209–66.

1986   "Rule-following and Constructivism," in C. Travis (ed.), *Meaning and
            Interpretation*, Oxford: Blackwell, 271–97.

1984   "Kripke's Account of the Argument Against Private Language," *Journal of
            Philosophy* LXXI (12), 759–78.

1981   "Rule-following, Objectivity and the Theory of Meaning," in C. Leich
            and S. Holtzman (eds.), *Wittgenstein: To Follow a Rule*, London: Routledge,
            99–117.

Notes on the Contributors

GEORGE S. BOOLOS taught at the Massachussetts Institute of Technology from 1969 until his death in 1996. His books include *The Logic of Provability* (Cambridge University Press 1993), and (with Richard Jeffrey and John Burgess) *Computability and Logic* (fifth edition, Cambridge University Press 2007). Many of his papers are collected in *Logic, Logic and Logic* (Harvard University Press 1999).

WILLIAM DEMOPOULOS taught at the University of Western Ontario from 1972 to his retirement in 2012. His publications include *Frege's Philosophy of Mathematics* (Harvard University Press 1997) and *Logicism and its Philosophical Legacy* (Cambridge University Press 2013) as well as numerous articles on the philosophy of science, philosophy of language and logic, the history of analytic philosophy, and the philosophy of mathematics.

JIM EDWARDS is Honorary Research Fellow at the University of Glasgow, where he taught philosophy for many years. He is the author of numerous articles on the philosophy of logic and language and related areas, in journals such as *Mind*, *Philosophical Quarterly, Analysis, Synthese*, and *Notre Dame Journal of Formal Logic*.

BOB HALE was emeritus Professor of Metaphysical Philosophy at the University of Sheffield until his death in December 2017, having held previous posts at Lancaster, St Andrews, and Glasgow. His published work includes numerous articles on philosophy of mathematics and the philosophy of logic and language, and related areas of metaphysics and epistemology, and numerous books including *Abstract Objects* (Blackwell 1987), *The Reason's Proper Study: Essays towards a Neo-Fregean Philosophy of Mathematics* (Oxford University Press 2001, jointly written with Crispin Wright), and, most recently, *Necessary Beings: an Essay on Ontology, Modality, and the relations between them* (Oxford University Press 2013, paperback 2015).

RICHARD KIMBERLY HECK has been Professor of Philosophy at Brown University since 2005. Previously, he taught at Harvard University from 1991 to 2005. His publications include *Frege's Theorem* (Oxford University Press 2011), *Reading Frege's Grundgesetze* (Oxford University Press 2012), and many articles on the philosophy of logic, language, and mathematics.

ALEXANDER MILLER is Professor of Philosophy at the University of Otago, New Zealand. He works mainly on the philosophy of language and mind, metaphysics, and metaethics. His books include *Contemporary Metaethics: An Introduction* (second edition, Polity Press 2013) and *Philosophy of Language* (third edition, Routledge 2018). He is co-editor (with Crispin Wright) of *Rule-Following and Meaning* (Acumen 2002), co-editor (with Bob Hale and Crispin Wright) of *A Companion to the Philosophy of Language* (revised and expanded second edition, Wiley-Blackwell 2017), and guest editor of *Australasian Philosophical Review* Volume 2 Number 3 (2019).

GIDEON ROSEN is Stuart Professor of Philosophy at Princeton University, where he has taught since 1992. Prior to that he taught at the University of Michigan, Ann Arbor. His publications include a co-authored book (with John. P. Burgess) *A Subject With No Object* (Oxford University Press 1997), and numerous articles on topics in metaphysics, philosophy of language, philosophy of mathematics, and related areas.

IAN RUMFITT is Senior Research Fellow of All Souls College, Oxford. He works mainly in the philosophy of language, philosophical logic, and the philosophy of mathematics. His book *The Boundary Stones of Thought* (Oxford University Press 2015) investigates conflicts between rival logic systems and how they might be rationally resolved.

STEPHEN SCHIFFER is Silver Professor of Philosophy at New York University. He works primarily in philosophy of language, philosophy of mind, and metaphysics. He is the author of numerous articles and of three books: *Meaning* (Oxford University Press 1972), *Remnants of Meaning* (MIT Press 1987), *and The Things We Mean* (Oxford University Press 2003). He is a Fellow of the American Academy of Arts and Sciences.

SANFORD SHIEH is Professor of Philosophy at Wesleyan University. He is the author of a number of papers on the philosophy of mathematics, philosophy of language and logic, and related areas, and co-editor (with Alice Crary) of *Reading Cavell* (Routledge 2006) and (with Juliet Floyd) of *Future Pasts: The Analytic Tradition in Twentieth Century Philosophy* (Oxford University Press 2001).

NEIL TENNANT is Arts & Sciences Distinguished Professor of Philosophy at the Ohio State University and has held positions at the University of Edinburgh, the University of Stirling, and the Australian National University. His books include *Core Logic* (Oxford University Press 2017), *Introducing Philosophy* (Routledge 2015), *Changes of Mind: An Essay on Rational Belief Revision* (Oxford University Press 2012), *The Taming of the True* (Oxford University Press 1997), *Natural Logic* (Edinburgh University Press 1990), and *Anti-Realism and Logic* (Clarendon Press 1987).

STEPHEN YABLO is David W. Skinner Professor of Philosophy at the Massachusetts Institute of Technology, where he has taught since 1997. He taught previously at the University of Michigan, Ann Arbor. His publications include two volumes of collected papers, *Thoughts* (Oxford University Press 2005) and *Things* (Oxford University Press 2010), as well as *Aboutness* (Princeton University Press 2014).

PART I
Frege and Neo-Logicism

1

Generality and Objectivity in Frege's *Foundations of Arithmetic*

William Demopoulos.

Frege's (1884) *Foundations of Arithmetic* is celebrated as the first formulation and defense of the philosophy of arithmetic known as logicism: the thesis that arithmetic is a branch of logic. The work is noted for its masterful deployment of polemics against earlier views, its centrality to the analytic tradition, and since *Frege's Conception of Numbers as Objects* (Wright 1983) it has been generally recognized as containing the first mathematically significant theorem of modern logic. I hope to show that there are misconceptions concerning the nature and significance of Frege's achievement that reflection on "Frege's theorem" is capable of correcting. I address two such misconceptions. The first is that arithmetic is a species of a priori knowledge for Frege only if it can be shown to be analytic, or more generally, that there can be nothing answering to a Fregean account of the apriority of arithmetic without the success of logicism. The particular form of this misconception that will be addressed has its basis in the idea that only the generality Frege associated with logic can support his account of the apriority of arithmetic. The second misconception is that any account of the natural numbers that deserves to be called Fregean must single out *the* natural numbers; this misconception is based on the assumption that a Fregean account of the objectivity of arithmetic must be derived from the thesis that numbers are particular objects.

My contentions are principally two: (*i*) *Foundations of Arithmetic* has a compelling reconstruction which addresses the *apriority* of arithmetic without in any way relying on the thesis that numbers are logical objects, or that arithmetic is analytic or a part of logic; moreover, this reconstruction respects a distinctive feature of Frege's definition of a priori knowledge—namely, that apriority requires a justification based exclusively on general laws. (*ii*) The *objectivity* of arithmetic can be secured by adherence to principles which, though unquestionably Fregean, are independent of the thesis that numbers are logical objects, and involve only a minimalist interpretation of the thesis that they are objects at all. Although both of these contentions mark points of divergence from neo-Fregeanism, the present study would hardly have been possible without Crispin's pathbreaking work. It is a special pleasure to be able to contribute it to this Festschrift for him.

William Demopoulos, *Generality and Objectivity in Frege's* Foundations of Arithmetic In: *Logic, Language, and Mathematics: Themes from the Philosophy of Crispin Wright*. Edited by: Alexander Miller, Oxford University Press (2020). © William Demopoulos. DOI: 10.1093/oso/9780199278343.003.0001

Frege's Desiderata

The beginning sections of Frege's *Foundations of Arithmetic* (hereafter, *Fdns*) are largely devoted to the criticism of other authors' attempts to address the question, "What is the content of a statement of number?" A little reflection suggests a relatively simple classification of Frege's objections to earlier views based on their failure to account for one or more of the following:

Statements of number exhibit generality. Anything can be counted. Mill and Kant fail to observe this, though in very different ways.

Statements of number are objective. The number of petals of a flower is as objective a feature as its color (*Fdns* §26). Berkeley who comes closest to the right answer in other respects, fails to accommodate this observation.

The grammar of statements of number is peculiar. Compare the difference between 'The apostles were fishers of men' with 'The apostles were 12'; also, numeral names are proper—they don't pluralize.

Frege's own celebrated answer, given in §46, is that a statement of number involves the predication of something of a concept. However, what I wish to focus on—at least initially—is not Frege's answer to this question, but the desiderata or constraints that this classification of his criticisms of earlier views imposes on the philosophy of arithmetic: A successful account of arithmetical knowledge must explain its *generality* and *objectivity*, and it must preserve its characteristic *logical form*.

One thing that is noteworthy about Frege's desiderata is the absence of two conditions that have come to dominate the contemporary philosophy of mathematics literature: the truths of arithmetic are *necessary* truths, and numbers are *abstract* objects.

Anyone familiar with the thesis that arithmetic is part of logic but with only a passing acquaintance with Frege's writings might expect that, for Frege, the necessity of arithmetic is inherited from the necessity of the laws of logic. But there is a simple consideration that argues against this expectation. Frege barely addresses the question of what characterizes a truth as logical, and when he does, necessity plays no role in his answer. Frege's great contribution to logic was his formulation of polyadic logic with mixed generality; but he contributed very little to our understanding of what constitutes a logical notion or a logical proposition, and still less to our understanding of logical necessity. For Frege the laws of logic are marked by their universal applicability and by the fact that they are presupposed by all of the "special" sciences. Because of logic's universality, a reduction of arithmetic to it would address Frege's generality constraint, but given Frege's conception of logic, it would shed no light on the matter of arithmetic's necessity.

The closest Frege comes to explicitly considering the necessity of arithmetic is in connection with his belief that it is a species of a priori knowledge. On Frege's definition of *a priori* this means that it is susceptible of a justification solely on the basis of general laws that neither need nor admit of proof.[1] The hint of Frege's

---

[1] §3 of *Fdns* appears to offer this only as a sufficient condition, not as a necessary and sufficient condition as a definition would require. Here I follow Burge (2005: 359–60) who argues, convincingly in my view, that the condition is intended to be both necessary and sufficient.

interest in necessity comes with the nature of the warrant that attaches to the general laws which are capable of supporting a proof of apriority. Such a law neither needs nor admits of proof because no premise of any purported proof of it is more warranted than the law itself. But to the extent to which this is a notion of necessity at all, it is a wholly epistemic one. By contrast, the contemporary concern is with the metaphysical necessity of arithmetical truths, and it came to prominence only with Ramsey's celebrated 1925 essay on the foundations of mathematics.[2]

Frege's understanding of the epistemological significance of a derivation of arithmetic from logic is subtle. Certainly such a derivation would make it clear that arithmetic is not *synthetic* a priori, which is something Frege certainly seeks to establish. But the derivation of arithmetic from logic should not be needed for the simpler thesis that arithmetical knowledge is encompassed by Frege's definition of a priori knowledge. To suppose otherwise would be to imply that our appeal to logical principles is necessary because there are no self-warranting *arithmetical* principles to sustain its apriority. This could only be maintained if arithmetical principles are less warranted than the basic laws of logic or if they lack the requisite generality. Discounting, for the moment, the second alternative, but supposing arithmetical principles to be less warranted than those of logic, the point of Frege's logicism would be to provide a justification for arithmetic in logic. But Frege did not maintain that the basic laws of arithmetic—by which I mean the second order Peano axioms[3] (hereafter, PA for *Peano Arithmetic*)—are significantly less warranted than those of his logic. Although the textual evidence is not unequivocal, Frege says on more than one occasion that the primary goal of his logicism is not to secure arithmetic, but to expose the proper dependence relations of its truths on others.

It is important always to bear in mind that *Fdns* was not written in response to a "crisis" in the foundations of mathematics; above all *Fdns* seeks to illuminate the character of our knowledge of arithmetic and to address various misconceptions, most notably the Kantian misconception that arithmetic rests on intuitions given a priori. The early sections of *Fdns* are quite explicit in framing this general epistemological project of the work as the following passages illustrate:

The aim of proof is, in fact, not merely to place the truth of a proposition beyond all doubt, but also to afford us insight into the dependence of truths on one another. (§2)

[T]he fundamental propositions of arithmetic should be proved...with the utmost rigour; for only if every gap in the chain of deductions is eliminated with the greatest care can we say with certainty upon what primitive truths the proof depends...

[I]t is above all Number which has to be either defined or recognized as indefinable. This is the point which the present work is meant to settle. On the outcome of this task will depend the decision as to the nature of the laws of arithmetic. (§4)

---

[2] See Ramsey (1925: 3–4).

[3] I do not regard the equation of the basic laws of arithmetic with the second order Peano axioms as at all tendentious. Each of the familiar Peano axioms or a very close analogue of it occurs in the course of the mathematical discussion of §§74–83 of *Fdns*, where the implicit logical context is the second order logic of *Begriffsschrift* which, as customarily interpreted, assumes full comprehension. For additional considerations in favor of this equation, see Dummett (1991: 12–13).

Although Frege's attempted demonstration of the analyticity of arithmetic would indeed show that the basic laws of arithmetic can be justified by those of logic, the principal interest of such a derivation is what it would reveal regarding the dependence of arithmetical principles on logical laws. This would be a result of broadly epistemological interest, but its importance would not necessarily be that of providing a warrant where one is lacking or insufficient.[4]

As for the idea that numbers are abstract objects, this also plays a relatively minor role in *Fdns*, and it is certainly not part of a desideratum by which to gauge a theory of arithmetical knowledge. Frege's emphasis is rarely on the positive claim that numbers are abstract, but is almost always a negative one to the effect that numbers are not ideas, not collections of units, not physical aggregates, not symbols and neither intuitions nor objects of intuition. The only positive claim of Frege's regarding the nature of numbers is that they are extensions of concepts, a claim which—at least in *Fdns*—he seems to assign the character of a convenience (§§69 and 107); nor does he pause to explain extensions of concepts, choosing instead to assume that the notion is generally understood (fn. 1 to §68).

Frege's mature—post *Fdns*—view of the characterization of numbers as classes or extensions of concepts is decidedly less casual. That it too is almost exclusively focused on the epistemological role of classes—they facilitate the thesis that arithmetic is recoverable by analysis from our knowledge of logic—is clearly expressed in his correspondence with Russell:

> I myself was long reluctant to recognize...classes, but saw no other possibility of placing arithmetic on a logical foundation. But the question is, How do we apprehend logical objects? And I found no other answer to it than this, We apprehend them as extensions of concepts.... I have always been aware that there are difficulties connected with [classes]: but what other way is there? (Frege to Russell, 28.vii.1902, in McGuinness 1980: 140–1)

The foregoing considerations suggest that we should distinguish two roles that sound or truth-preserving derivations are capable of playing in a foundational investigation of the kind Frege is engaged in. Let us call derivations that play the first of these roles *proofs*, and let us distinguish them from derivations that constitute *analyses*. *Proofs* are derivations which enhance the justification of what they establish by deriving them from more securely established truths in accordance with logically sound principles. *Analyses* are derivations which are advanced in order to clarify the logical dependency relations among propositions. The derivations involved in analyses do not purport to enhance the warrant of the conclusion drawn, but to display its basis in other truths. Given this distinction, the derivation of arithmetic from logic would not be advanced as a *proof* of arithmetic's basic laws if these laws were regarded as established, but as an *analysis* of them. The general laws, on the basis of which a proposition is shown to be a priori, neither need nor admit of proof in the sense of 'proof' just explained: Such laws may admit of analyses, but the derivation such an analysis rests upon is not one that adds to the epistemic warrant of the derived proposition. Unless this distinction or some equivalent of it is admitted, it would be difficult to maintain that the basic laws of arithmetic neither need nor admit of proof

---

[4] The nature of Frege's epistemological concerns is discussed more fully in Demopoulos 1994.

and advance the thesis of logicism, according to which they are derivable from the basic laws of logic. As we will see, there are difficulties associated with the notion of neither needing nor admitting of proof, but they are not those of excluding the very possibility of logicism merely by conceding that the basic laws of arithmetic neither need nor admit of proof.

The Problem of Apriority

Suppose we put to one side the matter of arithmetic's being analytic or synthetic. Does there remain a serious question concerning the mere apriority of its basic laws? This is actually a somewhat more delicate question than our discussion so far would suggest. Recall that truths are, for Frege, a priori if they possess a justification exclusively on the basis of *general* laws which themselves neither need nor admit of proof. An unusual feature of this definition is that it makes no reference to experience. A plausible explanation for this aspect of Frege's formulation is his adherence to the first of the three methodological principles announced in the introduction to *Fdns*: always to separate sharply the psychological from the logical, the subjective from the objective. Standard explanations of apriority in terms of independence from experience have the potential for introducing just such a confusion, which is why apriority and innateness became so entangled in the traditional debate between rationalists and empiricists. Frege seems to have regarded Mill's views as the result of precisely the confusion a definition in terms of general laws—rather than facts of experience—is intended to avoid. It is reasonable therefore for Frege to have put forward a formulation which, in accordance with his first methodological principle, avoids even the appearance of raising a psychological issue.

Although Frege's definition is nonstandard, it is easy to see that it subsumes the standard definition which requires of an a priori truth that it have a justification that is independent of experience: The justification of a fact of experience must ultimately rest on instances, either of the fact appealed to or of another adduced in support of it. But an appeal to instances requires mention of particular objects. Hence if a justification appeals to a fact of experience, it must also mention particular objects. Therefore, if a truth fails the standard test of apriority, it will fail Frege's test as well. However, the converse is not true: A truth may fail Frege's test because its justification involves mention of a particular object, but there is nothing in Frege's definition to require that this object must be an object of experience. This inequivalence of Frege's definition with the standard one would point to a defect if it somehow precluded a positive answer to the question of the apriority of arithmetic. Does it?

Thus far I have only argued that for Frege the basic laws of arithmetic are not significantly less warranted than those of his logic. But are they completely general? In a provocative and historically rich paper, Burge (2000: §V) argues that Frege's account of apriority prevents him from counting all the second order Peano axioms as a priori. Let us set mathematical induction aside for the moment. Then the remaining axioms characterize the concept of a natural number as Dedekind-infinite, that is, as in one-to-one correspondence with one of its proper subconcepts.

And among these axioms, there is one in particular which, as Burge observes, fails the test of generality because it expresses a thought involving a particular object:

> It is, of course, central to Frege's logist project that truths about the numbers—which Frege certainly regarded as particular, determinate, formal objects (e.g. *Fdns* §§13, 18)—are derivative from general logical truths. [...If] arithmetic is not derivable in an epistemically fruitful way from purely general truths [,...] then it counts as a posteriori on Frege's characterization. This would surely be a defect of the characterization.   (Burge 2005: 384–5)

As Burge observes, it matters little that for Frege zero is to be recovered as a "logical object," and that for this reason it is arguably not in the same category as the particular objects the definition of a priori knowledge is intended to exclude. The problem is that such a defense of Frege makes his argument for the apriority of arithmetic depend on his inconsistent theory of extensions. Hence the notion of a logical object can take us no closer to an account of this fundamental fact—the fact of apriority—concerning our knowledge of arithmetic. Thus, Burge concludes, unless his logicism can be sustained, Frege is without an account of the apriority of arithmetic.

There is, however, another way of seeing how the different components of *Fdns* fit together, one that yields a complete solution to what I will call *the problem of apriority*, namely, the problem of explaining the apriority of arithmetic in Frege's terms—that is, in terms of an epistemically fruitful derivation from general laws which do not depend on the doctrine of logical objects or the truth of logicism.

The reconstruction of *Fdns* that is suggested by the problem of apriority is so natural that it is surprising that it has not been proposed before. It is, however, a reconstruction, not an interpretation of Frege's views. Frege may have conceived of *Fdns* in the way I am about to explain, but there are at least two considerations that argue against such a supposition. First, Frege's main purpose in *Fdns* is to establish the analyticity of arithmetic, but on the proposed reconstruction, the goal of establishing arithmetic's apriority receives the same emphasis as the proof of its analyticity. Second, Frege frequently appeals to primitive truths and their natural order; by contrast the reconstruction uses only the notions of a basic law of logic or of arithmetic, and it uses both notions in entirely nontechnical senses; in particular I do not assume that the basic laws with which I am concerned reflect a "natural order of primitive truths." This idea is closely associated with another component of Frege's conception of apriority—that of neither needing nor admitting of proof. We identify those basic laws that are "primitive in the natural order of truths" as those which neither need nor admit of proof. But that this prescription is not so straightforward as it may seem can be seen by considering a recent attempt to understand it by an appeal to various notions of self-evidence.

Separating the Logical from the Psychological

In her interesting study, Jeshion (2001) attributes to Frege a "Euclidean rationale" according to which the

> primitive truths of mathematics have two properties. (i) They are *selbst-verständlich*: foundationally secure, yet are not grounded on any other truth, and, as such, do not stand in need of proof. (ii) And they are self-evident [generally signaled by Frege's use of *einleuchten*]: clearly

grasping them is a sufficient and compelling basis for recognizing their truth. [Frege] also thought that the relations of epistemic justification in a science mirror the natural ordering of truths: in particular, what is self-evident is *selbst-verständlich*. Finding many propositions of arithmetic non-self-evident, Frege concluded that they stand in need of proof.

(Jeshion 2001: 961)

It is clearly correct that for Frege some truths are *selbst-verständlich* or "foundationally secure," and that this is essential to their constituting a basis for arithmetic. But the idea that self-evident truths do not stand in *need* of proof, as on Jeshion's account they do not because they are not grounded on other truths, is potentially misleading. Parts II and III of *Begriffsschrift* are constructed around the derivation of a single arithmetical proposition, Proposition 133 of the work. As Frege explains in his completed but never published paper, 'Boole's Logical Calculus and the Begriffsschrift,' his purpose in choosing to set forth the derivation of this proposition within *Begriffsschrift*'s system of second order logic is not to rectify a failure of self-evidence—he disputes neither its self-evidence nor its truth—but, by the presentation of a complete and gap-free derivation, to show it to be analytic (to use the terminology of *Fdns*). To revert to our earlier terminology, a self-evident truth may not stand in need of a *proof*, but it may require a derivation to clarify the source of our knowledge of it.[5]

Frege takes for granted that arithmetic's basic laws are not only true, but are *known* to be true. But Frege's various appeals to self-evidence do not, by themselves, reveal a commitment to the notion that self-evidence is the correct explanation of the warrant basic laws enjoy. By contrast, Jeshion believes that the notion of self-evidence she has isolated can explain why we are justified in believing the primitive truths of mathematics, and that it can do so without forcing the content of such judgements to be mentalistic or otherwise dependent on our grasping them. Nor, according to Jeshion, does it demand that we interpret normative notions like truth and understanding by reference to our mental activity. But these considerations, important as they are, sidestep the decisive issue, which is whether self-evidence violates Frege's strictures against psychologism in connection with questions of justification. Jeshion admits that the notion of self-evidence must assign epistemic significance to obviousness, and that we "must rely on what we find obvious to judge propositions as self-evident" (p. 967). So even though self-evidence avoids many of the pejorative aspects of psychologism, its principal indicator is a thoroughly psychological notion. Basing primitive truths or basic laws—whether of logic or arithmetic—on such a notion of self-evidence, depending as its application does on obviousness, clearly runs counter to Frege's insistence on separating the psychological from the logical.

---

[5] [T]o show that I can manage throughout with my basic laws...I chose the example of a step by step derivation of a principle which, it seems to me, is indispensable to arithmetic, although it is one that commands little attention, being regarded as self-evident. The sentence in question is the following:

If a series is formed by applying a many-one operation to an object (which need not belong to arithmetic) and then applying it successively to its own results, and if in this series two objects follow one and the same object, then the first follows the second in the series or vice versa, or the two objects are identical.   (Hermes et al. 1979: 38)

I believe Jeshion is led to this notion of self-evidence by her interpretation of §5 of *Fdns*, a section which receives an extended discussion in her paper. This section concerns the self-evidence of particular arithmetical facts involving large numbers. Although I will not attempt to establish this in detail here, I believe that what is said of self-evidence in §5 is naturally understood as having a purely "dialectical" function, and that Frege should be understood as appealing to whatever notion of self-evidence is assumed by the proponents of the view he is opposing. By contrast, Jeshion takes Frege's remarks to reflect the notion of self-evidence implicit in his own positive view. Now it is generally recognized that §5 raises a criticism of the Kantian thesis that facts about particular numbers are grounded in intuition, and that Frege's criticism of this thesis hinges on the premise that when a truth fails to qualify as self-evident, any account of our knowledge of it as intuitive must also fail. But §5's general point against Kantian intuition doesn't really depend on this appeal to the connection between intuition and self-evidence: even if Frege's premise is rejected and it is necessary to concede that such facts are intuitive, they would still be unsuitable as a basis for arithmetic. For since facts about particular numbers are infinitely numerous, to take them as the primitive truths of arithmetic "conflicts with one of the requirements of reason, which must be able to embrace all first principles in a survey" (*Fdns* §5). So whether or not such facts are intuitively given, they cannot form the basis of our arithmetical knowledge, and we may conclude that the Kantian notion of intuition is not a useful guide to uncovering arithmetic's primitive truths. If I am right, and this is the structure of the argument of §5, it would be surprising if the section were a reliable guide to Frege's positive view of self-evidence. In any event, a reasonable condition to impose on a reconstruction of *Fdns* is that it should address questions of warrant without descending to the psychological and subjective level. This would seem to preclude relying on a notion of self-evidence like the one Jeshion attributes to Frege.

What of the idea that clearly grasping a self-evident proposition is a sufficient and compelling basis for recognizing its truth? An argument for a self-evident proposition would then consist in an explanation of its sense; acceptance of it as a truth would follow from the understanding such an explanation would facilitate. Such an argument does not in any way compromise the idea that the proposition neither needs nor admits of proof, because it merely makes explicit what is involved in grasping the proposition.

There is ample evidence that Frege was at times highly sympathetic to the idea that fundamental principles are sometimes justified on the basis of their sense. In *Function and concept* (Frege 1891: 11) he famously endorsed this methodology in connection with Basic Law V:

For any concepts F and G, the extension of the Fs is the same as the extension of the Gs if, and only if, all Fs are Gs and all Gs are Fs.

This lends plausibility to the relevant interpretative claim, but the fact that Law V arguably *does* capture the notion of a Fregean extension poses insurmountable difficulties in the way of accepting this methodology as part of a credible justification of it. Being analytic of the notion of a Fregean extension does not show Basic Law V to be analytic, true, or even consistent. If therefore it is a mark of primitive truths that our grasp of them suffices for the recognition of their truth, then some at least of Frege's basic laws are not primitive truths. Although grasping the sense of a basic law does not always suffice for the recognition of its truth, it is doubtful that Frege had a more considered methodology for showing that we are justified in believing his basic laws. Frege seems to have taken it for granted that the basic laws of logic and arithmetic are self-warranting and that this is an assumption to which all parties to the discussion are simply entitled.

The Solution to the Problem of Apriority

The purpose of the present reconstruction is to isolate a solution to the problem of apriority, one that respects Frege's three desiderata without appealing to the analyticity of arithmetic. On the proposed reconstruction, the argument of *Fdns* divides into two parts. The first, and by far the more intricate argument, addresses the problem of apriority. The second argument, which I will ignore except insofar as it illuminates the argument for apriority, is directed at showing the basic laws of arithmetic to be analytic.

A principal premise of the argument for apriority—a premise which is established in *Fdns*—is Frege's theorem, that is, the theorem that the second order Peano axioms are recoverable as a definitional extension of the second order theory—called FA, for *Frege Arithmetic*—whose sole nonlogical axiom is a formalization of the statement known in the recent secondary literature as *Hume's principle*:

For any concepts F and G, the number of Fs is the same as the number of Gs if, and only if, there is a one-to-one correspondence between the Fs and the Gs.

For the purpose of the present discussion, it doesn't matter how Hume's principle is represented: it may be understood as a partial contextual definition of the concept of number, or of the cardinality operator (the number of (...)); or it may be taken to be a "criterion of identity" for numbers, that is, as a principle which tells us when the same number has been "given to us" in two different ways (as Frege suggests in *Fdns* §§62–3); alternatively, we could follow a suggestion of Ricketts (1997: 92) and regard Hume's principle not as a definition of number—not even a contextual one—but a definition of the second-level relation of *equinumerosity* which holds of first-level concepts.[6] However we regard it, since Hume's principle makes no mention of particular objects it possesses the kind of generality that is required of the premises of an argument for the apriority of a known truth.

But basing arithmetic on Hume's principle achieves more than its mere derivation solely from a principle whose form is that of a universally quantified statement: it effects an *analysis* of the basic laws of pure arithmetic by revealing their basis in the principle which controls the applications we make of the numbers in our cardinality judgements. Since Hume's principle is an arithmetical rather than a logical principle, the derivation of PA from FA is not the reductive analysis of logicism which Frege

---

[6] I am assuming that all of these construals acknowledge the existence and uniqueness assumptions which are implicit in the use of the cardinality operator.

sought. Nonetheless, it is an analysis of considerable epistemological interest. In addition to recovering PA from a general law, the derivation of PA from FA is based on an account of number which satisfies Frege's *generality constraint*: Since the cardinality operator acts on concepts, the application of number is represented as being as general in its scope as conceptual thinking itself. In particular, since numbers fall under concepts, they too can be counted. Such an analysis contrasts with *Principia*'s use of its Axiom of Infinity, which is advanced as the weakest postulate regarding the number of "concrete individuals" that are necessary for the development of Whitehead and Russell's theory. The axiom makes no pretense to being knowable a priori, but is at best a truth whose justification rests on instances, either of the axiom itself, or of some other empirical claim from which it follows. Such a claim might be a highly theoretical one, but its justification ultimately rests on reports of particular observations involving particular objects. It explains neither the generality nor the objectivity of our arithmetical knowledge, and it has no bearing on the logical form of statements of number.

Is Hume's Principle Ad Hoc?

It has been argued that the primacy which, in *The Basic Laws of Arithmetic* (Frege 1893–1903), Frege assigns to Basic Law V comes at the expense of the methodological soundness of Hume's principle. Thus, of Basic Law V, Ricketts writes that

> Frege's most compelling motivation for [it] is...the explicit general basis he believed [it] to give for mathematical practice. Basic Law V promised to be a codification of the means for the introduction of "new" objects into mathematics via abstractive definitions,...applicable...in arithmetic and analysis...and geometry. Simply postulating [Hume's] principle would have the same ad hoc character Frege finds in Dedekind's construction of the real numbers.
>
> (Ricketts 1997: 196)

To be sure, Hume's principle does not in Frege's view reveal the *logical* basis of arithmetic. And Ricketts is also correct to insist that Basic Law V incorporates a kind of generality that is indispensable to the argument for the analyticity of arithmetic. But the absence of "logical generality" hardly shows Hume's principle to be ad hoc, and none of this need affect the fact that it *is* appropriately general for the demonstration of the apriority of arithmetic.

I am unsure how to understand Ricketts's comparison of Hume's principle with Frege's criticisms of Dedekind's (1963) theory of the reals. Frege's account of the counting numbers in terms of Hume's principle is fully continuous with his theory of the real numbers. As Dummett (1991: ch. 20) has emphasized, the difficulty Frege finds in alternative accounts of the real numbers is that unlike his own development of their theory, other approaches fail to tie the definition of the reals to the applications we make of them in measurement. But the theory of the natural numbers that emerges when they are based on Hume's principle is one that expressly emphasizes their application in judgements of cardinality. Indeed it is precisely the emphasis his account of the natural numbers places on their applications that Frege seeks to mimic in his theory of the real numbers, and it is the neglect of this feature that underlies Frege's criticisms of the theories of others.

Apriority and Foundational Security

A subtlety in the logical form of Hume's principle makes it all the more compelling that the account of neither needing nor admitting of proof should not rest on a naïve conception of self-evidence. The difficulty is that the strength of Hume's principle derives from the logical form of the operator which is essential to its formulation. The cardinality operator is neither type-raising nor type-preserving, but maps a concept of whatever "level" to an *object*, which is to say, to a possible argument to a concept of *lowest* level. Were the operator not type-lowering in this sense, Frege's argument for the Dedekind infinity of the natural numbers would collapse. The fact that only the type-lowering form of the cardinality operator yields the correct principle argues against taking neither needing nor admitting of proof to be captured by self-evidence in any naïve sense; for it might be that only one of the weaker forms of the principle drives the conviction of "obviousness," "undeniability," or "virtual analyticity" that underlies what I am calling naïve conceptions of self-evidence.

Although it is highly plausible that the notion of equinumerosity implicit in our cardinality judgements is properly captured by the notion of one-to-one correspond-ence, a further investigation is needed to show that Hume's principle is "self-warranting," *selbst-verständlich* or whatever one takes to be the appropriate mark of neither needing nor admitting of proof. As we saw earlier, the closest Frege comes to an account of this idea is the flawed methodology of arguing from the grasp of the sense of a basic law to the recognition of its truth.

I have been concerned to show that Frege's emphasis on the *generality* of the premises employed in a proof of apriority—rather than their independence from experience—is not only sustainable independently of the truth of logicism, but is also productive of a fruitful analysis of number and our knowledge of its theory. Although Frege's analysis would preserve the notion that the principles of arithmetic express a body of truths that are known independently of experience, there is a respect in which it is independent even of the weaker claim that Hume's principle is merely a known truth. For, even if the traditional conception of the a priori were to be rejected, it would still be possible to argue that Hume's principle is a priori in the sense that it expresses the principal condition on which our application of the numbers rests.

Even if we set to one side the question of the *truth* of Hume's principle, there is a fact on which to base the claim that it is "foundationally secure," albeit in a weaker sense than is demanded by the traditional or Fregean notion of apriority. By the converse to Frege's theorem FA is recoverable from a definitional extension of PA, or equivalently, FA is interpretable in PA. As a consequence, FA is *consistent* relative to PA; a contradiction is derivable in it only if it is derivable in PA. As noted in Burgess (2005: 149) this observation goes back to Geach (1976). Its later formulation and proof by Boolos in his (1987) explicitly showed the interpretability of FA in PA in the course of showing how the consistency of FA can be demonstrated within the domain of the natural numbers. This fact is surprising to a non specialist because FA implies the existence of an infinite cardinal, while PA does not. Boolos's proof overcomes this difference by interpreting the cardinal of an infinite concept by zero and the cardinal of a finite concept by the successor of its ordinary cardinal. This provides a "cardinal number" for every concept in the range of the concept variables of Hume's principle, and it does so compatibly with the requirement that the "cardinals" be the same when the concepts with which they are associated are one–one correlated with one another.

Someone might grant that Hume's principle is foundationally secure in this sense, but object that, by admitting an infinite cardinal number, FA extends PA in a manner that makes it unsuitable as an analysis of the concept of number which is characterized by that theory. In my view the extension is entirely warranted, but this is not a point that needs arguing since, as we will see in the final section of the chapter, this objection may be accommodated by weakening Hume's principle without in any way compromising the recovery of PA.

Frege Arithmetic and Hilbert's Program

Since Frege regards the basic laws of arithmetic to be known truths, the interpretability of FA in PA would certainly count as showing that FA is foundationally secure as well; but Frege would regard such an argument as superfluous since Hume's principle was for him also a known truth, and therefore certainly foundationally secure. This shows why an appeal to the consistency strength of FA in support of Frege is altogether different from its use in the program which motivated the concept's introduction into the foundations of mathematics. The study of the consistency strength of subtheories of PA is an essential component of the program we associate with Hilbert—namely, to establish the consistency of higher mathematics within a suitably restricted "intuitive" or "finitistic" mathematical theory. Gödel's discovery of the unprovability of the consistency of first order Peano Arithmetic (PA^1) within PA^1 (by representing the proof of the incompleteness of PA^1 within PA^1) motivated an investigation of subtheories of PA^1 incapable of proving their own consistency, and of extensions of PA^1 capable of proving the consistency of PA^1. The theory Q known as Robinson Arithmetic is a particularly simple example of a theory incapable of proving its own consistency, and it forms the base of a hierarchy of increasingly stronger arithmetical theories.[7] But the study of this hierarchy is not integral to the logicist program of Frege for whom knowledge of the truth, and therefore the consistency of PA (and thus of PA^1), is simply taken for granted.

Of the two foundational programs, only Hilbert's holds out any promise of providing a foundation which carries with it any real justificatory force. Frege's foundational focus differs from Hilbert's in precisely this respect. The goals of Frege's logicism are epistemological, but they are not those of making the basic laws of arithmetic more secure by displaying their basis in Hume's principle, or indeed, in logic. To use our earlier terminology, Frege's derivation of PA from FA is part of an *analysis* of PA rather than a *proof* of it.

---

[7] As Kripke has observed, since in Q one cannot even prove that $x \neq x + 1$, a natural model for Q is the cardinal numbers with the successor of a cardinal x defined as $x + 1$. Kripke's observation is reported on p. 56 of Burgess 2005. Burgess's book describes where a variety of subtheories and extensions of FA are situated in relation to this hierarchy; it is recommended to anyone wishing to know the current state of art for such results.

Hilbert's proposal for securing PA and its set-theoretical extensions on an intuitive basis runs directly counter to Frege's goal of showing arithmetic to be analytic and hence independent of intuition. This is because Hilbert's views depend on taking as primitive the idea of iteration (primitive recursion) and our intuition of sequences of symbols:

> that the [symbolic objects] occur, that they differ from one another, and that they follow each other, or are concatenated, is immediately given intuitively, together with the object, as something that neither is reducible to something else nor requires such reduction.
>
> (Hilbert 1925: 376)

An attractive feature of the program of basing the consistency of PA on some particularly weak fragment of arithmetic is that it holds out the promise of explaining *why* the theory on which it proposes to base the consistency of PA is plausibly viewed as self-warranting. The notions which such a theory permit us to express—the intuitability of finite sequences of symbols and the iterability of basic operations involving them—are arguably presupposed in the formulation of *any* theory. But a theory that rests on no more and no less than what is demanded by the formulation and metatheoretical investigation of any mathematical theory is clearly as warranted as any such theory.

Although Frege's *Basic Laws of Arithmetic* contains a theory of general principles by which to explain the apriority of our knowledge of arithmetic, it requires going outside of arithmetic to its basis in logic. Rather surprisingly—and independently of the issue of its consistency—the nature and intended scope of the system of logic of *Basic Laws* prevent it from possessing the simple intuitive appeal of Hilbert's proposal. Were it not for Gödel's second incompleteness theorem, Hilbert's approach to the consistency of PA would have constituted a successful transcendental justification of it, since it would have shown that PA is consistent relative to what is demanded by a framework within which to inquire into the nature of mathematical theories. Hilbert therefore promised to carry the argument for the apriority of arithmetic in a direction that is genuinely minimalist in its use of primitive assumptions. This cannot be said of the system of *Basic Laws*. For this reason, Hilbert's account is more recognizably a putative *foundation* for arithmetic—indeed, for mathematics generally—than Frege's.

Our emphasis thus far has been on those basic laws of arithmetic which ensure the Dedekind infinity of the numbers. Early on, Hilbert expressed concern with the method by which the existence of a Dedekind infinite concept is proved in the logicist tradition.[8] But the principal divergence between logicist and non-logicist approaches to arithmetic arises in connection with the explanation of the remaining law—the principle of mathematical induction. Frege explains the validity of reasoning by induction by deriving the principle from the definition of the natural numbers as the class of all objects having all hereditary properties of zero. (A property is *hereditary* if whenever it is possessed by x it is possessed by x's immediate successor.) Anti-logicists explicitly reject this explanation, resting as it does on the questionable

---

[8] See especially the paragraphs devoted to Frege and Dedekind in Hilbert 1904: 130–1.

idea of the totality of *all properties* of numbers, and argue that in light of the paradoxes we have no entitlement to our confidence in this notion.

Although it was conceived in ignorance of the set-theoretic paradoxes and the analysis of their possible source, there is a sense in which the program of recovering arithmetic from FA retains its interest and integrity even in light of the paradoxes: Frege's definition of the natural numbers shows that there is also a *basis*—if not an entirely satisfactory *justification*—for characteristically arithmetical modes of reasoning in logical notions. It is a further discovery that pursuing arithmetic from this perspective is susceptible to a doubt that might be avoided by taking as primitive some idea of indefinite iteration. But it remains the case that there is a logical account of reasoning by induction, even if our confidence in it may be diminished by various analyses of the paradoxes. In order to pursue these and related issues more effectively, it will be worthwhile to consider recent developments of Fregean and other approaches to the structure of the natural numbers.

Characterizing the Structure of the Natural Numbers

Let me begin with an unduly neglected recent development of the Fregean perspective. Bell (1999a) contains the following theorem:

Let v be a map to E with domain $dom(v)$ a family of subsets of E satisfying the following conditions:

  (i) $\emptyset \in dom(v)$
  (ii) $\forall U \in dom(v) \ \forall x \in E{-}U, \ U \cup \{x\} \in dom(v)$
  (iii) $\forall U,V \in dom(v) \ (v(U) = v(V) \text{ iff } U \approx V)$,

where by $U \approx V$ is meant that there is a bijection from U on to V.
Then we can define a subset N of E which is the domain of a model of Peano's axioms in their canonical set-theoretic formulation.

In his (1999b) Bell calls a pair (E, v) satisfying the above conditions a *Frege structure*. The connection of this notion with Frege's account of arithmetic derives of course from condition (*iii*), which is evidently a form of Hume's principle. Indeed, if v is required to be the smallest map satisfying the conditions of the theorem, then (*iii*) expresses what I will call *Finite Hume*, that is, Hume's principle restricted to weakly finite sets in the sense of conditions (*i*) and (*ii*).

It is evident that Bell's theorem bears on our earlier discussion of the suitability of FA as an analysis of PA.[9] Its significance for the reconstruction of Frege is that it not only characterizes models of PA in terms of a cardinality map which satisfies a form of Hume's principle, but it also isolates, in general terms, minimal conditions on the domain of such a map for the characterization to be successful. It also accords with Frege's "fundamental thought"—that a statement of number is an assertion about a concept—insofar as it recovers the numbers from an assumption concerning the

---

[9] The set-theoretic framework is entirely incidental, since there is obviously no impediment to reformulating the theorem and its proof in a higher-order logical setting.

structure of the family of (extensions of) concepts to which cardinality judgements apply: it is because the domain of our cardinality judgements is as extensive as it is and has the character that it does that a model of PA is recoverable from a Frege structure. The idea then is that a Fregean account of the infinite object which constitutes a domain for a model of PA is given in terms of a domain of finite sets over an arbitrary but fixed set E, ordered by inclusion and admitting a cardinality map which satisfies the conditions imposed by the theorem. The argument in favor of such a Fregean analysis is not that it affords greater security for our arithmetical beliefs, but that it explains our knowledge of pure arithmetic, which is encapsulated in PA, in terms of our knowledge of what underlies its application—which is what is expressed by FA.

Bell's discussion bears comparison with the ideas expounded by Feferman and Hellman in their (1995) and (2000). As we will see, their discussion stands to aspects of the work of Dedekind and Hilbert, as Bell's does to that of Frege. The sets mentioned in the hypothesis of Bell's theorem, those belonging to $dom(v)$, satisfy the following two finiteness axioms of Feferman and Hellman's "Elementary theory of finite sets" (EFS), namely

FS-I   Ø exists and is finite.
FS-II   If X is finite, so is $X \cup \{x\}$.

But Feferman and Hellman's work has no analogue of condition *(iii)* of the definition of a Frege structure. In place of *(ii)* and Finite Hume, Feferman and Hellman have a pairing axiom which asserts that pairing is one–one, and another asserting the existence of an urelement with respect to pairing. One sees immediately from the pairing axioms that although Feferman and Hellman are more circumspect than Dedekind in their choice of urelement and "transformation," their inspiration is evidently traceable to the account of *Was sind und was sollen die Zahlen?*; this is a point to which I will return.

There are important methodological differences between Bell's work and Feferman and Hellman's that have nothing to do with the Fregean idea that motivates condition *(iii)*, but are more in line with the emphasis on finiteness and the avoidance of the "logical totalities" that were mentioned in our discussion of Frege and Hilbert. Bell makes extensive use of *inductive properties* of elements of $dom(v)$. These are defined as properties φ such that $\varphi(\emptyset)$ and for any X if $\varphi(X)$ and $v(X) \notin X$, then $\varphi(X^+)$, where $X^+ = (X \cup \{v(X)\})$. A subfamily A of $dom(v)$ is an *inductive subfamily* if being a member of A is an inductive property. N is defined as the intersection of all inductive subfamilies of $dom(v)$; from this definition, Bell derives the following *Principle of Induction* for N:

For every property φ defined on members of N, if φ is inductive, then φ holds of every member of N.

Many of Bell's proofs appeal to this principle of induction, which is evidently as strong as induction on finite sets. By contrast, none of Feferman and Hellman's proofs appeal to finite set induction, let alone to a derivation of it from a definition of the character of the definition of N. This is essential to their methodology, since Feferman and Hellman are concerned to establish their results in a predicative setting based on a notion of finiteness; an appeal to an object defined as **N** is—as a certain minimal closure—would violate the spirit of such an investigation.

In any event, assuming induction on **N**, Bell in effect establishes that $N = \{v(X): X \in \mathbf{N}\}$ is the domain of what Feferman and Hellman call a *pre-N structure* (for *pre-natural number* structure)—namely, a triple *(M, a, g)* where *M* is a class in the sense of their "Elementary theory of finite sets and classes" (EFSC), $a \in M$ and *g* is a map $g: M \rightarrow M$ such that

$$\text{N-I} \quad \forall x \in M\, (g(x) \neq a)$$

$$\text{N-II} \quad \forall x, y \in M(g(x) = g(y) \rightarrow x = y).$$

The connection between Frege structures and pre-N structures is very simple: If we put $0 = v(\emptyset)$ and define $s(v(X)) = v(X \cup \{v(X)\}) = v(X^+)$, then *(N, 0, s)* is a pre-N structure.

An *N-structure* is defined by Feferman and Hellman as a pre-N structure which satisfies the following weak induction axiom—"weak" because the context is Feferman and Hellman's EFSC, whose *Comprehension axiom* is weak second order in the sense that the only classes it countenances are defined by formulas not containing bound class variables:

$$\text{N-III} \quad \forall X \subseteq M[(a \in X \,\&\, \forall x\, (x \in X \rightarrow g(x) \in X)) \rightarrow X = M].$$

Bell is able to derive full induction for Frege structures given the definition of **N** as the *minimal* inductive family contained in *dom(v)*. This is of no methodological advantage for an investigation like Feferman and Hellman's, since their goal is to avoid the technique employed in the definition of **N** and by proceeding predicatively, in a context of elementary facts about finiteness, to recover the existence and categoricity of the structure of the natural numbers.[10]

It is worthwhile to compare Feferman and Hellman's procedure with Frege's use of the ancestral of a relation and Dedekind's closely similar use of chains. Call a property φ *hereditary*—strictly, *hereditary with respect to a binary relation f*—if whenever $\varphi(x)$ and *xfy*, $\varphi(y)$. For Frege, the *ancestral* f^* of a binary relation *f* holds between *x* and *y*—*xf*y*—just in case *y* has every hereditary property of *x*. The ancestral is thus defined in terms of the notion of *all properties*. Frege often has occasion to use the ancestral in connection with properties of the form

$$[x : a\, f^*\, x],$$

or allowing for the possibility that *a = x*, with properties associated with the *weak ancestral* $f^*_=$, namely

$$[x : a\, f^*_=\, x],$$

A fact about finiteness which I have not mentioned is the axiom of EFS that Feferman and Hellman call *Separation*:

(Sep)   *Every definable subset of a finite set is finite.*

The axiom is historically interesting because when, in the 1925 *Principia*, Russell attempted to recover induction without the Axiom of Reducibility, it was in the proof of a lemma to the effect of (Sep) that his argument failed. (The error was first noted by Gödel 1944.)

which is read, "the property of being the same as a or following a in the f-series."[11] Dedekind's chains have the same character, and are, in effect, just the *extensions* of Frege's properties for the case where f is one–one. Thus $\leq_F$ is the weak ancestral (in the sense of Frege's definition—hence the subscript) of the relation *immediate successor*, and its associated chains are classes of the form

$$\{y : x \leq_F y\}.$$

The Frege–Dedekind definition of the class N of natural numbers can thus be formulated as

$$N = \{y : 0 \leq_F y\}.$$

The definition succeeds because the property *being a natural number* is both hereditary with respect to immediate succession and a property of 0, and so falls within the range of the quantifier in the definition of $\leq_F$.

Feferman and Hellman distinguish between characterizing a natural number structure and characterizing the natural numbers in a way that supports *full* induction. The construction of Frege, which we have just described, was principally conceived in aid of the second task, since Frege sought to explain iterative or inductive reasoning in terms of logical reasoning with all properties. But the characterization of the natural numbers requires *less* than is assumed in the explanation of full induction.

Feferman and Hellman's innovation is two fold. First, they recover the domain N of a natural number structure by consideration of the closures of sets under the *converse* of the relation of immediate succession: A is *closed* if whenever x's immediate successor belongs to A, x belongs to A. Feferman and Hellman's relation $\leq$ is an analogue of Frege's ancestral $\leq_F$: $y \leq x$ says that y belongs to every *finite* set which contains x and is closed with respect to immediate predecessor. The set

$$\{y : y \leq x\}$$

does not rest on the notion *all classes* (or using Frege's framework, *all properties*—the choice is not important in the present context), but on the notion *all finite sets*. This is in keeping with the primacy, in foundational work after Frege, of the notion of finiteness over the unrestricted use of *all properties*. Second, since Feferman and Hellman's definition of $\leq$ involves quantification over the totality of all finite sets, and N is not finite but infinite, N does not lie in the range of the quantifier occurring in the definition of $\leq$.

Following a simplification suggested by Aczel, Feferman and Hellman's preferred definition of N is

$$N = \{x : \mathrm{Fin}(\mathrm{Pd}(x)) \,\&\, 0 \leq x\},$$

where $\mathrm{Pd}(x)$ is the set $\{y : y \leq x\}$ of x's predecessors, and $\mathrm{Fin}(\mathrm{Pd}(x))$ asserts that $\mathrm{Pd}(x)$ is finite. It can be established in EFSC that N is the domain of an N-structure

---

[11] I am here borrowing a notational device of Boolos (1987) by using [:] to form a term for a Fregean concept or property.

and that all such structures are isomorphic. Feferman and Hellman thus avoid the kind of impredicativity that afflicts the classical Frege–Dedekind style of definition.

Let me return to Bell's results. Bell's proofs are all carried out within "constructive" or "intuitionist" set theory, which is to say ordinary set theory ZFC with intuitionist logic as its underlying logic; none of the proofs use excluded middle or intuitionistically equivalent principles. Bell stresses the constructiveness, in this sense, of his proof that every Frege structure has an associated pre-N structure. The domain of the cardinality map v of (E, v), and therefore the notion of a Frege structure itself is characterized independently of any concept of natural number. While Bell's proofs are carried out in constructive set theory, Feferman and Hellman's are carried out in one or another of their predicatively acceptable theories EFS or EFSC. Thus, Bell's proofs depend on a principle of induction equivalent to induction on finite sets, whereas Feferman and Hellman consciously avoid such an assumption, depending as it does "on an explicit understanding of the even more complex infinite structure of finite sets ordered [for example] by inclusion" (2000: 322). Feferman and Hellman's construction does, of course, depend on the notion of an arbitrary finite set, but they argue that this is a "self-standing" notion, rather than one that is characterized in terms of membership in a structure which includes all finite sets.

As already mentioned, given N, Bell can prove not only N-III, but *full* induction for $(N, 0, s)$. But the foundational interest of this is somewhat mitigated, since it can be argued that by proving the existence and categoricity of N-structures in EFSC one is able to keep separate issues that are genuinely distinct: There is, *first*, the problem of establishing the existence and nature of the structure of the natural numbers; and, *second*, that of justifying reasoning by induction. The solution to the first problem depends only on a notion of finite set, which can be explained independently of the notion of natural number itself. In fact, to the extent that a solution to this problem can be achieved without recovering induction from the definition of the property *being a natural number*, the notion of a finite set of individuals is justifiably regarded as "prior" to the notion of natural number, and Feferman and Hellman's methodology is vindicated.

Bell's procedure is much closer to Frege's, whose definition of *successor* (in *Fdns* §76) is adapted by Bell for the definition of X^+: X^+ and the subsequent definition of N in terms of v's values in E reflect Frege's demand that any successful account of the numbers must acknowledge the fact that anything can be counted, including the numbers themselves. The generalized form of induction on which Bell relies is *not* induction on the numbers, but induction on certain finite sets. Since *finite set* does not presuppose *natural number*, there is nothing circular about proceeding in this way. But since the definition of N appeals to *all* inductive families, it raises the same general concern that was urged against the classical logicist justification of mathematical induction in terms of the Frege–Dedekind definition of N. To highlight the parallel with Bell's account, we can express this definition in the following terms: A *property* is *inductive* (with respect to immediate successor) if whenever it is possessed by x it is possessed by x's immediate successor, and a *set* is *inductive* if membership in the set is an inductive property. Then the set N of natural numbers is the *minimal* inductive set containing *zero*, and this involves quantification over a totality to which N itself belongs. The issues raised by constructivist criticisms of logicism have focused on the alleged circularity of such a definition—whether of N or of $\mathbf{N}$—not on excluded middle. The constructivity of constructive set theory is therefore largely orthogonal to the philosophical discussion. So if the goal is to proceed constructively in a sense that bears on the conceptual issues relevant to this debate, then Feferman and Hellman would appear to have a better claim to having provided a constructivist account of the structure of the natural numbers. For a Fregean to make a comparable claim, the use of the minimal closure of the families of inductive sets to derive the Principle of Induction would have to be avoided by assuming this principle as a primitive truth about reasoning with finite sets. To proceed in this way would be to concede more to constructivism than is compatible with the goals of logicism.

It is important therefore to appreciate that the argument in favor of a Fregean approach has nothing to do with constructivity, but with its success in providing an explanation of the basic laws of arithmetic that connects them with applications. By contrast, Feferman and Hellman's pairing axioms,

P-I $\exists u\, \forall x, y\ (u \neq (x, y))$

P-II $(x,\ y) = (u, v)$ *if and only if* $x = u$ *and* $y = v$,

fulfill the role that for a Fregean is played by Hume's principle, or by Finite Hume. P-II has the form of a criterion of identity for pairs, just as Hume's principle is a criterion of identity for number. But since pairing is a type-preserving map—comparable to Cantor's map injecting the rationals into the natural numbers—it is evident that it cannot lead to paradoxes of the kind we associate with Basic Law V. The type-lowering character of Hume's principle is the feature that makes the question of its consistency a substantial one. But though intuitively plausible, the pairing axioms tell us little about the numbers, and say nothing about what the numbers are *for*. They are concretely realizable by systems of notation, and as such, are suggestive of a *psychological* explanation of our knowledge of arithmetic in terms of our acquaintance with symbolic systems. Indeed, this is little more than a linguistic transcription of Dedekind's idea that pairing is realized by the association of a thought with its object, and that one's Ego, which is not a (thought, object of thought)-pair, is an urelement with respect to pairing. One can see how, on this basis, an account of arithmetical knowledge might be forthcoming by treating it as derivative from our capacity for knowledge of language. Frege's concern is altogether different, attempting as it does, to address the *epistemological* basis for arithmetical knowledge in judgements of cardinality. So the difference in starting point that emerges is a fundamental one having to do with the goals one imposes on a theory of arithmetical knowledge and the methodological strictures by which such a theory is constrained.

The advantage which I am claiming for Frege's theory of arithmetic is easily separated from the project of identifying *which* objects the numbers are. It is essential to the success of the Fregean program that it recover N-structures from Frege structures, but to achieve this it is not necessary to single out the numbers uniquely. *Any E* which supports a family—including the smallest such family—of weakly finite subsets satisfying the conditions of Bell's theorem can serve as the domain of a Frege structure. Frege's original question, "What are the numbers?," has, in the process, been replaced by another—namely, "What is a natural number type structure?" This latter question is reminiscent of Dedekind and Hilbert, but the answer in terms of the notion of a Frege structure remains thoroughly Fregean.[12]

Here then is the bearing of our reconstruction on the *objectivity* of arithmetic. Finite Hume is not an arbitrary postulate, but is the principle which controls our ordinary judgements of cardinality. But these judgements are objective; hence so also is Finite Hume, since it is the principle on which their evaluation depends. The basic laws of arithmetic thus inherit *their* objectivity from being founded on the principle that controls our ordinary cardinality judgements. The objectivity the analysis imparts to arithmetic is altogether independent of Frege's conception of numbers as objects except in the very weak sense that they must be arguments to concepts of first level. Such an account of arithmetic's objectivity is entirely compatible with the fact that neither FA nor PA single out the domain of a particular natural number structure as *the* numbers; and neither the objectivity of arithmetic nor its generality rests on an appeal to the idea that numbers are logical objects.[13]

---

[12] It is therefore entirely open to a Fregean to concur with Feferman's assessment that

> [t]he first task of any general foundational scheme for mathematics is to establish the number systems. In both the extensional and intensional approach this is done from the modern structuralist point of view. The structuralist viewpoint as regards the basic number systems is that it is not the specific nature of the individual objects which is of the essence, but rather the isomorphism type of the structure of which they form a part. Each structure is to be characterized up to isomorphism by a structural property which, logically, may be of first order or of higher order.
>
> (Feferman 1985: 48; quoted by Feferman and Hellman 2000: 322)

This is reminiscent of Hilbert's remark to Frege that

> it is surely obvious that every theory is only a scaffolding (schema) of concepts together with their necessary connections, and that the basic elements can be thought of in any way one likes. For example, instead of points, think of a system of love, law, chimney-sweep ... which satisfies all the axioms; then Pythagoras' theorem also applies to these things. Any theory can always be applied to infinitely many systems of basic elements.
>
> (Excerpt from a draft or letter of Hilbert to Frege, 29.xii.99; in McGuinness 1980: 42; translation by Hans Kaal)

[13] This chapter was originally presented to a REHSEIS workshop in the philosophy of mathematics; my thanks to the organizer Marco Panza, my co-contributors Michael Hallett and Stewart Shapiro, and my audience for their stimulating questions. I wish to thank David Boutillier, Peter Clark, Janet Folina, and Edward Stabler, all of whose comments significantly improved my earlier formulations. I am especially indebted to Anil Gupta, Peter Koellner, and Erich Reck for their critical remarks and generous advice on the matters dealt with here. My research was supported by the Social Sciences and Humanities Research Council of Canada.

References

Bell, J. L. (1999a). "Frege's Theorem in a Constructive Setting," *Journal of Symbolic Logic* 64, 486–8.

Bell, J. L. (1999b). "Finite Sets and Frege Structures," *Journal of Symbolic Logic* 64, 1552–6.

Boolos, G. (1987). "The Consistency of Frege's *Foundations of Arithmetic*," in J. J. Thompson (ed.), *On Being and Saying: Essays in Honor of Richard Cartwright*, Cambridge, MA: MIT Press, 3–20.

Burge, T. (2005 [2000]). "Frege on Apriority," in T. Burge, *Truth, Thought, Reason*, Oxford: Oxford University Press, 356–89.

Burgess, J. (2005). *Fixing Frege*, Princeton: Princeton University Press.

Dedekind, R. (1963). *Essays on the Theory of Numbers*, W. W. Beman (trans.), New York: Dover.

Demopoulos, W. (1994). "Frege and the Rigorization of Analysis," *Journal of Philosophical Logic* 23, 225–46.

Dummett, M (1991). *Frege: Philosophy of Mathematics*, Cambridge, MA: Harvard University Press.

Feferman, S. (1985). "Intensionality in Mathematics," *Journal of Philosophical Logic*, 14, 41–55.

Feferman, S., and G. Hellman (1995). "Predicative Foundations of Arithmetic," *Journal of Philosophical Logic*, 24, 1–17.

Feferman, S., and G. Hellman (2000). "Challenges to Predicative Foundations of Arithmetic," in G. Sher and R. Tieszen (eds.), *Between Logic and Intuition: Essays in Honor of Charles Parsons*, New York: Cambridge University Press, 317–38.

Frege, G. (1884). *The Foundations of Arithmetic: A Logico-Mathematical Enquiry into the Concept of Number*, J. L. Austin (trans.), Evanston: Northwestern University Press, 1980, second revised edition.

Frege, G. (1891). *Function and Concept*, in B. McGuinness (ed.), *Gottlob Frege: Collected Papers on Mathematics, Logic and Philosophy*, P. Geach (trans.), Oxford: Blackwell, 1980, 137–56.

Frege, G. (1893–1903). *The Basic Laws of Arithmetic: Exposition of the System*, M. Furth (trans.), Berkeley: University of California Press, 1964.

Geach, P. (1976). "Review of Michael Dummett, *Frege: Philosophy of Language*," *Mind* 84, 436–49.

Gödel, K. (1944). "Russell's Mathematical Logic," in P. A. Schilpp (ed.) *The Philosophy of Bertrand Russell*, New York: Harper Torchbooks, 123–54.

Hermes, H. et al. (eds.) (1979). *Gottlob Frege: Posthumous Writings*, Peter White and Roger Long (trans.), Oxford: Blackwell.

Hilbert, D. (1904). "On the Foundations of Mathematics," S. Bauer-Mengelberg (trans.), in Jean van Heijenoort (ed.), *From Frege to Gödel*, Cambridge, MA: Harvard University Press, 1967, 129–38.

Hilbert, D. (1925). "On the Infinite," in Jean van Heijenoort (ed.), *From Frege to Gödel*, S. Bauer-Mengelberg (trans.), Cambridge, MA: Harvard University Press, 1967, 367–92.

Jeshion, R. (2001). "Frege's Notions of Self Evidence," *Mind* 110, 937–76.

McGuinness, B. (ed.) (1980). *Philosophical and Mathematical Correspondence of Gottlob Frege*, H. Kaal (trans.), Oxford: Blackwell.

Ramsey, F. P. (1925). "The Foundations of Mathematics," reprinted in R. B. Braithwaite (ed.), *The Foundations of Mathematics and Other Logical Essays*, London: Kegan Paul, 1931.

Ricketts, T. (1997). "Truth Values and Courses of Values in Frege's *Grundgesetze*," in W. W. Tait (ed.), *Early Analytic Philosophy: Frege, Russell and Wittgenstein*, Chicago: Open Court, 187–211.

Wright, C. (1983). *Frege's Conception of Numbers as Objects*, Aberdeen: Aberdeen University Press.

2

A Logic for Frege's Theorem

Richard Kimberly Heck

1. Opening

As is now well-known, axioms for arithmetic can be interpreted in second-order logic plus 'Hume's Principle,' or HP:

$$Nx:Fx = Nx:Gx \text{ iff } \exists R[\forall x \forall y \forall z \forall w (Rxy \wedge Rzw \rightarrow x = z \equiv y = w) \wedge$$

$$\forall x(Fx \rightarrow \exists y(Rxy \wedge Gy)) \wedge$$

$$\forall y(Gy \rightarrow \exists x(Rxy \wedge Fx))]$$

This result is *Frege's Theorem*. Its philosophical significance has been a matter of some controversy, most of which has concerned the status of HP itself. To use Frege's Theorem to reinstate logicism, for example, one would have to claim that HP was a logical truth. So far as I know, no one has really been tempted by that claim. But Crispin Wright claimed, in his book *Frege's Conception of Numbers as Objects* (1983), that, even though HP is not a logical truth, it nonetheless has the epistemological virtues that were really central to Frege's logicism. Not everyone has agreed.[1] But even if Wright's view were accepted, there would be another question to be asked—namely, whether the sorts of inferences employed in deriving axioms for arithmetic from HP preserve whatever interesting epistemological property HP is supposed to have. Only then would the axioms of arithmetic have been shown to have that interesting property.

The problem is clearest for a logicist. If the axioms of arithmetic are to be shown to be logical truths, not only must HP be a logical truth, the modes of inference used in deriving axioms of arithmetic from it must preserve logical truth. They must, that is to say, be logical modes of inference. For Wright, the crucial question is less clear. It would be enough for his purposes if these modes of inference preserved whatever interesting epistemological property HP was supposed to have. But Wright has, nonetheless, typically been content to claim that second-order reasoning is logical reasoning and to suppose, reasonably enough, that, if that claim is good enough for the logicist, it is good enough for his purposes, too.

---

[1] There's a nice back-and-forth about this between Wright (2001) and George Boolos (1998a).

Richard Kimberly Heck, *A Logic for Frege's Theorem* In: *Logic, Language, and Mathematics: Themes from the Philosophy of Crispin Wright*. Edited by: Alexander Miller, Oxford University Press (2020). © Richard Kimberly Heck.
DOI: 10.1093/oso/9780199278343.003.0002

The claim that 'second-order logic is logic,' as it is often put, has had both defenders and detractors.[2] I am not going to enter that debate here. What I want to argue here is that a neo-logicist does not need to commit herself to any claims about second-order logic.

In a typical proof of Frege's Theorem, axioms for arithmetic are derived from HP in second-order logic, but not all of the power of second-order logic is needed for the proofs of the axioms. The power of second-order logic derives from the so-called comprehension axioms, each of which states, in effect, that a given formula defines a 'concept' or 'class'—something in the domain of the second-order variables. These axioms take the form:[3]

$$\exists F \forall x [Fx \equiv A(x)].$$

In full second-order logic, one has such an axiom for every formula $A(x)$ in which 'F' does not occur free.[4] So-called 'predicative' second-order logic has comprehension only for formulae containing no bound second-order variables. Predicative second-order logic is weak in a well-defined sense: Given any first-order theory Θ, adding predicative second-order logic to Θ yields a conservative extension of it.[5] Full second-order logic, on the other hand, is extremely powerful, and it is that power that underlies much of the skepticism about the appropriateness of the term '*second-order logic*.'[6]

Between predicative second-order logic and full second-order logic are systems of intermediate strength, each admitting a different set of comprehension axioms. In principle, any set of comprehension axioms will do, and there are many that have been considered.[7] What are perhaps the most natural intermediate systems arise from syntactic restrictions on the formulae appearing in the comprehension axioms. Say that a formula containing no bound second-order variables is Π_∞^0. Then

---

[2] Quine (1986) was famously skeptical. Boolos (1998b,e) was an early proponent. Even still, Boolos was not at all sure that second-order reasoning would preserve analyticity (Boolos 1998d). And at the end of his life, Boolos claimed no longer to understand the question whether second-order logic is logic, stated so baldly.

[3] There are similar axioms for many-place predicates, of course.

[4] At the other extreme, one could consider a system that had no comprehension axioms at all, but this adds essentially nothing to first-order logic.

[5] A theory Γ' in a language L' conservatively extends a theory Γ stated in $L \subseteq L'$ if every theorem of Γ' statable in L is already a theorem of Γ. And we can show that Γ' does conservatively extend Γ by showing that every model of Γ can be extended to a model of Γ' by adding interpretations of the primitives in $L' \backslash L$. For suppose so and suppose A is not a theorem of Γ. By completeness, let $\mathcal{M}$ be a model of Γ in which A is false. We can extend $\mathcal{M}$ to a model $\mathcal{M}'$ of Γ', and then A will also be false in $\mathcal{M}'$ and so not a theorem of Γ'.

Now, given any model for Θ, let the second-order domain contain exactly the subsets of the first-order domain definable (with first-order parameters) in the language of Θ so interpreted. The result is a model of Θ plus predicative comprehension, which is thus a conservative extension of Θ.

There is some need for care here when Θ contains axiom-schemata: The schemata must not come to have new instances as a result of the addition of second-order vocabulary.

[6] Boolos expresses this worry (1998a). It also plays a role in some of John Etchemendy's discussions (1990). Peter Koellner (2010) has developed an extremely sophisticated version of this objection.

[7] The standard reference on second-order logic is Stewart Shapiro's *Foundations without Foundationalism* (Shapiro 1991: esp. chs 3–4). For a shorter, more accessible overview, see his piece in the *Blackwell Guide to Philosophical Logic* (Shapiro 2001). There are many intermediate systems other than those we shall consider here, some of which are not based upon comprehension principles but various sorts of choice principles, and so forth.

where φ is Π_∞^0, formulae of the form $\forall F_1 \ldots \forall F_n \varphi$ and $\exists F_1 \ldots \exists F_n \varphi$ are Π_1^1 and Σ_1^1, respectively. If φ is Σ_n^1 (Π_n^1), then $\forall F_1 \ldots \forall F_n \varphi$ ($\exists F_1 \ldots \exists F_n \varphi$) is Π_{n+1}^1 (Σ_{n+1}^1). Second-order logic with Π_n^1 comprehension has only those comprehension axioms in which $A(x)$ is Π_n^1 (or simpler).

It is important to note that, as I have formulated the Π_n^1 comprehension scheme, free second-order variables are allowed to occur in the comprehension axioms. As a result, there is no significant difference between Π_n^1 comprehension and Σ_n^1 comprehension. If $A(x)$ is a Σ_n^1 formula, then its negation is (trivially equivalent to) a Π_n^1 formula. Hence, Π_n^1 comprehension delivers a concept F such that:

$$\forall x[Fx \equiv \neg A(x)].$$

But then predicative comprehension delivers a concept G such that:

$$\forall x[Gx \equiv \neg Fx],$$

And so we have:

$$\exists G \forall x[Gx \equiv A(x)].$$

We might as well therefore regard Π_n^1 comprehension as Π_n^1-or-Σ_n^1 comprehension.

More significantly, consider the formula:

$$\forall F[Fa \wedge \forall x(Fx \wedge Pxy \rightarrow Fy) \rightarrow Fb].$$

This formula defines the so-called 'weak ancestral' of the relation P. It is obviously Π_1^1, so Π_1^1 comprehension delivers a concept $\mathbb{N}$ such that:

$$\mathbb{N}n \equiv \forall F[Fa \wedge \forall x(Fx \wedge Pxy \rightarrow Fy) \rightarrow Fn].$$

If we take a to be o and P to be the relation of predecession—read Pxy as: x is the number immediately preceding y—then that is Frege's definition of the concept of a natural number. And Π_1^1 comprehension delivers the existence of this concept even if P itself has been defined by a Σ_1^1 formula, as it normally is in Fregean arithmetics:

$$Pab \equiv \exists G \exists y[b = Nx : Gx \wedge Gy \wedge a = Nx : (Gx \wedge x \neq y)].$$

The existence of the relation P is guaranteed by Σ_1^1—equivalently, Π_1^1—comprehension.

We may seem to be cheating here: Won't such a method end up reducing all comprehension to Π_1^1 comprehension? That would indeed be disastrous, but no such result is forthcoming. Chaining instances of comprehension together works in this case only because the variable F does not occur within the scope of the quantifier $\exists G$ that appears in the definition of P. The method will allow us to apply Π_1^1 comprehension twice to a formula of the form:

$$\forall F[\ldots F \ldots \rightarrow \exists G(\ldots G \ldots)],$$

but not to one of the form:

$$\forall F[\ldots F \ldots \rightarrow \exists G(\ldots G \ldots F \ldots)].$$

But one might still think such 'chaining' impermissible, even if coherent. Comprehension, so formulated, collapses Π_1^1 and Σ_1^1 comprehension and, moreover, fails to distinguish Π_1^1 sets from sets that are Π_1^1 in Π_1^1 sets. Is that really wise? Obviously, I am not suggesting that these distinctions do not matter, and if one wishes to use second-order logic to investigate problems to which these distinctions are relevant, then comprehension should be formulated so as to prohibit such 'chaining': One need only prohibit free second-order variables from appearing in the comprehension scheme. But it is not clear that these distinctions matter in the present context. I shall discuss the matter further below (see §3, page 34). For the moment, I appeal to authority: Solomon Feferman formulates the comprehension axioms this way in his classic paper 'Systems of Predicative Analysis' (1964).

Both the concept of predecession and the concept of natural number are thus delivered by Π_1^1 comprehension: That should make it plausible that the standard proof of Frege's Theorem requires only Π_1^1 comprehension, a conjecture that can be verified by working through the proof in detail while paying careful attention to what comprehension axioms are used.[8] There is a sense in which this result is best possible. I have mentioned several times that axioms for arithmetic can be derived from HP in second-order logic, but I have not yet said which such axioms I have in mind. There are, of course, many equivalent axiomatizations—I shall present one such axiomatization below—but what is important at the moment is that standard presentations of Frege's Theorem do not include a derivation of the usual first-order axioms for addition and multiplication. The reason is that, in a standard second-order language, the recursive definitions of addition and multiplication can be converted into explicit definitions in a way due (independently) to Dedekind and to Frege. The recursion equations themselves—and these just are the first-order axioms—can then be recovered from the definition. Unsurprisingly, however, the derivation of the recursion equations from the explicit definition needs more than predicative comprehension. The proof that addition and multiplication are well-defined and satisfy the recursion equations is by induction, and the induction is on a predicate containing the definition of addition or multiplication. The legitimacy of the induction thus presumes that the predicate in question defines a relation. Since the formula that defines addition is Π_1^1, we will need at least that much comprehension even to interpret first-order PA.[9]

One can at least imagine a view that would regard Π_1^1 comprehension axioms as logical truths but deny that status to any that are more complex—a view that would, in particular, deny that full second-order logic deserves the name. In light of what has been said, such a view would serve the purposes of a neo-logicist such as Wright. I do not expect it to be obvious at this point how such a view might be motivated, and it is in fact no part of the view I want to defend here that, say, Δ_3^1 comprehension axioms are *not* logical truths. What I am going to suggest, however, is that there is a special case to be made on behalf of Π_1^1 comprehension. Or something like it.

---

[8] The mentioned fact was first noted in my paper 'Counting, Cardinality, and Equinumerosity' (Heck 2000). Øystein Linnebo gives a detailed proof, and he also proves the converse: PA with Π_1^1-comprehension is interpretable in FA with Π_1^1-comprehension (Linnebo 2004). See also John Burgess's *Fixing Frege* (2005: 151ff).

[9] Assuming we add no other axioms. It turns out that there is a way to interpret PA in ramified predicative Frege arithmetic, if we add a weak form of reducibility (Heck 2011).

2. Predecession

As it happens, the only comprehension axioms one actually needs for the proof of Frege's Theorem—besides a handful of instances of predicative comprehension—are these:

$$\exists P\{Pab \equiv \exists G\exists y[b = Nx:Gx \wedge Gy \wedge a = Nx:(Gx \wedge x \neq y)]\}$$

$$\exists R\{Ran \equiv \forall F[Fa \wedge \forall x(Fx \wedge Pxy \rightarrow Fy) \rightarrow Fn]\}$$

The latter, of course, defines the relation that would usually be written: $\xi P^{*=}\eta$. That is, it defines the relation that is the ancestral of predecession.

Øystein Linnebo suggests that there is something seriously wrong with Frege's definition of predecession (2004: 172–3). It simply does not seem reasonable to suppose that a notion as simple as that of predecession should logically be so complex. Consider, for example, the proof of the familiar fact that every number other than zero has a predecessor. This proposition, $x \neq 0 \rightarrow \exists y(x = Sy)$, is one of the axioms of Robinson arithmetic, Q, but it is redundant in PA, since it is provable in PA and, in fact, in the much weaker theory known as $I\Delta_0$, which has induction only for bounded formulae.[10] In Frege arithmetic, however, a formalization of that proof would require Σ_1^1 comprehension, since the induction must now be on $x \neq 0 \rightarrow \exists y(Pyx)$.[11] Are we really to believe that such strong logical resources are needed for the proof of such a simple statement? The more plausible view is the one enshrined in the usual treatment of arithmetic: Predecession is a *primitive* notion.

Linnebo's concern is a sensible one, but I think it can be answered. Although the definition of predecession is undeniably Σ_1^1 in form, it is not, I want to suggest, Σ_1^1 in spirit. The definition one would really like to give is this one:

(P-lite) $P(Nx:Gx, Nx:Fx) \equiv \exists y(Fy \wedge Nx:Gx = Nx:(Fx \wedge x \neq y))$.

To be sure, (P-lite) is not a proper definition. It does not tell us when *Pab* but only when *P(Nx:Gx,Nx:Fx)*: Nothing in (P-lite) tells us whether Julius Caesar, that familiar conqueror of Gaul, precedes 0 or not. But the obvious reply is that it was supposed to be implicit in (P-lite) that *only numbers have or are predecessors*. If Caesar is a number, then he is the number of Fs, for some F, in which case (P-lite) will determine which numbers he precedes and succeeds. If he is not a number, then he neither precedes nor succeeds any number. Hence, the question which numbers Caesar precedes and succeeds is equivalent to the question whether Caesar is a number and, if so, which one he is. Well, if that isn't a familiar problem! Maybe it is even a serious problem. But it is a problem the neo-logicist had anyway.

Suppose that the Caesar problem has either been solved or been justifiably ignored. (Maybe it isn't a serious problem, just an amusing one.) Then (P-lite) tells one everything one needs to know about predecession. How would that allow

---

[10] The induction can be carried out on $x \neq 0 \rightarrow \exists y < x(x = Sy)$.

[11] A different proof can be given that would not require comprehension at all, but there are other examples of this same form.

the neo-logicist to avoid appealing to Σ_1^1 comprehension? Well, the neo-logicist might regard predecession as *primitive* and regard (P-lite) as analytic of that notion. Moreover,

(P-imp) $Pab \rightarrow \exists F(a = Nx:Fx) \wedge \exists G(b = Nx:Gx),$

simply makes explicit the implicit requirement that only numbers can be or have predecessors, so it too is analytic of predecession. No appeal to comprehension is then needed to guarantee that the relation of predecession exists, any more than in the usual formulation of second-order arithmetic.

One might worry that this strategy makes everything too easy. Why can't the neo-logicist just regard the ancestral as primitive and take the usual definition of the concept of natural number to be analytic of it? Then no appeal to comprehension would be needed! It will become clear that I am in a way sympathetic with this suggestion, but the arguments just offered on behalf of the claim that (P-lite) is analytic do not generalize to the case of the ancestral. Those arguments apply only to certain sorts of explicit definitions—namely, those that can be resolved into something of the form:

$$Ra_1 \ldots a_n \overset{df}{\equiv} \exists F_1 \ldots \exists F_n[a_1 = \Phi x:F_1x \wedge \cdots \wedge a_n = \Phi x:F_nx \wedge \Re_x(F_1x, \ldots, F_nx)],$$

which is equivalent to the conjunction of

(R-lite) $R(\Phi x:F_1x, \ldots, \Phi x:F_nx) \equiv \Re_x(F_1x, \ldots, F_nx)$

and

(R-imp) $Ra_1 \ldots a_n \rightarrow \exists F_1(a_1 = \Phi x:F_1x) \wedge \ldots \wedge \exists F_n(a_n = \Phi x:F_nx).$

The arguments presented above purport to show that (R-lite) is already an adequate definition of R, *modulo* an instance of the Caesar problem. But they apply only to this sort of case.

The case of the ancestral is not such a case,[12] but there is a different such case that is important. Consider the so-called predicative fragment of *Grundgesetze*, which consists of predicative second-order logic plus a schematic form of Frege's Basic Law V:

$$\hat{x}A(x) = \hat{x}B(x) \equiv \forall x(A(x) \equiv B(x)).$$

This theory is known to be consistent (Heck 1996). What saves the system from inconsistency is the fact that membership is defined in terms of a Σ_1^1 formula:

---

[12] It is not such a case because we know that the ancestral cannot be defined by a Σ_1^1 formula. Proof: Suppose it can. Then there is a formula $\exists F_1 \ldots \exists F_n \varphi(a,x,R,F_1,\ldots,F_n)$ with φ being Π_∞^0 such that $\exists F_1 \ldots \exists F_n \varphi(a,x,R,F_1,\ldots,F_n)$ holds if, and only if, there is a finite sequence $a = a_0,a_1,\ldots,a_n = x$ where Ra_ia_{i+1}. Now consider the first-order formula $\varphi(a,x,R,F_1,\ldots,F_n)$ and consider the formulae $\neg Rax$, $\neg \exists x_0$ $(Rax_0 \wedge Rx_0x)$, $\neg \exists x_0 \exists x_1(Rax_0 \wedge Rx_0x_1 \wedge Rx_1x)$, etc. Obviously, there are models of $\varphi(a,x,R,F_1,\ldots,F_n)$ and any finite subset of these. By compactness, there is thus a model in which all of them hold. But that is not a model in which there is a finite sequence connecting a to x, and yet $\varphi(a,x,R,F_1,\ldots,F_n)$ holds in the model and so *a fortiori* $\exists F_1 \ldots \exists F_n \varphi(a,x,R,F_1,\ldots,F_n)$ holds as well.

$$a \in b \equiv \exists F(b = \hat{x}Fx \land Fa),$$

and we do not have comprehension for such formulae in the predicative fragment. So, crucially, we cannot prove the following instance of naïve comprehension:

$$a \in \hat{x}(x \notin x) \equiv a \notin a,$$

whence the paradox that threatens to arise when we take a to be $\hat{x}(x \notin x)$ is averted. But it is averted only at the cost of our inability to prove the formula just displayed, and that has always seemed to me to be deeply counterintuitive. I can now give some content to the intuition thus countered.

The definition of membership is of precisely the form we have been discussing. The definition of membership one would really like to give is this one:

$(\in\text{-lite}) \quad a \in \hat{x}B(x) \equiv B(a).$

That is not a proper definition. It does not tell us when $a \in b$ but only when $a \in \hat{x}B(x)$, and so on and so forth. But modulo the Caesar problem, or so I would argue, ($\in$-lite) is a perfectly good definition. Any neo-Fregean who is prepared to countenance Basic Law V ought to regard membership as primitive and characterized by ($\in$-lite) and

$(\in\text{-imp}) \quad a \in b \rightarrow \exists F(b = \hat{x}Fx).$

But then Russell's paradox reappears, since ($\in$-lite) immediately implies:

$$\hat{x}(x \notin x) \in \hat{x}(x \notin x) \equiv \hat{x}(x \notin x) \notin \hat{x}(x \notin x).$$

And that seems to me an intuitively satisfying result. There is nothing *truly* impredicative about the definition of membership.

The predicative fragment of *Grundgesetze* may be consistent, then, but it is not really *coherent*.[13]

3. Ancestral Logic

As noted above, the only impredicative instances of comprehension needed for the proof of Frege's Theorem are these:

$$\exists P\{Pab \equiv \exists G \exists y[b = Nx : Gx \land Gy \land a = Nx : (Gx \land x \neq y)]\}$$

$$\exists R\{Ran \equiv \forall F[Fa \land \forall x(Fx \land Pxy \rightarrow Fy) \rightarrow Fn]\}$$

The arguments of §2 purported to establish that the former has no significant epistemological cost. If that is accepted, then we may draw the following intermediate

---

[13] Note that this argument does not even purport to show that predicativity restrictions are not otherwise justified. It is entirely specific to the case of Basic Law V. Note further that it does not really depend upon the assumption that Basic Law V and the definition of membership are given in schematic form, rather than in the forms:

$$\forall F \forall G[\hat{x}Fx = \hat{x}Gx \equiv \forall x(Fx \equiv Gx)]$$
$$\forall F[a \in \hat{x}Fx \equiv Fa].$$

If $\in$ is taken as primitive, then $x \notin x$ is predicative and it can then be used to instantiate Fx.

conclusion: The question whether the logic used in the proof of Frege's Theorem is epistemologically innocent reduces to the question what our attitude should be to Frege's definition of the ancestral.

The assumption that the ancestral of an arbitrary relation exists is much weaker than full Π_1^1 comprehension. In fact, there is a logic known as *ancestral logic* that characterizes the logic of the ancestral in an otherwise first-order language (Shapiro 1991: 227).[14]

We begin with an ordinary first-order language $\mathcal{L}$ and form a new language $\mathcal{L}_*$ by adding an operator $*_{xy}$ that forms a relational expression from a formula with two free variables, these being bound by the operator. So we have formulae of the form: $*_{xy}(\varphi xy)(a, b)$, where φxy is a formula. Let an interpretation of $\mathcal{L}$ be given. We expand it to an interpretation of $\mathcal{L}_*$ as follows:[15] Suppose that φxy is satisfied by exactly the ordered pairs in some set Φ; then $*_{xy}(\varphi xy)(a, b)$ is true if, and only if, there is a finite sequence $a = a_0, \ldots, a_n = b$ such that each $\langle a_i, a_{i+1} \rangle \in \Phi$. Less formally: It is true just in case a can be linked to b by a finite sequence of φ-steps. We require that there should be at least one such step: $*_{xy}(\varphi xy)$ is therefore the *strong* ancestral of φ, so-called because we do not always have: $*_{xy}(\varphi xy)(a, a)$. The weak ancestral of φ, denoted $*^=_{xy}(\varphi xy)$, may be defined in the usual way as: $*_{xy}(\varphi xy)(a, b) \vee a = b$. We shall use the more familiar notation $\varphi^* ab$ and $\varphi^{*=} ab$, omitting the bound variables when there is no danger of confusion.

It is easy to see that ancestral logic is not completely axiomatizable: It permits the formulation of a categorical theory of arithmetic (Shapiro 1991: 228).[16] But, of course, that need not prevent us from partially axiomatizing the logic. One way to proceed would be to take as introduction rules

$$\varphi ab \vdash \varphi^* ab$$

$$\varphi^* ab, \varphi bc \vdash \varphi^* ac$$

and as an elimination rule:

$$\varphi^* ab \vdash \forall x\Big(\varphi ax \to A(x)\Big) \wedge \forall x \forall y\Big(A(x) \wedge \varphi xy \to A(y)\Big) \to A(b).$$

Call this system *weak ancestral logic*. Its introduction rules reflect the 'inductive' character of the ancestral: Taking φab to mean: b is a's *parent*, they tell us that one's parents are one's ancestors and that any parent of an ancestor is an ancestor. The elimination rule is a principle of induction, in schematic form.

Weak ancestral logic incorporates, in its elimination rule, one half of Frege's definition of the ancestral. But it does not incorporate the other half of Frege's definition:

$$\forall F[\forall x(\varphi ax \to Fx) \wedge \forall x \forall y(Fx \wedge \varphi xy \to Fy) \to Fb] \to \varphi^* ab,$$

---

[14] There is also some work by Arnon Avron (2003) on such logics.

[15] This specification is less precise that it would really need to be, since it does not allow for additional free variables in φ. But let us not be too pedantic.

[16] Add to a first-order formulation of *PA* the axiom: $\forall n[*^=_{xy}(y = Sx)(0, n)]$.

and that, to my mind, is an important weakness. Consider the following argument.[17]

Suppose that b is a's ancestor and that c is b's ancestor. Suppose further (i) that all of a's parents are, say, blurg and (ii) that blurghood is hereditary—that is, that any parent of someone who is blurg is also blurg. Since b is a's ancestor, b is blurg, by the elimination rule. But then, by (ii), all of b's parents are blurg and so, since c is b's ancestor, c is blurg, again by the elimination rule. That is, if (i) and (ii), then c is blurg. And so, by Frege's definition of the ancestral, c is a's ancestor.

That, obviously, is an argument for the transitivity of the ancestral and, so far as I can see, nothing like it can be formalized within weak ancestral logic. That is not to say that the transitivity of the ancestral cannot be proved in weak ancestral logic. It can be, though in a different way—namely, in roughly the way Frege proves it in *Begriffsschrift*.[18] But that is a different argument, one whose formalization in standard second-order logic requires the use of Π_1^1 comprehension. No comprehension at all is needed for the formalization of the argument just given (Boolos and Heck 1998: 319). That we can formalize the more complicated argument in weak ancestral logic but not the less complicated one suggests to me that weak ancestral logic has at least one thing upside down.

One may have been wanting to ask what the nonsense term 'blurg' is doing in the above argument, and that is a perfectly reasonable question. But such reasoning is very common. At least it is very common for me to engage in such reasoning, especially when I am teaching logic to undergraduates. Perhaps that would be more obvious if I were to replace 'blurg' with 'F,' but one does not have to use letters to engage in such reasoning. And there is no reason to dismiss it out of hand. In many cases, such reasoning can be understood as tacitly semantic. We may take 'blurg' to be a variable that ranges over expressions and construe the argument as tacitly invoking semantic notions, such as satisfaction. A related proposal would construe the argument substitutionally. On either construal, however, this particular argument would only establish something about concepts we can *name*, whence it is surely invalid. But it seems to me a perfectly good argument, so some other way of understanding such reasoning is needed.

A fan of second-order logic might suggest that 'blurg' is a second-order variable and that the argument as a whole tacitly involves second-order quantification, its intuitive force revealing the extent to which second-order reasoning is intuitively compelling (see Boolos 1998e: 59–60). But there are two aspects to this suggestion, and they can be disentangled: We can interpret 'blurg' as the natural language

---

[17] The argument is even more natural if we use plurals: Suppose that whenever a's parents are among some things and any parent of one of those things is one of those things. . . It would be a simple matter to reformulate everything we are doing here in such terms. Indeed, no reformulation is needed: One may, as Boolos (1998e) noted, simply regard variables like F as being plural variables and take Fx to mean: x is among the Fs. The point would then be that we can prove Frege's Theorem in a language that allows for plural reference, even if it does not allow for plural *quantification*.

[18] Frege's proof is by induction on $\varphi^*a\xi$. The elimination rule yields

$$\varphi^*bc \rightarrow [\forall x(\varphi bx \rightarrow \varphi^*ax) \wedge \forall x\forall y(\varphi^*ax \wedge \varphi xy \rightarrow \varphi^*ay) \rightarrow \varphi^*ac]$$

The second conjunct follows immediately from the second introduction rule; the first follows from φ^*ab and the second introduction rule. Hence, $\varphi^*bc \wedge \varphi^*ab \rightarrow \varphi^*ac$. It would be cleaner if I had a nice example of a theorem whose proof is easily formalized using Frege's definition of the ancestral but which cannot be formalized in weak ancestral logic. There must be one.

correlate of a *free* second-order variable and simultaneously deny that second-order *quantification* is involved in the argument at all.

Given a first-order language, add to it a stock of second-order variables. We do not permit these variables to be bound by quantifiers: They occur only free. Thus, there are formulae in the language such as:

$$\forall x(\varphi ax \rightarrow Fx) \wedge \forall x \forall y(Fx \wedge \varphi xy \rightarrow Fy) \rightarrow Fb,$$

where F is a free second-order variable. An interpretation of such a formula is simply a first-order interpretation. Free second-order variables, that is to say, are to be treated just as predicate-letters are in first-order logic: They are to be assigned subsets of the domain. Implication is then defined as usual: A set of formulae Γ implies a formula A if, and only if, every interpretation that makes all formulae in Γ true also makes A true. A formula is valid if it is implied by the empty set of formulae.[19]

The proof-theory is also straightforward. I shall take us to be working in a system of natural deduction. Such a system will have some mechanism or other for keeping track of the premises used in the derivation of a given formula. I assume that we have some natural set of rules for first-order logic already in place. No special rules that govern free second-order variables are being introduced at this point. Call the resulting system *minimal schematic logic* (minimal SL). It should be clear that minimal SL is sound with respect to the semantics mentioned above. It is also complete, since, at this point, the free second-order variables differ from predicate-letters only in what they are called.[20]

We can now reformulate ancestral logic. The elimination rule, which we may call (*−), remains one direction of Frege's explicit definition of the ancestral, though it now need not be formulated as a schema but can be formulated using a free second-order variable:

$$\varphi^* ab \vdash \forall x(\varphi ax \rightarrow Fx) \wedge \forall x \forall y(Fx \wedge \varphi xy \rightarrow Fy) \rightarrow Fb$$

The introduction rule (*+) is the other direction of Frege's definition of the ancestral:

$$\forall x(\varphi ax \rightarrow Fx) \wedge \forall x \forall y(Fx \wedge \varphi xy \rightarrow Fy) \rightarrow Fb \vdash \varphi^* ab,$$

where F may not be free in any premises on which the premise of this inference itself depends.[21] Call the resulting system *minimal schematic ancestral logic* (minimal SAL). It should again be clear that this logic is sound if the ancestral is interpreted as indicated above. It is not, of course, complete with respect to that semantics, since no recursive axiomatization can be.

What we have done is transcribe Frege's explicit definition of the ancestral into the framework of schematic logic. Why does the transcription work? Consider the introduction rule, (*+). If we can derive

---

[19] A formula of schematic logic is therefore valid only if its universal closure is a valid second-order formula.

[20] The point of calling them 'variables' would then be only to distinguish them from predicate constants: A schematic language would be regarded as interpreted even if it assigns no interpretations to free second-order variables.

[21] It is here that the additional expressive power provided by the presence of free second-order variables makes itself felt: No such rule could possibly be formulated in a purely first-order language.

(†) $\forall x(\varphi ax \rightarrow Fx) \land \forall x\forall y(Fx \land \varphi xy \rightarrow Fy) \rightarrow Fb$,

from premises in which F does not occur free, then, in standard second-order logic, we can use universal generalization to conclude that

(††) $\forall F[\forall x(\varphi ax \rightarrow Fx) \land \forall x\forall y(Fx \land \varphi xy \rightarrow Fy) \rightarrow Fb]$

and then use Frege's definition of the ancestral to conclude that φ^*ab. Similarly in the case of the elimination rule: φ^*ab and Frege's definition together imply (††) which in turn implies (†). But then (††) is just a layover and Frege's explicit definition is a ladder we can kick away.[22] Goodbye, ladder.

The transitivity of the ancestral can be proven in minimal SAL, thus:[23]

[1]	(1)	φ^*ab	Premise
[2]	(2)	φ^*bc	Premise
[3]	(3)	$\forall x(\varphi ax \rightarrow Fx) \land \forall x\forall y(Fx \land \varphi xy \rightarrow Fy)$	Premise
[1]	(4)	$\forall x(\varphi ax \rightarrow Fx) \land \forall x\forall y(Fx \land \varphi xy \rightarrow Fy) \rightarrow Fb$	(1,*−)
[1,3]	(5)	Fb	(3,4)
[1,3]	(6)	$\forall x(\varphi bx \rightarrow Fx)$	(3,5)
[2]	(7)	$\forall x(\varphi bx \rightarrow Fx) \land \forall x\forall y(Fx \land \varphi xy \rightarrow Fy) \rightarrow Fc$	(2,*−)
[1,2,3]	(8)	Fc	(3,6,7)
[1,2]	(9)	$\forall x(\varphi ax \rightarrow Fx) \land \forall x\forall y(Fx \land \varphi xy \rightarrow Fy) \rightarrow Fc$	(3,8,→+)
[1,2]	(10)	φ^*ac	(9,*+)

So minimal SAL does not suffer from the problem that plagues weak ancestral logic.

That said, minimal SAL is still a very weak logic. The transitivity of the ancestral can be proven in minimal SAL because it can be proven in second-order logic without any appeal to comprehension. But the proof of theorem (124) of *Begriffsschrift*:

$$\varphi^*ab \land \forall x\forall y\forall z(\varphi xy \land \varphi xz \rightarrow y = z) \land \varphi ac \rightarrow \varphi^{*=}ac,$$

breaks down, as a little experimentation will show. The reason is that Frege's proof uses Π_1^1 comprehension, and there is nothing in minimal SAL that gives us the power of Π_1^1 comprehension.[24]

How are we to get that power without second-order quantifiers? Easily. There are no explicit comprehension axioms in the formal systems of *Begriffsschrift* and

---

[22] Thanks to Stewart Shapiro for the allusion.

[23] I have not included all the steps that would be required to make this argument formally precise, only enough to make it clear that it could be.

[24] Zoltan Gendler-Szabó has proved that comprehension is, in fact, required. It is also required for the proof of theorem (133):

$$\varphi^{*=}ab \land \varphi^{*=}ac \land \forall x\forall y\forall z(\varphi xy \land \varphi xz \rightarrow y = z) \rightarrow \varphi^*bc \lor b = c \land \varphi^*cb,$$

which thus cannot be proven in minimal SAL.

Grundgesetze, either. Rather, Frege has a rule of substitution: Given a theorem of the form. . .*F*. . ., infer. . .*φ*. . ., for any formula *φ* (subject to the usual sorts of restrictions). The substitution rule is, as is well known, equivalent to comprehension.[25] What we need here is thus a rule of substitution:

Suppose we have derived A from the premises in Γ, and let $A_{F/\varphi}$ be the result of replacing all occurrences of F in A by the formula φ (subject to the usual sorts of restrictions, again). Then, if F is not free in Γ, we may infer $A_{F/\varphi}$.

As a special case, of course, if A is provable (from no assumptions), then we may infer any substitution instance of it. And, given this rule, theorem (124) of *Begriffsschrift* can now be proven. (See the Appendix for the proof.)

The substitution rule is clearly sound given the semantics sketched above, but it is another question whether we should regard it as *justified* and, if so, on what sort of ground. This question will be considered below, in section 7.

We thus have two kinds of systems: There are the systems without the substitution rule—minimal SL and minimal SAL—and there are the systems with the substitution rule—what we may call full schematic logic and full schematic ancestral logic. In fact, there are further distinctions to be drawn, since the substitution rule can be restricted in various ways. We might, for example, require the formula that replaces F not to contain *. Adding this rule to minimal SAL would give us the effect of predicative comprehension, so we might call the resulting system *predicative* schematic ancestral logic. Such subsystems will not be of much interest here, however.

4. Schemata in Schematic Logic: A Digression

One reason second-order languages are so appealing is that principles that have to be formulated, in first-order languages, as axiom-schemata can be formulated in second-order languages as single axioms. It would be nice if schematic languages had a similar appeal—if, for example, the axiom of separation could be expressed by the single axiom:

$$(Sep) \quad \forall z \exists y \forall x (x \in y \equiv x \in z \land Fx),$$

from which its various instances could then be inferred by substitution. But it can't be, not if we characterize the rule of substitution as we did above. A *theory*, after all, is a set of formulae, and its theorems are the formulae that are deducible from the sentences in that set.[26] A deduction must thus assume some of the sentences of the theory as premises and then derive a theorem from them. If (Sep) is taken as a premise in a deduction, however, the variable F will obviously be free in that premise, whence it cannot be substituted for. Indeed, it is easy to see that (Sep) does not imply

---

[25] Substitution implies comprehension: Trivially, we have $\forall x(Fx \equiv Fx)$, so existential generalization yields: $\exists G \forall x[Gx \equiv Fx]$; by substitution: $\exists G \forall x[Gx \equiv \varphi x]$, for each formula φ. The proof of the converse is messier but not difficult: It is by induction on the complexity of the formula. . .*F*. . . .

[26] Sometimes the term 'theory' is used in a different sense—a theory is a deductively closed set of sentences, and its theorems are just the members of that set—but that is not the sense that is relevant here.

all (or even most) other instances of separation, not if 'implies' is defined as it was above. So it is a very good thing that such instances cannot all be deduced from it.

There is really a more basic problem here: I've yet to say what it might mean to assert something like (Sep); I've yet, that is, to say what the truth-conditions of (Sep) are. One might reasonably want to deny that (Sep) *has* truth-conditions. Since it contains a free second-order variable, one cannot speak of it as being true or false absolutely but only as being true or false under this or that assignment of a value to F. But there is an alternative. One can give free second-order variables the so-called 'closure interpretation,' effectively taking (Sep) to be true just in case its universal closure is true. We do not actually need to consider the universal closure, of course, for we can define truth for formulae of schematic languages directly: A formula is true if, and only if, it is true under all assignments to its free variables.[27]

This definition of truth would solve our problem concerning separation. Unfortunately, it brings a whole host of other problems with it.[28] To make further progress, we need to distinguish two sorts of assumptions that occur in argument. Sometimes, one makes an assumption 'for the sake of argument.' For example, one might assume the antecedent of a conditional and try to prove its consequent in order to prove that conditional. So, for example, if one were trying to prove $F0 \rightarrow F0 \vee G0$, one might begin by assuming $F0$: 'Suppose 0 is blurg' one might say. One is not to continue with, 'So 0 is odd. And prime. And, for that matter, even.' One does not, that is to say, expect to be understood as having assumed that zero has every property there is. If we call an assumption made 'for the sake of argument' a *supposition*, then what has been shown is that suppositions are not to be understood in terms of their universal closures, the reason being that a supposition is relevantly like the antecedent of a conditional: No one would suppose that $F0 \rightarrow G0$ should be understood as $\forall F(F0) \rightarrow \forall G(G0)$, however tempted she was by the thought that it should be understood as $\forall F \forall G(F0 \rightarrow G0)$.

But this observation does not make the closure interpretation of (Sep) any less available. The reason is that suppositions are mere tools of argument: They are not put forward as true in their own right. In a sense, that is obvious, since one sometimes makes a supposition only for *reductio*, but I am suggesting something stronger: that suppositions are not even *assumed* to be true; that is not the role they play in argument. Still, it ought nonetheless to be possible to assume that something is true, if only to investigate its consequences. Let us reserve the term *hypothesis* for an assumption of this kind. Then there is no bar to our understanding hypotheses in terms of their universal closures.

Formally, then, we distinguish between premises that are suppositions and premises that are hypotheses. An inference is valid if every interpretation that makes all of

---

[27] Thanks, long after the fact, to George Boolos for impressing this point upon me.

[28] Shapiro considers a logic $L2K-$, which is similar to the systems we have been discussing: Free second-order variables are given the closure interpretation. It turns out, however, to be surprisingly difficult to define a notion of implication with respect to which any reasonable set of deductive principles is sound (Shapiro 1991: 81). In the end, Shapiro does define such a notion, but it is not really consistent with the closure interpretation. On Shapiro's definition, Fx does not imply Gx. But if the former really means $\forall F(Fx)$ and the latter really means $\forall G(Gx)$, it should. If one wanted to say that it is perhaps best if Fx doesn't imply Gx, I'd happily agree, but that intuition isn't consistent with the closure interpretation.

its suppositions and hypotheses true—where truth for hypotheses is understood in terms of the closure interpretation—also makes its conclusion true. The distinction must be tracked in the proof-theory as well, and rules of inference that discharge premises, such as *reductio* and conditional proof, will need some modification: The premise discharged must be a supposition, not a hypothesis. But the crucial observation, for our purposes, is that the rule of substitution can now be relaxed: One can infer $A_{F/\varphi}$ from A so long as F is not free in any *supposition* on which A depends; it may be free in *hypotheses* on which A depends. This rule is clearly sound: Since, semantically speaking, hypotheses are treated as if they were universally closed, it is as if there aren't any free variables in the hypotheses, at least as far as the definition of implication is concerned.

There are other issues concerning the formulation of ancestral logic that we could discuss, but let me set them aside: How they are resolved does not really bear upon the philosophical issues that are motivating this discussion.[29] The question with which we started was whether we can regard (Sep) as a formulation of separation. The answer is that we can if we regard a theory as a set of *hypotheses* from which theorems are to be deduced. As a *hypothesis*, (Sep) does imply all other instances of separation, including those containing other free variables.

One might wonder why I chose separation as my example rather than induction, since the same issues will, of course, arise with respect to induction in the context of schematic logic. They do not, however, arise in the context of *ancestral* logic, since we can formulate induction in ancestral logic as the single sentence:

$$\forall z^* \; \overset{=}{_{xy}}(y = Sx)(0, z).$$

As we shall see, this point can be generalized: Any principle expressed in first-order logic by an axiom scheme can be expressed in Arché logic,[30] to be introduced next, by a single sentence.

5. Arché Logic

The methods used in §3 allowed us to transcribe Frege's explicit definition of the ancestral into schematic logic. A brief review will reveal, however, that the methods used presume only that the formula defining the ancestral is Π^1_1: They presume nothing about what formula it is. We can thus generalize that construction.

Consider first the simplest case. Let $\varphi_x(Fx, y)$ be an arbitrary formula containing no free variables other than those displayed. In standard second-order logic, we can explicitly define a new predicate $\mathcal{A}_\varphi$ as follows:

---

[29] Of course it matters that the technical issues *can* be resolved. I'll leave that as an exercise. One nice thing to do is to add structural rules that allow a hypothesis freely to be converted to a supposition and a supposition to be converted to a hypothesis so long as none of its free variables are free in any of the suppositions on which it depends. The distinction I am drawing here is very close to a distinction I drew elsewhere, for an ostensibly quite different purpose, between what I called 'rules of inference' and 'rules of deduction' (Heck 1998). In fact, however, the formal situation is almost identical, since modal formulae are there interpreted as if they were always proceeded by a universal quantifier over accessible worlds. The techniques developed there can therefore be used here.

[30] I hope it's not too presumptuous to take this name for the logic. Obviously, it is intended in tribute.

$$\mathcal{A}_\varphi(y) \equiv \forall F \varphi x(Fx, y).$$

The definition is licensed, in effect, by Π_1^1 comprehension, which guarantees that $\mathcal{A}_\varphi$ exists. But the trick used with the ancestral can also be used here. We have an introduction rule, $(\mathcal{A}_\varphi +)$:

$$\varphi_x(Fx, y) \vdash \mathcal{A}_\varphi(y),$$

where, as usual, F may not be free in any premises on which $\varphi_x(Fx,y)$ depends, and an elimination rule, $(\mathcal{A}_\varphi -)$:

$$\mathcal{A}_\varphi(y) \vdash \varphi_x(Fx, y).$$

In effect, these rules define $\mathcal{A}_\varphi(y)$ as equivalent to $\forall F \varphi_x(Fx,y)$ without using an explicit universal quantifier to do so. The extension of the new predicate $\mathcal{A}_\varphi$ is then the set of all those y such that $\varphi_x(Fx,y)$ is true for every assignment to F.

Generalizing, we may allow more than one predicate variable to occur in φ;[31] we allow the predicate variables to be of various adicities; and we allow additional free first-order variables, in which case what is defined is a relation rather than just a predicate. So, in general, φ may be of the form

$$\varphi_{x_1 \ldots x_{max(k_i)}} \left(F_1(x_1, \ldots, x_{k_1}), \ldots, F_n(x_1, \ldots, x_{k_n}), y_1, \ldots, y_m \right)$$

and, simplifying notation, we introduce a new predicate $\mathcal{A}_\varphi(\bar{y})$ subject to the rules:

$$\varphi_{\bar{x}}(\bar{F}, \bar{y}) \vdash \mathcal{A}_\varphi(\bar{y})$$

$$\mathcal{A}_\varphi(\bar{y}) \vdash \varphi_{\bar{x}}(\bar{F}, \bar{y})$$

It is a more serious question whether we wish to allow additional free *second*-order variables to occur in φ. I'll return to this issue below (see §7). For now, we do *not* allow additional free second-order variables.

Call what was just described the *scheme of schematic definition*. It allows us to transcribe what we would normally regard as an explicit definition of a new predicate or relation in terms of a Π_1^1 formula into schematic logic: If we can prove $\psi_x(Fx,a)$, then we can, in standard second-order logic, use universal generalization to conclude that $\forall F \psi_x(Fx,a)$ and so that $\mathcal{A}_\psi(a)$; if we have $\mathcal{A}_\psi(a)$, then by definition, $\forall F \psi_x(Fx,a)$ and so, by universal instantiation, $\psi_x(Fx,a)$. But the explicit definition of $\mathcal{A}_\psi(a)$ in terms of $\forall F \psi_x(Fx,a)$ simply mediates the transitions between $\psi_x(Fx,a)$ and $\mathcal{A}_\psi(a)$. It can be eliminated in favor of a schematic definition of $\mathcal{A}_\psi(a)$ in terms of those same transitions. Goodbye, ladder.

---

[31] If we do not, we will get a logic in which only Π_1^1 formulae with a single universal quantifier can be defined. This is, I believe, enough for the proof of Frege's Theorem. If we have pairing, then we can collapse adjacent universal quantifiers to a single one, but we do not have pairing in Fregean arithmetics, the reason being that we do not know what else might be in the domain besides numbers. Still, if we know that the matrix formula is true only of numbers, we can use this familiar trick, so the arithmetical strength of the theory is not affected.

If we add the scheme of schematic definition to (minimal) SL, we thus get a system in which new predicates co-extensional with Π^1_1 formulae can be introduced by schematic definition. Here again, however, the scheme of schematic definition is, by itself, deductively very weak: To exploit its power, we need a rule of substitution. We thus have three sorts of systems, depending upon the strength of the substitution principle we assume. Call the system without substitution *minimal Arché logic* (minimal AL). *Predicative Arché logic* contains a restricted substitution principle: The formula replacing F may not contain any of the new predicates $\mathcal{A}_\varphi$. *Full Arché logic* allows unrestricted substitution.[32] These systems obviously include minimal, predicative, and full schematic ancestral logic, respectively, and the logical strength of minimal, predicative, and full AL are, I suspect, that of second-order logic with no comprehension, predicative comprehension, and Π^1_1 comprehension, respectively. But it should be equally clear that the language in which these systems are formulated has nothing like the expressive power of a second-order language.

The axiom scheme of separation can be expressed in full AL by a single axiom. Let $\sigma_x(Fx,w)$ be the formula:

$$\forall z \exists y \forall x (x \in y \equiv x \in z \land Fx) \land w = w.$$

Then the scheme of schematic definition gives us a new predicate $\mathcal{A}_\sigma(w)$ subject to the rules:

$$\forall z \exists y \forall x (x \in y \equiv x \in z \land Fx) \land w = w \vdash \mathcal{A}_\sigma(w)$$

$$\mathcal{A}_\sigma(w) \vdash \forall z \exists y \forall x (x \in y \equiv x \in z \land Fx) \land w = w$$

or equivalently:

$$\forall z \exists y \forall x (x \in y \equiv x \in z \land Fx) \vdash \forall w \mathcal{A}_\sigma(w)$$

$$\forall w \mathcal{A}_\sigma(w) \vdash \forall z \exists y \forall x (x \in y \equiv x \in z \land Fx)$$

So separation is expressed by the single sentence: $\forall w \mathcal{A}_\sigma(w)$. A similar technique plainly applies to any axiom scheme.[33]

The scheme of schematic definition also allows us to define new predicates that are co-extensional with Σ^1_1 formulae. Let $\varphi_x(Fx,a)$ be a formula. We want to define a new predicate that is equivalent to $\exists F \varphi_x(Fx,a)$. Use the scheme of schematic definition to introduce a new predicate $\mathcal{A}_{\neg\varphi}$, subject to the rules:

---

[32] In full AL, the elimination rule ($\mathcal{A}_\varphi$ −) effectively takes the form

$$\mathcal{A}_\varphi(a) \vdash \varphi_x\Big(B(x),a\Big),$$

where $B(x)$ is an arbitrary formula (subject to the usual restrictions). In predicative AL, $B(x)$ is not permitted to contain new predicates of the form $\mathcal{A}_\varphi$.

[33] A more elegant way to proceed is to extend the scheme of schematic definition to allow a new zero-place predicate (i.e., a sentential variable) to be defined in terms of a formula $\varphi_x(F_1x,\ldots,F_nx)$, in which case we have:

$$\varphi_x(F_1x,\ldots,F_nx) \vdash \mathcal{A}_\varphi$$
$$\mathcal{A}_\varphi \vdash \varphi_x(F_1x,\ldots,F_nx)$$

Then separation is expressed by the zero-place predicate thus defined when we take $\varphi_x(Fx)$ to be (Sep).

$$\neg\varphi_x(Fx, a) \vdash \mathcal{A}_{\neg\varphi}(a)$$

$$\mathcal{A}_{\neg\varphi}(a) \vdash \neg\varphi_x(Fx, a)$$

So $\mathcal{A}_{\neg\varphi}(a)$ is equivalent to $\forall F\neg\varphi_x(Fx,a)$. We now regard $\neg\mathcal{A}_{\neg\varphi}(a)$ as degeneratively of the form $\psi_x(Fx,a)$ and introduce a new predicate $\mathcal{A}_{\neg\mathcal{A}_{\neg\varphi}}$, which I shall write: $\mathcal{A}^{\varphi}$, subject to the rules:

$$\neg\mathcal{A}_{\neg\varphi}(a) \vdash \mathcal{A}^{\varphi}(a)$$

$$\mathcal{A}^{\varphi}(a) \vdash \neg\mathcal{A}_{\neg\varphi}(a)$$

So $\mathcal{A}^{\varphi}(a)$ is equivalent to $\forall F(\neg\mathcal{A}_{\neg\varphi}(a))$, that is, to $\neg\mathcal{A}_{\neg\varphi}(a)$, that is, to $\neg\forall F\neg\varphi_x(Fx,a)$ and so to $\exists F\varphi_x(Fx,a)$, as wanted.

This argument obviously depends upon our allowing predicates defined using the scheme of schematic definition to appear in formulae used to define yet further new predicates using that same scheme. As already noted, we can, formally speaking, restrict the scheme so as not to allow such iteration, the result being predicative AL. But this restriction has no motivation in the context of this investigation. The scheme of schematic definition formalizes a certain mode of *concept-formation*. Once one has used it to form a certain concept, one has that concept, and there is simply no reason one cannot iterate the process of concept-formation in the way we have allowed.

That said, there is a more elegant way to define predicates that are equivalent to Σ_1^1 formulae. As we have seen, the scheme of schematic definition, as currently formulated, in effect characterizes $\mathcal{A}_{\varphi}$ in terms of introduction and elimination rules that mirror those for the universal quantifier that appears in its explicit second-order definition. That suggests that we should characterize $\mathcal{A}^{\varphi}$ in terms of introduction and elimination rules that mirror those for the existential quantifier that appears in *its* explicit second-order definition. So we may take the introduction rule for $\mathcal{A}^{\varphi}$, $(\mathcal{A}^{\varphi}+)$, to be:

$$\varphi_x\Big(A(x), a\Big) \vdash \mathcal{A}^{\varphi}(a),$$

where, in this case, variables free in $A(x)$ may be free in premises on which $\varphi_x(A(x),a)$ depends. In full AL, $A(x)$ may be any formula; in predicative AL, it may not contain schematically defined predicates; in minimal AL, it must not contain schematic variables, either.

The elimination rule, $(\mathcal{A}^{\varphi}-)$, is more complex, but only because the elimination rule for the existential quantifier is itself more complex, involving as it does the discharge of an assumption. Suppose we have derived a formula B from formulae in some set Δ together with $\varphi_x(A(x),a)$, where none of the free variables occurring in $A(x)$, other than x itself—which is actually bound in $\varphi_x(A(x),a)$—occur free in B or in Δ. Suppose further that we have derived $\mathcal{A}^{\varphi}(a)$ from the formulae in some set Γ. Then we may infer B, discharging $\varphi_x(A(x),a)$, so that B depends only upon Γ and Δ.[34] Symbolically:

---

[34] If one wants to make the distinction between suppositions and hypotheses here, then $\varphi_x(A(x),a)$ must be a supposition.

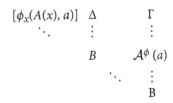

This rule simply parallels the relevant instance of the usual elimination rule for the second-order existential quantifier, except that we have replaced $\exists F\varphi_x(Fx,a)$ with $\mathcal{A}_\varphi(a)$. It is convenient to expand the scheme of schematic definition to allow schematic definitions of this form, too, since doing so adds no additional strength to the logic.

I intend the term 'scheme of schematic definition' to be taken seriously: I regard the introduction and elimination rules $(\mathcal{A}_\varphi+)$ and $(\mathcal{A}_\varphi-)$ as *defining* the new predicate $\mathcal{A}_\varphi$ and therefore regard the rules themselves as effectively self-justifying, since they are consequences of (indeed, components of) a definition. In particular, then, I am proposing that we should regard (*+) and (*−)[35] as defining the ancestral. Perhaps that would be a reason to regard the ancestral as a logical notion and to regard these rules are logical rules. I am not sure, because I am not sure what the word 'logical' is supposed to mean here. But the crucial issue for the neo-logicist is epistemological. The proof of Frege's Theorem makes heavy use of the ancestral and of inferences of the sort (*+) and (*−) describe. A neo-logicist must therefore show that she is entitled both to a grasp of the concept of the ancestral and to an appreciation of the validity of (*+) and (*−), and this entitlement must be epistemologically innocent in the sense that it does not itself import epistemological presuppositions that undermine the neo-logicist project: It must not, for example, presuppose a grasp of the concept of finitude, and it is a common complaint that our grasp of the concept of the ancestral presupposes precisely that. But if we regard the ancestral as schematically defined by (*+) and (*−), we may dismiss this complaint.

To be sure, one cannot simply introduce a new expression and stipulate that it should be subject to whatever introduction and elimination rules one wishes: Inconsistency threatens, as Arthur Prior (1960) famously showed. A complete defense of the position I am developing here would thus have to contain an answer to the question when such stipulations are legitimate,[36] and to many others besides. But my purpose here is more modest. I am trying to argue that a certain position is available and worth considering. Whether it is true is a question for another day.

---

[35] In schematic ancestral logic, * is an operator, and so its logic can be characterized by the pair of rules (*+) and (*−). In Arché logic, there is no such operator. Rather, we have to define the ancestral of each relation separately, using the scheme of schematic definition. But this point does not affect the present discussion, so I shall ignore it.

[36] See the discussion by Hale and Wright (2000).

6. Frege's Theorem

If we are to prove Frege's Theorem in some form of schematic logic, we must be able to formalize HP in schematic logic. HP is, of course, neither a definition nor an instance of comprehension, but the techniques developed above may none-theless be applied to it. We may represent HP as a pair of rules. Let $Fx \approx^{Ryz}_{xyx} Gx$ abbreviate:

$$\forall x \forall y \forall z \forall w (Rxy \wedge Rzw \rightarrow x = z \equiv y = w) \wedge$$

$$\forall x (Fx \rightarrow \exists y (Rxy \wedge Gy)) \wedge \forall y (Gy \rightarrow \exists x (Rxy \wedge Fx)),$$

for any formula Ryz, so that $Fx \approx^{Ryz}_{xyz} Gx$ says: R correlates the Fs one-one with the Gs. Then the introduction rule $(N+)$ is easy enough to state:

$$Fx \approx^{Ryz}_{xyz} Gx \vdash Nx : Fx = Nx : Gx$$

The elimination rule $(N-)$ is more complicated, but it simply parallels $(\mathcal{A}^\varphi -)$:

$$
\begin{array}{ccc}
\Delta & [F_x \approx^{Rxy}_{xyz} G_x] & \Gamma \\
\ddots & \vdots & \vdots \\
B & & Nx : Fx = Nx : Gx \\
& & \ddots \quad \vdots \\
& & B
\end{array}
$$

That is: If we have derived a formula B from assumptions in some set Δ together with the assumption that $Fx \approx^{Ryz}_{xyz} Gx$ and we have derived $Nx{:}Fx = Nx{:}Gx$ from the assumptions in some set Γ, then we may infer B, discharging $Fx \approx^{Ryz}_{xyz} Gx$. *Arché arithmetic* is full Arché logic plus these two rules.

I should emphasize before continuing that I am not claiming that $(N+)$ and $(N-)$ schematically define the cardinality operator. I am not even claiming that the possibility of formulating HP in schematic logic should do anything to ease any concerns one might have had about its epistemological status. My point here concerns only the logical resources needed for the proof of Frege's Theorem.

Frege's definitions of arithmetical notions can all be formalized in Arché arith-metic. Zero may be defined in the usual way:

$$0 = Nx : x \neq x$$

Frege's definition of predecession becomes a schematic definition of a new relation-symbol P subject to the rules $(P+)$:

$$\exists y [Nx : Gx = b \wedge Gy \wedge Nx : (Gx \wedge x \neq y) = a] \vdash Pab$$

and $(P-)$:

$$\exists y[Nx : Gx = b \wedge Gy \wedge Nx : (Gx \wedge x \neq y) = a] \qquad \Delta \qquad \Gamma$$

$$\ddots \qquad\qquad \vdots \qquad \vdots$$

$$B \qquad Pab$$

$$\vdots \qquad \vdots$$

$$B$$

where Gx is an arbitrary formula. The weak ancestral of this relation is schematically defined as subject to the two rules:

$$(P^*+) \quad \forall x(Pax \rightarrow Fx) \wedge \forall x \forall y(Fx \wedge Pxy \rightarrow Fy) \rightarrow Fb \vdash P^*ab$$

$$(P^*-) \quad P^*ab \vdash \forall x(Pax \rightarrow Fx) \wedge \forall x \forall y(Fx \wedge Pxy \rightarrow Fy) \rightarrow Fb$$

subject to the usual restrictions. The definition of the concept of natural number is then:

$$Nn \equiv P^*0n \vee 0 = n.$$

Frege's proofs of axioms of arithmetic can then be formalized straightforwardly.

We may take the axioms of arithmetic to be as follows:[37]

1. $N0$
2. $\forall x \forall y(Nx \wedge Pxy \rightarrow Ny)$
3. $\forall x \forall y \forall z(Nx \wedge Pxy \wedge Pxz \rightarrow y = z)$
4. $\forall x \forall y \forall z(Nx \wedge Ny \wedge Pxz \wedge Pyz \rightarrow x = y)$
5. $\neg \exists x(Nx \wedge Px0)$
6. $\forall x(Nx \rightarrow \exists x(Pxy))$
7. $A(0) \wedge \forall x \forall y(Nx \wedge A(x) \wedge Pxy \rightarrow A(y)) \rightarrow \forall x(Nx \rightarrow A(x))$

For convenience, induction has been formulated as a schema. As we saw above, this can be avoided, but let us work with the schema, for simplicity.[38]

In Arché arithmetic, just as in Frege arithmetic (second-order logic plus HP), axioms (1) and (2) follow easily from the definition of N: (1) is immediate, and (2) follows from the transitivity of the ancestral. The proofs of axioms (3), (4), and (5) in Frege arithmetic all appeal to HP, but they use only predicative comprehension and so are easily formalized in Arché arithmetic.

---

[37] Arithmetic and multiplication can be defined using the scheme of schematic definition, so we need no special axioms governing them.

[38] Consider the formula: $F0 \wedge \forall x \forall y(Nx \wedge Fx \wedge Pxy \rightarrow Fy) \rightarrow Fa$. The scheme of schematic definition yields a new predicate Ja subject to the two rules:

$$F0 \wedge \forall x \forall y(Nx \wedge Fx \wedge Pxy \rightarrow Fy) \rightarrow Fa \vdash Ja$$

$$Ja \vdash F0 \wedge \forall x \forall y(Nx \wedge Fx \wedge Pxy \rightarrow Fy) \rightarrow Fa$$

As before, Ja is then equivalent to $\forall F[F0 \wedge \forall x \forall y(Nx \wedge Fx \wedge Pxy \rightarrow Fy) \rightarrow Fa]$, and induction is: $\forall x(Nx \rightarrow Jx)$.

Axiom (7) is stronger than what the definition of the ancestral by itself delivers. Simple manipulations give us:

$$A(0) \land \forall x \forall y (A(x) \land Pxy \rightarrow A(y)) \rightarrow \forall x (\mathbb{N}x \rightarrow A(x)).$$

But (7) is stronger than this. The second conjunct of its antecedent is $\forall x \forall y (\mathbb{N}x \land A(x) \land Pxy \rightarrow A(y))$, which is weaker than $\forall x \forall y (A(x) \land Pxy \rightarrow A(y))$, since it contains the additional conjunct $\mathbb{N}x$ in its antecedent. But the instances of (7) can be proven, as usual, by induction on $\mathbb{N}\xi \land A(\xi)$.[39]

And finally, as was first noticed by Boolos (1998c), axiom (6) can be derived from axioms (3), (4), and (5), the definitions, and the following very weak consequence of HP:

(Log) $\forall x(Fx \equiv Gx) \vdash Nx : Fx = Nx : Gx.$

The proof uses only Π_1^1 comprehension, so it, too, can be formalized in full Arché logic.[40] See the Appendix for the details.

A form of Frege's Theorem can thus be proven in Arché arithmetic. Exactly how strong the resulting fragment of second-order arithmetic is, I do not know for sure. It seems exceedingly likely that it is equivalent to second-order arithmetic with Π_1^1 comprehension. It is certainly stronger than first-order PA, since it contains ancestral logic, and ancestral logic plus PA is, as noted earlier, categorical. For another illustration of the strength of Arché arithmetic, note that the explicit definition of satisfaction for the language of first-order arithmetic is Π_1^1, so that definition can be converted into a schematic definition in Arché arithmetic. The usual induction will then establish the consistency of first-order PA, so Arché arithmetic is a non-conservative extension of first-order PA.

7. Philosophical Considerations

A close examination of the proofs in the Appendix will show that, if one regards predecession as primitive and subject to the rules $(P+)$ and $(P-)$, as suggested in §2, then Frege's Theorem can be proven in full schematic ancestral logic. If one regards the ancestral too as primitive and subject to the rules $(*+)$ and $(*-)$, then Frege's Theorem can be proven in *predicative* schematic ancestral logic. Predicative systems are generally regarded as epistemologically innocent. So if both predecession and the

---

[39] It is not often noted that, for this reason, the proof of induction in Frege arithmetic requires Π_1^1 comprehension. That is yet another reason one should not expect to get by with much less if one is trying to derive the axioms of PA from HP. As it happens, however, if one is willing to forego induction and interpret a weaker theory, such as Robinson arithmetic, then one can do so in predicative second-order logic (Heck 2011).

[40] Boolos's proof uses axioms (1), (2), and (7), as well, but these can be derived simply from the definitions, as we have seen. As we have also seen, however, the proof of axiom (7) uses Π_1^1 comprehension, which is also used in the proofs of certain other, very general facts about the ancestral that are used in the proof of axiom (6). All such uses can, however, be avoided in ramified Frege arithmetic (Heck 2011): The proof of axiom (6) therefore does not *require* impredicative reasoning. The lesson is that the place impredicative reasoning is required is in the proof of *induction*—just as Poincaré famously suspected.

ancestral could be regarded as primitive, the mentioned rules being analytic of these notions, the logic needed for the proof of Frege's Theorem would be epistemologically innocent, and uncontroversially so.

But the question which notions are primitive does not seem to me to be the right question to ask here: It is far too slippery. It is better, I think, to regard both predecession and its ancestral as defined by means of the scheme of schematic definition and to regard $(P+)$, $(P-)$, (P^*+), and (P^*-) as analytic on the ground that they are consequences of, because components of, those definitions.[41] The proof of Frege's Theorem, in that case, needs full Arché logic. In particular, the proof needs the unrestricted rule of substitution. A philosopher with principled concerns about impredicativity might therefore be tempted to say—and might, indeed, long have been wanting to say—that the foregoing is largely beside the point if the question at issue is whether the logic required for the proof of Frege's Theorem is epistemologically innocent. The unrestricted substitution rule is impredicative, and the only question, really, was where the impredicativity would ultimately surface. The bump has been pushed around a fair bit, to be sure, but Frege's carpet is no flatter now than it was before.

I disagree. The scheme of schematic definition allows us to introduce a new predicate $\mathcal{A}_\varphi$ subject to the rules:

$(\mathcal{A}_\varphi +)$ $\varphi_x(Fx, a) \vdash \mathcal{A}_\varphi(a)$

$(\mathcal{A}_\varphi -)$ $\mathcal{A}_\varphi(a) \vdash \varphi_x(Fx, a)$

In the presence of the unrestricted substitution rule, the elimination rule is equivalent to:

$(\mathcal{A}_\varphi \text{sub})$ $\mathcal{A}_\varphi(a) \vdash \varphi_x(B(x), a)$,

where $B(x)$ is now a formula rather than a variable. In particular, in the case of the ancestral, the elimination rule is equivalent, in the presence of unrestricted substitution, to:

$(\varphi^*\text{sub})$ $\varphi^*ab \vdash \forall x(\varphi ax \to B(x)) \land \forall x \forall y(B(x) \land \varphi xy \to B(y)) \to B(b)$.

Such a rule can be understood in two ways.[42] One way takes the set of formulae $B(x)$ that can appear in the rule to be determined by reference to some fixed language. That is how axiom-schemata, such as the induction scheme in PA:

$$A(0) \land \forall x(A(x) \to A(Sx)) \to \forall x A(x),$$

---

[41] As noted earlier (note 37), addition and multiplication can be similarly defined.

[42] Feferman (1991) discusses this distinction at some length. Regarding his historical remarks in §1.5, it is perhaps worth noting that the first accurate formulation of the sort of substitution rule that is needed here is due to Frege: See Rule 9 in §48 of *Grundgesetze*. As Feferman notes, citing Church, such a rule is missing from *Begriffsschrift*.

are usually understood. The induction scheme is usually regarded as abbreviating an infinite list of axioms, one for each formula of the language of arithmetic. When we consider expansions of the language of PA, then—say, the result of adding a truth-predicate T—that expansion does not, in itself, result in any new axioms' being added to the original theory. In particular, sentences containing T, such as:[43]

$$T0 \land \forall x \Big(Tx \to T(Sx) \Big) \to \forall x (Tx),$$

do not automatically become axioms of the new theory, though such sentences do have the form of induction axioms. That is why adding a truth-predicate to PA, and even adding the Tarskian clauses for the truth-predicate, yields a conservative extension: One can't do much with the truth-predicate if it doesn't occur in the induction axioms.

Formally, of course, one can proceed how one likes, but this way of thinking of the induction scheme is not obviously best. Even if our theory of arithmetic is formulated in a first-order language, one would have thought that the induction scheme should be regarded as one that *does*, as it were, automatically import new instances of that scheme into our theory as our language expands. In everyday mathematics, we do not so much as ask, as our language expands, whether the new instances of induction that become available should be accepted as true.[44] As it is ordinary understood, that is to say, the induction scheme expresses an open-ended commitment to the truth of all sentences of a certain form, both those we can presently formulate and those we cannot. The rule (φ*sub) is to be understood in this way, too: It expresses an open-ended commitment to the validity of all inferences of a certain form.[45]

The unrestricted substitution rule thus expresses the open-ended nature of the commitments we undertake when we define a new predicate *via* an instance of the scheme of schematic definition. And that, simply, is how the scheme is intended to be understood. Someone whose understanding of the ancestral was completely constituted by her grasp of the rules (φ*+) and (φ*−) would, it seems to me, be quite surprised to hear that these rules do not license the sort of inference required for the proof of theorem (124) of *Begriffsschrift*. I am not saying that it would be *incoherent* to refuse to accept that inference. I am simply saying that it would not be a natural reaction.

One might object that this justification of the substitution rule, if it is defensible at all, ought to apply just as well in the context of full second-order logic. And perhaps it does so extend. But in the case of second-order logic, there is another and more fundamental problem with which we must contend: We must explain the second-order quantifiers. Absent such an explanation, we do not so much as understand second-order languages, and the question how the substitution rule should be

---

[43] Or, more importantly, the instance of induction you get if you take $A(x)$ to be: $\mathrm{Bew}_{PA}(x) \to T(x)$.

[44] One might suggest that vagueness and, in particular, the Sorites paradox refute this suggestion, since accepting induction for predicates like 'heap' leads to problems. But the common wisdom nowadays is that induction is irrelevant, since the paradox can be reformulated without it.

[45] I borrow the term 'open-ended' from Vann McGee (1997).

justified doesn't arise. Now, to understand the second-order universal quantifier, one must understand what it means to say that all concepts are thus-and-so. But to understand that sort of claim, or so it is often argued, one must have a conception of what the second-order domain comprises. One must, in particular, have a conception of (something essentially equivalent to) the full power-set of the first-order domain, and many arguments have been offered that purport to show that we simply do not have a definite conception of $\wp(\omega)$. It is not my purpose here to evaluate such arguments. Maybe they work, and maybe they do not. My purpose here is to identify an epistemologically relevant difference between second-order logic and Arché logic. Here it is: Since there are no second-order quantifiers in schematic languages, the problem of explaining the second-order quantifier simply does not arise.

A similar point arises in connection with a question set aside earlier: whether second-order parameters should be permitted in schematic definitions, that is, whether one should be allowed to define what Frege would have called a 'relation of mixed level' in the following sort of way:

$$A_{yz}(x, Fy, Gz) \vdash \Phi_y(x, Fy)$$

$$\Phi_y(x, Fy) \vdash A_{yz}\left(x, Fy, B(z)\right)$$

The first of these is the introduction-rule, of course, and $A_{yz}(x,Fy,Gz)$ is a formula with the free variables shown; what is being defined here is the 'mixed level' relational expression $\Phi_y(x,Fy)$.

Permitting this sort of definition leads to a system with the expressive power of full second-order logic.[46] A friend of second-order logic might welcome this news, of course: If Arché logic really is epistemologically innocent, and if full second-order logic can be obtained simply by admitting second-order parameters in schematic definitions, then the naturalness of that extension might be taken to suggest that second-order logic too is epistemologically innocent. But, of course, my opponent will see things differently: If it is only because we have disallowed second-order parameters that Arché logic is any weaker than full second-order logic, and if allowing them is indeed so natural, then the resources needed for an understanding of Arché logic might seem to be sufficient for an understanding of full second-order logic, and there is no epistemologically relevant difference between them to be found.

---

[46] The proof is by induction. Suppose that, for each Σ_n^1 formula $\varphi(x_1,\ldots,x_n,F_1,\ldots,F_m)$, where the F_i are second-order parameters, we can schematically define a formula $\Phi(x_1,\ldots,x_n,F_1,\ldots,F_m)$ that is equivalent to it. (The other case is similar.) Now consider a Π_{n+1}^1 formula $\forall F_1\varphi(x_1,\ldots,x_n,F_1,F_2,\ldots,F_m)$. Since $\varphi(x_1,\ldots,x_n, F_1,F_2,\ldots,F_m)$ is Σ_n^1, we can schematically define a formula $\Phi(x_1,\ldots,x_n,F_1,F_2,\ldots,F_m)$ that is equivalent to it. We then schematically define $\Psi(x_1,\ldots,x_n,F_2,\ldots,F_m)$ in the obvious way:

$$\Phi(x_1,\ldots,x_n,F_1,F_2,\ldots,F_m) \vdash \Psi(x_1,\ldots,x_n,F_2,\ldots,F_m)$$

$$\Psi(x_1,\ldots,x_n,F_2,\ldots,F_m) \vdash \Phi(x_1,\ldots,x_n,F_1,F_2,\ldots,F_m),$$

where the former is, of course, subject to the usual restrictions. By earlier remarks, $\Psi(x_1,\ldots,x_n,F_2,\ldots,F_m)$ is then equivalent to $\forall F_1\varphi(x_1,\ldots,x_n,F_1,F_2,\ldots,F_m)$.

But this response overlooks an important difference between Arché logic and this extension of it, namely: Before we can even consider whether schematic definitions containing such parameters should be permitted, we must first enrich the *language*. The language of Arché logic is the language of first-order logic, and the only sorts of relations that are definable in it are first-order relations, that is, relations between objects. The language of the extension just mentioned is not even the language of second-order logic: Except for the second-order quantifiers, the primitive expressions of second-order logic are just those of first-order logic. What we need in the language if we are to permit parameters in schematic definitions are primitives like $\Phi_y(x,Fy)$ that express relations not just between objects but between objects and concepts (to use Frege's language). Such expressions therefore sit at roughly the same level as the second-order quantifiers themselves: They are free *third*-order variables. It would not be at all surprising, then, if the resources needed to understand expressions of this kind were similar to the resources needed to understand the second-order quantifiers. But let me not pursue that question now.[47] I am, once again, just trying to establish that there are issues worth discussing here.

A similar concern is that our understanding of the introduction rule for the ancestral involves a conception of the full power-set of the domain. How else, it might be asked, are we to understand

$$\forall x(\varphi ax \rightarrow Fx) \wedge \forall x \forall y(Fx \wedge \varphi xy \rightarrow Fy) \rightarrow Fb,$$

as it occurs in the premise of the rule (*+), except as involving a tacit initial second-order quantifier? Does it not say, explicit quantifier or no, that all concepts F that are thus-and-so are so-and-thus? Doesn't understanding that claim therefore require the disputed conception of the full power-set? No, it does not. A better reading would be: A concept that is thus-and-so is so-and-thus. What understanding this claim requires is not a capacity to conceive of *all* concepts but simply the capacity to conceive of *a* concept: to conceive of an arbitrary concept, if you like. The contrast here is entirely parallel to that between arithmetical claims like $x + y = y + x$, involving only free variables, and claims involving explicit quantification over all natural numbers. Hilbert famously argued that our understanding of claims of the former sort involves no conception of the totality of all natural numbers, whereas claims of the latter sort do, and that there is therefore a significant conceptual and epistemological difference between these cases. I am making a similar point about claims involving only free second-order variables as opposed to claims that quantify over concepts (or whatever).

But one might still want to insist that, if we do not have a definite conception of the full power-set of the domain—if, in particular, there is nothing in our understanding of free second-order variables that guarantees that they range over the full power-set of the domain—then the meanings of the predicates we introduce by schematic definition will be radically underdetermined, at the very least. It was stipulated earlier that $\mathcal{A}_\varphi(a)$ is true if, and only if, $\varphi_x(Fx,a)$ is true for every assignment of a subset of

---

[47] If second-order logic is interpreted *via* plural quantification, as suggested by Boolos (1998e), then expressions of this kind are 'plural predicates,' as discussed by Agustín Rayo (2002).

the first-order domain to F. But why? If we have no conception of the full power-set, why not take the domain of the second-order variables to be smaller? Why not restrict it to the definable subsets of the domain? Surely the axioms and rules of Arché logic do not require the second-order domain to contain every subset of the first-order domain.

Obviously, there is a technical point here that is incontrovertible: The existence of non-standard models is a fact of mathematics, and a very useful one at that. But the philosophical significance of this technical point is not so obvious. It seems to me that there is something about the axioms and rules of Arché logic that requires the second-order domain to be unrestricted, and '*unrestricted*' is the crucial word. The difference between the standard model and the various non-standard models is to be found not in what the standard model *in*cludes but in what non-standard models *ex*clude: Since the second-order domain of a standard model includes all subsets of the first-order domain, a non-standard model must *ex*clude certain of these subsets. That, however, is incompatible with the nature of the commitments we undertake when we introduce a new predicate by schematic definition. Those commitments are themselves *unrestricted* in the sense that we accept no restriction on what formulae may replace $B(x)$ when we infer $\varphi_x(B(x),a)$ from $\mathcal{A}_\varphi(a)$.[48] One might be tempted to object that, if so, we must somehow conceive of the totality of all such formulae in advance. But that would simply repeat the same error: No such conception of the totality of all formulae is needed; what is needed is just the ability to conceive of *a* formula—an arbitrary formula, if you like.

A neutral referee might well declare a draw at this point, although both sides have moves remaining. But that is enough for my purpose here. What I am trying to do is not to convince the reader of any particular position. I am trying, rather, to convince the reader of the *interest* of a certain position—namely, the position that full Arché logic is epistemologically innocent in whatever way such positions as the neo-logicist's need logic to be. It is no part of my position that full second-order logic is not epistemologically innocent in that sense. It is my position that there is *enough of a difference* between Arché logic and second-order logic that it would not be unreasonable to regard them as epistemologically unequal in this same sense. I take myself to have accomplished that much. The resources deployed above in the defense of full Arché logic against the predicativist skeptic are not resources that are obviously sufficient to defend full second-order logic. Perhaps they can be built upon for that purpose. I don't necessarily say otherwise. But perhaps they cannot be.

The critical difference between Arché logic and second-order logic thus turns out to lie not so much in the logical principles that distinguish them but, rather, in the expressive power of the underlying languages. What we have seen is not just that the full deductive strength of second-order logic is not needed for the proof of Frege's Theorem: That has been known for some time. What we have seen is that not all of the expressive power of second-order languages is needed, either. Quine's view was

---

[48] And $B(x)$ might be demonstrative: Those ones. Definability considerations are therefore out of place here, too.

that second-order quantification is not even a logical *notion*: If second-order variables range over sets, then second-order quantification is quantification over sets, and second-order logic is set-theory in sheep's clothing, quite independently of its proof-theoretic strength (1986).[49] Even if one interprets second-order quantifiers in terms of plurals, as suggested by Boolos (1998e), however, one might have other reasons to suppose that plural quantifiers are non-logical constants (see, e.g., Resnik 1988), perhaps reasons connected with the expressive power of plural quantifiers. Again, it is no part of my view that plural quantifiers are not logical constants. What I am claiming is that the question whether axioms for arithmetic can be derived, purely logically, from HP does not depend upon how such issues are resolved. The language of Arché arithmetic has a significantly stronger claim to be a logical language than the language of second-order logic does.[50]

Appendix

Here, as above, I am not including all the steps that would be necessary to make the arguments completely formal. In particular, standard first-order moves will be repressed for the most part and only briefly indicated where they are not. My intention is simply to make it clear that these results can be proven in the relevant systems.

Proof of *Begriffsschrift,* Theorem (124)

The proof is in full schematic ancestral logic. It could also, of course, be carried out in full Arché logic.

---

[49] That is not to say that Quine didn't make a great deal of the proof-theoretic strength of second-order logic. He seems, in particular, to have regarded completeness as an essential property of anything rightly regarded as a 'logic.' But, so far as I know, Quine never says what is supposed to justify this restriction.

[50] This material was presented to the Philosophy of Mathematics Workshop at Arché, the AHRC Research Centre for the Philosophy of Logic, Language, Mathematics and Mind, at the University of St Andrews, in February 2005. Thanks to Arché for its support, which is much appreciated, and to Crispin Wright for arranging another visit. I thank everyone who attended for their comments, but special gratitude is due to Crispin and to Stewart Shapiro for their enthusiasm about this material, which is what led me to write it down.

Earlier versions of these ideas were presented in a graduate seminar given at Harvard University in fall 2004. Thanks to the members of that seminar for their reaction. Thanks too to Øystein Linnebo for his comments on an early draft.

It should be obvious that my work owes a great deal to Crispin's. That is true not only of my work on Frege's philosophy of mathematics, but also of my work on vagueness and of my work on philosophy of language. But my debt to him is far greater than that. Although I first met Crispin in the summer of 1993 at a conference organized by Matthias Schirn, the proceedings of which were published as *Philosophy of Mathematics Today* (Schirn 1998), he had already been generous with his time, discussing philosophy over email with a distant graduate student. Since then, I have many times had the privilege of spending time with Crispin, whether in St Andrews or elsewhere, discussing philosophy, football, and our families, and I am honoured now to call him a friend. He has been a reliable supporter, both of my work and of me, so much so that I am quite certain that my career would have been far different if not for his presence in my life.

Thank you, Crispin, for everything.

[1]	(1)	φ^*ab	Premise
[2]	(2)	φac	Premise
[3]	(3)	$\forall x \forall y \forall z(\varphi xy \wedge \varphi xz \rightarrow y = z)$	Premise
[1]	(4)	$\forall x(\varphi ax \rightarrow Fx) \wedge \forall x \forall y(Fx \wedge \varphi xy \rightarrow Fy) \rightarrow Fb$	(1,*—)
[1]	(5)	$\forall x(\varphi ax \rightarrow \varphi^{*=}cx) \wedge \forall x \forall y(\varphi^{*=}cx \wedge \varphi xy \rightarrow \varphi^{*=}cy) \rightarrow \varphi^{*=}cb$	(4,subst)
[6]	(6)	φax	Premise
[2,3,6]	(7)	$x = c$	(2,3,6)
[2,3,6]	(8)	$\varphi^{*=}cx$	def $\varphi^{*=}$
[2,3]	(9)	$\forall x(\varphi ax \rightarrow \varphi^{*=}cx)$	(6,7)
[]	(10)	$\forall x \forall y(\varphi^{*=}cx \wedge \varphi xy \rightarrow \varphi^{*=}cy)$	transitivity
[1,2,3]	(11)	$\varphi^{*=}cb$	(5,9,10)

The substitution rule, applied at line (5), is essential to this proof, which therefore collapses in minimal SAL. Since the substituted formula contains *, the proof cannot be replicated in predicative SAL either.

Proof of Axiom (6)

This proof is in full schematic ancestral logic plus the following restricted form of HP, which George Boolos dubbed **Log**:

$$\forall x(Fx \equiv Gx) \rightarrow Nx : Fx = Nx : Gx$$

We will need the 'roll-back theorem': $P*ab \rightarrow \exists y(P^{*=}ay \wedge Pyb)$. Its proof is straightforward. We will show that axiom (6) follows from the other axioms of arithmetic and Frege's definitions of arithmetical notions, to which we freely appeal.

We start by proving Theorem (145) of *Grundgesetze*: $P^{*=}0x \rightarrow \neg P*xx$.

[]	(1)	$P^{*=}0n \wedge \neg P^*00 \wedge \forall x \forall z(P^{*=}0x \wedge \neg P^*xx \wedge Pxz \rightarrow \neg P^*zz) \rightarrow \neg P^*nn$	Axiom (7)
[]	(2)	$P^*00 \rightarrow \exists y(P^{*=}0y \wedge Py0)$	roll-back
[]	(3)	$\neg \exists y(P^{*=}0y \wedge Py0)$	Axiom (5)
[]	(4)	$\neg P^*00$	(2, 3)
[5]	(5)	$P^{*=}0x \wedge \neg P^*xx \wedge Pxz$	Premise
[6]	(6)	P^*zz	Premise
[6]	(7)	$\exists y(P^{*=}zy \wedge Pyz)$	roll-back
[8]	(8)	$P^{*=}zy \wedge Pyz$	Premise
[5,8]	(9)	$x = y$	5, 8, Axiom (4)
[5,8]	(10)	$Pxz \wedge P^{*=}zx$	5, 8, 9
[5,8]	(11)	P^*xx	10, transitivity
[5,6]	(12)	P^*xx	11, [8] $\exists$—
[5]	(13)	$\neg P^*zz$	5, 12; [6] $\neg$+
[]	(14)	$\forall x \forall z[P^{*=}0x \wedge \neg P^*xx \wedge Pxz \rightarrow \neg P^*zz]$	13, [5] $\rightarrow$+; $\forall$+
[]	(15)	$P^{*=}0n \rightarrow \neg P^*nn$	1, 4, 14

We now prove the existence of successor by proving: $P^{*=}on \rightarrow P(n,Nx:P^{*=}xn)$. We shall need the following simple fact twice:

$$Fa \rightarrow P[Nx:(Fx \wedge x \neq a), Nx:Fx].$$

That is Theorem 102 of *Grundgesetze* and shall be cited as such. It follows immediately from (P+).

[]	(1)	$P^{*=}0n \wedge P(0, Nx : P^{*=}x0) \wedge \forall y \forall z[P^{*=}0y \wedge P(y, Nx : P^{*=}xy) \wedge Pyz \rightarrow$	Axiom (7)
		$P(z, Nx : P^{*=}xz)] \rightarrow P(n, Nx : P^{*=}xn)$	
[]	(2)	$P^{*=}00 \rightarrow P[Nx : (P^{*=}x0 \wedge x \neq 0), P^{*=}x0)$	Gg 102
[]	(3)	$P[Nx : (P^{*=}x0 \wedge x \neq 0), P^{*=}x0)$	2, def $P^{*=}$
[]	(4)	$P^{*=}x0 \wedge x \neq 0 \rightarrow P^{*}x0$	def $P^{*=}$
[]	(5)	$P^{*}x0 \rightarrow \exists u(P^{*=}0u \wedge Pu0)$	roll-back
[]	(6)	$\neg \exists u(P^{*=}0u \wedge Pu0)$	Axiom (5)
[]	(7)	$\neg(P^{*=}x0 \wedge x \neq 0)$	4, 5, 6
[]	(8)	$\forall x[(P^{*=}x0 \wedge x \neq 0) \equiv x \neq x]$	7, $\forall$+
[]	(9)	$Nx : (P^{*=}x0 \wedge x \neq 0) = Nx : (x \neq x)$	8, **Log**
[]	(10)	$Nx : (P^{*=}x0 \wedge x \neq 0) = 0$	9, def 0
[]	(11)	$P(0, Nx : P^{*=}x0)$	9, 10

So that establishes the basis step. We now prove the induction step to complete the proof.

[12]	(12)	$P^{*=}0y \wedge P(y, Nx : P^{*=}xy) \wedge Pyz$	Premise
[]	(13)	$P^{*=}zz \rightarrow P[Nx : (P^{*=}xz \wedge x \neq z), Nx : P^{*=}xz]$	Gg 102
[]	(14)	$P[Nx : (P^{*=}xz \wedge x \neq z), Nx : P^{*=}xz]$	13, def $P^{*=}$
[12]	(15)	$z = Nx : P^{*=}xy$	12, Axiom (3)
[12]	(16)	$Nx : (P^{*=}xz \wedge x \neq z) = Nx : P^{*=}xy \rightarrow P(z,Nx : Nx : P^{*}xz)$	14, 15
[12]	(17)	$\forall x[(P^{*=}xz \wedge x \neq z) \equiv P^{*=}xy] \rightarrow$	
		$Nx : (P^{*=}xz \wedge x \neq z) = Nx : P^{*=}xz$	**Log**
[12]	(18)	$\forall x[(P^{*=}xz \wedge x \neq z) \equiv P^{*=}xy] \rightarrow P(z,Nx :Nx :P^{*=}xz)$	16, 17

We now need only establish the antecedent of (18) to complete the proof. First, right-to-left:

[19]	(19)	$P^{*=}xy$	Premise
[19]	(20)	$P^{*=}xz$	12, 19, transitivity
[21]	(21)	$x = z$	Premise
[12,21]	(22)	$P^{*=}xy \wedge Pyx$	12, 19, 21
[12,21]	(23)	$P^{*=}0y \wedge P^{*}yy$	12, 22, transitivity
[12]	(24)	$x \neq z$	23, Gg 145; [12]$\neg$+
[12]	(25)	$P^{*=}xy \rightarrow P^{*=}xz \wedge x \neq z$	20, 24; [19]$\rightarrow$+

Now left-to-right:

[26]	(26)	$P^{*=}xz \wedge x \neq z$	Premise
[26]	(27)	$P^{*}xz$	def $P^{*=}$
[26]	(28)	$\exists u(P^{*=}xu \wedge Puz)$	27, roll-back

[29]	(29)	$P^{*=}xu \wedge Puz$	Premise
[12,29]	(30)	$u = y$	12, 29, Axiom (4)
[12,29]	(31)	$P^{*=}xy$	29, 30
[12,26]	(32)	$P^{*=}xy$	28, 31; [29]∃−
[12]	(33)	$P^{*=}xz \wedge x \neq z \rightarrow P^{*=}xy$	32, [26]→+
[12]	(34)	$\forall x[(P^{*=}xz \wedge x \neq z) \equiv P^{*=}xy]$	25, 33, ∀+
[12]	(35)	$P(z, Nx : Nx : P^{*=}xz)$	18, 34
[]	(36)	$\forall y \forall z[P^{*=}0y \wedge P(y, Nx : Nx : P^{*=}xy) \wedge Pyz \rightarrow P(z, Nx : Nx : P^{*=}xz)]$	35, [12]→+, ∀+

That completes the proof of the induction step. So we may conclude:

[]	(37)	$P^{*=}0n \rightarrow P(n, Nx : P^{*=}xn)$	1, 11, 36

References

Avron, A. (2003). 'Transitive Closure and the Mechanization of Mathematics,' in F. Kamareddine (ed.), *Thirty Five Years of Automating Mathematics*, New York, Kluwer Academic Publishers, 149–71.

Boolos, G. (1998a). 'Is Hume's Principle Analytic?,' in *Logic, Logic, and Logic*, Cambridge, MA: Harvard University Press, 301–14.

Boolos, G. (1998b). 'On Second-Order Logic,' in *Logic, Logic, and Logic*, Cambridge, MA: Harvard University Press, 37–53.

Boolos, G. (1998c). 'On the Proof of Frege's Theorem,' in *Logic, Logic, and Logic*, Cambridge, MA: Harvard University Press, 275–91.

Boolos, G. (1998d). 'Reading the *Begriffsschrift*,' in *Logic, Logic, and Logic*, Cambridge, MA: Harvard University Press, 155–70.

Boolos, G. (1998e). 'To Be Is to Be a Value of a Variable (or to Be Some Values of Some Variables),' in *Logic, Logic, and Logic*, Cambridge, MA: Harvard University Press, 54–72.

Boolos, G. and Heck, R. K. (1998). '*Die Grundlagen der Arithmetik* §§82–83,' in *Logic, Logic, and Logic*, Cambridge, MA: Harvard University Press, 315–38. Originally published under the name 'Richard G. Heck, Jr.'

Burgess, J. P. (2005). *Fixing Frege*, Princeton, NJ: Princeton University Press.

Etchemendy, J. (1990). *The Concept of Logical Consequence*, Cambridge, MA: Harvard University Press.

Feferman, S. (1964). 'Systems of Predicative Analysis,' *Journal of Symbolic Logic* 29, 1–30.

Feferman, S. (1991). 'Reflecting on Incompleteness,' *Journal of Symbolic Logic* 56, 1–49.

Hale, B. and Wright, C. (2000). 'Implicit Definition and the *A Priori*,' in P. Boghossian and C. Peacocke (eds.), *New Essays on the A Priori*. Oxford: Clarendon Press, 286–319.

Heck, R. K. (1996). 'The Consistency of Predicative Fragments of Frege's *Grundgesetze der Artithmetik*,' *History and Philosophy of Logic* 17, 209–20. This and subsequent papers were originally published under the name 'Richard G. Heck, Jr.''

Heck, R. K. (1998). 'That There Might Be Vague Objects (So Far as Concerns logic),' *The Monist* 81, 277–99.

Heck, R. K. (2000). 'Counting, Cardinality, and Equinumerosity,' *Notre Dame Journal of Formal Logic* 41, 187–209.

Heck, R. K. (2011). 'Ramified Frege Arithmetic,' *Journal of Philosophical Logic* 40, 715–35.

Koellner, Peter (2010). 'Strong Logics of First and Second Order,' *Bulletin of Symbolic Logic* 16, 1–36.

Linnebo, Ø. (2004). 'Predicative Fragments of Frege Arithmetic,' *Bulletin of Symbolic Logic* 10, 153–74.

McGee, V. (1997). 'How We Learn Mathematical Language,' *Philosophical Review* 106, 35–68.

Prior, A. (1960). 'The Runabout Inference-Ticket,' *Analysis* 21, 38–9.

Quine, W. V. O. (1986). *Philosophy of Logic*, Cambridge, MA: Harvard University Press, 2nd edition.

Rayo, A. (2002). 'Words and Objects,' *Noûs* 36, 436–64.

Resnik, M. (1988). 'Second-Order Logic Still Wild,' *Journal of Philosophy* 85, 75–87.

Schirn, M. (ed.) (1998). *Philosophy of Mathematics Today*, Oxford: Oxford University Press.

Shapiro, S. (1991). *Foundations without Foundationalism: A Case for Second-order Logic*, Oxford: Oxford University Press.

Shapiro, S. (2001). 'Classical Logic II—Higher-Order Logic,' in L. Goble (ed.), *The Blackwell Guide to Philosophical Logic*, Malden, MA: Blackwell Publishing, 33–54.

Wright, C. (1983). *Frege's Conception of Numbers as Objects*, Aberdeen: Aberdeen University Press.

Wright, C. (2001). 'Is Hume's Principle Analytic?,' in *The Reason's Proper Study*, Oxford: Clarendon Press, 307–32.

3

Logicism and Logical Consequence

Jim Edwards

1. Introduction

Crispin Wright relaunched Frege's programme for showing that numbers are objects and that the truths of arithmetic are logical truths. There are differences between Frege's original programme and Wright's (1983) neo-Fregean logicism. Frege had sought to derive the truths of arithmetic from what he took to be purely logical axioms by purely logical rules of inference. Had the programme succeeded, the truths of arithmetic could in principle have been rewritten in basic logical vocabulary, and would have been shown to be proof-theoretic consequences of logical axioms by logical rules of inference. Frege's project foundered on Russell's discovery of paradox in Frege's axioms, and subsequently Gödel showed that no finitistic proof-theory can deliver all the truths of arithmetic. Wright starts from second-order proof theory, generally assumed to be consistent, but with an added axiom, called Hume's Principle, which itself contains an explicit operator 'Nv:...v...' forming a singular term for a number from a predicate: $\forall F \forall G(Nx:Fx = Nx:Gx \leftrightarrow \exists R(F_1\text{-}1_R G))$. Here '$F_1\text{-}1_R G$' abbreviates a claim, which can be made in basic vocabulary, that R one–one correlates the extensions of the predicates F and G. Wright claims that whilst Hume's Principle is not a truth of logic per se, and is not an explicit definition allowing the elimination of numerical terms from all sentences, it is constitutive of our conception of natural number, and so an analytic truth. The idea is then to derive Peano's axioms for arithmetic in this system. In the light of Gödel's result not all the truths of arithmetic are derivable in this or any other finitely codifiable system. But in response to Gödel and following Tarski, the primary notion of logical consequence is now semantic and not proof-theoretic. Wright can appeal to the semantic result that Peano's axioms are categorical with respect to arithmetic sentences in what are known as full models. Hence, *provided we agree that full models measure the semantic relation of logical consequence among the sentences of arithmetic*, the derivation of Peano's axioms shows that the truths of arithmetic are logical consequences, in the semantic sense, of second-order logic augmented with Hume's Principle. Hence they are analytic truths, being the logical consequences of an analytic axiom.

Subsequent work by Wright and Bob Hale in support of neo-Fregean logicism has focused upon defending the Fregean ontology of numbers as objects, upon the status of Hume's Principle, upon fending off the 'Caesar problem' that led Frege to introduce explicit definitions of numbers and led from there to paradox, and upon

Jim Edwards, *Logicism and Logical Consequence* In: *Logic, Language, and Mathematics: Themes from the Philosophy of Crispin Wright.* Edited by: Alexander Miller, Oxford University Press (2020). © Jim Edwards.
DOI: 10.1093/oso/9780199278343.003.0003

extending the programme from arithmetic to analysis—see the papers in Wright (2000), Hale and Wright (2001), and for an overview Hale and Wright (2005). But attention is now turning to the semantic relation of second-order logical consequence. It is crucial to neo-Fregean logicism that the semantic relation of logical consequence among the sentences of arithmetic be proof-theoretically *in*complete. For otherwise Gödel's result would defeat neo-Fregean logicism, by generating arithmetic truths which are not analytic truths, not being proof-theoretic consequences of the *ex hypothesi* proof-theoretically complete logic applied to Hume's Principle. I shall argue that the semantic relation of second-order logical consequence most naturally suited to the practice of arithmetic *is* proof-theoretically complete. Hence Wright's programme of neo-Fregean logicism is threatened. Gödel's result shows that there are arithmetic truths which are not derivable in Wright's proof theory augmented by Hume's Principle, and hence, given the completeness of that second-order proof-theory, are not logical consequences of Hume's Principle, I shall claim.

First, some preliminaries to the main argument to come in later sections—rather a lot of preliminaries, I'm afraid. In the remainder of this section I shall (i) sketch a second-order language and its attendant proof-theory, (ii) give full and Henkin semantics for that language, (iii) frame the resulting full and Henkin definitions of logical consequence for that language, (iv) give a *philosophical* conception of logical consequence to which, I think, such definitions should answer, and, finally, (v) I shall relate the full and Henkin definitions to my philosophical conception of logical consequence.

(i) Stewart Shapiro (2000a) has presented an extended and detailed defence of a view of second-order logical consequence favourable to Wright's neo-Fregean project.[1] We can use Shapiro's formal language, attendant proof theory and associated semantics to frame our discussion. Shapiro's language L2K comprises an unspecified but denumerable set of non-logical constants K, the logical constants '$\neg$,' '$\rightarrow$,' and '$\forall$,' an infinite but denumerable set of first-order variables 'x_1,' 'x_2,' etc., an infinite but denumerable set of n-adic second-order predicate variables 'X_1^n,' 'X_2^n,' etc., for each $n \geq 1$, and similarly of function variables 'f_1^n,' 'f_2^n,' etc., for each $n \geq 1$. The formation rules for open and closed sentences are as expected, and other constants are introduced by the usual definitions. Defined on this language we have a proof theory D2. Taking 'φ' and 'ψ' as meta-variables for any open or closed sentence of L2K, we have:

Axioms of D2:
Any open or closed sentences of the forms:
$\varphi \rightarrow (\psi \rightarrow \varphi)$
$(\varphi \rightarrow (\psi \rightarrow \chi)) \rightarrow ((\varphi \rightarrow \psi) \rightarrow (\varphi \rightarrow \chi))$

---

[1] Well, favourable up to a point, Shapiro himself thinks—see Shapiro and Weir (2000). Favourable in that Peano's axioms are derivable in a somewhat extended second-order proof-theory from Hume's Principle, and favourable in that Peano's axioms are categorical with respect to Shapiro's notion of logical consequence, as Wright requires. However, Shapiro and Weir doubt that the epistemic aims of neo-Fregean logicism are well served. They argue that the theory delivers Peano's axioms only because the underlying second-order proof theory is heavily committed to an ontology of sets prior to the introduction of Hume's Principle.

$(\neg\varphi \rightarrow \neg\psi) \rightarrow (\psi \rightarrow \varphi)$

$\forall x\varphi(x) \rightarrow \varphi(t)$, where t is any term free for x in φ.

$\forall X^n\varphi(X^n) \rightarrow \varphi(R^n)$, where R^n is either an n-place relational variable free for X^n in φ, or an n-place relation constant.

$\forall f^n\varphi(f^n) \rightarrow \varphi(F^n)$, where F^n is either an n-place function variable free for f^n in φ, or an n-place function constant.

Comprehension schema: $\exists X^n\forall\langle x\rangle_n(X^n\langle x\rangle_n \leftrightarrow \varphi\langle x\rangle_n)$, provided X^n does not occur free in φ.

Axiom of Choice: $\forall X^{n+1}(\forall\langle x\rangle_n\exists y X^{n+1}\langle x\rangle_n y \rightarrow \exists f^n\forall\langle x\rangle_n X^{n+1}\langle x\rangle_n f^n\langle x\rangle_n)$.[2]

Rules of inference of D2:

For any open or closed sentences:

From φ and $\varphi \rightarrow \psi$ infer ψ

From $\varphi \rightarrow \psi(x)$ infer $\varphi \rightarrow \forall x\psi(x)$, provided that x does not occur free in φ or in any premise of the deduction.

From $\varphi \rightarrow \psi(X^n)$ infer $\varphi \rightarrow \forall X^n\psi(X^n)$, provided that X^n does not occur free in φ or in any premise of the deduction.

From $\varphi \rightarrow \psi(f^n)$ infer $\varphi \rightarrow \forall f^n\psi(f^n)$, provided that f^n does not occur free in φ or in any premise of the deduction.

(ii) We turn now to semantics for L2K. Shapiro takes semantics for L2K to consist of three elements: a set of models, a set of assignment functions, and a satisfaction relation. Different triads of different types of models, with their attendant assignment functions and satisfaction relations will yield different types of semantics for L2K. Shapiro offers full semantics, two versions of Henkin semantics, and first-order, multi-sorted semantics. We'll consider here just full semantics and what he calls 'faithful' Henkin semantics—though we'll return briefly to first-order, multi-sorted semantics in §3. Since we will only consider one species of Henkin semantics, I'll drop the sobriquet 'faithful.' Briefly before we get to detail, a *model* is an ordered n-tuple of a domain (full models) or domains (Henkin models) for the quantifiers, and a function assigning entities of the appropriate categories arising from the domain (or domains) to the non-logical constants in K. An *assignment function* assigns values from the domain (or domains) to variables. Lastly, a *satisfaction relation* between a model, an assignment function and an open or closed sentence defines what it is for that sentence to be satisfied by that model relative to that assignment function. A closed sentence is true of a given model iff it is satisfied by that model relative to all assignment functions. In somewhat more detail:

Full Semantics

A *full model* for L2K is a structure $\langle d,I\rangle$, where d is a non-empty set, the domain of the first-order variables, and I is an interpretation function assigning suitable set-theoretic entities taken or constructed from d to any non-logical constants in K.

---

[2] For perspicuity comprehension and choice are expressed using the defined symbols '$\exists$' and '$\leftrightarrow$.'

A *variable assignment* s with respect to a model <d,I> is any function that assigns a member of d to each x_i, a subset of d^n to each X_i^n, and a function from d^n to d to each f_i^n. (Thus s may assign to the variable X_i^n any member of the full *powerset* of d^n as its extension, that is any n-adic relation taken in extension among the members of d; and s may assign to f_i^n any function from d^n to d as its extension.)

A relation of *satisfaction* of an open or closed sentence of L2K by a variable assignment s relative to a model <d,I> is then defined straightforwardly as expected.

A closed sentence of L2K is *true* of a model <d,I> iff it is satisfied by all variable assignments relative to <d,I>.

Henkin Semantics

A *Henkin model* for L2K is a structure <d,D,F, I>, with d as before. The additional member D is a sequence $<D_1, D_2, \ldots, D_n, \ldots>$ such that each D_n is a non-empty subset of the powerset of d^n, and similarly F is a sequence $<F_1, F_2, \ldots, F_n, \ldots>$ such that each F_n is a non-empty subset of the functions from d^n to d. (In effect these are the relations and functions recognized by that model of L2K, and they may fall short of the relations and functions recognized by the corresponding full model.) However, every instance of the axiom of comprehension is true, and the axiom of choice is true. That is, every open sentence $\varphi<x>_n$ of L2K determines a member of D_n, and if a sentence of L2K of the form $\forall<x>_n \exists y X^{n+1}<x>_n y$ is true of a model, then F_n for that model contains a function making $\exists f^n \forall<x>_n X^{n+1}<x>_n f^n<x>_n$ true. The interpretation function I assigns to any singular constant of K a member of d, to any n-adic relational constant of K a member of D_n, and to any n-adic functional constant of K a member of F_n.

A *variable assignment* s with respect to a model <d,D,F,I> is any function that assigns a member of d to each x_i, of D_n to each X_i^n, and of F_n to each f_i^n.

The relation of *satisfaction* of an open or closed sentence of L2K by a variable assignment s relative to a model <d,D,F,I> is defined as for full semantics.

A closed sentence of L2K is *true* of a model <d,D,F,I> iff it is satisfied by all variable assignments relative to <d,D,F,I>.

By examination, each full model <d,I> is in effect a Henkin model in which D_n = the powerset of d^n, for each $n \geq 1$, and F_n = all functions from d^n to d, for each $n \geq 1$. But not vice versa, there are Henkin models which are not full models.

The proof theory D2 does not include any axioms or rules dealing especially with the identity symbol '=' used with first-order terms, and neither semantics gives special status to the relation assigned by I to '=.' That is to say '=' is not a logical constant of L2K. This will become relevant when we extend L2K by adding a term-forming operator 'Nv:...v...,' semantically a function from D_1 to d, to formulate Hume's Principle: $\forall F \forall G (Nx:Fx = Nx:Gx \leftrightarrow \exists R(F1-1_R G))$. Neo-Fregean logicism proposes to derive Peano's Axioms from Hume's Principle, but to do that we will need to add to D2 the standard axioms of identity:

$$\forall x(x = x)$$
$$\forall x \forall y \Big(\big(\varphi(x) \, \& \, x = y \big) \rightarrow \varphi(y) \Big)$$

The letter of neo-Fregean logicism as described above would need these additional axioms to be logical truths on a par with D2, and hence '=' to be a logical constant, since officially Hume's Principle is to be the only non-logical axiom. The issue of whether or not '=' is a logical constant is moot. One influential view makes '=' a logical constant. In the view of Tarski (1986) and elaborated by Sher (1991) '=' is a logical constant par excellence. Logical constants are taken to be expressions whose extensions are invariant under all one–one permutations of the domain d, and clearly {<x,y>: x=y} is so invariant. But my own broadly Davidsonian view is that '=' is not a logical constant, and the standard axioms for identity are therefore not logical truths—see Edwards (2002). According to my view, logical constants are dealt with by the recursion axioms of an optimal theory of meaning, and non-logical constants by the lexical axioms. An optimal theory of meaning for L2K will deal with '=' as an ordinary binary relational expression by a lexical axiom: $\forall x \forall y (<x,y>$ satisfies '=' $\leftrightarrow x=y$). Hence in intended models <d,I> and <d,D,F,I> of L2K '=' is a member of K and in intended interpretations I('=')={<x,y>:x=y}. However, although my view of the status of '=' conflicts with the letter of neo-Fregean logicism it does not offend the spirit. Hume's Principle is not taken to be a logical truth, but is taken to be analytic, constitutive of the meaning of the operator 'Nv:...x....' A neo-Fregean logicist could add the standard axioms for identity to D2 claiming they are similarly analytic, not logical truths but constitutive of the meaning of '='.[3] It seems to me that the epistemic and ontological ambitions of neo-Fregean logicicsm would not thereby be compromised.

(iii) Now let us turn to relations of logical consequence among the sentences of L2K under full and Henkin semantics, respectively. A Tarskian definition takes the form:[4]

(⊨)   A sentence φ is a *logical consequence* of a set of sentences Γ iff for every model M and assignment s on M, if for every ψ in Γ, ψ is satisfied by M under s, then φ is satisfied by M under s.

Since we have two conceptions of models in play, we have two putative definitions of logical consequence on offer.

(⊨F)   A sentence φ is a *full logical consequence* of a set of sentences Γ iff for every *full* model M and assignment s on M, if for every ψ in Γ, ψ is satisfied by M under s, then φ is satisfied by M under s.

(⊨H)   A sentence φ is a *Henkin logical consequence* of a set of sentences Γ iff for every *Henkin* model M and assignment s on M, if for every ψ in Γ, ψ is satisfied by M under s, then φ is satisfied by M under s.

---

[3] Although Davidson himself did not do so, Edwards (2007) shows how a Davidsonian can recognize such a notion of the analytic.

[4] Tarski (1935) defines logical consequence in terms of truth, not satisfaction: A sentence φ is a *logical consequence* of a set of sentences Γ iff for every model M, if every ψ in Γ is true in M, then φ is true in M. The definition given in the text allows for relations of logical consequence between open sentences too, as required by D2.

As is well known, ($\models_F$) and ($\models_H$) differ in extension.

(iv) When Tarski presented his semantic definition of logical consequence in his (1935) he did not see himself as laying down by definition a new concept. Rather, he sought to capture in mathematically tractable form a prior conception of logical consequence prevalent among logicians of his day. Thus his opening remark was:

The concept of logical consequence is one of those whose introduction into the field of strict formal investigation was not a matter of arbitrary decision on the part of this or that investigator; in defining this concept, efforts were made to adhere to the common usage of the language of ordinary life.   (Tarski 1935: 409)

I take from Tarski the key idea that a model-theoretic definition of logical conse-quence should answer to a prior *philosophical* conception of logical consequence. As a first pass, here is my prior philosophical conception of logical consequence for closed sentences:

> (PC)   A sentence φ is a logical consequence of a set of sentences Γ
> iff
> it is impossible for φ to be false and all the members of Γ true, where the meaning of the logical constants in φ and in the members of Γ, and otherwise their logical forms, are sufficient in context to generate this impossibility.

(PC) raises immediate prior philosophical questions. What are the logical constants of a language? What are the logical forms of its sentences? But they need not delay us, since in L2K the logical constants are agreed to be '∀,' '→,' and '¬'—setting aside controversy over the status of '='—and the logical forms of sentences of L2K are their surface forms. By saying that the logical forms and the meaning of the logical constants are sufficient *in context* to generate the impossibility of φ being false and all the members of Γ true, I am gesturing at the platitude that in general the truth or falsity of a sentence depends upon the meanings of its non-logical constants too, on the sentence meaning that so-and-so, and upon the facts, upon it being the case or not that so-and-so. For example, 'Snow is white v ¬Snow is white' is a logical consequence of the null set of premises (a logical truth), but the meaning of the non-logical expression 'snow is white' and the fact of snow being white together play a part in making it true by making the first disjunct true and therefore the whole disjunction true. Still, it is also the case that 'Snow is white v ¬Snow is white' would be true whatever interpretation we gave to 'snow is white,' even one making it false. Further, by saying that the logical forms and the meanings of the logical constants are sufficient in context *to generate* the impossibility of φ being false and all the members of Γ true, I am gesturing at a distinction between why it is impossible for 'Snow is white v ¬Snow is white' to be false on the one hand, and why it is impossible, for example, for 'Water is H_2O v ¬Snow is white' to be false, on the other, where the latter depends not only upon the meaning of the non-logical expression 'Water is H_2O' but, crucially, on the fact that *necessarily* water is H_2O. As will be seen in §2, I intend (PC) to apply not only to L2K whose logical forms and logical constants are relatively uncontroversial, but to an extended language L2K⁺ of which L2K is only a fragment, whose logical constants and logical forms are as yet undetermined. Thus (PC) is inchoate insofar as our conception of logical constants and of logical form in the wider language $L2K^+$ is inchoate. And (PC) is also shallow because it appeals to an unexamined generative relation between necessity, truth, meaning, and logical form. But guiding prior philosophical conceptions are often inchoate and shallow. My conception is, I hope, a familiar enough philosophical conception of logical consequence, so far as it goes, and one whose linking of logic to meaning should be at least prima facie congenial to neo-Fregeans, given their claim that Hume's Principle is constitutive of the meaning of numeric terms and therefore analytic.

(v) Logical consequence, according to (PC), turns on logical form and on the meaning of the logical constants. Full and Henkin semantics generate different extensions for the relation of logical consequence among the open and closed sentences of L2K, but they assume the same logical forms for the sentences of L2K. So, on my view, they presuppose different meanings for the logical constants. A semantics gives at least the referential skeleton, the *Bedeutung* if not the *Sinn* in Frege's terms, of the meaning of the sentences of L2K. Each semantics assigns to each logical constant what I will call an 'outer' function. In the case of full semantics and '$\neg$,' for example, this is a function from assignments s, domains d, interpretations I, and sentences φ either to $\{<s,d,I,\varphi>:$ s satisfies $\ulcorner\neg\varphi\urcorner$ with respect to d and I$\}$ or to $\{<s,d,I,\varphi>:$ s does not satisfy $\ulcorner\neg\varphi\urcorner$ with respect to d and I$\}$. But this is not the function a user of '$\neg$' need understand as its reference, since in using L2K she is given one particular interpretation and domain, and need not think of generalizing over other domains or interpretations. She will need to grasp a derived 'inner' function, which takes particular values of d and I as fixed, and goes from assignments s and sentences φ only either to $\{<s,\varphi>:$ s satisfies $\ulcorner\neg\varphi\urcorner\}$ or to $\{<s,\varphi>:$ s does not satisfy $\ulcorner\neg\varphi\urcorner\}$. These inner functions are the referents of '$\neg$' assigned by the semantics, and they differ in extension from model to model of course. But what allows us nonetheless to say that the meaning of '$\neg$' does not differ from model to model is that each inner function (distinct in its extension) is generated, presupposing given values for d and I, by the *same* rule understood by a speaker in intension:

For all s, φ, s satisfies $\ulcorner\neg\ \varphi\ \urcorner$ iff s does not satisfy φ.

These different inner functions are the references of '$\neg$' in each full model, and the common rule its single sense.

Since both full and Henkin semantics employ the same clauses for '$\neg$' and '$\rightarrow$,' the different relations of logical consequence they generate among the sentences of L2K are due to the different meanings they respectively assign to '$\forall$.' For each model $<d,I>$ full semantics assigns a different triplet of inner functions to '$\forall$,' but for each model those functions are generated by the same three rules:

For any function s assigning to x_i a member of d, to X_i^n a subset of d^n, and to f_i^n a function from d^n to d:

s satisfies $\ulcorner\forall x_i\varphi\urcorner$ iff any assignment s*, which differs from s at most in the member of d assigned to x_i, satisfies φ,

s satisfies $\ulcorner\forall X_i^n\varphi\urcorner$ iff any assignment s*, which differs from s at most in the subset of d^n assigned to X_i^n, satisfies φ, and s satisfies $\ulcorner \forall f^n_i \varphi \urcorner$ iff any assignment s*, which differs from s at most in the function from d^n to d assigned to f^n_i, satisfies φ.

For each given Henkin model $\langle d,D,F,I \rangle$, Henkin semantics assigns an inner function according to the following three rules. Only the first is common to Henkin and to full semantics.

> For any function s assigning a member of d to x_i, a member of D_n to X^n_i and a member of F_n to f^n_i:
>
> s satisfies $\ulcorner \forall x_i \varphi \urcorner$ iff any assignment s*, which differs from s at most in the member of d assigned to x_i, satisfies φ,
>
> s satisfies $\ulcorner \forall X^n_i \varphi \urcorner$ iff any assignment s*, which differs from s at most in the member of D_n assigned to X^n_i, satisfies φ,
>
> s satisfies $\ulcorner \forall f^n_i \varphi \urcorner$ iff any assignment s*, which differs from s at most in the member of F_n assigned to f^n_i, satisfies φ,
>
> *and* d, D and F are so related that the axiom of choice—$\forall X^{n+1}(\forall \langle x \rangle_n \exists y X^{n+1} \langle x \rangle_n y \rightarrow \exists f^n \forall \langle x \rangle_n X^{n+1} \langle x \rangle_n f^n \langle x \rangle_n)$—comes out true, and each instance of the comprehension schema—$\exists X^n \forall \langle x \rangle_n (X^n \langle x \rangle_n \leftrightarrow \varphi \langle x \rangle_n)$, provided X^n does not occur free in φ—comes out true.[5]

An alternative account of the meaning of '$\forall$' in Henkin semantics would remove the restriction on d, D, and F due to Comprehension and Choice from the specification of the inner function, the reference of '$\forall$,' and take the restriction as given, as built into the specification of a Henkin model. However, this would be a specification of the reference of '$\forall$' according to 'unfaithful' Henkni semantics. For there would be nothing in the *meaning* of '$\forall$' to exclude its application to *unrestricted* models in which Comprehension and Choice fail. On my view of the relation between logical consequence and the meaning of the logical constants, the restrictions must be embodied in the meaning of '$\forall$,' in the rule generating the inner functions, the references of '$\forall$,' if Comprehension and Choice are to be logical truths.

In §3 we will consider the use of L2K to practise elementary arithmetic and analysis. To do so we will take L2K to be furnished with the appropriate *full* models. But as we noted, full models are a proper subset of Henkin models, a proper subset in which D_n = the powerset of d^n, for each $n \geq 1$, and F_n = all functions from d^n to d, for each $n \geq 1$. The inner functions assigned to '$\forall$' by full and by Henkin semantics will be the same in extension, but they presuppose different meanings for '$\forall$,' different intensions, different rules by which that common extension is generated, and with these different meanings for '$\forall$' go extensionally different relations of logical consequence ($\vDash_F$) and ($\vDash_H$). Principles of interpretation of the use of L2K to practise arithmetic or analysis will be needed to determine which relation of logical consequence that use manifests. This will be the topic of §3.

---

[5] The full and Henkin rules for '$\forall$' are recursively specified, since φ may itself contain '$\forall$.' And obviously so in the case of Henkin meaning, where a condition is laid on d, D, and F to make true certain sentences employing '$\forall$.'

2. Comparing ($\models_F$) and ($\models_H$) as Putative Definitions of Logical Consequence

Tarski claimed that his ($\models$) 'agrees quite well with common usage' (1935: 417). What kind of agreement are we after? Not, I think, conceptual analysis. (PC) is modal, and requires a certain explanation of that modality—it is impossible for φ to be false and all the members of Γ true, where the meanings of the logical constants in φ and in the members of Γ, and otherwise their logical forms, are sufficient in context to generate this impossibility. But ($\models$) is non-modal: all models of Γ are models of φ. The agreement we are looking for is agreement in extension between ($\models$) and (PC), and therefore agreement in extension between ($\models_F$) or ($\models_H$) and (PC), when '∀' is given full or Henkin meaning respectively. Agreement in extension need not be perfect. For my prior conception is inchoate and shallow, whereas ($\models_F$) and ($\models_H$) are relatively clear and precise. But we should require that they agree over *definite* cases. If the prior conception definitely determines that φ is or is not a logical consequence of Γ, then Γ $\models_F$φ should match that verdict, and Γ $\models_H$φ similarly.[6] However, to accept ($\models_F$) or ($\models_H$) as *definitions* of logical consequence we require more than that they *in fact* capture in this way the extension of (PC), we need to *know* that they so capture the extension of (PC). Reasonable conjecture is not good enough. Conjecture leaves open the epistemic possibility of them not coinciding. But *definition* closes off that avenue of inquiry. It is an epistemic solecism to say 'I define X as Y, and in this case Y, and I conjecture that in this case X also,' even if I am quite confident of my conjecture. If X is defined as Y, there is no epistemic room to conjecture that this given Y is X.[7] Nonetheless, a definition may be proposed against a presupposed background theory Z, which may itself be only a conjecture. So it is in order to say: assuming Z, I define X as Y, where given Z, I would then know of any Y that it is X. So to take ($\models_F$) (or ($\models_H$)) as a definition of logical consequence, perhaps against a background assumption, we need to *know*, perhaps to know only relative to a presupposed background, that ($\models_F$) (or ($\models_H$)) coincides in extension with (PC) applied to the language L2K, at least over cases where (PC) is definite. If we do not, even relative to whatever is our background presupposition, know that they so coincide, we can of course take ($\models_F$) (or ($\models_H$)) as definitions of *something*—full consequence or Henkin consequence, say. But then it is a conjecture that full consequence or Henkin consequence *coincides* with logical consequence.

Shapiro's position is somewhat complicated by the fact that he is dealing with (at least) two languages: the artificial formal language L2K and the natural language in which arithmetic, and mathematics generally, is conducted. Shapiro intends L2K to *model*, only, the natural language—'model' in the scientific sense in which point

---

[6] I intend to allow that (PC) may be vague relative to ($\models_F$) and to ($\models_H$) by requiring agreement of extension only over cases (PC) definitely determines. Clearly, we should *not* take (PC) to be effectively decidable, even for definite cases. A case may be definite even though we cannot discover that it is so. So, although logical form and meaning are things we grasp, those logical forms and meanings may definitely determine that φ is a logical consequence of Γ, or definitely determine that φ is not a logical consequence of Γ, without our having an effective procedure to discover which.

[7] Cf. I define a bachelor as an unmarried male, Tom is an unmarried male, and I conjecture that he is a bachelor!?!

masses model the behaviour of a gas.[8] The purpose of ($\models_F$) and ($\models_H$) is then to model, only, the relation of logical consequence among the sentences of the natural language. As such we should *not* require that ($\models_F$) or ($\models_H$) coincide in extension with the relation of logical consequence holding among the sentences of the natural language. For models may be more or less good, and may involve idealization so that the definition of logical consequence in the model need not, and even for good reason should not, coincide with that of the natural language. But we should not confuse this issue with the requirement that ($\models_F$) or ($\models_H$) coincide in extension with (PC) *when (PC) is applied to the sentences of L2K itself.* We should not think that (PC) applies only to the sentences of the natural language, being our intuitive conception of logical consequence, and not require that ($\models_F$) or ($\models_H$) themselves answer to (PC). Such confusion might be encouraged by the following remark by Shapiro.[9] He is discussing the influence of the axioms of set theory (adopted in the metalanguage) upon the output of ($\models_F$) among the sentences of L2K. In particular he is discussing the axiom of infinity and he remarks:

> We don't need to claim that the axiom of infinity is true on logical grounds, since we are out to *model* the intuitive notion of consequence, not to give some sort of philosophical analysis of it.
> (Shapiro 1998: 151, his emphasis)

This suggests, at least, that the intuitive conception of (logical) consequence is not to be applied to L2K. But that would be a mistake. When L2K is taken to model in the scientific sense the natural language in which we do arithmetic, L2K is itself an *interpreted* language, albeit the interpretation is stipulated in a metalanguage. As an interpreted language, a relation of logical consequence answering to (PC) holds among the sentences of L2K. My claim is that ($\models_F$) (or ($\models_H$)) can be taken as definitions of logical consequence *among the sentences of L2K,* perhaps subject to background assumption, only if we know that ($\models_F$) (or ($\models_H$)) coincides in extension with (PC) over definite cases when (PC) is applied *to the sentences of L2K themselves.* What relation of logical consequence holds among the sentences of the natural language of arithmetic we are trying to model in L2K, and how ($\models_F$) or ($\models_H$) compare, are separate issues.

However, unlike Shapiro I do not intend L2K, furnished with suitable semantics, to be a *model,* in the scientific sense, of a natural *language* of arithmetic. Rather, I take it, so furnished, to be a language for expressing the *facts* of arithmetic and a language in which one could *do* arithmetic. Of course, those facts are expressed in, and the practice of arithmetic conducted in, natural language too, and I take the ontology of that natural language at face value; an attitude Shapiro shares and calls 'working realism.' But my objective is to get L2K to express the arithmetic facts and to be a vehicle for doing arithmetic. And as we shall see in §3, Shapiro's complaint against L2K furnished with Henkin semantics is that it cannot unambiguously express the arithmetic facts, and is not itself a suitable vehicle for doing arithmetic. Officially, I presume, this is an argument that it is therefore not a good model, in the scientific sense, of the natural language in which we express arithmetic facts and do arithmetic.

---

[8] See Shapiro (2000b: Part 1 *passim*).   [9] I do not claim that Shapiro himself is thus confused.

But it seems to me that Shapiro's complaints, if well grounded, count directly against the adequacy of L2K, when furnished with Henkin semantics, for expressing and doing arithmetic. And this last is the claim we shall examine in §3. The modelling of an unspecified natural language is an idle wheel.

Let us return to applying (PC) to the sentences of L2K itself. I shall argue that we can take ($\models_F$) to *define* logical consequence among the sentences of L2K when '∀' has full meaning, but only against a background presupposition, a presupposition whose epistemic status is that of a *conjecture* merely. In contrast, I shall argue, we can *know* that ($\models_H$) does indeed capture the extension of logical consequence when '∀' has Henkin meaning, without need of a parallel conjecture. In this sense, ($\models_H$) is more *philosophically secure* than ($\models_F$) as a definition of logical consequence.

For ($\models_F$) to succeed in defining a relation which answers to my prior conception of logical consequence, when '∀' has full meaning, we require to *know*, perhaps only relative to a presupposition, that the following holds:

(1)   In cases which (PC) definitely determines, and in which '∀,' '→,' and '¬' have the meaning assigned by full semantics, $\Gamma \models_F \varphi$ iff φ is a logical consequence of Γ according to (PC).

We can break this down into:

(1A)   In cases which (PC) definitely determines, and in which '∀,' '→,' and '¬' have the meaning assigned by full semantics, if $\Gamma \models_F \varphi$ then φ is a logical consequence of Γ according to (PC).

(1B)   In cases which (PC) definitely determines, and in which '∀,' '→,' and '¬' have the meaning assigned by full semantics, if φ is a logical consequence of Γ according to (PC) then $\Gamma \models_F \varphi$.

(1B) is trivial. If there is a full model in which Γ are all true and φ false, then that would provide the referential basis for an interpretation of L2K in which Γ are all true and φ false, and hence show that φ is not a logical consequence of Γ in the prior philosophical sense (PC). Such a model would be a definite counterexample to (PC).[10] Hence, by contraposition, if φ is definitely a logical consequence of Γ according to (PC), then $\Gamma \models_F \varphi$.

However, we know (1A) holds only relative to a certain presupposition. We can frame ($\models_F$) in a suitably extended L2K itself, L2K⁺. L2K⁺ may be used as a metalanguage to express set theory, second-order ZFC say, define sentencehood, full models, etc., and hence to frame the proposed definition ($\models_F$) for an object language of the kind L2K.[11] The *truth* of second-order ZFC in this metalanguage requires only the

---

[10] We do not require these counterexamples to be languages of practical interest, or ones we could actually use or even understand were they not specified via a metatheory. Their purpose is only, so to speak, to 'filter out' cases where there is no impossibility of φ being false and all the members of Γ true due to, etc.

[11] Our earlier informal sketch of semantics of L2K and definition ($\models_F$) cannot be *directly* formalized in L2K⁺. L2K⁺ has only second-order functions and predicates: 'f^n' symbolizes a function from d^n to d, 'X^m' a property satisfied by a subset of d^n. But our informal semantics used a function, a *variable assignment* s, assigning a subset of d^n to each predicate 'X^n,' and a function from d^n to d to each function symbol 'f^n.' So far so good, as the syntax of the object language L2K will be in the domain of the metalanguage L2K⁺. But *satisfaction* was a third-order relation: for example, '$X^n x_1, \ldots, x_n$' is satisfied by $\langle X^n, x_1, \ldots, x_n \rangle$

cumulative hierarchy up to but not including the first inaccessible cardinal. So, if that is how the metalanguage is interpreted, the range of full models quantified over in ($\vDash_F$) framed in this metalanguage is confined to those in which d is a set of accessible cardinality. But this interpretation of L2K$^+$ is itself not represented by any model quantified over in ($\vDash_F$). It is represented by a model in which d has the cardinality of the entire cumulative hierarchy of accessible cardinals, itself the first inaccessible cardinal.

Thus we have a further interpretation of L2K, taken as metalanguage, not among those interpretations quantified over in defining ($\vDash_F$) for L2K taken as object language. But so what? That L2K bears yet another interpretation is not itself significant to the status of ($\vDash_F$). It becomes significant however when we note that the new interpretation assigns the same logical forms to the sentences of L2K and the same meaning to the logical constants '$\forall$,' '$\rightarrow$,' and '$\neg$.' A semantic theory of the metalanguage would assign the same logical forms to its sentences and employ the same clauses for the logical constants. The only difference is that the domain is a bigger set, the first inaccessible, than any provided for the quantifier in the object language. (PC) looks only at logical form and the meaning of the logical constants, so $\Gamma \vDash_F \varphi$ is hostage to (PC), when applied to the bigger domain of the metalanguage, definitely making Γ all true but φ false. If so, ($\vDash_F$) fails to capture the extension of the relation of logical consequence even among the sentences of L2K the *object* language.[12]

It will not help to increase the range of models quantified over in ($\vDash_F$) by adding to second-order ZFC an axiom of infinity sufficient for, say, up to the nth inaccessible. For instance, let the metalanguage quantify over all sets up to but not including the second inaccessible, let Γ be second-order ZFC, and φ be 'There exists a first inaccessible,' all expressed in L2K$^+$. Then Γ and φ are both true in the metalanguage, since the domain contains a first inaccessible, but $\Gamma \vDash_F \neg\varphi$ is also true, since the first inaccessible is the biggest domain the metalanguage has to offer ($\vDash_F$), and all *its* members are themselves smaller than the first inaccessible. The point generalizes. If the domain of the metalanguage includes all sets up to but not including the nth inaccessible and φ is 'There is an (n-1)th inaccessible,' then Γ and φ are both true in the metalanguage, but so is $\Gamma \vDash_F \neg\varphi$. ($\vDash_F$) so understood produces false positives.[13]

The key point does not depend upon any particular belief about the size of the set-theoretic universe. Of course we will want our definition of ($\vDash_F$) to quantify over absolutely all models and will furnish our metatheory with whatever axioms we think

---

iff X^n= s('X^n') & x_1=s('x_1') & ... & x_n=s('x_n') & $<x_1, ..., x_n>\varepsilon$s('$X^n$'). However, we can frame the semantics using second-order L2K$^+$ as the metalanguage if we first arithmetize the syntax of the object language L2K so as to represent it by members of d, and extend d to include suitably rich sets—see Shapiro (1991: 134-7).

[12] Wright has doubts about the existence of uncountable transfinite sets—see Wright (1985) and the concluding remark of Hale and Wright (2005). However, such scepticism does not nullify the essential point being made. Suppose the interpretation of L2K$^+$ used to frame ($\vDash_F$) quantifies over only all countable sets. Suppose there is no set of all countable sets. Even so, this interpretation of L2K is both one on which (PC) bears, and not one represented by the models quantified over in formulating ($\vDash_F$). So we can accept ($\vDash_F$) as a definition of logical consequence only subject to the presupposition that this extra interpretation of L2K will not provide a definite instance where all the members of Γ are true and φ is false.

[13] Thanks to an anonymous referee for pointing out these examples.

ensure this. And of course we want these models between them to represent every interpretation of L2K in which '∀,' '→,' and '¬' retain their meaning. Earlier in this paragraph I seemingly generalized in informal English over all cardinals at once. Some authors take their cue from this common practice and its intuitive intelligibility. They argue that we should not accept restrictions in formal semantics which we clearly ignore, and seemingly successfully ignore, in natural language. Hence they argue that for a formal language a domain d may contain absolutely everything, whilst seeking other ways to avoid paradox. If this project is accomplished, ($\models_F$) framed in L2K when the first-order quantifiers are taken to range over absolutely everything will quantify over absolutely every full model. But now the bump in the carpet appears elsewhere. We now have an interpretation of L2K, the metalanguage, which is not itself identified with any full model M = <d,I>. There is no set of all sets, so the domain of this interpretation would be 'too big' to be a set at any level in the hierarchy. As Shapiro remarks, commenting on Williamson's advocacy in his (2003) of a domain which includes absolutely everything:

> To speak roughly, for Williamson, 'absolutely everything' is a proper class *par excellence*. There can be *no* language which contains the word 'set' governed by the axioms of ZFC (with urelements) and which contains a set of absolutely everything. Such a set would have no powerset. We would encounter genuine contradiction if we did a Burali-Forti reflection on that language, producing a successor language. So for Williamson, it is decidedly not true that the proper classes (so to speak) in the present language become sets in some future language or some successor to that language [as per the hierarchy of inaccessibles]. They can *never* form a set (in the sense of ZFC).   (Shapiro 2003: 481, his emphases)

Nonetheless, in such an interpretation of L2K the logical constants, we suppose, retain their customary meaning and the sentences their customary logical forms. But ($\models_F$) with its invocation of models <d,I> requires d to be a set. So we would still have to conjecture that, given Γ $\models_F$φ, the metalanguage itself does not provide a definite counterexample to φ being a logical consequence of Γ.

Thus we can accept ($\models_F$) as a *definition* of logical consequence among the sentences of L2K, but only subject to the *presupposition* that no definite counterexample lurks among any further relevant interpretation of L2K not generalized over in the definition. The epistemic status of this presupposition, I think, is that of a *conjecture*. In the case of ($\models_H$) no such presupposition is needed. The point was made by Kreisel (1967). We can show, and therefore *know*, the following:

(2)   In cases which (PC) definitely determines, and in which '∀,' '→,' and '¬' have the meaning assigned by Henkin semantics, Γ $\models_H$ φ iff φ is a logical consequence of Γ according to (PC).

Again, break this down into:

(2A)   In cases which (PC) definitely determines, and in which '∀,' '→,' and '¬' have the meaning assigned by Henkin semantics, if Γ $\models_H$ φ then φ is a logical consequence of Γ according to (PC).

(2B)   In cases which (PC) definitely determines, and in which '∀,' '→,' and '¬' have the meaning assigned by Henkin semantics, if φ is a logical consequence of Γ according to (PC) then Γ $\models_H$ φ.

As before, (2B) is trivial. But this time we can also know (2A), although again we know that the models generalized over in ($\models_H$) do not include all the relevant interpretations of L2K. The proof theory D2 is complete with respect to Henkin semantics. Hence:

(3)   If $\Gamma \models_H \varphi$ then $\Gamma \vdash_{D2} \varphi$.

The axioms and rules of D2 are intuitively sound with respect to any interpretation of L2K in which '∀,' '→,' and '¬' have the meaning assigned by Henkin semantics. That is, a proof $\Gamma \vdash_{D2} \varphi$ shows us how the meaning of '∀,' '→,' and '¬' together with the logical forms of $\Gamma \cup \{\varphi\}$, as exploited by the rules of D2, make it impossible for all of Γ to be true and φ false. Hence, if $\Gamma \vdash_{D2} \varphi$, then (PC) is definitely satisfied: no interpretation of L2K in which '∀,' '→,' and '¬' have the meaning assigned by Henkin semantics will make all the members of Γ true and φ false, whether or not that interpretation is represented in the particular totality quantified over in ($\models_H$). Hence we have:

(4)   If $\Gamma \vdash_{D2} \varphi$ then φ is definitely a logical consequence in the prior philosophical sense (PC) of Γ, with '∀,' '→,' and '¬' having the meaning assigned by Henkin semantics.

Hence, by (3) and (4):

(5)   If $\Gamma \models_H \varphi$ then φ is definitely a logical consequence in the prior philosophical sense (PC) of Γ, with '∀,' '→,' and '¬' having the meaning assigned by Henkin semantics.

Thus we have (2A). Thanks to the intuitive soundness of D2 relative to the meaning assigned by Henkin semantics, and to the completeness of D2 relative to Henkin semantics, we know that (2) holds. Thus we can accept ($\models_H$) as a definition of logical consequence even if the range of models is limited to those provided by the accessible hierarchy. We know that no counterexample lurks among the further interpretations not generalized over in ($\models_H$). ($\models_H$) is in this way more *philosophically secure* than ($\models_F$).

There is another dimension along which we can compare ($\models_F$) and ($\models_H$) as putative definitions of logical consequence. L2K is given as a restricted language with just '∀,' '→,' and '¬' as its logical constants. These have been chosen because L2K is then, when given an appropriate domain and interpretation for its non-logical constants, a vehicle for doing arithmetic, or analysis, or set theory. In fact we practise these not in L2K but in English, or an extension of English. And the analogues of '∀,' '→,' and '¬' in English are not confined to mathematical topics, but extend their influence throughout the language. English also has further semantically significant devices without analogue in L2K: modal, deontic, alethic, temporal operators; adverbs and adjectives of various iterable kinds. Our default assumption is that the natural counterparts of '∀,' '→,' and '¬' mean the same in the fragment in which we do mathematics as they do across the extended whole. It should be a working assumption that L2K too can be extended syntactically beyond its customary mathematical vocabulary to L2K$^+$, where K$^+$ includes what are, intuitively, modal, deontic, alethic, temporal operators; adverbs and adjectives of various iterable kinds. L2K$^+$ has the syntactic resources to build by recursion predicates and sentences of arbitrary complexity, where any occurrences of '∀,' '→,' and '¬' may or may not be within the scope of a member of K⁺.[14] Our default assumption, parallel to that we make for the natural language, is that the meaning of '∀,' '→,' and '¬' is uniform as between the fragment L2K and the extended L2K⁺, and that logical forms of L2K are at least a proper subset of those of L2K⁺—being both the forms of the fragment L2K and forms exemplified in L2K⁺ at large.[15] There is, to say the least, as yet no consensus among philosophers as to how to extend semantics for L2K to semantics for L2K⁺. We do not know whether we can any longer treat the syntactic surface forms of the new sentences of L2K⁺ as their logical forms, as we do the sentences of L2K, or whether they will need to undergo various transformations before logical forms are revealed to which a compositional semantic theory can be applied. We don't know which, if any, of the new vocabulary are new logical constants. And we don't know what entities we will need to assign as the semantic values of the new vocabulary: possible worlds, propositions, thoughts, or entities from the realm of *Sinn*, perhaps. Still, it remains a long-term ambition of philosophy to construct such an L2K⁺ and to give its semantics, a project to which much literature has contributed.[16]

($\models_F$) and ($\models_H$) each purport to give the relation of logical consequence due to the meaning of just '∀,' '→,' and '¬' in L2K⁺, both the relation of logical consequence holding in the fragment L2K and the relation among the sentences of L2K⁺ at large due solely to those constants.[17] It may be that in the course of devising semantics for L2K⁺ we have to revisit and revise the semantics we initially gave to L2K. If so, it seems it is all bets off for ($\models_F$) or ($\models_H$) defining logical consequence even for the fragment L2K. However, subject to two relatively modest presuppositions ($\models_H$) may survive such a semantic sea change. Suppose that in the hoped for, perhaps revised semantics for L2K⁺:

(6)   The rules and axioms of D2 remain sound for the language L2K⁺.

(7)   Henkin models for the fragment L2K remain at least a proper subclass of the models provided by the hoped for semantics of L2K⁺ for the semantics of its fragment L2K.

---

[14] For example '□(∀xFx)' and 'a believes (∀xFx),' where '∀' seemingly occurs within the scope of '□' and 'a believes.'

[15] For example, '□(∀xFx) → ∀xFx' exemplifies the form 'P → Q' found in L2K. It also exemplifies the further form '□(P) →P,' not found in L2K.

[16] As Shapiro takes L2K to model the natural language of arithmetic, so I take L2K⁺ to model the richness of natural English. If we ever do achieve a compositional semantics for L2K⁺, I take no view on whether we will thereby have revealed the logical form of English manifest in its 'surface' syntax (Montague), or revealed the hidden logical form (Davidson), or have regimented the unclear semantics of English (Quine), or produced a semantically sound language which achieves what semantically defective English can only aspire to (Frege, Tarski). Nor do I have a view on whether the ontological commitments of English would be laid bare by the semantics for L2K⁺. Though I would hope that L2K⁺ and its semantics would shed light on all these questions.

[17] I mean that we require of ($\models_F$) not only for example ∀x(Fx → ¬Gx), Fa $\models_F$ ¬Ga which involves only sentences of the fragment L2K, but also for example ∀x(Fx → ¬◊Gx), Fa $\models_F$ ¬◊Ga, which shares the same form but employs vocabulary from L2K⁺ at large. And of course we require the same of ($\models_H$).

It follows:

(8) For all sentences $\{\varphi\}\cup\Gamma$ of L2K$^+$ belonging to the fragment L2K, $\Gamma \vDash_{\text{L2K}+}\varphi$ iff $\Gamma \vDash_{\text{H}}\varphi$.

(9) For all sentences $\{\varphi\}\cup\Gamma$ of L2K$^+$ belonging to the fragment L2K, if $\Gamma \vDash_{\text{H}}\varphi$, then $\Gamma^* \vDash_{\text{L2K}+}\varphi^*$ for all uniform substitutions on $\{\varphi\}\cup\Gamma$ employing the vocabulary of L2K$^+$ at large to generate $\{\varphi^*\}\cup\Gamma^*$.

Left to right of (8) holds because, if there is no model in the extended semantics of L2K$^+$ in which all the members of Γ are true and φ false, then given (7), there is no such Henkin model, and hence $\Gamma \vDash_{\text{H}}\varphi$. Right to left of (8) also holds: if $\Gamma \vDash_{\text{H}}\varphi$ then by completeness we have $\Gamma \vdash_{\text{D}2}\varphi$, and given (6) we have $\Gamma \vDash_{\text{L2K}+}\varphi$. (9) also follows. If $\Gamma \vDash_{\text{H}}\varphi$ then, by completeness we have $\Gamma \vdash_{\text{D}2}\varphi$, and hence by uniform substitution $\Gamma^* \vdash_{\text{D}2}\varphi^*$. Hence by (6) we have $\Gamma^* \vDash_{\text{L2K}+}\varphi^*$. Thus, subject to the conjectures (6) and (7) about the hoped-for semantics, we can accept ($\vDash_{\text{H}}$) as a definition of logical consequence among the sentences of L2K$^+$ due solely to the meanings of '$\forall$,' '$\rightarrow$,' and '$\neg$' and the logical forms they generate.[18]

Conjectures (6) and (7) will not sustain the claim of ($\vDash_{\text{F}}$) to define the relation of logical consequence for the fragment L2K of L2K$^+$. Suppose we had initially assigned full semantics to L2K. Suppose on extending to L2K$^+$ we find we have to revisit and assign Henkin semantics to the fragment L2K instead. (6), we may suppose, is satisfied, since D2 is sound for Henkin semantics. And the analogue of (7) for full semantics would be satisfied, since full models are a proper subclass of the Henkin models now assigned to the fragment L2K. Yet the relation of logical consequence among the sentences of the fragment L2K would have changed from ($\vDash_{\text{F}}$) to ($\vDash_{\text{H}}$).

Given an initial position in which we assign full semantics to L2K, and therefore ($\vDash_{\text{F}}$) as its relation of logical consequence, I can think of no *conjecture* concerning the hoped-for semantics of L2K$^+$ which will clearly sustain ($\vDash_{\text{F}}$) as the *definition* of logical consequence for the now fragment L2K. At the start of this section I argued that a definition could not itself be a conjecture, a definition requires knowledge but a conjecture leaves open the epistemic possibility of falsity. But a definition can be given presupposing a conjecture, if subject to that conjecture we would have the required knowledge: in this case know that ($\vDash_{\text{F}}$) coincides with ($\vDash_{\text{L2K}+}$) for the sentences of the fragment L2K, or more basically that ($\vDash_{\text{F}}$) coincides in extension with (PC) applied to the fragment L2K of the sentences of L2K$^+$ under the hoped-for semantics. Call what we need to know (K). We can of course conjecture that '$\forall$,' '$\rightarrow$,' and '$\neg$' continue to have full meaning, or equivalently that full semantics continue to apply to the fragment L2K in the hoped-for semantics for L2K$^+$. Call this conjecture (C). However, ($\vDash_{\text{F}}$) itself explicitly quantifies over full models:

---

[18] Multi-sorted first-order semantics, which we will visit briefly in §3, are an example of an extended semantics for L2K such that Henkin models are a proper subclass of multi-sorted first-order models for L2K, and yet ($\vDash_{\text{H}}$) remains the relation of logical consequence among the sentences of L2K. This need not make multi-sorted first-order semantics for the fragment L2K semantically otiose. It may be, for example, that L2K$^+$ employs what appears to be third-order quantification, to which multi-sorted first-order models for the fragment L2K pave the way.

($\vDash_F$) $\Gamma \vDash_F \varphi$ iff for every full model M and assignment s on M, if for every ψ in Γ, ψ is satisfied by M under s, then φ is satisfied by M under s.

Our conjecture (C) is thus perilously close to conjecturing directly that (K), that ($\vDash_F$) continues to coincide in extension with (PC) applied to the fragment L2K. In contrast, the conjectures (6) and (7) stand at a substantial epistemic distance bridged by the completeness result from our knowledge of (8) and (9) subject to presupposing (6) and (7). It is doubtful that we have enough of an inferential gap between the epistemic uncertainty of the conjecture (C) (that full semantics continue to reign in the fragment) and what we need to know (K) (that ($\vDash_F$) coincides with (PC) for the fragment) for (C) to ground knowledge of (K). Enough gap, that is, to sustain the epistemic difference between a conjecture and a definition. So it is not clear that ($\vDash_F$) can be taken as a *definition* of logical consequence among the sentences of L2K, although we may *conjecture* that it coincides in extension with logical consequence, if L2K is regarded, not as a restricted language bounded by just '∀,' '→,' and '¬' but is thought of as open to expansion towards L2K⁺.

3. On Using L2K to Practise Arithmetic or Real Analysis

In this section I shall argue that L2K with Henkin semantics *can* be used to express and do arithmetic, and that Henkin semantics *should be* the preferred option when taking L2K as a language for doing arithmetic. And similarly when taking L2K as a language for doing real analysis. It is uncontroversial that L2K with appropriate K and given suitable semantics can express arithmetic, and also real analysis. Shapiro claims that it can express all of mathematics, except perhaps category theory. But which semantics should we furnish it with to do arithmetic, or to do analysis? In this section we will concentrate upon arithmetic, but parallel remarks apply to analysis. We shall proceed by reverse engineering. Let us suppose a community who use L2K (K = {'o,' 's,' '+,' '.'}) as an autonomous language to practise arithmetic—by 'autonomous' I mean that the meaning of L2K is determined by their usage, not stipulated by our semantics. Do arithmetic with just these meagre syntactic resources?! Well no, but the rest can be explicitly defined (by us) in terms of them, and these few will be enough for our philosophical purposes. We will adopt the pose of radical interpreters, but of radical interpreters whose job is nearly done. Suppose we have interpreted their vernacular language, have parsed L2K, interpreted '→,' '¬,' and the members of K, and thereby discovered, at least tentatively, that they use L2K to do arithmetic. Two matters remain: Does '∀' have full or Henkin meaning, and if the latter, is their intended interpretation of L2K a full model or not? I assume for the moment, a matter to be discussed later in this section, that either would suffice for them to practise arithmetic. We, the radical interpreters, are to express our theory by stipulating semantics for L2K in our home language. As well as selecting semantics, we also need to assign an intended interpretation. Thus as radical interpreters we need to fix *two* parameters. Our stipulation of meaning for '∀' *fixes a range of models* thereby yielding a relation of logical consequence. Since full models are a proper subclass of

Henkin models, it is possible that the semantics for L2K picked out by radical interpretation will have *Henkin* logical consequence, but a *full* intended model. I shall argue in this section that this is indeed the case.

Shapiro seemingly underplays the need to determine two parameters, and this leads him to think that radical interpretation may be more straightforward than it is. He considers the writings of logicians from the first part of the last century, before model theory was standard, where they discuss in informal vernacular semantics for their proposed formal languages and attendant proof theories. Shapiro remarks:

> The author may simply discuss the intended range of the predicate variables directly. In such cases, the situation may be straightforward. We only need to see if the *stated* range of the predicate variables is the entire powerset of the domain, or includes *all* its properties. If so, the logic is higher-order and [full] semantics is an appropriate model of its logical relations.
>
> (Shapiro 1991: 174, Shapiro's emphasis)

What is straightforward here is only that the author *intended* a full model. It does not follow that full semantics are appropriate for the logic of the author's formal language. When the author says that the range of the predicate variables is the entire powerset of the domain (or all properties), and we take him at his word, there is no reading off whether he was specifying the *meaning* of the quantifier (thereby making full semantics appropriate), or merely setting the *extension* (thereby allowing that Henkin semantics also may be appropriate). To labour the point: the author has only told us which model of his formal language we should take as his intended model, and if that model belongs to both semantic families, the issue of which relation of logical consequence is appropriate remains open.

My example supposes that we have already interpreted their vernacular language, as well as much of L2K. I have set things up this way to avoid certain contentious issues. David Lewis (1974) saw philosophical discussions of radical interpretation as a device for probing how non-intentional broadly physical facts determine, insofar as they do determine, intentional facts about meaning and the mental. I doubt that the method of radical interpretation which Lewis preferred would get even as far as assigning meaning.[19] In my example we have mental facts to draw on from the start since we already understand our subjects at least as far as their vernacular language goes. Given that much, why not simply *ask* them in their vernacular to say whether they assign '∀' full or Henkin meaning, and if the latter whether their intended model is full or not? However, as I intend the example, I would not take their answer as authoritative. I am supposing they did not set up L2K by explicit semantics couched in their vernacular, but rather L2K has developed as an autonomous language alongside their vernacular, and I suppose they have never considered our questions before. My assumption is that it is their usage of L2K to practise arithmetic which determines its meaning and interpretation. If either will serve—full meaning or

---

[19] Lewis thought that the physical facts determine propositional attitudes, principally the subject's beliefs and desires, *prior to* determining facts about the meaning of the subject's sentences. I doubt it. Our beliefs are too sensitive to the testimony of our fellows. Thus on safari our subject Karl may hold-true the utterance of his guide given that morning: 'Hereit bionlay nicht cheetlay.' On later seeing a paw print and a pile of fresh ordure, does Karl believe that a lion is around, or that a cheetah is around? No saying, until we learn that he understood his guide to mean that there are lions about, but not cheetahs.

Henkin meaning with a full intended model—then their opinion is not authoritative if they had not considered the matter, though it may be authoritative for later practice with L2K. Prior to the question being raised for them, it is settled, if it is settled, by a general principle of meaning-parsimony: given the purpose for which a language is used, meaning is no more specific than is needed to achieve that purpose. I take this as a metaphysical principle concerning how meaning is related to use: not 'use' construed behaviouristically or just physically, but use in the service of purpose, in the case of human language use for recognizably human purposes.

Hence I have a somewhat different account of the source of the intuitions we draw on when discussing examples of radical interpretation from Lewis or Davidson. Despite what some have claimed,[20] we are not experienced radical interpreters, so the intuitions we draw on in philosophical discussions of radical interpretation must have some other source. In Lewis (1974) and Davidson (1974) this is our general commonsense interpretative theory of persons. Roughly, an interpretation of a human language user is correct only if it makes its subject out to be another rational human believer in truth and seeker after good. I agree this is one source, but I would add also a limiting principle from the metaphysics of meaning: there cannot be more to meaning than is needed to do this job. This metaphysical limiting principle is far from being of a piece with our commonsense interpretative theory of persons which Davidson and Lewis appeal to. Commonsense cleaves to some form of the myth of the museum: meaning is 'in the head' in a proprietry sense that informs usage but may transcend even all possible usage. The limiting principle, and hence any intuitions about radical translation which it may drive, arise therefore from relatively esoteric metaphysics, although it is a principle that would be agreeable to Davidson and Lewis.

This metaphysical limiting principle—that there is no more to meaning than is needed to do its job—spawns three methodological principles bearing on radical translation. The first is that we should not read a more specific meaning into usage than is needed to explain that usage in its context:

(Min)   Do not read more specific meaning into use than is needed to explain that usage, in context, as serving the purposes of the users, as those purposes would be apparent to fellow human initiates of the same culture.[21]

---

[20] For example Davidson (1973) takes radical interpretation to underlie even understanding fellow speakers of our home language. If so, the intuitions that we bring to bear on abstruse philosophical thought experiments would just be the intuitions employed in our ordinary practice.

[21] Does (Min) open the door to Quinean indeterminacy? Assigning 'gavagai' a rabbitish meaning somehow indeterminate between rabbit, undetached rabbit part, etc., is less specific than assigning it the meaning rabbit, and, Quine claimed, is equally justified by native usage. I think not. As Wright has effectively argued in his (1997) citing Evans (1975), the requirement of a compositional semantics will narrow the field of plausible interpretations. So also will Davidson's requirement that interpretations make sense of human communication in terms of human interests and purposes, and reflect what is salient to humans. Perhaps an alien might properly wonder whether 'gavagai' means rabbit or undetached rabbit part. But it is enough that we fellow humans find the latter interpretation contrived, unless a special purpose is in play—how to conceal the whole of a live rabbit within the conjurer's hat, perhaps. Somewhat similarly, it is a familiar point that red things are, at the level of physical science, a disparate lot. So an alien lacking human visual sensibility might be baffled by the things we group under 'red' in contrast to those we group under 'yellow.' But it comes naturally to us, because colour saliences reflect features of our visual apparatus.

(Min) talks of human purposes, but to determine human purposes we need, inter alia, to interpret their utterances. The interpretative circle is broken in general terms, *pace* Lewis but as Davidson has urged, by achieving a reflective equilibrium in which those others we can understand appear, generally speaking, as rational individuals, more or less like ourselves. When we apply (Min) to the interpretation of our supposed community practising arithmetic with L2K, it constrains the meaning we stipulate in our interpretative semantics for '∀.' (Min) becomes:

> (Min*)   Do not assign a more specific meaning to '∀' than is needed for L2K to serve the practice of arithmetic.

There is a clear sense in which full meaning is more specific than Henkin meaning. Full models are a proper subclass of Henkin models, so full meaning has more expressive power than Henkin meaning. Thus assigning full meaning to '∀' restricts the interpretation of L2K more than does assigning Henkin meaning. So (Min*) favours assigning Henkin semantics to L2K—but only, of course, if so doing enables the practice of arithmetic in L2K.

My second methodological principle for radical interpretation concerns the ranges of quantifiers. It requires that we do not place contextual restrictions which are neither indicated nor warranted by the purposes of the communication we are interpreting.

> (Max)   Do not restrict the context, when there is nothing in the circumstances of usage to warrant this restriction in the light of the purposes of the users, as those purposes would be apparent to fellow human initiates of the same culture.

Although it is a requirement to maximize the domains of any quantifiers, (Max) follows, I think, from the metaphysical principle of minimalizing meaning. The less meaning we endow our utterances with the wider the range of their application. More meaning renders them more specific and narrows their range. Of course, there are often contextual restrictions which are understood but not voiced. To borrow an example from Timothy Williamson, if you arrive at the airport saying 'I've got everything in the one bag,' I will not take you to mean the kitchen sink too, but only to mean items needed for the holiday. But in addition I will not take you to be claiming, falsely, that you have in the bag the clothes you are actually wearing; nor your passport, nor wallet, nor flight ticket, which I cannot currently see but would expect you to keep about your person. However, if there is a first-order domain which includes absolutely everything in the manner of Williamson (2003), (Max) enjoins us to take it as the default domain. Suppose the default is in force when someone says 'Donkeys can't talk.' I raise the possibility of talking donkeys on Alpha Centauri. The speaker may respond that she was not intending them. In my view, whether or not she intended them she did mean them. Her response illustrates the prevalence of the myth of the museum: if she didn't intend the donkeys on Alpha Centauri she didn't mean them. I think she did mean them whether she realizes it or not, since the default was in force. Intuitions to the contrary reflect, I think, the fact that the default seldom is in force. (Max) sanctions the default only when taking the domain to include absolutely everything does not impede the purposes of the discourse. And normally it would: taking the quantifiers of common or garden empirical statements to range over absolutely everything would play havoc with their epistemology. 'Concentrated caustic soda is inimical to life' seems a well-attested generalization to me, relevant to my project of sterilizing the drain. But it wouldn't be well attested if life on Alpha Centauri which thrives on caustic soda is relevant. Issues around relevance and meaning are complex. The seas of language run high here, but fortunately we need paddle only in the shallows. The following application of (Max) to interpreting L2K seems independently plausible to me:

(Max*)   Do not stipulate a restriction to the domain of '∀' in the intended model for L2K, if there is nothing in the practice of arithmetic to warrant such a restriction.

Together (Min) and (Max) make a prima facie case for assigning Henkin semantics to L2K but with a full model as the intended interpretation.[22,23]

Shapiro would agree that there is nothing in the practice of arithmetic, or in the practice of real analysis, to warrant restricting the domain of the higher-order variables. But he gives reasons why he thinks, ($\models_H$) is too weak a relation of logical consequence for L2K to express the facts of arithmetic, or the facts of real analysis, and too weak to support the epistemology of arithmetic or of analysis. If these reasons hold, (Min*) will require us to impose the stronger relation ($\models_F$) on the sentences of L2K by stipulating a full meaning for '∀.' His argument centres upon the upward and downward Löwenheim–Skolem theorems for Henkin models, and upon the contrasting categoricity results for full models. Let us start with his claim that Henkin semantics are too weak for L2K to express the facts of arithmetic and analysis. In summary form and transposed to our position as radical interpreters, his reasons are:[24]

(a)   An upward Löwenheim–Skolem theorem applies to Henkin models of L2K. The second-order Peano axioms in L2K have uncountable models in Henkin semantics. Hence if L2K, so understood, were taken as a vehicle for a practice, that practice would be radically indeterminate as between countable and uncountable interpretations. Intuitively our own arithmetic practice is not radically indeterminate between countable and uncountable interpretations:

---

[22] My third methodological principle for radical interpretation will appear later, when we consider first-order, multi-sorted semantics for L2K.

[23] (Min) and (Max) are reminiscent of Paul Grice's conversational maxim of Quantity: make your contribution as specific as the purpose of the conversation requires, and no more specific than the purpose of the conversation requires. Certainly there are connections, see for example the discussion of Jennifer Saul (2002) and the development of formal pragmatics described in Craige Roberts (2003). But there are also differences: Grice's maxims concern the complex interplay between literal meaning, context of utterance and communicative purposes, non-philosophical common or garden communicative purposes, where we take the literal meaning as a given. (Min) and (Max) are concerned with finding literal meaning given an esoteric philosophical limiting principle.

[24] Shapiro's project was to make a case for second-order logic with full semantics when reconstructing mathematics in general, and his main target was first-order semantics. Our interest is comparatively narrow. Logicists take arithmetic considered in isolation as their entry point into mathematics, and are moving on towards analysis. Our interest is therefore confined to arithmetic and analysis, and to the contrast between Henkin and full semantics at second-order. Much of Shapiro's interesting and detailed discussion passes us by.

its objects are paradigmatically countable. And there is no reason to suppose that they are in a worse position than we are when doing arithmetic. So stipulating Henkin semantics for their use of L2K is not a felicitous interpretation. In contrast, when L2K is interpreted by full semantics, the second-order Peano axioms are categorical, and have only countable models. Hence stipulating full semantics for their use of L2K to practise arithmetic is a felicitous interpretation.

(b) A downward Löwenheim–Skolem theorem applies to Henkin models of L2K. Hence, any second-order axioms of real analysis given in L2K have a countable model in Henkin semantics. Hence if L2K, understood according to Henkin semantics, with suitable K and axioms of real analysis, is taken as a practice, then that practice would be radically indeterminate between uncountable and countable interpretations. But intuitively our own practice of analysis is not radically indeterminate between uncountable and countable interpretations: its objects are paradigmatically uncountable. And there is no reason to suppose that they are in a worse position than we are when practising analysis. So stipulating Henkin semantics for their use of L2K is not a felicitous interpretation. In contrast, when L2K is interpreted by full semantics, the customary axioms of analysis are categorical, and have only uncountable models. Hence stipulating full semantics for their use of L2K to practise analysis is a felicitous interpretation.

(c) As a consequence of the upward and downward Löwenheim–Skolem theorems, schemas of L2K which attempt to specify cardinalities of domains of various apparent sizes, from Dedekind infinity upwards,[25] fail to do so when L2K is understood by Henkin semantics, since then all have countable Henkin models, and all have Henkin models of all uncountable cardinalites. So again, if L2K so interpreted is taken as a practice, then that practice would fail to distinguish between countable and various uncountable cardinalities. But intuitively, the cardinality schemas of analysis do uniquely pick out specific infinite cardinalities, and we have no reason to think that they are in a worse position than us. So again L2K interpreted by Henkin semantics is not a felicitous interpretation. And again, when L2K is given full semantics, these schemas succeed in specifying each their respective infinite cardinality, from Dedekind infinity upwards, for all full models. Hence, stipulating full semantics for their use of L2K to practise arithmetic and real analysis is a felicitous interpretation.

The response to these objections is clear. Objections (a) and (b) fail. The Natives' use of L2K to practise arithmetic, or to practise analysis, is an interpreted

---

[25] At second-order, schemas identify the various cardinalities, and there are no explicit definitions of those cardinalities. For example, the schema for Dedekind infinity:

$$INF(X): \quad \exists f(\forall x \forall y(fx = fy \rightarrow x = y) \,\&\, \forall x(Xx \rightarrow Xfx) \,\&\, \exists x(Xx \,\&\, \forall y(Xy \rightarrow \neg fy = x)))$$

If we had third-order predicates, we could replace the schema by an explicit definition:

$$\forall X(INF(X) \leftrightarrow \exists f(\forall x \forall y(fx = fy \rightarrow x = y) \,\&\, \forall x(Xx \rightarrow Xfx) \,\&\, \exists x(Xx \,\&\, \exists y(Xy \rightarrow \neg fy = x))))$$

language. Thus we need to assign *intended* models to their use of L2K as well as a meaning for '∀.' These intended models are, by (Max*), full models. The second-order Peano axioms are categorical for full models, as are the corresponding axioms for analysis. So these models impute no indeterminacy, as between countable and uncountable interpretations, to L2K when used to practise arithmetic or to practise analysis. Reference is determinate[26] in each case, to countable natural numbers and to uncountable reals respectively. That there are unintended uncountable models of arithmetic, and unintended countable models of analysis is irrelevant, since, precisely, these interpretations are not those intended. Consequently, (c) fails: their schemas attempting to specify various cardinalities in L2K are *interpreted*, and those schemas specify those cardinalities in whatever model is their intended model, since (Min*) and (Max*) yield as their intended model a Henkin model which is also a full model. So as required, Henkin semantics permits these cardinalities to be specified in practice. Thus, *pace* Shapiro's objections (a) to (c), L2K with Henkin semantics is adequate to express the *facts* of arithmetic and of real analysis. It expresses *the same* arithmetic facts, or *the same* facts of real analysis, as it would were '∀' given full meaning. As to the relations of logical consequence defined on Henkin and full semantics respectively, objections (a) to (c) would tell against ($\models_H$) and in favour of ($\models_F$) only if there were no way of identifying the intended model of L2K as a full model. What the objections really show is *not* that practising arithmetic or analysis in L2K requires ($\models_F$), but that it requires the intended model in each case to be a full model. Of course, if ($\models_F$), L2K only has full models. But the neglected option is to combine ($\models_H$) with a full intended model. In short, we don't need to restrict the meaning of ' $\models$ ' to ward off interpretations of L2K which are counterintuitive and esoteric.

In §2 we considered the epistemic nature of definitions. We turn now to their semantics. Our practitioners will want to formulate non-trivial definitions by minimal closure. For example, they may introduce the standard definition of natural numbers by minimal closure.

$$\text{(DN)} \quad \forall x(Nx \leftrightarrow \forall X((X0 \,\&\, \forall y(Xy \rightarrow Xsy)) \rightarrow Xx))$$

Of course, since we interpret '∀' as having Henkin meaning, the upward Löwenheim–Skolem theorem applies and even in the presence of the axioms of arithmetic (DN) will have unwanted models: models in which the smallest member of the domain of 'X' containing 0 and all its successors is of uncountable cardinality because it does not stop with the finite cardinals but proceeds to $\aleph_0$ and beyond. Nonetheless, since the intended model is a full model, the extension of 'Nx' in that model will be just the natural numbers.

---

[26] Determinate only up to isomorphism? Structuralism raises its head at this point, as in Benacerraf (1965). But I ignore structuralism for two reasons: first, because my target is logicism and logicists take the semantics of arithmetic practice at its objectual face value; and second, because Henkin semantics with a full intended model is no more or less vulnerable to structuralist attack than is full semantics.

John Corcoran (1971) gives as the semantic requirement for a successful definition:
D is a definition of an n-adic predicate P in terms of prior vocabulary K relative to prior theory T

iff

for every model <d,I> of T there is a *unique* $r \subseteq d^n$ such that if <d,I*> is a model of T∪D, where I* interprets K∪P, and where I(α)=I*(α) for all α∈K, then I*(P)=r.

That is, a definition should *uniquely* extend the language. Corcoran remarks that the rationale for his definition of a definition is that a definition of a new predicate should uniquely determine its meaning in terms of the meanings of the prior vocabulary. He also appeals, as does Shapiro, to the result that such definitions are semantically non-creative and eliminable. D is non-creative with respect to K and T iff for each sentence φ of employing only K such that T∪D $\models$ φ, T $\models$ φ. And D is eliminative iff for each sentence φ containing P there is a sentence φ* employing only K such that T∪D $\models$ φ↔φ*. It is simple to adapt Corcoran's definition to Henkin models:

D is a definition of an n-adic predicate P in terms of prior vocabulary K relative to prior theory T

iff

for every model <d,D,F,I> of T there is a unique $r \in D^n$ such that if <d,D,F,I*> is a model of T∪D, where I* interprets K∪P, and where I(α)=I*(α) for all α∈K, then I*(P)=r.[27]

It is easy to see that definitions of predicates by non-trivial minimal closure do uniquely extend L2K when L2K is understood in terms of Henkin semantics. Such a definition takes the form: $\forall x_1 \ldots \forall x_n (P x_1 \ldots x_n \leftrightarrow \Phi x_1 \ldots x_n)$, where $\Phi x_1 \ldots x_n$ contains only prior vocabulary. The comprehension schema in our version of T ensures that there is at least one member of D^n for I* to assign to P. Suppose, for reductio, that <d,D,F,I> can be extended in two ways <d,D,F,I*> and <d,D,F,I**>, such that I(α)=I* (α)=I**(α) for all α∈K, but I*(P)≠I**(P), and where both <d,D,F,I*> and <d,D,F,I**> are models of T∪D. Then there would be a third extension <d,D,F,I⁺> possible of <d,D,F,I>, an extension introducing *two* new predicates P* and P** and in effect combining the two earlier models. That is, two new predicates P* and P** would be introduced by the axioms: D*: $\forall x_1 \ldots \forall x_n (P^* x_1 \ldots x_n \leftrightarrow \Phi x_1 \ldots x_n)$, and D**: $\forall x_1 \ldots \forall x_n (P^{**} x_1 \ldots x_n \leftrightarrow \Phi x_1 \ldots x_n)$. And <d,D,F,I⁺> would be such that I(α) =I⁺(α) for all α∈K, but I⁺(P*)≠I⁺(P**), and <d,D,F,I⁺> would be a model of T∪D*∪D**. But it is easy to see that there is no such model as<d,D,F,I⁺>. Given premises of the following forms:

$$D^* \qquad \forall x_1 \ldots \forall x_n (P^* x_1 \ldots x_n \leftrightarrow \Phi x_1 \ldots x_n)$$

$$D^{**} \qquad \forall x_1 \ldots \forall x_n (P^{**} x_1 \ldots x_n \leftrightarrow \Phi x_1 \ldots x_n)$$

We have, as a *proof- theoretic* consequence:

$$\forall x_1 \ldots \forall x_n (P^* x_1 \ldots x_n \leftrightarrow P^{**} x_1 \ldots x_n).$$

---

[27] 'd^n' in Corcoran's original definition referred to the nth Cartesian product of the domain d. 'D^n,' in contrast, refers to the nth member of the sequence, D^1,D^2,...which make up D in a Henkin model.

So, since D2 is sound for Henkin models, there is no Henkin model extending <d,D,F,I> and making D* and D** both true where $I^+(P^*) \neq I^+(P^{**})$. Hence definitions by non-trivial closure in Henkin models satisfy a straightforward extension of Corcoran's definition of a definition to Henkin models. They uniquely extend the language, and so meet Corcoran's rationale for his definition of a definition. It is interesting to note that although a definition by minimal closure, for example (DN) above, will assign to 'Nx' unintended extensions in unintended models of L2K, it qualifies as a definition in all models, whether intended or not, since Corcoran's definition of a definition generalizes over all models and (DN) satisfies Corcoran's definition. What (DN) is a definition *of* varies from Henkin model to Henkin model. In Henkin models which are full models it defines the natural numbers, and that is all our practitioners of arithmetic with L2K require.

Turning from semantics to epistemology, Shapiro emphasizes the centrality of categoricity results to the *epistemology* of mathematical practice. As well as proving results, arithmetic results for example, by calculation, or by deduction from Peano's axioms, mathematicians also prove arithmetic results by embedding Peano's axioms in a richer theory, set theory perhaps, deducing the result in question in the richer theory, and then transferring the result back to elementary arithmetic, where it can't[28] be deduced from Peano's axioms alone.[29] Call the embedded theory Γ, the embedding theory Σ, and the result in question Φ. Shapiro claims that the procedure shows that Φ is a *logical consequence* of Γ, where logical consequence is distinct from provability. He describes the process in general terms thus:

In symbols (and roughly), the mathematician established (Γ+Σ) ⊢Φ, and this shows that Γ ⊨Φ. However, it does not show that Γ ⊢Φ and perhaps Φ is not justifiable on the basis of Γ alone. Perhaps not-(Γ ⊢Φ).

(Shapiro 1999: 49, with an insignificant change of format in the last sentence)

Shapiro points out, (Γ+Σ) ⊢Φ shows that Γ ⊨Φ *only if* the theory Γ is categorical. Φ is a sentence of the language of Γ; that is any model of Γ assigns a truth value to Φ. If Γ is categorical, then every model of Γ assigns the same truth value to Φ, and given that Φ is true in the rich model of Γ provided by Γ+Σ, Φ is true in all models of Γ, and therefore Γ ⊨Φ. Second-order Peano's axioms and the axioms of analysis are each categorical for full models. So embedding either of these axioms in say set theory may enable a mathematician to prove Γ ⊨_F Φ where we do not have Γ ⊢Φ. A famous example of embedding is the proof of the Gödel sentence for Peano's axioms by arithmetic embedding. Shapiro takes the mathematician to (rightly) claim that Φ is a

---

[28] Either can't in practice, if the proof would be just too long for humans, or can't in principle, for example a Gödel sentence.

[29] Embedding is not strictly relevant to my project. It presupposes some richer theory in which to embed, and I am supposing that arithmetic is the *only* theory the practitioners express in L2K. Embedding is relevant to Shapiro's project because he is concerned to model in formal language mature mathematics, where the various branches of mathematics—arithmetic, analysis, set theory, etc.—may interact one with another as in embedding, and where mathematicians move from theory to metatheory and back. My interest is Wright's logicism. Wright supposes a community innocent of mathematics who then acquire the concepts to begin to practise first just arithmetic, and then arithmetic and analysis, and then.... So my community use L2K to practise just arithmetic: they have not yet got further mathematical theories which will one day provide the tools to enrich their knowledge of arithmetic by such techniques as embedding.

logical consequence of Γ, in this case that the Gödel sentence is a logical consequence of Peano's axioms. Shapiro makes it a requirement of a successful axiomatization of arithmetic for, as he puts it 'the modal/semantic purpose' that that the axiomatization be categorical. Then, the axiomatization being categorical, it will determine all arithmetic truths. And those we may be unable to discover by direct deduction from the axioms, we may hope to discover by ingenious embeddings.

An axiomatization A of, say arithmetic, is successful for the modal/semantic purpose if the natural numbers are a model of A and if every model of A is isomorphic to the natural numbers. Now, a sentence Φ is a consequence of A if and only if Φ holds in all models of A. It follows that if A is successful for the modal/semantic purpose then Φ is a consequence of A if and only if Φ is true of the natural numbers. In the light of Gödel's incompleteness theorem, then, completeness is emphatically not desired for the model/semantic consequence relation.

(Shapiro 1999: 56)

In my view mathematicians should not claim so much for embedding. Taking arithmetic as our example, the unembedded language of arithmetic is an *interpreted* language. Suppose (Min*) and (Max*) assign a full model as its interpretation. Then since Γ is categorical for full models, Φ has the same truth value in *all* full models in which Γ is true. Σ need not be itself an interpreted language, provided there is *a* model of Σ ∪ Γ whose submodel for the fragment dealing with Γ ∪ Φ is a *full* model—not necessarily the same full model as the intended model of the unembedded language. Since Σ ∪ Γ ⊢Φ, Φ is true in the model of Σ ∪ Γ. It follows that Φ is true in the intended model of the unembedded language—since Φ has the same truth value in *all* full models. But it is not required that Φ be a logical consequence of Γ. There may be non-standard models in which Γ and not-Φ. That is to say, we have established that Φ is a *material consequence* of Γ in the unembedded interpretation, but not, as Shapiro supposed, Γ ⊨Φ. Good enough. The important thing the mathematician wants to establish is that Φ is a *truth* of arithmetic (or analysis); that is, that Φ is true given Γ, not, I suggest, that Φ is a logical consequence of Γ. In general, to know that Φ is true given Γ we require that:

(i) All models which satisfy Γ settle the truth value of Φ one way or another, but perhaps without assigning the same value in all models.
(ii) Γ is categorical for some subclass of these models, perhaps a proper subclass of all the models.
(iii) This subclass of models includes the intended model.
(iv) At least one model which satisfies Σ, perhaps the intended model, is included in the subclass, or its sub-model dealing with the vocabulary of Γ∪Φ is in the subclass.
(v) Φ is a logical consequence of Γ∪Σ—which we know since we have (Γ∪Σ) ⊢Φ.

This epistemic process establishes that Φ is true given Γ, a material consequence of Γ. It does not require either Γ ⊨_F Φ or Γ ⊨_H Φ. Note that we don't need to know that Σ is itself true, only that some model which satisfies Γ∪Σ is either itself in the subset for which Γ is categorical, or its sub-model is.

Let us return to the metaphysical principle of parsimony of meaning—do not assign a more specific meaning to L2K than is needed for the job, viz. practising arithmetic, or practising analysis. The principle led via (Min*) and (Max*) to our favouring Henkin semantics over full semantics for L2K, but with a full model as the practitioners' intended model. This was because the intended full model that does the job is a member of both families, and, full models being a proper subset of Henkin models, Henkin semantics assign less specific meaning to L2K than do full semantics. Thus Henkin semantics were favoured by the principle. This position looks unstable. Henkin models are themselves in effect a proper subset of first-order, multi-sorted models, so stipulating first-order, multi-sorted semantics for L2K would assign a still less specific meaning. Moreover, the axioms of arithmetic so understood are categorical for a subset of first-order, multi-sorted models, viz., the subset of first-order, multi-sorted models which are the analogues of the full models, and the intended model is a member of this subset. Similarly for the axioms of analysis. Thus the case saying that L2K with Henkin semantics and a full intended model can do the job carries over to a case for saying that L2K with first-order, multi-sorted semantics and with an intended model corresponding to a full model can also do the job. And now, it seems, the metaphysical principle of meaning parsimony will favour the latter, since the meaning assigned thereby is less specific.

I think this line of thought fails. But to show why I need first to sketch first-order, multi-sorted semantics (hereafter 1MS semantics). 1MS semantics read an atomic sentence '$X^n x_1, \ldots, x_n$' of L2K not as an n-adic predicate symbol 'X^n' followed by n singular terms 'x_1,' $\ldots$,'x_n,' but as '$P^{n+1} X^n, x_1, \ldots, x_n$' with an elided n+1-adic *predicate* symbol 'P^{n+1}' followed by n+1 *singular* terms 'X^n,' 'x_1,'$\ldots$, 'x_n.' Similarly, 1MS semantics reads a function term '$f^n x_1, \ldots, x_n$' not as an n-adic function symbol 'f^n' followed by n singular terms 'x_1,'$\ldots$, 'x_n,' but as '$a^{n+1} f^n, x_1, \ldots, x_n$' with an elided n+1-adic *function* symbol 'a^{n+1}' followed by n+1 *singular* terms 'f^n,' 'x_1,'$\ldots$,'x_n.' The values of each new first-order variable 'X^n' are sets of n-tuples of entities from D^n. The idea is that in the *intended* model of L2K, I('P^{n+1}') is set membership between $<x_1, \ldots, x_n>$ and X^n, so that the fully expressed '$P^{n+1} X^n, x_1, \ldots, x_n$' is satisfied iff the n-tuple assigned to 'x_1' through 'x_n' is a member of the set assigned to 'X^n.' And the values of a new first-order variable 'f^n' are functions taken in extension from D^n to d. The idea is that in the *intended* model of L2K, I('a^{n+1}') is the n+1 adic function such that $a^{n+1} f^n, x_1, \ldots, x_n$ is the value of the function f^n for the arguments x_1 through x_n. However, 'P^{n+1}' and 'a^{n+1}' are implicit non-logical terms of L2K, and so may be assigned other (n+1)-adic relations and functions in other models. Since this semantics is first-order, the predicate and functional symbols 'P^{n+1}' and 'a^{n+1}' are not open to quantification; only singular terms, now including 'X^n' and 'f^n,' are open to quantification. The domains for the new singular terms 'X^n' and 'f^n' are D_n and F_n respectively, as for Henkin semantics. Thus in some models D_n is only a proper subset of the powerset of d^n, and F_n similarly. Clearly, each Henkin model corresponds to a 1MS model in which 'P^{n+1}' and 'a^{n+1}' are given their intended interpretations. Given a Henkin model $<d,D,F,I>$ of L2K, there is an equivalent 1MS model $<d,D,F,I^*>$ in which $I(\varphi) = I^*(\varphi)$ for all φ in K, and $<X^n, x_1, \ldots, x_n> \varepsilon I^*('P^{n+1}')$ iff $<x_1, \ldots, x_n> \varepsilon X^n$, and $I^*('a^{n+1}')$ is the function from $(f^n, x_1, \ldots, x_n)$ to $f^n(x_1, \ldots, x_n)$. Clearly also, Henkin models correspond to only a proper subclass of 1MS models, since 'P^{n+1}' and 'a^{n+1}' may be given other interpretations.

Let L2K$^+$ be the language generated from L2K by making the elided predicate and functional symbols explicit We can now see that 1MS semantics does *not* after all assign a less specific meaning to the sentences of L2K$^+$ than did Henkin semantics to the sentences of L2K, even though Henkin models for L2K are equivalent to only a proper subclass of 1MS models for L2K$^+$. Each Henkin model for L2K is equivalent to a *single* 1MS model for L2K$^+$ in which the non-logical constants 'P^2,''P^3,'...,'P^n,'...and 'a^2,''a^3,'...,'a^n,'...have been given particular meanings. The extra 1MS models are generated by giving these non-logical constants other particular meanings. But once these new constants are given particular meaning, the meaning assigned to the original constants in L2K$^+$ under 1MS semantics is no more or less specific than the meaning assigned to them in L2K under Henkin semantics. In contrast, given an interpretation function I fixing the meaning of the non-logical constants of L2K, the meaning assigned by Henkin semantics is less specific than that assigned by full semantics, since each full model <d,I> corresponds to just one of a plethora of Henkin models <d,D,F,I>, the one in which D_n = the powerset of d^n, for each n≥1, and F_n = all functions from d^n to d, for each n≥1.

Still, 1MS semantics for L2K$^+$ are no more specific in the meaning assigned to the constants of L2K than Henkin semantics for L2K. So do we have a tie, both interpretations are equally justified? I think not. The metaphysical principle of usage generating minimal meaning to do the job spawns a principle of minimal syntax required for meaning to do the job. Suppose K of L2K is augmented with the usual decimal numerals, and consider the sentence '2+2=4' which L2K regards as a simple syntactic transformation of '=^{2}(+^{2}(2,2),4)' Here the superscripts give the acidity of the predicate and function, and the syntactic transformation is to remove brackets and commas by replacing the first comma by its governing function symbol and the second by its governing predicate symbol. 1MS sees a more complex logical form in '2+2=4': viz., P^3(=^{2},a^3(+^{2},2,2),4). This having been syntactically transformed to yield perhaps '2+^{2}2$_{a3}$=$^2_{P3}$4' and the symbols 'a^3' and 'P^3' elided. Clearly Henkin semantics sees '2+2=4' as closer to its logical form than does 1MS.

As radical interpreters we assign logical forms to the sentences of a language so as to yield a compositional account of the meaning of those sentences, and such that those forms and that meaning capture a plausible relation of logical consequence among the sentences, sufficient for the purposes of the users, in this case the practice of arithmetic. The metaphysical principle of meaning parsimony requires us not to assign more form than is necessary.

(MinS)   Do not read more syntax into use than is needed to explain that usage, in context, as serving the purposes of the users, as those purposes would be apparent to fellow human initiates of the same culture.

Applied to the case in hand this becomes:

(MinS*)   Do not read their usage as L2K$^+$ rather than L2K if that is not needed for their practice of arithmetic.

Thus the metaphysical principle of meaning parsimony favours Henkin semantics over 1MS semantics for arithmetic. And similarly for analysis.[30]

If the Henkin relation of logical consequence among the sentences of L2K is all that is needed to practise arithmetic or analysis, then it will be all that is needed in my home language to express semantics for L2K. But now there is a problem of a familiar shape. I applied (Min*), (Max*), and (MinS*) formulated in my home language to assign Henkin semantics with a full model as the intended model of L2K. However, (Min*), etc., will assign a full model only if the intended interpretation of my home language is itself a full model, otherwise they will assign only the 'nearest' to a full model that the intended interpretation of my home language provides, and that model will fall short of being genuinely a full model if the intended model of my home language is not itself a full model. And now my appeal to the categoricity results for full models fails to establish what I took it to establish in response to Shapiro. I took categoricity to establish that the intended model of the axioms of arithmetic is genuinely the natural numbers, or at least isomorphic to the natural numbers, and not some larger set; and that the intended model of the axioms of analysis is genuinely the reals, or at least isomorphic to the reals, and not some smaller set. But the proofs of the categoricity results will not show this if the intended model of the home language in which those proofs are conducted is not itself a full model. We would still have it that all models $m_1, m_2, \ldots$ of the axioms of arithmetic in L2K are isomorphic one with another. But if my home language in which the categoricity result is derived were given instead a full interpretation, then the models of the axioms of arithmetic in L2K might be revealed to be $m^*_1, m^*_2, \ldots$, where m^*_1, $m^*_2, \ldots$ are isomorphic one with another, but *not* isomorphic with $m_1, m_2, \ldots$. Indeed, the sets $d_1, d_2, \ldots$ which were assigned to the natural numbers in m_1, $m_2, \ldots$ may be revealed by the full interpretation to be of *uncountable* cardinality. The sets $d_1, d_2, \ldots$ were merely the smallest sets containing the natural numbers provided by the intended interpretation, but, because that interpretation was not a full model, they contained much else besides and were in fact of uncountable cardinality. Similarly for the axioms of analysis in L2K: given that the intended interpretation of my home language is not a full model, the models provided for the axioms, though isomorphic, may actually be of countable cardinality. The upshot is that (Min*), (Max*), and (MinS*) select full models as the interpretations of the axioms of arithmetic in L2K, and of the axioms of analysis, only if the intended interpretation of my home language is itself a full model; that is, only if

---

[30] The relation of Henkin logical consequence among the sentences of L2K is exactly equivalent to the relation of logical consequence among the corresponding sentences of L2K⁺. Let Δ be a set of sentences of L2K, let Δ^* be the corresponding set of sentences of L2K⁺—that is, of the members of Δ with P^{n+1} and a^{m+1} added where required. Shapiro proves that for any set of sentences Δ of L2K, there is a Henkin model satisfying Δ iff there is a 1MS model satisfying Δ^*. Consequently, $\Gamma^* \models_{1MS} \varphi^*$ holds among the sentences of L2K⁺ iff $\Gamma \models_H \varphi$ holds among the sentences of L2K. Prima facie, this is surprising. Henkin models are a proper subclass of 1MS models, so we might have expected the relation of logical consequence to weaken, as it did when we extended the class of full models to that of Henkin models. It doesn't because for each adicity n, we have introduced just *one* predicate constant P^{n+1}, and just *one* function constant f^{n+1}. The picture would change if L2K⁺ were to include more than one of either, and we did not require uniform insertion of these new constants to get $\Gamma^* \models_{1MS} \varphi^*$ from $\Gamma \models_H \varphi$.

my second-order quantifier ranges over the full powerset of the domain of my first-order quantifier.

Of course, my *saying* in my home language that my second-order quantifier ranges over the full powerset of the first-order domain, over *all* the subsets of the first-order domain, does not help, since the issue is precisely the range of that quantifier in the intended interpretation of my home language. And I cannot claim that you should understand my words so because that is what I intend, for I have eschewed appeals to myths of the museum. My usage must show it, or nothing does. And my usage does show it, through the application of (Min), (Max), and (MinS) to my home language. Not (Min), (Max), and (MinS) formulated *in* my home language, that would be bootstrapping. But (Min), (Max), and (MinS) formulated in whatever the language it is in which the worry about the intended interpretation of my home language is raised. In *that* language, when we consider a sentence of my home language 'The practitioners use "∀" at second-order to range over *all* the subsets of d,' (Min), (Max), and (MinS) assign to my 'all' Henkin meaning with a full intended interpretation, given that my purpose is apparent as requiring the widest range. That someone, in philosophical vein, may feign not to see that purpose is, in the end neither here nor there: one can always be misunderstood.

That response may seem lame. The worry can be raised again about the language in which the worry about my home language was raised and in which (Min), etc., are formulated to purportedly settle the original worry about my home language. And this meta-worry can be settled again by the application of (Min), etc., formulated in yet another language. Clearly, the worry and response iterate ad inf. I don't think this regress is vicious. Much has been written on this kind of issue, especially since Hiliary Putnam's seminal (1980). I don't want to say much in defence of my claim that the regress is not vicious in this case. Nor do I want to claim that regresses of this general shape are never vicious: a theory which is introduced by formulating axioms and supplying an explicit semantic interpretation of those axioms is hostage to ambiguity in the language expressing those explicit semantics since it is dependent for whatever meaning it has on those explicit stipulations. But the issue now is the semantics of my home language, the language in which I explicitly formulate (Min*), (Max*), and (MinS*), and in which guided by (Min*), etc., I formulate semantics for the practitioners' use of L2K. That language, my home language, is not dependent for its semantic properties upon stipulations formulated in some further language used to formulate what is purportedly the semantics of my home language, and about whose semantics the question of its intended interpretation can again be raised. My home language is an autonomous language which gets its meaning from the human purposes it is used to serve. If the question of the semantics of my home language is raised and a theory formulated in a language with an unsuitable intended interpretation, then that semantic theory of my home language will simply be wrong. Enough said? Undoubtedly not, but I leave it there.

Assume now that when I say 'all subsets' of some given set I do mean the entire powerset, and therefore that (Min*), etc., applied by me to the practitioners of arithmetic will select a full model as their intended model. Still a question arises as to what, more precisely, that full model is? I assume the practice I am interpreting is not itself dependent upon my view of the cardinality of the continuum, but only my view of their practice is so dependent. That view I express in my home language. There I may assert the Continuum Hypothesis, or any one of indenumerably many alternatives of the form: $|R| = \aleph_i$, for i>0 where i continues into the transfinite ordinals. We know each of these alternatives is a consistent addition to my home theory. Assume that in my home language I adopt $|R| = \aleph_1$, then the set of all subsets of natural numbers will, according to me, have cardinality $\aleph_1$, as will the full model I assign as the practitioners' intended model of analysis, and by the categoricity proof, all full models of the axioms of analysis, according to me. But had I instead adopted, say, $|R| = \aleph_{23}$ then the set of all subsets of natural numbers would, according to me, have cardinality $\aleph_{23}$, as would the full model I assign as the practitioners' intended model, and by the categoricity proof, all full models of the axioms of analysis, according to me. In short, the categoricity proof for the axioms of analysis settles that all models have the same cardinality, but not what that cardinality is. In contrast to such views, Paul Cohen speculates that the Continuum Hypothesis and all its rivals of the form $|R| = \aleph_i$ may simply be false. The powerset axiom, he speculates, may take us to a cardinality beyond the reach of the operations which generate the alephs—beyond the reach, that is, of successor, union, and replacement.

A point of view which the author feels may eventually come to be accepted is that [the Continuum Hypothesis] is *obviously* false... $\aleph_1$ is the set of countable ordinals and this is merely a special and the simplest way of generating a higher cardinal. The set C [= the set of subsets of the set of natural numbers] is, in contrast, generated by a new and more powerful principle, namely the Power Set Axiom. It is unreasonable to expect that any description of a larger cardinal which attempts to build up that cardinal from ideas deriving from the Replacement Axiom can ever reach C. Thus C is greater than $\aleph_n$, $\aleph_\omega$, $\aleph_\alpha$, where $\alpha = \aleph_\omega$, etc., This point of view regards C as an incredibly rich set given to us by one bold new axiom, which can never be approached by any piecemeal process of construction.

(Cohen 1966: 151, his emphasis)

In view of the above various (epistemic) possibilities open up, some but not all of which threaten the idea that there is such a thing as *the* intended model to be selected by (Min*), (Max*), and (MinS*). First, if the Continuum Hypothesis or one of its alternatives is actually true and the rest false, then, whether I know it or not, the set of all subsets of the natural numbers has that cardinality and (Min*), etc., will select a full model of that cardinality as the practitioners' intended model, whether they know it or not. Indeed, let me decree they have no view as to the cardinality of the set of reals. But, second, the idea that one of these alternatives is actually true and the rest false, though we do not know and may never be able to know which or offer even a speculative explanation as to why it is that one, may seem a metaphysical surd too far. Hartry Field concludes that there simply is no fact of the matter; full models are simply indeterminate on the cardinality of the continuum (Field 1998, reprinted in Field 2001 with postscript). If it is indeterminate as to what a full model is, even to the point of the cardinality of the reals, then (Min*), etc., applied to our practitioners cannot select a model of determinate cardinality as their intended model. Now, as Field argues, the categoricity proof for the axioms of analysis in full models loses its metaphysical force. The proof shows that all full models of the axioms of analysis have the same cardinality, but it is indeterminate what that cardinality is. Still, it is determinate that it is not $\aleph_0$, so the downward Löwenheim–Skolem theorem is still blocked by (Min*), etc., selecting a full model. But for any $\aleph_i > 0$, there is no argument that no model of the axioms has the cardinality of $\aleph_i$. Finally, if Cohen's speculation is correct, then the issue of the size of the continuum is no bar to (Min*), etc., selecting a determinate full model, but the size is beyond any reached by the series of alephs. These are open questions to which it would be pointless to offer a guess.

Radical interpretation is no more than a dramatic device. If their use of L2K to practise arithmetic or analysis calls for Henkin semantics and a full intended model, then so would ours. We would be in the same position regarding our usage of L2K were we to use it that they are with theirs. Radical interpretation inoculates us against the myth of the museum: if their usage for those purposes generates that meaning, then our usage for those purposes would do the same for us. It also inoculates us against taking the meaning of the mature[31] usage to be determined by whatever training games or preliminary semantic stipulations lie in the psychological, anthropological history of our mature usage. The analogy here is of evolutionary precursors rather than cumulative development. The current usage to practise arithmetic determines meaning, not the history of how we came to that usage. Suppose we had introduced L2K to practise arithmetic by explicit stipulation of full meaning for '$\forall$' and explicit stipulation of a semantics of full models. Still, as those stipulations fade into the distant history of our practice of arithmetic, and if they are not kept alive like a swearing of allegiance to the flag from time to time, then metaphysical parsimony in the form of (Min*), (Max*), and (MinS*) will trim the meaning down to Henkin semantics with a full intended model.

(Min), (Max), and (MinS) speak of the *purposes* in using L2K which would be *apparent* to fellow human initiates of the same culture, and I have contrasted such apparent purposes with whatever *intentions* may lie *hidden* in a user's private museum. This separation of 'purpose' and 'intention' is merely a piece of convenient temporary labelling. Of course purposes can be covert, and intentions manifest; and I know of no important distinction between purpose and intention. Rather, I am labelling the distinction between what is manifest in the use of language to fellow human initiates of the same culture and necessary to understanding what has been said, and what may be covert. The distinction between manifest and covert is not sharp: our nearest and dearest may read more of us than is comfortable. Nor can a grasp of the speaker's purposes which are necessary to understanding what has been said be specified in some illuminating way independently of specifying what has been said. This is a difficult area to which Wright has contributed much. But for the case in hand I have simply assumed in (Min*), (Max*), and (MinS*) that the manifest purpose is to practise arithmetic (or analysis), and what stays in the museum is whatever is not needed to practise arithmetic (or analysis).

---

[31] Maturity is relative. The state of maturity I have in mind is that we have left whatever training games behind and reached the theory of pure arithmetic encapsulated in Peano's axioms. But, bearing in mind that Wright's logicism takes arithmetic to be its independent point of entry into the mathematical world, that is as far as our current mathematical knowledge and practice extends, we suppose. Adolescence perhaps, rather than maturity.

Let me sum up so far. In §1 I set out my intuitive notion of logical consequence (PC). (PC) gave rise to a pluralist view of logics. The relation of logical consequence among the sentences of L2K differs according to which meaning, full or Henkin, we give '∀.' In §2 I argued that the proof-theoretic completeness of ($\models_H$) bestows two philosophical advantages over ($\models_F$) as a putative definition of logical consequence. The first advantage was that even though the set of models quantified over in the definition ($\models_H$) is not the proper class of absolutely all models of L2K relevant to Henkin logical consequence among the sentences of L2K, proof-theoretic completeness enabled us to know that ($\models_H$) reflects absolutely all models, so we could accept ($\models_H$) as a *definition* of the relation of Henkin logical consequence among the sentences of L2K, even though the models quantified over in the definition ($\models_H$) had, say, domains of only accessible cardinality. In contrast we could accept ($\models_F$) as a definition of logical consequence due to '∀' having full meaning only subject to the *conjecture* that no counterexample lurks among models not quantified over in the definition ($\models_F$). The second advantage of proof-theoretic completeness was that, subject to certain modest conjectures, we know that ($\models_H$) will continue to define logical consequence among the sentences of L2K due to '∀' continuing to have Henkin meaning even if L2K becomes merely a fragment of an extended language able to do all the things that natural English can. We could take no such position regarding ($\models_F$) and full meaning for '∀' in advance of such development. The issue in §2 was *not* whether '∀' having full meaning would generate a bona fide relation of logical consequence among the sentences of L2K. The issue was whether we *know* ($\models_F$) measures that relation of logical consequence, as we do *know* that ($\models_H$) measures the relation due to '∀' having Henkin meaning.

In this section, §3, I have argued that to express the facts and to practise arithmetic, or analysis, we need no more than the Henkin relation of logical consequence among the sentences of L2K, and that even if a practice were initially set up say by stipulating full meaning for '∀,' the metaphysical principle of meaning parsimony in the form of (Min*), (Max*), and (MinS*) would, if over time L2K were used *just* for the practice of arithmetic or analysis, trim the meaning of '∀' down to Henkin meaning with a full intended model, because nothing stronger is needed to do the job. Of course, ($\models_H$) lacks much of the mathematical and philosophical interest of ($\models_F$), but I don't see why the *logic* of arithmetic, or of analysis, should be mathematically interesting.

4. Neo-Fregean Logicism

Let's call it just 'logicism.' The language of arithmetic assumed so far is not suitable for logicism. We have supposed L2K equipped with just arithmetic constants and used to practise just pure arithmetic, or equipped with just the constants needed to practise real analysis. In contrast, arithmetic logicism assumes a prior language without arithmetic concepts, used to talk of cabbages, kings, and whatever, into which the concept of natural number is then introduced by Hume's Principle.

$$(HP) \quad \forall F \forall G(Nx\!:\!Fx = Nx\!:\!Gx \leftrightarrow \exists R(F1\!-\!1_R G))$$

Logicism claims that if (HP) is taken as a stipulation implicitly defining the new operator, then 'Nx:...x' will naturally be interpreted as taking the members of the second-order domain onto the cardinal number of their members. Ordinals will follow when a zero and a predecessor relation (or a successor function) are chosen. Logicism then shows that the Peano axioms can be deduced in second-order logic from (HP) as the sole premise via definitions of zero and predecessor.[32] The proof-theory used will be an extension of D2, since it will need axioms/rules to deal with identities involving the operator 'Nv:...v...'—see Shapiro and Weir (2000). And, Peano's axioms will not be derived in a 'pure' form: they will need to be relativized to natural numbers to exclude transfinite cardinals, whose identity is also given by (HP), and any non-numbers in the domain supplied by the prior language. In his (1983) §xix Wright sketches a route essentially discovered by Frege which takes (HP) as its sole non-logical premise and employs the following definitions:

Def 0: $0 = Nx{:}\neg x=x$

Def P, the predecessor relation: $\forall x\forall y(Pxy \leftrightarrow \exists F\exists z(Fz \,\&\, y=Nu{:}Fu \,\&\, x=Nv{:}(Fv \,\&\, \neg z=v)))$

Def P*, the ancestral of P: $\forall x\forall y(P^*xy \leftrightarrow \forall F[(\forall z(Pxz \rightarrow Fz) \,\&\, \forall u\forall w((Fu \,\&\, Puw) \rightarrow Fw)) \rightarrow Fy])$

Def Nat, natural numbers: $\forall x(Natx \leftrightarrow (x=0 \lor P^*0x))$

Wright sketches out a demonstration of Peano's axioms taken in the form:

1. Nat0
2. $\forall x(Natx \rightarrow \exists y(Naty \,\&\, Pxy))$
3. $\forall x\forall y\forall z\forall w((Pxw \,\&\, Pyz) \rightarrow (x=y \leftrightarrow w=z))$
4. $\neg\exists x(Natx \,\&\, Px0)$
5. $\forall X[(X0 \,\&\, \forall x(Xx \rightarrow \forall y(Pxy \rightarrow Xy))) \rightarrow \forall x(Natx \rightarrow Xx)]$

(HP) does not qualify as a definition by Corcoran's definition of a definition. Corcoran's idea was that a definition should fix the sense of a new term such that there is only *one* way of extending any model of the prior language to become a model which satisfies the definition in question. (HP) does not meet this condition. Given a model of the prior language, there are many ways of extending it to a model satisfying (HP). Assigning 'Nx:...x' *any* function from D_1 to d will do provided it takes equinumerous members of D_1 onto the same member of d and non-equinumerous members of D_1 onto different members of d. The definitions of '0,' 'P,' and 'P*' will then order the objects so assigned into an ω-sequence.[33,34]

---

[32] There is no suggestion that our tyro arithmeticians will discover this deduction any time soon. The point of the deduction is to show the logical power latent in (HP).

[33] An attractive feature of assigning the von Neumann numbers to Frege's numerals is that each N_i is itself an i-membered set. So if $Nx{:}Fx = N_i$, then $\exists R(Fx1{-}1_Rx\varepsilon N_i)$. But this is not required by (HP), and is not true of Frege's own interpretation of his numerals, nor Russell's, nor Zermelo's.

[34] There is a set of theorems of (HP) which might make it *look* as if the interpretation of 'Nv:...v...' *must* be a function from the extension of a predicate to specifically the number of its members. Wright proves in an appendix to his (1999) that every instance of $\exists_n xFx \leftrightarrow n^*=Nx{:}Fx$ may be deduced from (HP), where $\exists_n$ is a numeric quantifier—there are exactly n—and n* is the nth Fregean numeral in the

Although many models satisfy (HP) and therefore via the definitions also satisfy Peano's axioms, Wright imposes contextual conditions which eliminate some models as unintended models. He emphasizes that if (HP) is adopted to first introduce a concept of number, then it is to be understood as providing sortal criteria for these newly introduced objects, the numbers. And he invokes Frege's 're-carving' analogy: that the sense of the left hand side of (HP) is a re-carving of the sense of the right— Wright (1983) and Hale and Wright (2001). Let us suppose that these (and perhaps other) contextual features will yield as the intended interpretation, primarily[35] the natural numbers, and return to our main topic, logical consequence.

In §3, I had supposed L2K equipped just with arithmetic concepts and used to practise just arithmetic. There was no attempt to explain how the supposed practitioners came by their arithmetic concepts, nor what other linguistic skills they might have had. Logicism, in contrast, offers an idealized conceptual genealogy of arithmetic concepts, an idealized explanation of how a community might have acquired them: by adopting (HP) in a language otherwise innocent of arithmetic concepts. And, following what Wright (2000) calls 'Frege's constraint,' logicism's idealized genealogy of pure arithmetic readily extends to an idealized explanation of applications such as counting oranges in a bowl, and comparing their number to that of apples in a bag. Clearly, in the light of these aims, they see it as a virtue of (HP) that the domain of its initial quantifiers '$\forall F \forall G$' includes non-arithmetic sets—oranges in the bowl, apples in the bag—available in the language to second-order quantifiers prior to the introduction of 'Nx:...x.' Hence the logicist's startpoint is a language for talking of cabbages and kings but free of arithmetic concepts, into which 'Nx:...x' and resulting arithmetic concepts are introduced by (HP).

It appears that the case made against ($\models_F$) and for ($\models_H$) in §3 does not carry over to arithmetic as envisaged by logicism. I had supposed that L2K was a language used to practise just pure arithmetic, and I argued that this called for no more than ($\models_H$) with a full intended interpretation. But the prior language presupposed by logicism to talk of cabbages and kings, might *already* have the stronger relation of logical consequence ($\models_F$), it might *already* assign '$\forall$' full meaning, having had this *before* the introduction of 'Nv:...v...' and the initiation of its users into the practice of arithmetic. '$\forall$' might *retain* this full meaning because of the other non-arithmetic purposes to which the language is put. If so, the truths of arithmetic will indeed be logical consequences of the newly introduced (HP). So the case against ($\models_F$) and for ($\models_H$) made in §3 is undercut, hostage to an account of the relation of logical consequence extant prior to and surviving beyond the introduction of (HP). However, this is not the conclusion I think we should draw. Certainly, if logicism

---

series: $0^* = $ Nx:¬x=x, $1^* = $ Nx:x=Nx:¬x=x, $2^* = $ Nx:(x=Nx:¬x=x v x=(Nx:x=Nx:¬x=x)), etc. If we were to read this as $\exists_n xFx \leftrightarrow n = $ Nx:Fx, then these theorems would require that the value of Nx:Fx be the number of Fs. But the proof goes via the definitions of the Fregean numerals. So to take an example, $\exists_2 xFx \leftrightarrow 2^* = $ Nx: Fx, the content of the theorem is, when we eliminate the numeral 2^* via its Fregean definition: $\exists_2 xFx \leftrightarrow $ Nx: (x=Nx:¬x=x v x=(Nx:x=Nx:¬x=x))=Nx:Fx. All this requires of the value of Nx:Fx is that it be the same as that of Nx:(x=Nx:¬x=x v x=(Nx:x=Nx:¬x=x)) iff there are exactly two Fs. But it does not require their common value be the number 2.

[35] 'Primarily' because $\exists x(x = $ Nx:Natx) is provable from (HP) via the definitions, and if the Frege numerals are assigned the natural numbers, 'Nx:Natx' will be assigned $\aleph_0$.

is required to explain how a community might come to acquire specifically arithmetic concepts, then a prior language is rightly presupposed. And clearly with it will come a prior relation of logical consequence. But there is a second question logicism should answer. What relation of logical consequence is required to express the facts of pure arithmetic and to sustain the epistemology of pure arithmetic once this practice is up and running? To answer this question we should bracket off how the community came by their arithmetic concepts, and bracket off any other non-purely arithmetic purposes they put the language to which might influence the meaning of '∀.' We should simply assume a mastery of arithmetic concepts employed in a practice of pure arithmetic, bracketing off whatever background was needed to get that practice going,[36] and ask what relation of logical consequence is needed, generated from what meanings of the logical constants, if that practice is to express the ontology of pure arithmetic and sustain the epistemology of pure arithmetic? Hence we posed in §3 as radical interpreters who knew that L2K was used just to practise pure arithmetic, but were blind to the context and history of that practice. We want to know what relation of logical consequence pure arithmetic *requires*, rather than what relation was *inherited* from other sources, or is sustained by other purposes. Thus, I claim, the case for ($\models_H$) and a full intended model made in §3 does apply.

My discussion has been severely limited. I have considered only two contenders for the relation of logical consequence, arising either from assigning full or from assigning Henkin meaning to '∀.' The nature of particularly second-order logic is a contested topic, and a number of other very different views are current. Wright himself, without to my knowledge committing himself one way or the other, has shown sympathy for a constructivist approach, as witness the following remark:

The possibility must be taken seriously, it seems to us, that the most tractable philosophies of higher-order logic will have the side effect that the neo-Fregean treatment of analysis will be constrained to be *constructivist*, and that the classical continuum will be out of reach. (Whether that would be a cause for regret is, of course, a further question.)

(Hale and Wright 2005: 200, their emphasis)

Wright points out the modesty of the resources needed for the journey from (HP) to Peano's axioms, citing John Bell (1999):

J. L. Bell has shown that the higher-order logic necessary for the demonstration of Frege's Theorem on the basis of Hume's Principle [i.e., the demonstration of Peano's axioms from (HP)] is in fact quite weak.   (Hale and Wright 2001: 433)

And in a footnote to this passage:

Bell shows specifically that the proof of Frege's Theorem may go through without appeal to Excluded Middle and with the higher-order variables taken to range over just the *finite* concepts and relations on the domain in question. Thus no more than a countable population of higher-order 'entities' need be presupposed. Bell does not himself point out the corollary that if all the non-numerical objects in the domain are nameable in the language in question,

---

[36] Ladders to be climbed and thrown away, in Wittgenstein's recklessly unapt metaphor. Crutches are to be discarded; stabilizers removed from a child's bicycle when she has got the hang of it, boats burned if there is no going back. But a ladder is best left in place until safely down again.

then every concept presupposed by the proof of Fege's Theorem will be definable in that language and hence—if one were otherwise attracted to such a thing—that a *substitutional* interpretation of the higher-order quantifiers in Fregean Arithmetic is technically viable.

(Hale and Wright 2001: 433, n. 19, their emphasis)

Bell takes the values of the variable 'X' to be confined to Ø, the singleton of any member of the first-order domain d, and to be closed under unions of any two members. And similarly confines the values of 'R' to finite relations. The proof of Peano's axioms sketched in Wright's (1983) will still go through virtually unchanged.[37] Thus Bell provides a relatively sparse, countable, model for (HP), Peano's axioms and the proof theory needed to get from one to the other. However, I shall argue, a logicist needs to combine this with a relatively *strong* conception of logical consequence. Each of the Gödel sentences '¬*Bew*Nx: ¬x=x,' '¬*Bew*Nx: x=Nx:¬x=x,' etc., is provable in this proof-theory. So they are all true in Bell's model. Hence the Gödel sentence '∀x(Natx → ¬*Bew*x)' is true in Bell's model. Hence the logicist will require the Gödel sentence to be a *logical* consequence of the axioms. A weak conception of logical consequence like ($\models_H$) which is complete for this proof theory will not do. Since the Gödel sentence is not provable, ($\models_H$) permits models of the axioms and proof-theory in which the Gödel sentence is false, though '¬*Bew*Nx: ¬x=x,' '¬*Bew*Nx: x=Nx:¬x=x,' etc., remain true. So the moral, as I see it, is this: although the satisfaction of (HP) and derivation of Peano's axioms require only a countable model and a weak conception of logical consequence, the logicist's project of showing that *all* arithmetic truths are consequences of those axioms requires a strong conception of logical consequence.

Wright claims that Gödel's incompleteness result has no specific bearing on logicism:

If there is compelling reason to deny that second- or higher-order logic is logic, then logicism—at least in its most exacting form [= that all arithmetic expressions may be explicitly defined in purely logical terms and that when this is done, all arithmetic truths can be seen to be logical truths]—must be rejected anyway, even before considerations of incompleteness are brought to bear. If, on the other hand, the logicist is entitled to take logic as including second- and perhaps higher-order logic, then Gödel's incompleteness theorems raise no special problem for her. In short Gödel's incompleteness result has no specific bearing on the logicist project.

(Hale and Wright 2001: 4–5, n. 5)

The second horn of the dilemma is ambiguous. I concede that the logicist *is* entitled to take second-or higher-order logic as logic. Both ($\models_F$) and ($\models_H$) are conceptions of second-order logical consequence in good standing. But the logicist is *not* entitled to freely choose among the available conceptions of logical consequence. She needs to *justify* her choice in terms of what is required for the practice of arithmetic.

---

[37] Some minor adjustments will be needed. Lemma 3 '∀R∀x∀y(Rxy→R*xy)' and Lemma 4 '∀R∀x∀y∀z((R*xy & R*yz) → R*xz),' Wright (1983: 160), employ quantification over infinite relations R, since the instances wanted are 'Pxy' and 'P*xy' and these are infinite relations, applying to all numbers. But it is easy enough to prove the ungeneralized instances in their stead—'∀x∀y(Pxy→P*xy),' '∀x∀y∀z((P*xy & P*yz) → P*xz)'—and these are all that are actually needed in the proof of Peano's axiom 2.

Although our earlier discussion was limited to only two accounts of the semantics of L2K, full and Henkin, neither of which need be congenial to Wright, that discussion did show I think *a route* a logicist who is also a logical pluralist should take in quest of a favourable conception of logical consequence. A contrasting logical monist presumably gives her account of the one true relation of logical consequence, and then applies it to L2K. But my prior philosophical conception of logical consequence (PC) yields a pluralist view of logical consequence: logical consequence is a matter of the repertoire of logical forms and the meanings of the logical constants found in a language, and these may differ from language to language. We therefore need to ask what relation of logical consequence we require in arithmetic language. The logicist should start, I suggest, by postulating a formal language, say L2K augmented with 'Nx:...x...,' in which she supposes *only pure* arithmetic is practised. The point is that she needs to determine, by some plausible principles of radical interpretation or otherwise, what relation of logical consequence is needed to express the ontology of pure arithmetic and to support the epistemology of pure arithmetic, *uncontaminated* by any non-arithmetic purposes for which a language might also be used. Certainly, the logicist may also seek satisfy *other* epistemic and ontological aspects of her programme by rationally reconstructing the genesis of arithmetic concepts, postulating a prior non-arithmetic language into which arithmetic concepts are then introduced by stipulating (HP). But she should not allow the prior language to smuggle in a stronger relation of logical consequence than that needed in the purely arithmetic language.[38] Having thus identified weakest the relation of logical consequence necessary for expressing and for doing pure arithmetic, the logicist needs to show that Peano's axioms are demonstrable from (HP) in this logic. And finally she needs to show that Peano's axioms are suitably categorical for this relation of logical consequence, so that the remaining truths of arithmetic are not merely material consequences of the axioms in the intended interpretation, but are logical consequences of those axioms. In all a tall order.

Suppose that in the end the order cannot be fulfilled. What then would remain of the logicist's programme? Wright sums up what he requires of (HP) in four claims:

(i) that the vocabulary of higher-order logic plus the cardinality operator,... 'Nx:...x...,' provides a sufficient definitional basis for a statement of the laws of arithmetic [i.e., for Peano's axioms];

(ii) that when they are so stated, Hume's Principle provides for a derivation of those laws within higher-order logic;

(iii) that someone who understood a higher-order language to which the cardinality operator had been added would learn, on being told that Hume's Principle governs the meaning of the operator, all that it is necessary to know in order to construe any of the new statements that would then be formulable; and

---

[38] Suppose she finds that the introduction of arithmetic concepts somehow *requires* the prior language to have a stronger relation of logical consequence, say ($\models_F$), than that *needed*, say ($\models_H$), for the ontology and epistemology of pure arithmetic once arithmetic concepts have been acquired? Well, that would be an interesting consequence, revealing a fine-structure in logicism.

(iv) finally and crucially, that Hume's Principle may be laid down *without signifi-cant epistemological obligation*: that it may simply be stipulated as an explan-ation of the meaning of statements of numeric identity, and that—beyond the issue of the satisfaction of the truth-conditions it thereby lays down for such statements [i.e., subsidiary premises of the form $\exists R(F1-1_RG)$]—no compe-tent demand arises for an independent assurance that there *are* objects whose conditions of identity are as it stipulates. (Wright 1999; reprinted in Hale and Wright 2001: 321; Wright's emphasis)

These are in no way less plausible if the language which (HP) extends has Henkin logical consequence and a full intended model. (i) is satisfied because the definitions of '0,' 'Pxy,' 'P*xy,' and 'Natx' needed for the formulation of Peano's axioms are unaffected. (ii) is satisfied since the derivation of those axioms from (HP) is in an extended D2, which is sound for Henkin semantics. Even assuming a background of full logical consequence, (iii) and (iv) are of course highly controversial and controverted—for example, whether the right-hand side of (HP) can properly be seen as fixing the sense of the left-hand side, whether it is permissible for implicit definitions to be ontologically creative, whether that ontology is honoured in fact, whether (HP) can be properly separated from 'bad company'—that is, from incon-sistent stipulations of this general form such as Frege's Basic Law V, and from individually consistent but jointly inconsistent bedfellows, and more. But so far as I can see, to take the Hale–Wright side of these various controversies *over the role of (HP) as an implicit definition of the concept of number* does not require full logical consequence. But what would be lost, of course, is the claim that the truths of arithmetic are analytic, being all logical consequences of the stipulative definition (HP). The truths of arithmetic would be downgraded to being only material conse-quences of (HP) in the intended interpretation.

Still, it is premature to conclude that Wright's conditions will not be fulfilled. And, in any case, the extensive debate concerning the various aspects of logicism has been deep and illuminating, and it is ongoing. For the vigorous and detailed pursuit of these issues, for the insights that have resulted, both for the philosophy of mathem-atics and more widely given the centrality of that discipline, we should be grateful for the high philosophical quality of the major participants drawn to one side or the other of the debate that Wright initiated and continues to play a major part in. It is a pleasure to record my thanks in particular to the champions of the neo-Fregean side, Crispin Wright and his co-worker Bob Hale, and to look forward to more to come.[39]

References

Bell, J. L. (1999). "Frege's Theorem in a Constructive Setting," *Journal of Symbolic Logic* 64, 486–8.
Benacerraf, P. (1965). "What Numbers Could Not Be," *Philosophical Review* 74, 47–73.
Cohen, P. (1966). *Set Theory and the Continuum Hypothesis*, New York: W.A. Benjamin, Inc.

[39] Thanks to Bob Hale, Gary Kemp, Philip Percival, Alan Weir, and most especially to Adam Rieger. Sadly Bob Hale died after this chapter was written. I leave the final sentence as originally written in tribute to Bob.

Corcoran, J. (1971). "A Semantic Definition of Definition," *Journal of Symbolic Logic* 36, 366–7.

Davidson, D. (1973). "Radical Interpretation," *Dialectica* 27, 313–28.

Davidson, D. (1974). "Belief and the Basis of Meaning," *Synthese* 27, 313–28.

Edwards, J. (2002). "Theories of Meaning and the Logical Constants: Davidson versus Evans," *Mind* 111, 249–79.

Edwards, J. (2007). "Response to Hoeltje: Davidson Vindicated?," *Mind* 116, 131–41.

Etchemendy, J. (1990). *The Concept of Logical Consequence*, Cambridge, MA: Harvard University Press.

Evans, G. (1975). "Identity and Predication," *Journal of Philosophy* 72, 343–63.

Field, H. (1998). "Which Undecidable Mathematical Sentences Have Determinate Truth Values?," in G. H. Dales and G. Oliveri (eds.), *Truth in Mathematics*, Oxford: Oxford University Press. Reprinted with postscript in Field (2001).

Field, H. (2001). *Truth and the Absence of Fact*, Oxford: Oxford University Press.

Hale, R., and C. Wright (2001). *The Reason's Proper Study: Essays towards a Neo-Fregean Philosophy of Mathematics*, Oxford: Oxford University Press.

Hale, R., and C. Wright (2005). "Logicism in the Twenty-first Century," in S. Shapiro (ed.), *The Oxford Handbook of Philosophy of Mathematics and Logic*, Oxford: Oxford University Press, 166–202.

Henkin, L. (1951). "Completeness in the Theory of Types," *Journal of Symbolic Logic* 15, 81–91.

Jané, I. (2005). "Higher-Order Logic Reconsidered," in S. Shapiro (ed.), *The Oxford Handbook of Philosophy of Mathematics and Logic*, Oxford: Oxford University Press, 781–810.

Kreisel, G. (1967). "Informal Rigour and Completeness Proofs," in I. Lakatos (ed.), *Problems in the Philosophy of Mathematics*, Amsterdam: North Holland, 138–86.

Lewis, D. (1974). "Radical Interpretation," *Synthese* 23, 331–44.

McGee, V. (2000). "Everything," in G. Sher and R. Tieszen (eds.), *Between Logic and Intuition: Essays in Honor of Charles Parsons*, Cambridge: Cambridge University Press.

Putnam, H. (1980). "Models and Reality," *Journal of Symbolic Logic* 45, 464–82; reprinted in Putnam (1983), 1–25.

Putnam, H. (1983). *Realism and Reason*, Cambridge: Cambridge University Press.

Roberts, C. (2003). "Formal Pragmatics," in N. Nadel (ed.), *Encyclopaedia of Cognitive Science*, London: Macmillan.

Saul, J. (2002). "Speaker Meaning, What is Said and What is Implicated?," *Noûs* 36, 228–48.

Shapiro, S. (1991). *Foundations without Foundationalism: A Case for Second-Order Logic*, Oxford: Oxford University Press.

Shapiro, S. (1998). "Logical Consequence: Models and Modality," in M. Schirn (ed.), *The Philosophy of Mathematics Today*, Oxford: Oxford University Press.

Shapiro, S (1999). "Do Not Claim Too Much: Second-Order Logic and First-Order Logic," *Philosophia Mathematica* 7, 42–64.

Shapiro, S. (2003). "All Sets Great and Small: And I Do Mean All," *Philosophical Perspectives* 17, 467–90.

Shapiro, S. (ed.) (2005). *The Oxford Handbook of Philosophy of Mathematics and Logic*, Oxford: Oxford University Press.

Shapiro, S., and A. Weir (2000). "'Neo-Logicist' Logic Is Not Epistemically Innocent," *Philosophia Mathematica* 8, 160–89.

Sher, G. (1991). *The Bounds of Logic*, Cambridge, MA: MIT Press.

Skolem, T. (1970 [1941]). "Sur la Portée du Théorème de Löwenheim-Skolem," in Fenstad E. (ed.), *Selected Works in Logic*, Oslo: Universitetsforlaget, 455–82.

Tarski, A. (1935 [1956]). "On the Concept of Logical Consequence," in A. Tarski (ed.), *Logic, Semantics, Metamathematics*, Oxford: Oxford University Press, 409–20.

Tarski, A. (1986). "What Are Logical Notions?," in J. Corcoran (ed.), *History and Philosophy of Logic* 7, 143–54.

Williamson, T. (2003). "Everything," *Philosophical Perspectives* 17, 415–65.

Wright, C. (1983). *Frege's Conception of Numbers as Objects*, Aberdeen: Aberdeen University Press.

Wright, C. (1985). "Skolem and the Skeptic," *Proceedings of the Aristotelian Society Supplementary Volume* 59, 117–37.

Wright, C. (1997). "The Indeterminacy of Translation," in R. Hale and C. Wright (eds.), *A Companion to the Philosophy of Language*, Oxford: Blackwell, 397–426.

Wright, C. (1999). "Is Hume's Principle Analytic?," *Notre Dame Journal of Formal Logic* 40; reprinted in Hale and Wright (2001), 307–32.

Wright, C. (2000). "Neo Fregean Foundations for Real Analysis: Some Reflections on Frege's Constraint," *Notre Dame Journal of Formal Logic* 41, 317–34.

4

Logicism and Second-Order Logic

George S. Boolos
edited by *Richard Kimberly Heck*

Editor's Introduction

After George Boolos died, on 27 May 1996, I had the privilege of going through the papers and other effects that he had left in his office. There were a lot of letters (which went into the archives at the Massachusetts Institute of Technology); there was a huge collection of off-prints and papers from colleagues and students (which I distributed to people who would appreciate them); and there were lists of errata for his published books. To my surprise, however, there was not much unpublished paper-like material. It seemed that George had either had the good fortune to have most of what he wrote reach publishable form (as I, most certainly, have not), or else he had discarded what did not. For one reason or another, though, I found a few substantial drafts of unpublished material among his papers.

What follows is an edited version of one of these. It was discovered in a folder with a separate draft of what was published, in 1975, as "On Second-Order Logic" (Boolos 1975). The two drafts were typed on similar paper: Five-hole, wide-ruled. Some of the pages of the draft have square edges, and some round, and a number of changes are indicated in the text. It appears to be something on which George worked a fair bit.

The later parts of the draft, which begin with a section entitled "Quine's Criticisms of Second-Order Logic," overlap with "On Second-Order Logic" quite heavily. I have omitted that material, not just because of the overlap—quite substantial overlap remains—but because it had been so heavily reworked that it is impossible to be confident how George intended it to be reconstructed. The manuscript is littered with numerical indices that seem to indicate how paragraphs were supposed to be re-ordered. Indeed, it seems likely to me that, at some point, George decided that his most important points were contained in this material and that it needed a complete overhaul, so that he abandoned the present draft and began a new one, which would eventually become "On Second-Order Logic."

Still, it seems clear that George intended to return to this material at some point. The first footnote in "On Second-Order Logic" reads as follows:

My motive in taking up this issue [viz., whether second-order logic is logic] is that there is a way of associating a truth of second-order logic with each truth of arithmetic; this association

George S. Boolos *edited by* Richard Kimberly Heck, *Logicism and Second-Order Logic* In: *Logic, Language, and Mathematics: Themes from the Philosophy of Crispin Wright*. Edited by: Alexander Miller, Oxford University Press (2020).
© George S. Boolos with Postscript and notes by Richard Kimberly Heck.
DOI: 10.1093/oso/9780199278343.003.0004

can plausibly be regarded as a "reduction" of arithmetic to set theory. (It is described in Chapter 18 of *Computability and Logic*.)[a] I am inclined to think that the existence of this association is the heart of the best case that can be made for logicism and that unless second-order logic has some claim to be regarded as logic, logicism must be considered to have failed totally. I see the reasons offered in this paper on behalf of this claim as part of a partial vindication of the logicist thesis. I don't believe we yet have an assessment that is as just as it could be of the extent to which Frege, Dedekind, and Russell succeeded in showing logic to be the ground of mathematical truth. (Boolos 1975: 509, n. 1; 1998d: 37, n. 1)

Until I read the present paper, I had not paid much attention to this footnote. Much of the interest of this paper, then, lies in what it reveals about George's earliest thinking about logicism, a topic on which he would work throughout his career, but on which he would work especially intensively in the last decade of his life.

My editing has been fairly minimal. Obvious typos have been corrected; grammar has been repaired; material in parentheses has often been moved into footnotes; some references have been added. I have also added some editorial comments, which appear in square brackets in the text or else as footnotes, marked with superscripted letters.[b] What is reproduced here, though, is pretty much as I found it.

Let me say, finally, that, were George still with us, he would certainly have wanted to contribute to this volume. He had great respect for Crispin's work, even as he disagreed with it, often in fundamental ways. He was also fond of Crispin personally, and I have dear memories of a visit to St Andrews, in the fall of 1993, when George and I both gave papers at the Joint Session of the Mind Association and the British Society for the Philosophy of Science. I vividly remember running into George, shortly after I'd arrived, near the home of the Royal and Ancient Golf Club. George was walking around with Jason Stanley and laughing as Jason raved about just having seen James Bond (i.e., Sean Connery). I can only imagine how much fun we might all have had during the Arché conferences on neo-logicism had George been able to attend some. I can only imagine, too, what we might have learned from his participation.

Much thanks to Sally Sedgwick for permission to publish this paper.

1. Reductions of Arithmetic to Logic and Set Theory

I shall begin by discussing something that I do not consider to be a reduction of arithmetic to logic. It is a reduction to a "tiny" fragment of set theory.[1] The axioms of tiny set theory are those of extensionality, adjunction:[2]

$$\forall x \forall y \exists z \forall w (w \in z \rightarrow w \in x \lor w = y),$$

---

[a] This is of course chapter 18 of the first edition (Boolos and Jeffrey 1974). It has the same title, "Second-Order Logic," in the later editions. In the most recent edition, the fifth, it is chapter 22.

[b] Like this one.

[1] Terminology: Set theory is always Zermelo-Frankel set theory; elementary number theory is Hilbert-Bernays Z [i.e., PA—ed.], of which the axioms are the recursion axioms for successor, plus, and times, and the induction axioms; arithmetic is the set of all truths (under the standard interpretation) in the language of elementary number theory.

[2] Adjunction is not usually taken as an axiom of set theory, but it is a consequence of the axioms of pairing and union, which are.

and all the Aussonderungsaxioms.[c] Tiny set theory's consistency is beyond non-Cartesian doubt, for if '$x \in y$' is re-interpreted to mean 'there is a "1" at the xth place in the binary numeral for y counting from the right and starting with zero,' then tiny's axioms all come out true of the natural numbers. Tiny set theory assumes infinite classes in neither way in which a theory may do so: it has a model whose domain contains only finite classes (although infinitely many of them), for the hereditarily finite [pure] sets[3] form a model of tiny set theory. And no axiom of infinity is a theorem. Nor is the axiom of choice.

For all its weakness, though, tiny set theory is a theory to which elementary number theory can be reduced. There is a suitable translation of the language of set theory under which all theorems of elementary number theory can be proved in tiny set theory. Moreover, the translation preserves truth; truths about the natural numbers become truths about sets under the translation.[4]

But even though tiny set theory is an (interpreted) theory whose theorems assert only truths about sets and membership, it cannot be considered to be a system of logic, and the reduction of elementary number theory to it does not show arithmetic, or even elementary number theory, to be reducible to logic. It is not a system of logic, because some of its theorems are not logical truths, are not logically valid, that is, are false under some interpretations. The axiom of extensionality is not valid, for example. This point has been made, exceedingly well, by Paul Benacceraf:[d] '$\exists x \exists y (x \neq y)$,' though a truth about numbers, sets, or objects,[5] is not a logical truth, for it is not true in all interpretations, as it would have to be in order to count as a logical truth. And there are theorems of tiny set theory that are no better off than '$\exists x \exists y (x \neq y)$'—which, in fact, *is* a theorem of tiny set theory. And it certainly would appear to be reasonable to require that a reduction of arithmetic to logic not produce logically invalid statements as those that are supposed to express the "real content" of truths of arithmetic. The damaging point is not that truths about numbers are reduced to truths about what turn out to be, after all, only another sort of abstract object too much like the numbers. It is that they are reduced to statements that are not truths of logic. Weak though tiny set theory is, it is too strong to count as a system of logic.

The reducibility of elementary number theory to tiny set theory illustrates a danger in the use of the existence of a reduction to make an anti-skeptical claim, one that is reminiscent of the danger Mill warned of in the use of the syllogism. Suppose that a theory or system A has been shown to be reducible (whatever that may mean) to

---

[c] That is, all the instances of $\forall y \exists z \forall x [x \in z \equiv x \in y \wedge A(x)]$.

[d] There is no reference indicated here. The points Boolos mentions are made in Benacerraf's paper "Frege: The Last Logicist" (Benacerraf 1981, 1995), but that was not published until well after this paper was written. Boolos often did remark, though, that he greatly admired Benacerraf's dissertation, *Logicism: Some Considerations* (Benacerraf 1960)—he kept a copy in his library and once recommended I read it— and these points are also made there. So I think it likely that Boolos had Benacerraf's dissertation in mind. It's perhaps also worth noting that this point is one on which Boolos placed a good deal of weight: He mentioned it frequently in his later writings on logicism (Boolos 1998a,b,f).

[3] These are the finite sets all of whose members, members' members, etc., are finite.

[4] The natural numbers, incidentally, get reduced to their von Neumann counterparts.

[5] That is, it is a truth when its variables are taken to range over numbers, sets, or objects.

another, L. Suppose further that before the reduction A would have been supposed to be less "secure" than L. What "secure" will mean will vary from case to case. It may mean "certainly consistent," "analytic," "known," "logically necessary," "intelligible," "true,"[6] "well confirmed." After the reduction is achieved, it may or may not be advisable to take its existence as sufficient reason to think that A is at least as secure as L would have been thought to be, for, of course, in the absence of further reasons one way or the other, it is no more reasonable to do this than to think, for the same reason, that L is as little secure as A would have been thought to be. The reduction of elementary number theory to tiny set theory does not show that arithmetic is analytic; it does show that elementary number theory does not assume infinite classes. But the existence of a reduction does not by itself establish these two claims. What is needed in addition for the first is the fact that no one ought to maintain the analyticity of all statements of tiny set theory; for the second, that it was really just a guess that arithmetic assumed infinite classes. Frege was aware that his Basic Law V might be viewed as extra-logical and gave some considerations in support of it (Frege 2013: vol. I, p. vii). (Would that the problem had only been that concept and extension were not properly logical notions.)

There is a point of view—not, I believe, Frege's—according to which logic, essentially, treats of the relations between three sorts of things: sentences, interpretations, and truth-values, and provides an account or theory of the circumstances under which various sentences have various truth-values under various interpretations. A sentence is a logical truth if it has the truth-value "true" under every interpretation under which it has some truth-value.[e] In classical quantification theory, there are two truth-values, and an interpretation is taken to be an object "consisting of" a non-empty set, a domain, and an assignment of suitable objects to "non-logical" constants. Frege, it has been noted, seems nowhere to allow for the possibility of interpretations which differed in respect of their domains. It would appear that Frege would have maintained that the range of the (object-)variables was always the same: the totality of objects that there were.[f] Logical principles alone, he supposed, could guarantee that there were infinitely many objects. Frege's point of view seems foreign to us now. We are prepared to allow that '$\exists x \exists y (x \neq y)$' is not a logical truth even though we are prepared to grant that Parmenides was wrong.[7] And we are prepared to allow this even though we regard the sign of identity as a logical constant and interpret it uniformly, in the standard and familiar way.

In what follows, I am going to discuss certain "systems of logic," which, to the extent that they merit the term "logic," do so largely because of pronounced resemblances to classical quantification theory. They are two-valued systems; the notion of

---

[e] I am not sure why Boolos puts the point this way. The wording suggests that he means to allow for truth-value gaps, but that seems unlikely given his aversion to such things. Perhaps what he had in mind was just the simple fact that, on some treatments, anyway, the assignment $p \to T$, $q \to F$ counts as an interpretation, and on this interpretation $p \vee r$ has no truth-value. Of course, the same thing can happen with assignments of values to objectual variables.

[f] I'll note in passing that Boolos would later accept my arguments to the contrary (Heck 2012: 69ff).

[6] "A is less true than L" is to mean "L is true and A is not."

[7] We are not prepared to regard '$\exists x (x = x)$' as not a logical truth. Why is there something rather than nothing? Answer: social practice.

an interpretation is the same as that used in classical quantification theory; although more sequences of symbols count as (well-formed) sentences, each (classical) sentence receives the same truth-value under an interpretation according to any of these systems that it ordinarily receives; and the new sentences are interpreted in a manner strikingly analogous to the way in which the old ones are. I do not mean to deny—or affirm—that modal and intuitionist logics are systems of logic. I only wish to stress that we need not make a marked alteration in our notion of what logic is in order to admit that certain systems are systems of logic.[g] As we shall see, it will be perfectly compatible with this admission to maintain that set theory, or even tiny set theory, is not a system of logic. Whether a reduction of arithmetic to any such system [one that should be counted as a system of logic—ed.] can confer any interesting sort of security upon arithmetic is a question that we shall defer.

2. Some Objections to Logicism

It would seem that a logical discovery that was made after the logicist program had been formulated shows once and for all that no successful reduction of arithmetic to any system of logic can be had, at least not if that system meets a very mild requirement. The discovery is Church's theorem that there is no decision procedure for arithmetical truth, and the requirement is that it be decidable whether or not a sentence holds in an interpretation whose domain contains only one element. (Classical quantification theory certainly meets this requirement.) With these in the background, it might then be argued that if there is a reduction of arithmetic to logic, then there must be an effective procedure which, when given an arithmetical truth, delivers a logical truth to which it is 'reduced,' and when given an arithmetic falsehood, delivers a logical falsehood to which it is 'reduced.' The source of arithmetical falsity cannot be supposed to be different from that of arithmetical truth, and a reduction ought to treat truth and falsity alike. If arithmetical truth is really logical truth, arithmetical falsity must be logical falsity, and if a reduction must supply a logical truth that gives the content of each arithmetical truth, it must also supply a logical falsehood for each arithmetical falsehood. But of course there can be no such reduction, for then arithmetic would be decidable: to tell whether any given sentence of arithmetic is true, find the statement yielded by the reduction and determine whether it holds in a one-element interpretation (any one you please). The determination can be made effectively, and the statement will hold in the interpretation if and only if the given sentence was true. Corollary: No reduction commutes with the truth-functions.

It does not seem to me to be perfectly certain that this objection is a killer. The requirement on the reducing system of logic is certainly a very weak one. But the requirement that a reduction treat arithmetical truth and falsity symmetrically is not so very mild, and it is hard to see why a reduction must reduce arithmetical falsehoods to logical falsehoods in addition to reducing their negations to logical

---

[g] The original text here mentions infinitary logics and logics admitting quantifiers like 'finitely many,' as well as second-order logic, but these are not discussed further in this draft. It may be that Boolos originally intended to discuss these logics, too, but ended up focusing exclusively on second-order logic.

truths. Of course, we cannot have a reduction in which falsehoods are reduced to logical truths; but we might have one in which they are reduced to statements which are not logical truths, provided that arithmetical truths are reduced to logical truths. It might be argued that the reduced form of the joint denial of two statements must be the joint denial of the reduced forms of the statements, thence that a reduction must commute with truth-functions, and thence that arithmetical falsehoods must be reduced to logical falsehoods. The premise of this argument, however, needs more support than the obvious, simplistic, "simplicity" considerations that could be brought forth, and for now I shall simply assume that the objection is not decisive against the possibility of anything that could be called a reduction of arithmetic to logic.

Another objection is suggested by a remark of Russell's that a certain procedure of definition (which was Dedekind's, in fact) "does not enable us to know whether there are any sets of terms verifying Peano's axioms" (Russell 1919: 10). The objection is that a reduction of arithmetic to logic must guarantee that there is a model that is isomorphic to the natural number series,[h] but that anything which does guarantee the existence of such a model cannot count as a reduction to logic. A successful reduction must prove the existence of a progression;[i] but no reduction to logic can do this.

We noted, in connection with tiny set theory, the falsity of the first claim made in this argument. There are perfectly successful reductions of elementary number theory to set theories which do not prove the existence of any infinite set. That is, these theories do not entail any sentence that asserts that there is a set that contains infinitely many members. The denial of the existence of a progression is perfectly compatible with tiny set theory.[j] That any reduction of arithmetic must guarantee that there are progressions is false if it is an assertion, and entirely unmotivated if it is intended as a requirement upon the notion of a reduction.

It may be replied that a reduction must guarantee the existence of at least two objects, however, and then, again, there can be no reduction to logic, for that there are at least two objects, though true, is not logically true, and thus cannot be guaranteed by any system of logic.

Two things should be said in response to this modified objection. The first is that, although it may be agreed that no system of logic ought to certify '$\exists x \exists y(x \neq y)$' as a logical truth, it seems to be only a guess, though indeed a plausible seeming one, that a reduction must guarantee the existence of at least two objects: there are associations—not to beg the question by calling them 'reductions'—that assign to each truth of arithmetic (and only those) a statement compatible with '$\exists x \exists y(x \neq y)$.'

---

[h] By a 'model' here, Boolos means what one would normally mean in model theory: a set of a certain complex sort, including, among other things, a set that is the domain of the model. So, in particular, any model of PA will have to have, as one of its components, an infinite set that figures as domain. So, if the claim under discussion is correct, then any reduction of PA to logic will have to guarantee the existence of an infinite set.

[i] By a 'progression,' Boolos means an ordered set that is isomorphic to ω, that is, an ω-sequence.

[j] Indeed, tiny set theory is consistent with the negation of the axiom of infinity as, for that matter, is hereditarily finite set theory.

Whether these are reductions is best discussed after they are presented. The second is that a reduction to logic that does not guarantee that there is more than one object may well fail to provide certain kinds of important security for arithmetic, but nevertheless count as a reduction since it provides enough other kinds.

We shall discuss one last, and more powerful, argument against the possibility of a reduction of arithmetic to logic before presenting the "association" that would seem to have the best claim to be a reduction.

That arithmetic cannot be reduced to logic might be thought to be demonstrated in this way: If there is a reduction, then there is an effective procedure (which need not be assumed to be single-valued) which associates at least one logical truth with any arithmetical truth and associates no logical truth with any arithmetical falsehood. But if there is such a procedure, then there is an effective procedure for deciding whether any statement of arithmetic is a truth or a falsehood. To decide whether S is a truth, apply the procedure to both S and $\neg S$. (Exactly one of them will be an arithmetical truth.) *Ex hypothesi*, at least one statement will be associated with each of S and $\neg S$; of the associated statements, at least one will be a logical truth; and any logical truths will be associated with whichever of S and $\neg S$ is the arithmetical truth. Apply any sound and complete effective method for identifying logical truths to each of the sentences that is associated with either S or $\neg S$. (One may think of the procedure as successively generating statements associated with S and $\neg S$, and of the method being applied to each of the associated statements as they appear.) *Ex hypothesi*, again, at least one of these will eventually be identified by the method as a logical truth, and whichever of S and $\neg S$ the procedure associates with it is the logical truth. But as there is no effective method for deciding whether or not a given statement of arithmetic is a truth, there is no association procedure of the sort described, and consequently there is no reduction of arithmetic to logic, either.

It might be suggested that this argument only shows that there can be no *effective* reduction of arithmetic to logic. Non-effective functions abound which associate logical truths with arithmetical truths and associate logical falsehoods with arithmetical falsehoods. The interest of these would seem to be limited, since it is difficult to imagine how any non-effective association of logical truths with arithmetical truths could provide, or constitute, anything that could reasonably be called a reduction of arithmetic to logic. The reduced form of any arithmetical statement is supposed to be the truth of logic which the arithmetical statement, if true, "really is," or "is in disguise," or whatever. And it is sufficiently hard to see how any non-effective association could always yield the logical truth that any given arithmetical statement "really was" that we can feel justified in ignoring non-effective reductions: If arithmetic is logic disguised, the disguise must be effectively strippable. It is probably a serious mistake to say that a requirement of effectiveness is "part of the concept" of a reduction, but the attempted reductions of arithmetic to logic or set theory have all been effective associations, as is the one with which we shall mainly be concerned. Perhaps we might just say that if we are provided with something that is clearly a reduction of arithmetic to logic and demonstrably non-effective, we shall be no less surprised than we would be if provided with a non-recursive function that was clearly effectively calculable.

The argument convinces me, at any rate, that there is no reduction of arithmetical truth to logical truth, where the logic in "logical" is understood to be elementary, or first-order, logic, or indeed, any system of logic (such as axiomatic second-order logic,[k] or any of the familiar modal logics) whose theses form an effectively generable set.[8] (Godel's theorems of 1931 and the improvements and related results that followed soon afterwards meant the death of more philosophies of mathematics than just Hilbert's formalism.) The possibility of a significant reduction of arithmetic to something that might be called a system of logic is left open by our argument, however. What is excluded is that there be an effective method which identifies all and only the theses of the system: a proof procedure. It may seem that it is essential to a system of logic that it have a proof procedure, and that logics without proof-procedures are so called only laughingly or by courtesy. I think that this essentialist claim is wrong, however, and that at least two of modal logic, infinitary logic, second-order logic, and ω-logic deserve their names, in view of strong resemblances to classical elementary logic, which is entitled to the name 'logic' if anything is. One of them, second-order logic, does not seem to me to have been correctly appreciated. It is sometimes said to be set theory (in sheep's clothing, according to Quine) and sometimes said to be first-order logic (in elevator shoes). I think that a fair judgement of the accomplishments of Frege and Dedekind is impossible without a correct understanding of second-order logic, for it does seem to me that their work can be said, with some justice, to have effected a reduction of arithmetic to logic. "Logic" here means standard (or "full") classical, non-axiomatic second-order logic, which lacks a complete proof procedure. (Boy, does it lack it.) And I think that the effects of the disparagement, by Quine and others, of the term "second-order logic" might possibly be weakened by attending to the definition of validity in second-order logic (and in infinitary logic, too), for it is there that the analogy with elementary logic is sharpest.

3. A Short Course in Second-Order Logic

Syntax. In addition to the usual connectives and quantifiers, we have two distinctions of signs: into constants and variables, and into function letters and predicates of various degrees. 'Individual' means 'function letter of degree 0'; 'sentence,' 'predicate letter of degree 0.'[l] Rules of formation, and rules of freedom and bondage, are what you would expect. Constants never occur in quantifiers. All variables may. Two signs are of the same type if they are both predicate letters of the same degree or function letters of the same degree. A first-order formula is one containing no non-individual variable; the others are second-order.

---

[k] What Boolos means is the logical system whose validities are those of any standard axiomatic presentation of second-order logic, as opposed to the logical system whose validities are defined in terms of truth in all standard interpretations of second-order logic. The validities in the former case are effectively generable, but not in the latter case.

[l] This last quoted phrase is crossed out in the text, but nothing replaces it.

[8] A generalization of the argument shows that there is no reduction of arithmetical truth to truth in any system of logic whose theses form a set of arithmetical degree of unsolvability.

(Standard) Semantics. An interpretation (structure) is exactly the same sort of object it is in first-order logic: an ordered pair consisting of a nonempty set D, and a function which assigns to each constant β of degree n a function from D^n into a certain set S; S is D if β is a function constant and S is $\{\top, \bot\}$ if β is a predicate constant. (A function from D^0 into S is, as usual, just a member of S.) Two interpretations $\mathfrak{I}$ and $\mathfrak{I}'$ are β-variants if they are the same or differ only in what they assign to β. In explaining when a sentence σ is true in an interpretation $\mathfrak{I}$, we need only consider the case in which the sentence begins with a quantifier; the other cases are familiar. But so are the quantifier cases: if α is a variable, ϕ a formula, σ a sentence, and β a constant of the same type as α that does not occur in ϕ, then if $\sigma = \ulcorner \exists \alpha \phi \urcorner$ (resp., $\ulcorner \forall \alpha \phi \urcorner$), σ is true in $\mathfrak{I}$ iff $\phi_{\alpha/\beta}$ is true in some (every) β-variant of $\mathfrak{I}$.[9] Notice that in our account of the conditions under which a sentence of second-order logic is true in an interpretation, we have had to change neither the definition of an interpretation nor, really, the explanation of when a quantified sentence is true. Our account merely extends in the obvious way the one given in Benson Mates's *Elementary Logic* (1972, ch. 4) to cover the new sorts of sentences that arise in second-order logic. We can get first-order logic back from second- by permitting only individual variables to occur in sentences, and our account then becomes merely a summary rewording of the one given in Mates's book.

In first-order logic, the definition of truth in an interpretation $\mathfrak{I}$ is usually given by means of the recursively defined auxiliary notion of satisfaction in $\mathfrak{I}$ by a sequence of elements of the domain of $\mathfrak{I}$. Mates and others have noticed that a recursive definition of truth in $\mathfrak{I}$ (for variable $\mathfrak{I}$) can be given which does not mention satisfaction: the trick, if it is one, is in the assumption that interpretations may assign denotations to individual constants, which are of the same type as individual variables. Interpretations will, of course, typically assign denotations to some predicate letters, thereby enabling us to extend Mates's definition of truth of a first-order sentence in an interpretation in a perfectly straightforward way to second-order sentences containing predicate variables. Notice that we have not needed to introduce a separate non-empty domain for each sort of non-individual variable in order to give a satisfactory, "objectual" definition of truth in $\mathfrak{I}$ which agrees with the usual one that mentions satisfaction. It seems to me, therefore, that there is at least one sense in which second-order logic need not be regarded as "quantifying over sets" (or relations or functions). We do not need to assume that an interpretation associates a domain or a range with each kind of second-order variable in order to state when a second-order sentence is true in an interpretation.

Our account also provides a uniform explanation of quantification: an existentially (universally) quantified sentence is true in an interpretation if a certain other sentence is true in some (every) other (variant) interpretations. We do not need a separate clause in the account for each sort of variable. One statement can be made, without artificiality, to work for all.

The definitions of validity and consequence are what they always are: A sentence S is valid if it is true in all of its interpretations, and a sentence S is a consequence of a

---

[9] $\phi_{\alpha/\beta}$ is the result of substituting β for all free occurrences of α in ϕ.

set of sentences Γ if S is true in all of its interpretations under which all members of Γ are true. Mates lists nineteen laws of validity in *Elementary Logic* (1972: 65–6). Number 17 is the compactness theorem, which fails for second-order logic. The other eighteen hold for second-order logic if one makes the appropriate slight changes where necessary. And, of course, a first-order sentence is valid or a consequence of a set of first-order sentences according to the definition we have just given if and only if it is valid or a consequence of that set according to the usual one. There is a standard account of validity for first-order sentences, and there is an obvious, straightforward, non-*ad hoc* way of extending that account to second-order sentences.

Because every sentence has at least one interpretation, there are not two valid sentences each of which is the negation of the other. Any (axiomatic) system of second-order logic each of whose axioms is valid, and each of whose rules is validity-preserving, is therefore consistent.

Axiomatics. Axiom systems for second-order logic that are formulated in accordance with contemporary standards of rigor can be found in the books by Church (1956: ch. V) and Robbin (1969: ch. 6). These systems are sound but not complete with respect to the standard semantics. A simple syntactic consistency proof can be given for them (and indeed for axiomatic systems for the theory of types), as Gentzen showed.[10] All instances of the so-called "comprehension" axiom schema

$$\forall \alpha \exists X_n \forall x \left(X_n x \equiv \phi(x, \alpha) \right)$$

are theorems. As Quine and others have noted, the presence of the comprehension schema and the availability of the rule of substitution come to the same thing: via the rule of substitution we can prove each instance of the schema, and via the schema, we can prove anything we could prove with the help of the rule without its help. To repeat: these systems are demonstrably consistent.

Axiomatic second-order logic was first formulated by Frege in the *Begriffsschrift* (Frege 1967). Throughout §II of that book, a presentation of an axiomatic system of first-order logic with identity, Frege avails himself of a rule of substitution (which he never explicitly formulates), and he continues to do so in §III, where bound (unary) predicate variables are permitted to occur in formulas. In axiomatic formulations of first-order logic, substitution is not commonly taken as a primitive rule of inference; it can be shown to be a derived rule if the axioms of the system are given by means of schemata, which they usually are. The axioms of the *Begriffsschrift* are not schemata, but sentences, and Frege is obliged to use substitution in §II where he might have avoided its use by presenting his axioms schematically. §III, however, contains theses (such as (83)) whose derivations make essential use of substitution: its use is avoidable only at the cost of assuming some form of the comprehension schema. (Its use is inessential in the derivation, for example, of (98), the transitivity of the ancestral.)[m]

---

[m] This point is discussed in more detail in Boolos's later paper "Reading the *Begriffsschrift*" (Boolos 1998e: 159).

[10] The underlying idea is to suppose that exactly one individual exists; quantifiers can then be eliminated.

Although Frege did not formulate his rule of substitution explicitly and precisely, as van Heijenoort (1967: 3) notes, it is an easy matter to do this in a way which preserves Frege's derivations. Once this is done, the *Begriffsschrift* system can then be seen to be a presentation of axiomatic second-order logic which differs from current ones only in utilizing a rule of substitution instead of assuming a comprehension schema. (Frege never uses bound variables of degree greater than one, and perhaps a reconstruction of his system would distinguish it in this minor particular, as well.)[n] And, of course, Frege's *Begriffsschrift* system is also demonstrably consistent, once the rules of formation and substitution are formulated with the necessary precision.

Van Heijenoort (1967: 3) asserts that certain of Frege's derivations show him to be "on the brink of a paradox," the brink over which he had fallen by 1891. Van Heijenoort claims this because Frege substitutes $\mathfrak{F}$ for a in $f(a)$ in the derivation of (77), and substitutes it for f in the derivation of (91). I suppose that van Heijenoort thinks that Frege was on the brink of a paradox because his performing these two substitutions indicates that he would have allowed himself to perform them simultaneously, thereby obtaining $\mathfrak{F}(\mathfrak{F})$, a formula which figures in a well-known contradiction. If this is his reason for thinking Frege on the brink of an abyss, it seems to me excessively fearful. Nothing in *Begriffsschrift* suggests that Frege would have been willing to regard $\mathfrak{F}(\mathfrak{F})$ as well-formed or that the two substitutions Frege separately performed would have been performed simultaneously. Van Heijenoort (1967: 3) notes that Frege's derivations can be "unambiguously reconstructed." So can his (tacit) rules of formation. Examining (77) and (91) gives on no reason to think that '$\vdash \exists G \forall F(G(F) \equiv \neg F(F))$' was anywhere in the *Begriffsschrift's* offing.[o]

No sound axiomatic system of second-order logic will be complete with respect to the standard semantics for second-order logic described above. Since the theses of any axiom system will always form an effectively generable set, Leon Henkin has devised a kind of semantics with respect to which the usual axiom systems for second-order logic are complete: A formula is a thesis iff it holds in all Henkin models. The interest of Henkin models seems to me to be somewhat limited, however, for they are devised solely for the purpose of proving completeness.[11] That all instances hold in all Henkin models is something that is insured by the definition of a Henkin model: a "pre-structure" is an arbitrary system of non-empty domains and assignments of appropriate objects in those domains to constants; truth in a pre-structure is defined as one would expect. A Henkin model is then defined to be a pre-structure in which all instances of the comprehension schema hold. It is then

---

[n] As Boolos notes below, monadic second-order logic is decidable, so Frege's use just of monadic second-order variables in *Begriffsschrift* is perhaps not so minor a difference after all. (That said, Frege clearly intends to use relational variables in the *Foundations*, since these are needed for the definition of equinumerosity.) Maybe the most striking version of this sort of point, though, is that the monadic second-order theory of successor is decidable, too (Büchi 1962), whereas adding relational variables allows us to interpret second-order Peano arithmetic—which, indeed, is sometimes just formulated as the polyadic theory of successor.

[o] There is, in fact, a fair bit to say about this peculiar feature of Frege's system in *Begriffsschrift*. See my "Formal Arithmetic Before *Grundgesetze*" (Heck 2019: §3).

[11] That all instances of the comprehension schema are valid with respect to standard semantics can be proved (without any particular difficulty) by means of the Aussonderungs schema of set theory.

a small wonder that the usual axiom systems are complete with respect to Henkin models. No independent justification or essentially different justification is given for singling out those pre-structures in which comprehension holds, however.

Characterization. A sentence is sometimes said to characterize or express a property (or class) of interpretations if the interpretations in which the sentence is true are precisely those with the property (or in the class). Thus the sentence

$$\exists x \exists y (x \neq y) \tag{α}$$

characterizes the property of having a domain containing at least two members; '$\forall x(Ax \to Bx)$' can be said (loosely) to characterize the notion "subset of." It is easy to write down a first-order sentence which characterizes the notions of a linear ordering, of a one-one function, of the converse of a relation, and many other interesting notions. Certain interesting notions, however, such as well-ordering and enumerability, cannot be characterized by any first-order sentence. It sounds rather bizarre to say that one of the virtues of first-order logic is that it cannot characterize the notion of a well-ordering.[P] That no first-order sentence does so ought, it seems, to be regarded as something unfortunate about first-order logic, a failing and not an advantage. Although first-order logic's expressive capacity is occasionally surprising, second-order logic has it all over first-order logic as far as powers of characterization are concerned.

For example, if we conjoin the first two "Peano postulates," replace constants by variables, and existentially close, we obtain:

$$\exists z \exists S[\forall x(z \neq Sx \land \forall x \forall y(Sx = Sy \to x = y)] \tag{β}$$

a sentence that holds in an interpretation iff the domain is (Dedekind) infinite. And if we do the same for the induction postulate, we obtain

$$\exists z \exists S \forall F[Fz \land \forall x(Fx \to FSx) \to \forall x(Fx)] \tag{γ}$$

which is true iff the domain is countable. The notions of infinity and countability can be characterized by second-order sentences, though not by first-order sentences, as the compactness and upward Skolem–Löwenheim theorems show.

That there is a single second-order sentence $(\delta)^q$ that characterizes the class of interpretations isomorphic to the natural number series together with the operations of addition and multiplication is a discovery which is due to Dedekind (1902, Theorem 132). Frege, preoccupied with the task of finding specific references for

---

[P] That it cannot is a consequence of the compactness theorem. For suppose W_R is a sentence that is true in a model iff the extension of R is a set of ordered pairs that well-orders its field in that model. Let $\mathcal{M}$ be such a model in which the extension of R is infinite. Add infinitely many constants c_i and consider the set of sentences: $\{R(c_1,c_0),R(c_2,c_1),R(c_3,c_2),\dots\}$, etc. It is clear that every finite subset of this set is co-satisfiable with W_R. By compactness, then, the whole set is. But now, in any model of W_R plus that set, $\dots c_n R c_{n-1} R \dots R c_1 R c_0$ is an infinitely descending sequence.

[q] Boolos does not say exactly which sentence he has in mind, but the first footnote in "On Second-Order Logic," quoted above, says that the translation Boolos is about to describe is discussed in the chapter on second-order logic in *Computability and Logic*. Example 22.6 (of the latest, fifth, edition) states that the conjunction of the axioms of Robinson arithmetic and the second-order induction axiom is categorical. That conjunction, then, is presumably (δ).

the natural number words, never sought to characterize this class.[r] Frege did show how to characterize the notion of the ancestral, however, and this characterization can be seen as the heart of the hard part of Dedekind's accomplishment.

The existence of (δ) yields a reduction of arithmetic to second-order logic. Let us call the conditional whose antecedent is (δ) and whose consequent is σ 'the translation of σ,' σ^*. Then σ is a truth of arithmetic iff σ is true in all interpretations isomorphic to the natural number series, iff σ is true in all interpretations in which (δ) is true, iff $(\delta) \rightarrow \sigma$ is true in all its interpretations, iff σ^* is a (second-order) logical truth. Of course the translation of any sentence can be effectively found from that sentence.[12] Below we shall discuss whether this translation may be regarded as a reduction of arithmetic to (second-order) logic.

Metatheory. The completeness, compactness, and Skolem–Löwenheim theorems all fail for second-order logic, and their failure can be deduced quite directly from the existence of (δ) and related sentences. If a sound and complete proof-procedure existed for second-order logic, then arithmetic would be decidable. If $\exists_n$ is a sentence asserting the existence of at least n objects, then every finite subset, but not the whole, of $\{\neg(\delta), \exists_0, \exists_1, \exists_2, \ldots\}$ is satisfiable. The conjunction of (β) and the negation of (γ) holds in just those interpretations with uncountably infinite domains. Just about the only important metatheorem that does hold for both first- and second-order logic is the Craig Interpolation Lemma.

One response to these failures might be that they show that second-order logic is not properly so called. It might be said that since the central theoretical results about logic fail for second-order "logic," the usual designation is a misnomer and that "set theory" or "second-order something or other" would be preferable. There is a quick reply, viz., that these are central theoretical results about first-order logic. But this reply is too quick, for two reasons. The manner in which the completeness and Skolem–Löwenheim theorems fail reveals important likenesses between second-order validity and set theoretic truth. But let us first look at the failure of the compactness theorem.

There are infinite sets of statements which, one would have thought, were inconsistent, even though each of their finite subsets are consistent. Here are four examples:

1. "Smith is an ancestor of Jones," "Smith is not a parent of Jones," "Smith is not a grandparent of Jones," "Smith is not a great-grandparent of Jones,"...
2. "Not: there are infinitely many stars," "There are at least two stars," "There are at least three stars,"...
3. "R is a well-ordering," "$a_1 R a_0$," "$a_2 R a_1$," "$a_3 R a_2$,"...
4. "x is a natural number," "$x \neq 0$," "$x \neq S0$,"... (Tarski)[s]

---

[r] This is now known to be false (Heck 2012: ch.7), though that certainly was not known at the time.

[s] Presumably, Boolos has in mind here some (in)famous remarks Tarski makes in "On the Concept of Logical Consequence" (Tarski 1958: 410–11).

[12] σ may be a second- or higher-order sentence of arithmetic; nothing in the proof turns on σ's being first-order. The argument we gave earlier shows that no first-order sentence can serve in place of (δ); (first-order) arithmetic would then be decidable. (δ) is a conjunction of several sentences, only one of which, the induction principle '$\forall F[F0 \wedge \forall x(Fx \rightarrow FSx) \rightarrow \forall x(Fx)]$,' is a second-order sentence.

These sets of statements are inconsistent:

5. "Not: there are at least three stars," "Not: There are no stars," "Not: There is exactly one star," "Not: there are exactly two stars"
6. "R is a linear ordering," "a_0Ra_1," "a_1Ra_2," "$\neg a_0Ra_2$"

There is a translation into the notation of first-order logic, under which the sets of statements in the second group are formally inconsistent; moreover, the translation, together with an explanation of when the translated sentences are true in an interpretation, is a most important part of an explanation of the inconsistency of their originals. One would have hoped that the same sort of thing might be possible for the sets in the first group. It seems impossible, on reflection, that all of the statements in any one set should be true; it also seems that the reasons for this impossibility are of the sort which it has always been the province of logic to adduce. That the logic that is taught in courses like 24.251 (Philo 312, Phil 140)[t] cannot be used to represent the inconsistency of our four sets of sentences shows not that these sets are consistent after all, but that not all (logical) inconsistencies are representable by means of that logic. One may suspect that the second-order account of these inconsistencies is not the "correct" one and that some sort of infinitary logic might more accurately reflect the logical form of the sentences in question, but one really cannot praise first-order logic for muffing these cases altogether. Mathematically delightful though the compactness theorem for first-order logic may be, compactness must fail for any logic whose consequence relation is supposed to be useful in explicating the "common concept of consequence" (Tarski 1958: 409).

Löwenheim showed that any satisfiable first-order sentence is true in some interpretation whose domain is countable. Skolem extended this theorem by showing that "first-order sentence" could be replaced by "set of first-order sentences," and today we know that any intepretation with an infinite domain is an elementary extension (resp., submodel) of some interpretation whose domain has any desired smaller (greater) infinite cardinality. Now every logic is going to be limited in a way reminiscent of the Skolem–Löwenheim theorem. For any sentence of the logic is either going to be satisfiable or not; if S is satisfiable, let $h(S)$ be the smallest of the cardinal numbers of interpretations that satisfy it; if not, let $h(S) = 0$. If we assume that the sentences of a system of logic form a set, then the sets of sentences do, too, and so an axiom of replacement guarantees that $\{h(S)\}$ has a least upper bound. This least upper bound is called the Löwenheim number of the logic. Sentences such as '$\exists x\exists y(x \neq y)$,' '$\exists x\exists y\exists z(x \neq y \neq z \neq x)$,' etc., show that the Löwenheim number of a logic must be at least $\aleph_0$. Skolem's extension of Löwenheim's theorem thus shows that the Löwenheim number of first-order logic is as low as possible. It is a simple matter to write down second-order sentences that characterize the notions: of power $\aleph_1$, of power $2^{\aleph_0}$, etc.

It might be claimed that although it is the proper business of logic to characterize such notions as ancestral, identity, and (possibly) progression, it need not have to

---

[t] These are, or were, the course numbers for the introductory logic courses at MIT, Princeton, and Harvard, respectively.

characterize the lower higher cardinalities. Even so, the best light in which the limitation imposed by the Skolem–Löwenheim theorem can be seen is one in which it appears only as a (neutral) feature of first-order logic, and not a reason for esteeming or deprecating systems that have it or lack it. And the compactness and Skolem–Löwenheim theorems show first-order logic to be weak in one respect in which, it seems, a logic ought not to be so: that of characterizing the states of affairs in which its sentences are true or false.

The failure of the completeness theorem is more serious. For although it may plausibly be maintained that the failure of the Skolem–Löwenheim or the compactness theorem for a logic need not be regarded as a bad thing, the failure of the completeness theorem must be, it would seem.

"Gödel proved that (first-order) logic was complete and that (formal) arithmetic was incomplete." Does arithmetic then have a property that logic lacks? Yes and no. In one sense of "complete," a formal system is said to be complete if all sentences that are true in a certain class of interpretations are provable in the system.[13] But what that class of interpretations is may vary from occasion to occasion. In the case of arithmetic, it is the class containing just the standard interpretation; in the case of logic, it is the class of all interpretations. So not all truths of arithmetic are provable; all logical truths are, by any complete (!) system of proof.

To say then that first-order logic is complete is to say that there exist procedures by means of which all valid sentences can be proved. The fact of completeness would be of little interest if none of these procedures were sound, that is, capable of proving only valid sentences. But sound and complete proof-procedures exist for first-order logic.

A proof procedure for a formal system need not be a procedure which supplies proofs, if a proof of a sentence is supposed to be a sequence of statements, each of which is either an axiom or a consequence of earlier sentences by rules of inference, the last of which is the sentence. A proof procedure need not be an axiomatic proof procedure. A proof of a sentence need only be an arrangement of finitely many symbols which verifies (in some way, depending on the procedure) that the sentence is valid. In the most common sorts of procedure the relation which holds between an arrangement of symbols and a sentence iff the arrangement counts as a proof of the sentence is effectively decidable, but a proof of a sentence will not always be a sequence of valid sentences, of which the last is the sentence in question.[u] It is a pleasant fact that there do exist sound, complete axiomatic proof procedures for first-order logic.

But the mere existence of proof procedures for first-order logic guarantees that the set of valid sentences of first-order logic is effectively generable (recursively enumerable): there is an effectively calculable function from the natural numbers onto the set of first-order validities. A set may be effectively generable without being effectively decidable; there may be an effective procedure for enumerating all its members

---

[u] An example of the sort of alternative proof-procedure Boolos has in mind would be the tree method, which is what is used in the book Boolos used in the introductory logic course at MIT (Jeffrey 1981).

[13] In another sense, irrelevant here, a theory is said to be complete if every sentence of its language is either provable or refutable. Formal arithmetic is incomplete in this sense, too, of course.

without there being any such procedure for deciding whether or not any given object is a member. And indeed, the set of valid first-order sentences is such a set. According to Church's theorem, first-order logic is undecidable. An argument, due to Kleene, shows that a set is decidable just in case both it and its complement (relative to some decidable set) are effectively generable. We can thus conclude that there is no effective enumeration of the set of all invalid sentences of first-order logic.

All of this is of course well known, as is the fact that fragments of first-order logic exist which are decidable. The propositional calculus is one; monadic logic (in which no two- or more-place predicate letters are permitted to occur in sentences) is another. That monadic logic remains decidable if the equals-sign is permitted is slightly less well known. But decidability vanishes if even a single two-place predicate letter is allowed to appear in sentences (or if two-place function symbols are allowed). Now '$\forall x \forall y [x = y \equiv \forall F(Fx \equiv Fy)]$' is valid. Identity is definable in second-order logic: The definition of '$x = y$' contains a predicate variable 'F,' but it is a one-place predicate variable. We are thus led to consider the class of monadic second-order sentences, in which only individual and one-place predicate constants and variables may occur, and to ask whether it is decidable. For if it were, the decidability of monadic first-order logic with or without identity would be an immediate consequence, and a significant line between monadic and dyadic logic, whether first- or second-order, would be established.

The answer was discovered by Skolem, and can be found in second part of his paper "Untersuchungen ber die Axiome der Klassenkalkuls" (Skolem 1970). The answer, by the way, is yes, second-order monadic logic is decidable.

Quine, discussing the contrast between classical first-order quantification theory and an extension of it containing branching formulas, writes:

[T]here is reason, and better reason, to feel that our ... conception of quantification ... is not capriciously narrow. On the contrary, it determines an integrated domain of logical theory with bold and significant boundaries, designate it as we may. One manifestation of these boundaries is the following. The logic of quantification in its unsupplemented form admits of complete proof procedures for validity.... (Quine 1970: 90)

The extension is then noted not to admit of complete proof procedures.

A remarkable concurrence of diverse definitions of logical truth ... suggested to us that the logic of quantification as classically bounded is a solid and significant unity. Our present reflections on branching quantification further confirm this impression. It is at the limits of the classical logic of quantification, then, that I would continue to draw the line between logic and mathematics. (Quine 1970: 91)

We shall critically discuss the concurrence of diverse definitions of logical truth below;[v] suffice it for the present to say that the concurrence vanishes when the consistency of sets of sentences is considered.

It seems to me somewhat arbitrary to fasten on the fact of the completeness of first-order logic to draw a line between logic and mathematics. We have seen, first, that

---

[v] That is, in the portion of the paper that would become "On Second-Order Logic" (see Boolos 1975: 523ff). It is omitted here.

monadic first-order logic differs from dyadic first-order logic on the score of decidability, which is every bit as significant a property as completeness; we have further seen that this difference persists into second-order logic; and we have also seen that there is an account of the conditions under which a first-order sentence is true in an interpretation which may be extended to second-order sentences in a perfectly obvious and straightforward way, without change in the notion of an interpretation or the definition of validity as truth in all interpretations. To be sure, second-order logic possesses no complete proof procedure, but nor does first-order logic possess a decision procedure. We need not wonder at the suggestion that a line ought to be drawn between logic and mathematics, nor at the provenance of this suggestion, to be surprised that the semi-effectiveness of the set of first-order logical truths could be thought to provide a reason for distinguishing logic from mathematics. Why completeness rather than decidability or interpretation? Of course there is a big difference between second- and first-order logic; there are many. There are also big differences among fragments of first-order logic; there is a big difference between second- and third-order logic; and there are big differences between second-order logic and set theory. My objection is that the completeness theorem is not a sufficient ground for the essentialist claim about logic that Quine wishes to make.

We ought not, I think, to lose the sense that the completeness theorem is a surprising theorem: a priori, it is no more to be expected that arithmetical truth should outstrip formalization than that logical truth should do so. It doesn't seem as if anything in the idea of a logical truth should guarantee that there be at least one method of proof by means of which all logical truths can be demonstrated, any more than anything in the idea of an arithmetical truth insures that there be a method for demonstrating all arithmetical truths.

Postscript

A few readers have expressed some puzzlement about a claim Boolos makes in the discussion of 'Semantics' in §3 of this chapter. I will try here to fill in some details. Some of this is based upon my memories of lectures on second-order logic that Boolos gave when he and I co-taught a seminar on Frege in Spring 1993.

The claim in question is this one:

It seems to me...that there is at least one sense in which second-order logic need not be regarded as "quantifying over sets" (or relations or functions). We do not need to assume that an interpretation associates a domain or a range with each kind of second-order variable in order to state when a second-order sentence is true in an interpretation. (p. 104)

The important thing to appreciate is that Boolos is *not* saying here, as he famously would in later work (e.g., Boolos 1984), that the definition of truth in an interpretation of a second-order formula does not involve the notion of set *at all*. Rather, what he is saying here is it does not involve the notion of set *any more* than it does in the case of first-order logic. At this time, he would not have denied that the definition of interpretation in the case of first-order logic does involve the notion of set. So the later view really was a later view.

What's in an interpretation in the case of a first-order formula? Well, it's as follows:

- The domain is specified as some set D^w
- Each individual constant (or free variable) is assigned an element of D
- Each n-place predicate-symbol is assigned a set of n-tuples of elements of D, that is, a subset of D^n
- Each n-place function-symbol is assigned an function from n-tuples of elements of D into D, that is, a subset of D^{n+1} meeting existence and uniqueness conditions

Boolos's first claim is that the *exact* same thing counts as an interpretation of a second-order formula. Of course, in that case, we need to extend the clause for predicate- and function-letters also to cover free second-order variables. But that is the point: The values for second-order variables are just the same sorts of things as the values for predicate- and function-symbols in first-order logic (and we can permit those in the case of second-order logic, too, if we wish to distinguish predicate constants from predicate variables).

Actually, however, Boolos remarks, with a nod to Benson Mates, that we do not need to make use of free variables in defining truth, and therefore do not need to make use of assignments to such variables, either. Mates's semantics makes use instead of what I once called "auxiliary names" (Heck 2012: 58ff.). We add to whatever language we are considering an infinite stock of new constants, $c_0, c_1, \ldots$, whose interpretation is, as usual, an element of the domain. We then define truth for sentences of the original language in terms of truth in the expanded language. More precisely, $\forall v_i \, \Phi(v_i)$ is to be true in some interpretation J iff $\Phi(c_i)$ is true in all interpretations that are just like J except in what value they assign to c_i.[x] What Boolos observes is that this definition is entirely independent of what type of variable v_i is, in particular, whether it is first- or second-order. So Mates's semantics extends smoothly to the second-order case.

Boolos's conclusion, then, is that there is no significant difference between first- and second-order logic concerning what we need to say about the 'domain' of second-order expressions. In both cases, we need to say what the possible interpretations of such expressions are: Subsets of the (first-order) domain. Moreover, we quantify over such interpretations already in the first-order case, and we do not need any other resources to define truth in the second-order case.

There is, in fact, a difference that Boolos does not note. The definition of *truth* for first-order formulas, even done Mates's way, need only quantify over what we might call 'first-order variants' of a given interpretation: ones that differ in what they assign to *individual* (auxiliary) constants. It is only when we come to define *validity* for first-order formulas that we need to consider 'second-order variants.' But it's hard to see that this difference makes much of a difference, in particular, that it makes any difference to what sorts of set-theoretic assumptions we are making in the two cases (which is what was bothering Quine). Suppose there were some formula of first-order logic such that the only counter-model for it involved certain incredibly complex sets. In the worst possible case, one might imagine that we were only able to prove that the counter-model existed (say, using choice), but that we could not actually define the model concretely.[y] Then the 'standard' view would surely be that the formula in question is not valid. Every set is a possible domain for an interpretation, and every subset of the domain is

---

[w] To a significant extent, it was concerns about *this* part of the definition that led Boolos to his later views. In the case of the language of set theory, the intended interpretation is one in which the quantifiers range over all sets. But there is no set of all sets, so it looks as if the intended interpretation is not an interpretation.

[x] Interestingly enough, this treatment of quantification is what Gareth Evans (1985: 83ff.) called a "Fregean" one in his paper "Pronouns, Quantifiers, and Relative Clauses." He even cites Mates as an antecedent in note 5 of that paper.

[y] Think, for example, of the kinds of forcing constructions used to prove independence results in set theory. These sometimes make use of very powerful set theoretic techniques.

a possible interpretation of a predicate letter. If so, however, then it seems that our ordinary understanding of first-order validity quantifies over the 'full power set' of the first-order domain every bit as much as does Boolos's preferred understanding of second-order truth.

As it happens, we can show (via a careful proof of the arithmetized completeness theorem) that, if a first-order formula is invalid, then it has a counter-model of restricted logical complexity, namely, Δ_2: That is, the domain is either $\mathbb{N}$ or some finite subset thereof and all predicate-letters have values that are given by Δ_2 formulas. But surely no one is going to *define* validity in such terms, or even in terms of countable models. Validity is *defined* in terms of all possible subsets of D^n.

So, in short: The notion of an arbitrary subset of the (first-order) domain is already implicated in the usual definition of validity for first-order logic. The definition of truth for second-order logic needs no more; nor does the definition of validity for second-order logic.

References

Benacerraf, P. (1960). "Logicism: Some Considerations." PhD thesis, Princeton University.

Benacerraf, P. (1981). "Frege: The Last Logicist," *Midwest Studies in Philosophy* 6, 17–35.

Benacerraf, P. (1995). "Frege: The Last Logicist," in W. Demopoulos (ed.), *Frege's Philosophy of Mathematics*, Cambridge, MA: Harvard University Press, 41–67.

Boolos, G. (1975). "On Second-Order Logic," *Journal of Philosophy* 72: 509–27.

Boolos, G. (1984). "To Be Is to Be a Value of a Variable (or Some Values of Some Variables)," *Journal of Philosophy* 81, 430–49. Reprinted in R. Jeffrey (ed.), *Logic, Logic, and Logic*, Cambridge, MA: Harvard University Press, ch. 4.

Boolos, G. (1998a). "The Advantages of Honest Toil over Theft," in R. Jeffrey (ed.), *Logic, Logic, and Logic*, Cambridge, MA: Harvard University Press, 255–74.

Boolos, G. (1998b). "The Consistency of Frege's *Foundations of Arithmetic*," in R. Jeffrey (ed.), *Logic, Logic, and Logic*, Cambridge, MA: Harvard University Press, 183–202.

Boolos, G. (1998c). *Logic, Logic, and Logic*, ed. R. Jeffrey, Cambridge, MA: Harvard University Press.

Boolos, G. (1998d). "On Second-Order Logic," in R. Jeffrey (ed.), *Logic, Logic, and Logic*, Cambridge, MA: Harvard University Press, 37–53.

Boolos, G. (1998e). "Reading the *Begriffsschrift*," in R. Jeffrey (ed.), *Logic, Logic, and Logic*, Cambridge, MA: Harvard University Press, 155–70.

Boolos, G. (1998f). "The Standard of Equality of Numbers," in R. Jeffrey (ed.), *Logic, Logic, and Logic*, Cambridge, MA: Harvard University Press, 202–19.

Boolos, G. S., and R. C. Jeffrey (1974). *Computability and Logic*, first edition, New York: Cambridge University Press.

Büchi, J. R. (1962). "On a Decision Method in Restricted Second Order Arithmetic," in E. Nagel et al. (eds.), *Logic, Methodology, and Philosophy of Science: Proceedings of the 1960 International Congress*, Stanford, CA: Stanford University Press, 1–11.

Church, A. (1956). *Introduction to Mathematical Logic*, revised edition, volume 1, Princeton: Princeton University Press.

Dedekind, R. (1902). "The Nature and Meaning of Numbers," in *Essays on the Theory of Numbers*, W. W. Beman (trans.), Chicago: The Open Court Publishing Company, 31–115.

Evans, G. (1985). "Pronouns, Quantifiers, and Relative Clauses (I)," in *Collected Papers*, Oxford: Clarendon Press, 76–152.

Frege, G. (1967). "Begriffsschrift: A Formula Language Modeled upon that of Arithmetic, for Pure Thought," in J. van Heijenoort (ed.), *From Frege to Gödel: A Sourcebook in Mathematical Logic 1879–1931*, S. Bauer-Mengelberg (trans.), Cambridge, MA: Harvard University Press, 5–82.

Frege, G. (2013). *The Basic Laws of Arithmetic*, P. A. Ebert and M. Rossberg (trans.), Oxford: Oxford University Press.

Heck, R. K. (2012). *Reading Frege's Grundgesetze*, Oxford: Clarendon Press. Originally published under the name "Richard G. Heck, Jr."

Heck, R. K. (2019). "Formal Arithmetic before *Grundgesetze*," in P. Ebert and M. Rossberg (eds.), *Essays on Frege's* Basic Laws of Arithmetic, Oxford: Oxford University Press, 497–537.

Jeffrey, R. (1981). *Formal Logic: Its Scope and Limits*, second edition, New York: McGraw-Hill.

Mates, B. (1972). *Elementary Logic*, second edition, Oxford: Oxford University Press.

Quine, W. V. O. (1970). *Philosophy of Logic*. Englewood Cliffs, NJ: Prentice Hall.

Robbin, J. (1969). *Mathematical Logic: A First Course*, New York: W. A. Benjamin.

Russell, B. (1919). *Introduction to Mathematical Philosophy*, London: George Allen and Unwin.

Skolem, T. (1970). "Untersuchungen über die Axiome des Klassenkalkuls und über Produktations- und Summationsproblem, welche gewisse Klassen von Aussagen betreffen," in J. E. Fenstad (ed.), *Selected Works in Logic*, Oslo: Universitetsforlaget, 67–101.

Tarski, A. (1958). "On the Concept of Logical Consequence," in J. Corcoran (ed.), *Logic, Semantics, and Metamathematics*, Indianapolis: Hackett, 409–20.

van Heijenoort, J. (ed.) (1967). *From Frege to Gödel: A Sourcebook in Mathematical Logic 1879–1931*, Cambridge, MA: Harvard University Press.

5

Solving the Caesar Problem—with Metaphysics

Gideon Rosen and Stephen Yablo

1. Introduction

In the course of defining numbers Frege memorably digresses to discuss the definition of items he calls *directions* (Frege 1997: 110ff.). His first thought (§64) is that directions are things that lines a and b share if and only if they are parallel:

(DE) $\mathrm{dir}(a) = \mathrm{dir}(b)$ iff $a \parallel b$.

But then a worry occurs to him (§66). The goal was to define a certain kind of *object*, a kind including, for instance, the direction of the Earth's axis. And on reflection it is not clear that DE does that.

Certainly DE imposes a strong constraint on the direction-of *function*: it must associate parallel lines with the same object and non-parallel lines with distinct objects. And it constrains the *number* of directions too: there must be as many as there are non-parallel lines. But it is hard to see how DE constrains the directions themselves. As far as DE is concerned, directions might be almost anything. The direction of the Earth's axis, to take Frege's example, might be England. "[O]f course," Frege says, "no one is going to confuse England with the direction of the Earth's axis" (§66). But the lack of confusion here is "no thanks to our definition."

This sounds like a psychological point: if anyone were *inclined* to confuse directions with countries, DE would not get in the way. But the psychological point has a logical basis. Frege's real concern is that it would not *be* a confusion to identify England with the direction of the Earth's axis if there were no more to directions than is set out in the proposed definition.

What does it mean to say that "there is no more to directions" than is set out in the definition? That depends on how one thinks of definitions. The usual view is that definitions are *verbal* in nature. They convey verbal understanding by conveying what a word means, otherwise known as its sense or the concept it expresses. Explicit definitions do this by aligning the word with a phrase assumed to be already understood. Implicit definitions do it by stipulating "the truth of a certain sentence...embedding the definiendum and composed of otherwise previously understood vocabulary" (Hale and Wright 2000: 286).

Gideon Rosen and Stephen Yablo, *Solving the Caesar Problem—with Metaphysics* In: *Logic, Language, and Mathematics: Themes from the Philosophy of Crispin Wright*. Edited by: Alexander Miller, Oxford University Press (2020).
© Gideon Rosen and Stephen Yablo.
DOI: 10.1093/oso/9780199278343.003.0005

On this view, to say that "there is no more to directions than is set out in DE" is to say that DE exhausts the *concept* of direction (= the sense of "direction"). Frege's point would be that DE does *not* exhaust the concept, since DE allows us to identify the direction of the Earth's axis with England, whereas the full concept of direction does not allow that.

How might one attempt to answer Frege on this interpretation of his complaint? It would have to be shown that DE delivers a richer concept of direction than he supposes. Its overt content may not be up to the task, but perhaps it has latent content that he is overlooking. Anyone who thinks that it would be a misunderstanding of the proposed definition to suppose that England is (or might be) the direction of the Earth's axis must think that the definition imposes a constraint on the interpretation of *dir* that goes beyond the requirement that DE be true. A solution to our problem (the Caesar problem) must therefore involve two ingredients: (a) an explicit formulation of this further constraint, and (b) a demonstration that any function that satisfies this further constraint cannot possibly have ordinary concrete objects[1] like countries or Roman emperors in its range.

One approach to this problem is to insist that when a new function symbol like *dir* is introduced by means of an abstraction principle like DE, the real constraint imposed by the definition is that this new symbol express a *concept* or a *sense* such that

The thought/content/proposition that dir (a) = dir (b) is identical to the thought/content/proposition that $a \parallel b$.

This approach takes on large burdens. It must first explain the relevant notion of a thought/content/proposition, and second explain why concepts satisfying this sort of constraint cannot have ordinary objects in their ranges. Perhaps these challenges can be met, but it seems clear that they have not been yet (Hale 1997, 2001; Potter and Smiley 2001, 2002; Yablo 2008).

Another proposal is that the latent content of the definition is that *dir* is to express a function, the items in whose range fall under a *sortal concept* for which parallelism constitutes a *criterion of identity* (Wright 1983; Hale 1988; Hale and Wright 2001; Fine 2002). Here again the challenge is first to explain these technical notions, and then to show that ordinary objects cannot fall under a sortal concept for which parallelism constitutes such a criterion. Maybe this approach can be made to work and maybe not (Rosen 1993; Rosen 2003; Hale and Wright 2003); hopes of a simple, straightforward solution along these lines are now faded, however, and so we propose to try something different.[2]

2. Understanding a Definition

Consider an ordinary explicit definition. Suppose for instance that we introduce the word "grue" by laying it down that

---

[1] Or for that matter, ordinary, independently identifiable abstract objects like the key of E flat minor or the Declaration of Independence.

[2] *Apparently* different, anyway. As we note below (n.6), our proposal may represent a version of this more orthodox approach.

(GR)   For all x, x is grue iff x is green and observed or blue and unobserved.

What does the competent recipient of such a stipulation learn about the word "grue"? At a minimum he learns that its extension coincides with the extension of the open sentence on the right-hand side; but of course a competent recipient will learn much more. He will learn, for instance, that the word has a certain *intension*: that an object is grue *in a world w* iff it is green and observed or blue and unobserved *in w*. So part of the latent content of GR is given overtly by GR+:

(GR+)   Necessarily, for all x, x is grue iff x is green and observed or blue and unobserved.

Suppose now that a one place predicate like "grue" stands for a property; then we can say that someone who receives the definition GR in the intended spirit learns the intension of the new word's associated property. Does he learn anything else about this property? It would seem that he does. Consider someone who receives the definition and then says,

I see that a thing counts as grue (in a world) whenever it is either green and observed or blue and unobserved. But I am still not sure *what it is* to be grue. Perhaps for a thing to be grue just is for it to be green and observed or blue and unobserved. But it might also be that for a thing to be grue is for it to be *known by God* to be green and observed or blue and unobserved. Or perhaps to be grue is to be green and observed *and such that* $e^{i\pi} + 1 = 0$ or blue and unobserved and such that $e^{i\pi} + 1 = 0$.

More bizarrely yet, consider someone who hears the definition and says

Ah, I see what you're driving at: For a thing to be grue is for it to be either green and observed and such that that $e^{i\pi} + 1 = 0$ or blue and unobserved and such that $e^{i\pi} + 1 = 0$.

Both of these characters have failed to understand the definition, even though both have interpreted "grue" so as to make both GR and and GR+ true. They have failed to appreciate that the latent content of the definition—the real constraint it imposes on the interpretation of the new word—is that

(GR++)   For a thing to be grue *just is* for it to be green and observed or blue and unobserved.

More generally, we claim that in many cases (though not in all), when a new predicate F is introduced by means of an ordinary explicit definition of the form

For all x, Fx iff $\phi(x)$,

the real constraint imposed on the interpretation of F is the hyperintensional constraint that

To be F is to be ϕ.[3]

---

[3] There are significant exceptions to this principle, such as mere "reference fixing" stipulations. The word "acid" might be introduced by the stipulation

A substance is an acid iff it tends to turn litmus paper red.

But this sort of stipulation does not even purport to tell us what it is to be an acid.

The new predicate is constrained to pick out a property with a certain definition or analyis given explicitly in the *definiens*, and is thus precluded from picking out other properties that may coincide with the intended referent in extension or intension. The bizarre misunderstandings listed above come from failing to appreciate the fact that the verbal definition has a latent content of this form.

Our hypothesis is that something analogous goes wrong if one hears the implicit definition of *dir* and imagines that England either does or might fall within the associated function's range. Just as hearing the explicit definition of "grue" teaches us everything there is to know about what it is for a thing to be grue,[4] hearing the implicit definition of *dir* teaches us everything there is to know about what it is for a thing to be the direction of a line. We further submit that learning this puts us in a position to know that ordinary objects like England cannot possibly qualify as directions.

3. Formalities

The basic idea might be implemented in various ways, but the simplest seems to be this. Each entity t has a real definition, DEF_t. The real definition of t is a collection of structured propositions involving t, which together say all of what there is to be said about what it is to be t (for real definitions, see Fine 1994a, 1994b). Corresponding to t we have two sentential operators, Δ_t and ∇_t. The first is truly prefixable of a formula ϕ (t) just if the proposition $\phi(t)$ expresses is a member of (or trivial logical consequence of—a qualification henceforth omitted) DEF_t. The second is truly prefixable of $\phi(t)$ just if the propositions in DEF_t are conjuncts of (or trivial logical consequences of—another qualification henceforth omitted) the proposition expressed by $\phi(t)$. So, for instance, if *grue* is the color-like feature attributed by the use of "grue," then

$$\Delta_{grue} \, GR$$

says that it is definitive of *grue* that a thing is grue iff it is green and observed or blue and unobserved, while

$$\nabla_{grue} \, GR$$

says that nothing beyond GR is definitive of *grue*. The conjunction of these two claims can (and so might as well) be written

$$\maltese_{grue} \, GR$$

—in words, GR is exhaustively definitive of what it is to be grue. The reason that there is no room for the thought that another (unmentioned) part of what it is to be grue is to be known to the divine intellect, or such that $e^{i\pi} + 1 = 0$, is that anyone who appreciates the standard definition comes to know, not just that GR is true of grue, but that GR is exhaustively definitive of grue, and that there is nothing about *God* or *e* or *i* or π anywhere in it.

---

[4] More carefully: someone who hears the definition and *knows what it is for a thing to be blue, green, observed, etc.,* is thereby in a position to know all there is to know about what it is for a thing to be grue.

As the "grue" example suggests, the idea is meant to apply not just to definitions of terms but definitions of all kinds of expressions—predicates, function symbols, even perhaps connectives. The latent content of an explicit definition of the form

$a = $ the a,
for all $x_1 \ldots x_n$, $Px_1 \ldots x_n \leftrightarrow \psi(x_1 \ldots x_n)$,
for all $x_1 \ldots x_n$, $f(x_1 \ldots x_m) = \phi(x_1 \ldots x_m)$,
for all X, Y, $X*Y \leftrightarrow \ldots X \ldots Y \ldots$

is given explicitly by the corresponding formula prefixed by $\maltese_a$ or $\maltese_P$ or $\maltese_f$ or $\maltese_*$. Definitions of this sort assign an object/property/function/operator with a determinate nature to a name/predicate/functor/connective by specifying what it is for a thing to be the object in question or the function in question or etc.

4. The Proposal

So far we have been talking about explicit definitions, but the proposal is that implicit definitions like the direction equivalence and (to come finally to Frege's main concern) Hume's Principle may likewise be seen as real definitions of the functions they introduce. Hume's Principle

(HP) $\#F = \#G \leftrightarrow F \approx G$[5]

is normally taken as a verbal definition, and of course it is a verbal definition in part. But its ambitions are higher. Just as GR, advanced in the right definitional spirit, tells us not only that "grue" stands for a property that makes GR true, but also that

(GR++) $\maltese_{grue} \forall x (x$ is grue iff x is green and observed or blue and unobserved),

so HP advanced in the same spirit tells us not only that "#" stands for a function that makes HP true, but also that

(HP++) $\maltese_\# (\#F = \#G \leftrightarrow F \approx G)$

If we understand the definition correctly, we come away knowing that "#" can only stand for a function whose real definition is exhausted by the fact that it satisfies Hume's Principle. Of course, one can't be sure, to begin with, that there are functions like this. But if there are—call them *essential numerators*—then there is no question of what their natures are, since their natures flow from their definitions and their definitions are settled. To put the point in epistemic terms, if a function is an essential numerator then anyone who knows that it is an essential numerator and knows what it is for two bunches of things to be equinumerous is thereby in a position to know all there is to know about the function's nature.

---

[5] The right hand side, $F \approx G$, is shorthand for the second-order formula that asserts the existence of a one–one correspondence between the Fs and the Gs.

Now it is one thing to define the function-symbol "#," another (and prima facie easier) thing to define the predicate "is a Number." Assuming we understand "#," the predicate can be defined explicitly:

(N)   For all x, x is a Number iff for some F, $x = \#F$.

If this definition is understood along the lines sketched above, then anyone who understands it comes to know that "Number" picks out a property satisfying the following condition:

(N++) $\diamondsuit_{\text{Number}}$(For all x, x is a Number iff for some F, $x = \#F$)

Putting this all together, we conclude that when the neo-Fregean's definitions are properly understood, their recipient (assuming she knows what it is for two bunches of things to be equinumerous) comes away knowing everything there is to know about what it is to be a number.[6]

5. Three Questions

Suppose we are right that the neo-Fregean has it open to him to offer his stipulations in the spirit indicated.[7] Three questions then arise. First, how can we convince ourselves that there is at least one essential numerator, and hence that the stipulations serve to assign a function to the new function word? Second, how can we convince ourselves that there is at most one essential numerator, and hence that the

---

[6] Any property whose definition has this form will be a *sortal* property in one sense of the term: To be an instance of the property is to be the value of a function, whose nature is given by reference to an equivalence relation on items of a more basic kind. Grasp of (the nature of) any such property will then require a capacity to understand claims of identity and difference among the items in question, when those items are given as values of the relevant function. This is probably not an adequate account of the general notion of a sortal concept. *Person*, for example, is supposed to be a paradigmatic sortal concept, and yet it is implausible that the real definition of *person* makes reference in this way to an equivalence relation on more basic items (*person stages?*). We do not possess any adequate general analysis of the notion of a sortal. But our proposal seems clearly to *entail* that the properties associated with neo-Fregean predicates—that is, predicates defined in the manner illustrated here for *Number*—will be sortal. Hence the suggestion (n. 2 above) that our proposal may amount to an unorthodox version of Hale and Wright's approach to the Caesar problem.

[7] It is worth stressing that stipulations of a neo-Fregean sort *need not* be offered in this ambitious spirit. A neo-Hilbertian might introduce a function symbol φ by laying down the axiom

$$\phi(a) = \phi(b) \text{ iff } Rab$$

with the express intention that *any* interpretation of the new symbol that satisfies the overt axiom is as good as any other. ("It must always be possible to substitute in all geometric statements the words *table, chair, beer mug* for *point, line, plane*.") The challenge for neo-Fregean Platonism as we see it is to describe *an* understanding of the latent content of abstraction principles that would suffice to introduce concepts of the sort we in fact possess: concepts that apply more or less determinately to abstract objects of a certain sort. It is no objection if these same stipulations might also have been used for a different purpose.

stipulations confer *determinate* reference on the new function word and on complex functional terms constructed from it? Third, how can we show that the items in the range of the essential numerator are abstract mathematical objects of the sort with which arithmetic is intuitively concerned, and in particular, that Julius Caesar is not among them?

The first of these challenges is basically to answer the skeptic who doubts that any ontological ground is gained by the neo-Fregean stipulations. This is obviously a serious challenge, but it has nothing to do with the Caesar problem, so we propose to ignore it.[8] The second challenge is to answer a character we call the *libertine*, who thinks that so far as the neo-Fregean definitions are concerned, numbers and directions could be almost anything because the stipulations are multiply satisfiable.[9] The third challenge is to answer a character we call the *pervert*, who grants that the stipulations are uniquely satisfied but insists that numbers and directions could still (in an epistemic sense) be almost anything. What does it gain us to know that there is such an entity as *the* number of Martian moons if that entity might, for all we know, be Julius Caesar?

6. Simple Perversity

Consider the perverse hypothesis that the number of Martian moons = Julius Caesar. We instantly reject the notion. Why? There may be many reasons, but the following seems intuitively most fundamental. When it is proposed that JC might be the number of Martian Moons, we ask: Why him? Why not Caligula? Why not the key of E-flat minor? There is something *absurd* in the suggestion that JC might be the number of Martian moons precisely because the suggestion is impossibly *arbitrary*. And we are inclined to conclude that because the hypothesis would be absurd in this way, it cannot be true.

Our implicit reasoning seems to be: If JC were the number of Martian moons—if JC were 2—then there would have to be some account of why *he*, rather than some other thing, is 2—and, if this is different, some account of why 2, rather than some

---

[8] A complete theory of these matters would tell us when in general a formula of the form $\natural_f[\ldots f \ldots]$ is satisfiable. A maximally generous position would suppose that whenever $\ldots f \ldots$ has the form of an explicit definition there is always automatically an item with the requisite nature. But whatever the merits of generosity, the "bad company" objection to neo-Fregean Platonism (Boolos 1990) shows that this cannot be maintained when $\ldots f \ldots$ is an abstraction principle. A solution to the bad company objection will identify a class of *kosher* abstraction principles (presumably including the direction equivalence and HP) whose truth may be freely stipulated for the purpose of introducing a new function symbol. Conjecture: On any suitable account of this form, $\natural_f[\ldots f \ldots]$ will be satisfiable whenever $\ldots f \ldots$ is a kosher abstraction.

[9] The most debauched libertine maintains that the values of the function introduced by means of abstraction principles can be *anything at all*—or more precisely, that the admissible interpretations of the new function symbol include every function that satisfies the formal constraint on the right-hand side, including those whose ranges include ordinary concrete objects. A more restrained sort of libertine maintains that the values can be anything at all so long as they are abstract, or so long as they exist necessarily (Rosen 2003). We ignore this distinction in what follows.

other number, is JC. But it seems perfectly clear that there can be no such account. And so we conclude that the perverse hypothesis cannot be true.

The principle we seem to be relying on here is that facts of the form $[\#F = a]$ cannot be brute facts. (We write $[P]$ for the fact that P, and suppose that facts are structured complexes involving objects, relations, functions and the like as real constituents.) It is because this principle would be violated if 2 were JC that we reject the hypothesis as absurd. So we have two premises and a conclusion.

(P1)   Facts of the form $[\#F=a]$ cannot be brute facts. When they obtain there must be an account of why they obtain.

(P2)   On the perverse hypothesis some facts of that form would be brute and unaccountable.

(C)   So, the perverse hypothesis is false.

Before we consider the status of the premises we should say a bit more about the sort of explanation or account we have in mind. Obviously enough, we are not looking for causal or historical explanations. Our guiding assumption is that some facts are grounded in others, or hold in virtue of others. The clearest examples of this are disjunctive facts, which typically hold in virtue of their true disjuncts, and existentially general facts that typically hold in virtue of their true instances. But there are other examples, some potentially controversial. Thus it might be said that facts involving determinable properties hold in virtue of the corresponding determinate facts—The ball is blue in virtue of being (say) cobalt blue; that facts involving definable properties hold in virtue of facts involving the real definitions of those properties—The figure is a square in virtue of being an equilateral right quadrilateral;[10] that certain thin moral facts hold in virtue of some more concrete *wrong-making* feature—The act was wrong in virtue of the fact that it was an unjustified violation of the victim's right to privacy; and so on.

(P1) asserts that facts of the form $[\#F = a]$ must be grounded in more fundamental facts in this sense. There must be some other facts about a and about F and about $\#$ in virtue of which such identities obtain when they do.

Of course there is no point in pining after this sort of explanation if it is not to be had. So it may be helpful to show how there *could be* an explanation for why $\#F = a$. Suppose that there exist *Frege numbers*. A Frege number is a thing with a distinctive sort of essence. For each Frege number n there is a cardinality quantifier, $\exists_n$, such that

$$\lozenge_n \text{ For all } F \ (n = \#F \leftrightarrow \exists_n x \, Fx).$$

So, for instance, the Frege number 2 is an item that is, by definition, the number that numbers the Fs iff there is an F, x, and another F, y, and every F is either x or y. Also, though, and just as important, *all* essential facts about the Frege number 2 are implicit in the one just mentioned. This means that it is not essential to this item that, say, there are or could be human beings.

---

[10] This assumes that the fact that s is a square is distinct from the fact that s is an equilateral right quadrilateral, and hence that the properties in question are distinct.

If there are Frege numbers, then facts of the form $[\#F = a]$ have a straightforward explanation. Why is it that the number of eggs in the bowl = 6? Because there are six eggs in the bowl and 6 is, by its very Frege-numberish nature, the number that belongs to a concept iff six things fall under it. Once again, we do not insist for present purposes that there are Frege numbers. The claim is rather that *if* there are such things, then facts of the form $[\#F = a]$ will be explicable.

This suggestion about the nature of #'s values illustrates a larger claim about (we borrow the term from Goodman) "generating functions."[11] A generating function g is a function whose value on a given input x is *essentially* a value of g for arguments like x, and whose value on x has no more to its essence than that. Take for instance the *class of* function C—the function that takes plural arguments X and yields as output the class (if any) to which all and only the Xs belong. Few would dispute that

$$\text{For all } X, \text{ for all } y, \quad y = C(X) \rightarrow \maltese_y \#y = C(X).$$

It strikes us as wholly definitive of a class that it be the class to which all and only *these* items belong. In more complex cases, however, this simple formula $(y = g(x) \rightarrow \maltese_y\, y = g(x))$ will not hold. It does not lie in the nature of a given Frege number to be the number that belongs to some *particular* concept. Rather it lies in the nature of such a number to be the number that belongs to *any* concept with a certain higher-order feature (corresponding to the cardinality quantifier).

How should the notion be defined in general? A generating function g is associated with a partition, $\phi_1, \ldots \phi_i \ldots$ of its domain, in such a way that for each value y of g, there is a cell ϕ_y in this partition such that it lies in the nature of y to be the value of g for any member of ϕ_y. The cardinality quantifiers partition the domain of concepts, and each Frege number is associated, by definition, with a particular cell in this partition. Similarly, the various *orientation properties*—properties shared in common by classes of parallel lines—partition the domain of lines. And it lies in the nature of (what might be called) *Frege directions* to be the directions that belong to any line that possesses a certain such property. (The simple case mentioned above, where the values of the function are definitionally linked to some particular input, is the special case in which the relevant partition is the maximally fine grained unit partition.) The proper definition, then, seems to be this:

$g: X \rightarrow Y$ is a generating function iff there exist properties $\phi_1 \ldots \phi_i \ldots$ such that
(a) Every item in X instantiates exactly one ϕ_i,
(b) For all x, y, if $\phi_i x$ & $\phi_i y$ then $g(x) = g(y)$
(c) For each y in Y there is a ϕ_i such that $\maltese_y$ (for all x ($y = g(x) \leftrightarrow \phi_i x$)).

The point that matters is that facts of the form $[g(x) = a]$ will be explicable whenever g is a generating function. In each such case we will be able to say: The reason that $g(x) = a$ is that x is ϕ, and g by nature maps all ϕs to the same thing, and a is defined to be *that very thing*, the thing to which g by nature maps all ϕs.

---

[11] Goodman (1956).

7. Polymorphous Perversity

Now it is clear that on the perverse hypothesis, facts of the form $[\#F = a]$ cannot be explained in the manner indicated. For it is characteristic of the perverse hypothesis to suppose that the values of abstraction functions are not definitively values of such functions, much less exhaustively definitively values of such functions. They are objects like Julius Caesar or the key of E-flat minor whose natures make no reference to the number-of function or equinumerosity. So on the perverse hypothesis, this style of explanation is cut off.

Another style of explanation is, however, conceivable. For the pervert might reply as follows: I agree that facts of the form $[\#F = a]$ cannot be brute facts, and that they cannot be explained by reference to the nature of the object a. But perhaps they can be explained by reference to the nature of the abstraction function, in this case $\#$. Perhaps the answer to the question, "Why is Caesar the number of Martian moons?" is just this: There are two Martian moons, and it lies in the nature of $\#$ that Caesar = $\#F$ iff there are two Fs.

How can we respond to this sort of pervert? By reminding her that the $\#$ function is *exhaustively* defined by HP and that HP makes no mention of Caesar or of anything that might implicate him. To put it epistemically, anyone who appreciates the definition and knows what equinumerosity and identity are is in a position to know all there is to know about the nature of $\#$. But someone who appreciates the definition and knows what equinumerosity and identity are is *not* thereby in a position to know that Julius Caesar is the number of Martian moons iff there are two martian moons. So, our objector notwithstanding, it is not definitive of $\#$ that Julius Caesar is $\#F$ iff $\exists_n x\ Fx$. The same goes for the key of E-flat minor and every other independently identifiable object, abstract or concrete. It would therefore seem that if the perverse hypothesis is true, facts of the form $[\#F = a]$ cannot be explained by appeal to the definition of a or by appeal to the definition of $\#$.

It is not out of the question that these facts might be explicable in some other way—either by appeal to the definition of some third thing or without appeal to definitions at all. We cannot say with confidence that every constitutive explanation—every truth of the form P *obtains in virtue of* Q—must be mediated either implicitly or explicitly by a claim about the definitions or essences of the items that figure in P. So we shall only say this. It is hard to think what such an explanation would look like in the present case. If this is right, then it is reasonable to accept (P2). Facts of the form $[\#F = a]$ would be brute if the perverse hypothesis were true.[12]

8. Brute Perversity

So the pervert must say the following: I cannot tell you why Caesar is the number two. I cannot tell you what it is about him that suits him to the role, or what it is about

---

[12] We leave aside the possibility of explanations that reduce functional facts to *relational* facts with a uniqueness clause. It may be that $[\#F = a]$ obtains in virtue of the fact that a is the unique item that bears the number-of *relation* to F. But then the question will arise why *this* fact obtains. And it seems no more satisfactory that the relational fact should be brute than that the corresponding functional fact should be brute.

the role that suits him to it, or what it is about the two together that suits them to each other. But why shouldn't facts of the form $[\#F = a]$ be brute facts? There is nothing wrong with the supposition that *some* facts are brute facts. Indeed it is natural (why?) to suppose that *all* facts are somehow determined by a foundational array of brute facts—facts that do not obtain in virtue of anything more fundamental.[13] Some of these brute facts will presumably be relational, for example, the fact that one thing is at such and such a distance from another. If relational facts can be fundamental, why shouldn't some *functional* facts be fundamental? And if functional facts can be fundamental, why shouldn't facts of the form $[\#F = a]$ be among them?

One response is to remind the pervert that her hypothesis strikes us as absurd, and that this reflects how natural we find it to think that there would have to be some explanation of why Caesar was the number two, if in fact he were. It is no answer to say that brute facts are not *intrinsically* objectionable. For they are not intrinsically unobjectionable either. We need to consider our reaction to the supposed brutality of this supposed fact in particular. And it seems highly relevant that we react to the suggestion with incredulity. One might rest the case for premise (P1) on this intuition alone.

Alternatively, one might make a broadly theoretical case. Suppose we are comparing two hypotheses: the perverse hypothesis, on the one hand, and the hypothesis that there exist Frege-numbers on the other. We note that the former must count facts of the form $[\#F =a]$ inexplicable, while the latter offers natural, intuitively compelling explanations for them. The latter package, then, exhibits greater explanatory coherence. And this, one might think, constitutes a reason to believe that it is true.

There is much to be said for responses like this; but they are not in the spirit of neo-Fregean Platonism. It is more in the spirit of the neo-Fregean view to seek a general principle that would tell us when facts of a certain kind cannot be brute facts.[14] We mention one such principle that seems to do the trick in the present context while conceding that it is not as luminously self-evident as one might wish.

A fact A is non-basic or derivative, we said, if it holds in virtue of another fact (or other facts). Non-basic facts can be explained by pointing to the facts in virtue of which they obtain; and if they can be so explained, they should be, meaning just that they should not be treated as inexplicable.

Now one great source, perhaps the main source, of non-basic facts is definitions; for facts about a *definiendum* will in general hold in virtue of facts about its *definiens*. The definition of grue, for example, in telling us that

(GR++) $\#_{grue}$ (for all x, x is grue iff x is green and observed or blue and unobserved)

seems also to tell us that facts of the form [. . . grue . . .] are always grounded in facts about blue, green and observation. The definition of the Tri-State Area, in telling us that

---

[13] Of course the facts in this foundational array need not be altogether inexplicable. They may admit of causal explanation, for example.

[14] For a discussion of principles of roughly this sort, though in a different context, see Hale and Wright (1992).

(TS++) $\style{}{✿}_{TS}$ (for all x, x is the Tri-State Area iff New York, New Jersey, and Connecticut are parts of x, and every part of x overlaps New York, New Jersey, or Connecticut.)

seems to be telling us too that facts about the Tri-State area invariably derive from facts about New York, New Jersey, and Connecticut. How, though, do GR++ and TS++ have these results? To answer this, we need a notion of a non-basic *thing* to put alongside our earlier notion of a non-basic fact.

A thing is non-basic if it is *reductively definable*, in the sense that its real definition has the form of an equation (typically a universally quantified biconditional or identity) in which the item in question is totally absent from the right-hand side. So for example, a reductively definable *relation* R will be one whose real definition takes the form

$$✿_R \, \forall x_1...x_n (R(x_1...x_n) \leftrightarrow \phi(x_1...x_n)),$$

where R is totally absent from ϕ.[15] A reductively definable *object* a might be one whose real definition takes the form

$$✿_a \, (\forall x)(x = a \leftrightarrow \phi(x)),$$

where a is totally absent from ϕ, etc.[16] Perhaps the principle we need is just this:

(NB)   If a is a non-basic entity, then facts involving a are non-basic facts.

In assessing this principle it is important to note that not all items that admit of real definition are non-basic. At one extreme, we have items a whose real definition consists in non-biconditional propositions involve a, or biconditional propositions with a on both sides:

$$DEF_a = \{ Ca, \sim Da, Fa \leftrightarrow Ga, Rab, \text{ etc.} \}$$

Here there is no suggestion that the facts about a should hold in virtue of a-free facts. By contrast, and at the other extreme, when the real definition takes the form of an *explicit definition*, as with GR++ and TS++ above, then there does seem to be the implication that facts involving the defined item should be explicable in terms of more basic facts.

HP lies between these two extremes. It is a *reductive* definition of #, but not an *explicit* definition of #. Brute facts about reductively definable items seem to us about as objectionable as brute facts about explicitly definable items. NB thus strikes us as plausible, and we know of no compelling counterexamples. But we concede that NB is not self-evident. A more searching account would look for a derivation of NB, or some other rule for determining which facts demand explanation. Our main point is that for reasons we may or may not have identified, facts of the form $[\#F = a]$ clearly seem to require explanation, and that such explanations are unlikely to be

---

[15] Absent in the strong sense that R figures neither in ϕ nor in the definition of any of the constituents of ϕ, nor in the definitions of any of *their* constituents, etc.

[16] We could also allow for the case in which the definition takes the form of an identity, for example, $✿_a \, a = f(b)$.

forthcoming on the perverse hypothesis. If this is right, then numbers might well be Frege-numbers. But they cannot be Roman emperors.

9. Uniqueness

So far we have been assuming that definitions like HP have at most one solution. Let us now ask whether this assumption can be justified.

Note that it would not be a disaster if the assumption were mistaken. For we have already seen reason to think that any suitable candidate for # must be a generating function, and this means that its values must be essentially among its values, and that there is not much more to their nature than that. This already rules out functions with "ordinary" objects as values, since on the one hand, ordinary objects are not essentially values of abstraction functions, and on the other hand, even if they were essentially values of such functions, that would fall far short of exhausting their essence.

Suppose we are right that the only admissible solutions to HP are essential numerators ν—functions whose real definitions are exhausted by HP—whose values are exhaustively definable as ν-applied-to-so-and-so-many-things. If there are many such essential numerators, then singular terms formed with # are to some extent indeterminate in reference. But they are not wildly indeterminate. They do not divide their reference over absolutely everything, or over every necessarily existing object, or over every necessarily existing abstract object. The most radical and implausible versions of libertinism are thus refuted. We are left, at worst, with the sort of moderate indeterminacy implied by Benacerraf's discussion of set theoretic reductions of arithmetic (Benacerraf 1965).[17] And on reflection it would be neither surprising nor disturbing if the language of arithmetic turned out to exhibit this sort of indeterminacy.

That said, it seems possible to argue that there is at most one solution in the offing. We know that any candidate for the referent of # must have a certain real definition: it must be an item such that everything there is to be said about its nature is in some sense a consequence of the fact that, by its nature, it satisfies HP. The only way there could be two such functions is for there to be two functions with precisely the same real definition. One way to "solve" the uniqueness problem is therefore to appeal to a general principle that functions, or perhaps items in general, are individuated by their real definitions.[18]

---

[17] In fact the indeterminacy that threatens here is significantly milder than the indeterminacy that concerned Benacerraf. In Benacerraf's case the trouble was that there are countless ways of identifying the numbers with items of a very different sort, namely sets: items whose natures to not involve the relation of equinumerosity. In the worst case scenario for the present view, the numerals would divide their reference over items whose essences involve an essential numerator, and hence the notion of equinumerosity. So in our case, while there may be many candidates for the referent of '2,' the candidates are all in some sense clearly *numbers*.

[18] This principle needs to be stated carefully. As we have defined the notion, the real definition DEF_a of an item a is a set of propositions involving a itself, and given this, it may be trivial that distinct items never have the same real definition. The principle we wants holds that when $a \neq b$, DEF_a is distinct from the result of substituting a for b and b for a in DEF_b.

Now this general principle may be too strong. Consider the positive and negative square roots of -1, i, and $-i$. It may be that the only thing to be said about the natures of these items is that each is defined by the condition $x^2 + 1 = 0$. And yet the theory requires that these two items be distinct. And so we should be open to the possibility that there might be two *objects* with same essence or real definition. But now consider the corresponding claim about (say) *properties*. Above we saw that

(GR++) $\maltese_{grue}$ (for all x, x is grue iff x is green and observed or blue and unobserved)

leaves no room for the thought that another part of what it is to be grue, for some reason unmentioned, is to be such that God exists, or such that $e^{i\pi} + 1 = 0$. But GR++ would also seem to leave no room for the thought that there are two grue-type features both of which satisfy the formula (and are therefore alike in real definition). Consider how bizarre it would be to say: "I know exactly what it is to be grue: it is to be green and observed or blue and unobserved. And I know exactly what it is to be *groo*: it is to satisfy exactly the same condition. And yet grue and groo are distinct properties."

This suggests the following principle: While objects (or perhaps substances) may be the exception, in every other case things are individuated by their real definitions. In particular, if f and g are functions whose real definitions run exactly the same way, they are one and the same function. If this is right, there can be at most one essential numerator, and hence at most one solution to Hume's Principle properly understood.

10. Recap

As noted earlier, any solution to the Caesar problem must provide two things: (a) an explicit account the latent content of HP and hence of the constraint HP imposes on the interpretation of the function symbol #, and (b) a demonstration that the values of the function(s) that satisfy this constraint do not include "ordinary" objects like Julius Caesar.

As to (a), we have suggested that when a function symbol is introduced by a neo-Fregean abstraction principle, the symbol is constrained to denote a function that is exhaustively defined by that principle. This rules out the vast majority of functions that simply satisfy the principle. It is one thing to be a function that (as a matter of fact, or even as a matter of necessity) takes parallel lines into the same object and non-parallel lines into distinct objects. It is another thing to be a function whose nature is exhausted by the fact that it has this feature. We might call any function with this sort of nature an *abstraction function*.

As to (b), we have suggested, first, that when f is an abstraction function, facts of the form $[f(b) = a]$ must admit of a certain sort of explanation—there must be some account of why $f(b)$ is a rather than some other thing—and, second, that such facts are only explicable if the abstraction function is also a *generating function*: roughly, a function whose values are essentially among its values. In one good (though perhaps somewhat narrow) sense of the word, an *abstract object* is an object that is, by its very nature, the value of a certain abstraction function for a certain range of arguments. So if our principles are sound, it follows that the objects "introduced" by a neo-Fregean abstraction principle must be abstract objects in this sense. And this means that ordinary objects like Julius Caesar are excluded, since it is perfectly clear that such things are not abstract objects in this sense.[19]

Finally we have argued that given a plausible principle concerning the individuation of functions—no two functions with the same real definition—there is *at most one* function that satisfies the real content of any given neo-Fregean abstraction principle. And this means that functional terms of the form #F and dir(*a*) refer determinately if they refer at all.

What we have not done is show that there are in fact functions and objects satisfying these stringent conditions in the central cases. The initial appeal of neo-Fregean Platonism lay in large part in its claim to vindicate the thought that reference to abstract objects like numbers and directions is a relatively unproblematic business. All it takes is the free stipulation of an abstraction principle—a definition, of sorts—and then *presto*, a new thoroughly meaningful function symbol is introduced, and therewith a capacity to refer to its (abstract) values. The bad company objection and its descendants throw cold water on that prospect. On pain of contradiction we must admit that the neo-Fregean procedure for introducing a new function symbol sometimes fails. On our account, the real latent content of such principles is both richer and stranger than one might at first have been inclined to suppose, and so it may seem that our proposal "increases the risk" of failure associated with the procedure. We have conjectured (see n. 8) that this apparent aggravation of the problem is an illusion. This will be true if in the end there is a metaphysical guarantee that every *good* abstraction principle defines a function, in the sense that there exists a function whose nature is exhausted by the fact that it satisfies the principle in question. If *that* could be shown (and if the other principles upon which we have relied can be defended) we would possess a complete vindication of the central claims of neo-Fregean Platonism as we understand it.

References

Benacerraf, P. (1965). "What Numbers Could Not Be," *Philosophical Review* 74, 47–73.

Boolos, G. (1990). "The Standard of Equality of Numbers," in G. Boolos (ed.), *Meaning and Method: Essays in Honor of Hilary Putnam*, Cambridge: Cambridge University Press, 261–77. Reprinted in his *Logic, Logic and Logic*, Cambridge, MA: Harvard University Press, 1998, 202–19.

Fine, K. (2002). *The Limits of Abstraction*, Oxford: Clarendon Press.

Fine, K. (1994a). "Senses of Essence," in Walter Sinnott Armstrong (ed.), *Modality, Morality, and Belief*, New York: Cambridge University Press.

Fine, K. (1994b). 'Essence and Modality,' *Philosophical Perspectives* 8, 1–16.

[19] It also means the objects introduced by one abstraction principle are always distinct from the objects introduced by a distinct abstraction principle involving a distinct equivalence relation. So numbers, directions, shapes, etc., are all distinct. This principle has a number of surprising consequences. See Fine (2002: 46–54) for discussion.

Frege, G. (1884). "The Foundations of Arithmetic," reprinted in M. Beaney (ed.), *The Frege Reader*, Oxford: Blackwell, 1997, 84–129.

Frege, G., and M. Beaney (1997). *The Frege Reader*. Oxford: Blackwell Publishers.

Goodman, N. (1956). "A World of Individuals," *The Problem of Universals: A Symposium*, Notre Dame, IN: University of Notre Dame Press, 13–31. Reprinted in Paul Benacerraf and Hilary Putnam (eds.), *Readings in Philosophy of Mathematics*, New York: Prentice Hall, 1964, 197–210.

Hale, B. (1997). "Grundlagen Paragraph 64," *Proceedings of the Aristotelian Society* 97: 243–61.

Hale, B. (2001). "A Response to Potter and Smiley: Abstraction by Recarving," *Proceedings of the Aristotelian Society* 101, 339–58.

Hale, B., and C. Wright (1992). "Nominalism and the Contingency of Abstract Objects," *Journal of Philosophy* 89, 111–35.

Hale, B., and C. Wright (2000). "Implicit Definition and the A Priori," in P. Boghossian and C. Peacocke (eds.), *New Essays on the A Priori*, Oxford: Clarendon.

Hale, B., and C. Wright (2001). *The Reason's Proper Study: Essays towards a Neo-Fregean Philosophy of Mathematics*, Oxford: Clarendon Press.

Hale, B., and C. Wright (2003). "Responses to Commentators," *Philosophical Books* 44, 245–63.

Hale, B. (1988). *Abstract Objects*, New York: Blackwell.

Potter, M., and T. Smiley (2001). "Abstraction by Recarving," *Proceedings of the Aristotelian Society* 101, 327–38.

Potter, M., and T. Smiley (2002). "Recarving Content: Hale's Final Proposal," *Proceedings of the Aristotelian Society* 102, 351–4.

Rosen, G. (1993). "The Refutation of Nominalism?," *Philosophical Topics* 21, 149–86.

Rosen, G. (2003). "Platonism, Semi-Platonism and the Caesar Problem," *Philosophical Books* 44, 229–44.

Wright, C. (1983). *Frege's Conception of Numbers as Objects*, Aberdeen: Aberdeen University Press.

Yablo, S. (2008). 'Carving Content at the Joints,' *Canadian Journal of Philosophy* 38 (sup1), 145–77.

PART II
Vagueness

6

Vagueness and Intuitionistic Logic

Ian Rumfitt

Crispin Wright's papers and books have for forty years been read and admired by philosophers throughout the world. He has made, and continues to make, important contributions to debates about vagueness, rule-following, realism, scepticism, self-knowledge, the philosophy of mathematics, and the interpretation of Wittgenstein. Just as impressive, to those of us who know him personally, are Wright's qualities as a philosophical interlocutor. Whether in public discussion or private conversation he is quick on the uptake, acute in response, and seemingly incapable of descending to the footling or trivial. One's understanding of a philosophical issue is always deepened by debating it with him.

It is his passion for philosophical debate that led him to found two institutions which have contributed so much to philosophy in Britain and abroad. Philosophers would make the journey to the Arché Centre at St Andrews, and more recently to the Northern Institute of Philosophy at Aberdeen, in large part to get the benefit of his reaction to their ideas. The creation of those institutions bears witness to further qualities that are rarely found combined with philosophical distinction: considerable social and administrative energy, and an unquenchable optimism no matter how difficult the times.

It was Wright's essay 'On the Coherence of Vague Predicates' (1975) that first won him a wide readership, and his subsequent papers on vagueness have also made waves. I have no space to survey his work in this area, but one of his recent essays on the topic provides my point of departure. In '"Wang's Paradox,"'[1] his contribution to the *Library of Living Philosophers* volume that honoured our common teacher, the late Sir Michael Dummett, Crispin Wright presents the most powerful argument I know in favour of the thesis that we should employ intuitionistic logic, not classical logic, when reasoning with vague concepts.[2] For reasons that will emerge, I am not convinced that this thesis is true, but nor am I convinced that it is false. The aim of

---

[1] The topic of Wright's essay is not Wang's Paradox per se, but rather Dummett's paper 'Wang's Paradox' (Dummett 1975), hence the quotation marks in Wright's title.
[2] The idea that intuitionistic logic might help with the paradoxes of vagueness was floated by Hilary Putnam in his 'Vagueness and Alternative Logic' (1983). Stephen Read and Crispin Wright (1985) sharply criticized what they took to be Putnam's view, but Putnam (1985) protested that they had misread him. Wright's paper of 2007 may be read as developing the approach to the *Sorites* that Putnam had floated in 1983 and then presented more clearly in 1985. See also n. 4 below.

Ian Rumfitt, *Vagueness and Intuitionistic Logic* In: *Logic, Language, and Mathematics: Themes from the Philosophy of Crispin Wright*. Edited by: Alexander Miller, Oxford University Press (2020). © Ian Rumfitt.
DOI: 10.1093/oso/9780199278343.003.0006

this chapter, which I confess is inconclusive, is to bring into the debate some considerations that ought to inform any eventual decision on the question.

1. Wright's Paradox of Sharp Boundaries

The tangled skein that is the *Sorites* contains several distinct threads. A proper exploration of these and of their relations is impossible here, but one of Wright's contributions has been to focus attention on a form of the paradox that is especially stark and difficult to solve. The discussion in this essay is confined to this version of the problem.

Let us suppose that we have a sequence of a hundred transparent tubes of paint, $a_1, \ldots, a_{100}$, with the following properties: tube a_1 is clearly red; tube a_{100} is clearly orange and hence clearly not red; but for each n, tube a_{n+1} is only marginally more orange (and hence only marginally less red) than its predecessor a_n. Indeed, let us suppose that an observer with good eyesight, when viewing any pair of adjacent tubes a_n and a_{n+1} together in white light but without comparing them with other tubes, is unable to perceive any difference in colour between them. That is, we suppose that any two adjacent members of the sequence are *indiscriminable* in colour. In such a case, the claim that nowhere in the sequence is a red tube immediately followed by a non-red tube seems highly plausible. For if there were a number N for which a_N were red while a_{N+1} were not, we would have a pair of tubes which are indiscriminable in colour but where one member is red while the other is not red; this seems to conflict with the correct use of the predicate 'red.'[3] In a semi-formalized language, then, it seems that we may affirm:

(1) $\neg \exists n \, (a_n \text{ is red} \wedge \neg(a_{n+1} \text{ is red}))$.

In the situation described, however, we also have

(2) a_1 is red

and

(3) $\neg(a_{100} \text{ is red})$.

Now if we suppose

(4) a_{99} is red

then the rule of conjunction-introduction applied to (3) and (4) would yield

(5) a_{99} is red $\wedge \neg(a_{100} \text{ is red})$,

which, by $\exists$-introduction, yields

---

[3] Wright's leading example in his 2007 paper is a *Sorites* for 'looks red' rather than 'is red.' It is perhaps even clearer that the correct use of 'seems red' precludes the existence of a pair of indiscriminable tubes, one of which seems red while the other does not. However, claim (1) is also highly plausible, and the logical issues I want to address arise equally with 'is red,' which is why I have switched to the semantically simpler predicate.

(6) $\exists n\,(a_n$ is red $\wedge \neg(a_{n+1}$ is red$))$

which directly contradicts (1). Given (1) and (3), then, supposition (4) stands refuted, so by *Reductio* we may assert

(7) $\neg(a_{99}$ is red$)$.

By repeating this inferential sub-routine a further 98 times, we reach

(8) $\neg(a_1$ is red$)$

which contradicts (2). This, then, is the initial paradox. We have some reason to accept the trio of postulates (1), (2), and (3), but we also have an apparently valid deduction showing that the trio is inconsistent. It may be noted that the form of *Reductio* that is applied in reaching line (7)—and is reapplied at the corresponding later steps—is acceptable to an intuitionist. So the trio comprising (1), (2), and (3) is inconsistent in intuitionistic logic as well as in classical logic.

How should we react to this apparent demonstration of inconsistency? Since the case is one in which (2) and (3) are clearly true, it seems that we must take it as showing that (1) is false. In other words, we would appear to be entitled—indeed, compelled if the question of (1)'s truth arises—to make a further application of the relevant form of *Reductio* and infer the negation of (1), namely,

(9) $\neg\neg\exists n\,(a_n$ is red $\wedge\neg(a_{n+1}$ is red$))$.

In classical logic, however, (9) is equivalent to

(10) $\exists n\,(a_n$ is red $\wedge \neg(a_{n+1}$ is red$))$.

This, however, seems to land us in a yet more acute paradox, one that Crispin Wright has called the *Paradox of Sharp Boundaries*. For formula (10) says that at some point in the sequence a red tube is immediately followed by a non-red tube, and this seems to ascribe a sharp boundary to the concept *red*. Some philosophers take the vagueness of a predicate to consist in there not being a sharp boundary between those objects that satisfy it and those objects that do not. But even if the absence of a boundary is not constitutive of vagueness, it still seems absurd to say that one tube is red while a tube indiscriminable in colour from it is not red: little wonder that Wright calls formula (10) the 'Unpalatable Existential.'

Yet we have deduced this apparent absurdity by applying the rules of classical logic to the indisputably true premises (2) and (3).

One initial comment about the argument is in order. Although I have used an existential quantifier in formulating premiss (1) and some subsequent lines of the argument, its employment is entirely dispensable. Since the quantifier ranges over the natural numbers from 1 to 100, one could replace any statement of the form '$\exists n\,\varphi(n)$' by the corresponding disjunction '$\varphi(1) \vee \ldots \vee \varphi(100)$.' Thus a propositional logic— indeed, a logic of negation, conjunction, and disjunction—suffices to assess the validity of the argument.

The apparent antinomy here has long been a source of scepticism about the applicability of classical logic to deductions involving vague terms, but it is worth distinguishing between two sorts of sceptic. Some hold that the laws of some other logical system codify the standards for assessing the validity of those deductions, while others take a more radical position and deny that such assessments admit of codification at all. Among the radicals was Frege. On his view, statements can stand in law-governed logical relations only if each of their component expressions possesses a reference or *Bedeutung*. The reference of a predicate is a concept or *Begriff*. And yet:

A concept that is not sharply defined is wrongly termed a concept. Such quasi-conceptual constructions cannot be recognized as concepts by logic; it is impossible to lay down precise laws for them. The law of excluded middle is really just another form of the requirement that the concept should have a precise boundary.   (Frege 1902: §56, p. 69)

A few years earlier, he had elaborated the same message in a letter to Peano:

Logic can only recognize sharply delimited concepts. Only under this presupposition can it set up precise laws...Just as it would be impossible for geometry to set up precise laws if it tried to recognize threads as lines and knots in threads as points, so logic must demand sharp boundaries of what it will recognize as a concept unless it wants to renounce all precision and certainty.   (Frege 1896: 182–3 = Frege 1980: 114–15)

Indeed, Frege diagnoses the *Sorites* as arising from the application of precise laws to 'quasi-conceptual constructions': 'The fallacy known as the *Sorites* rests on this: that something (for example, *heap*) is treated as a concept when logic cannot acknowledge it as such because it is not properly circumscribed' (Frege 1897–8: 165 = Frege 1979: 155).

2. Intuitionism as the Logic of Vagueness

Unlike Frege, Wright does not despair of finding a precise logic for vague reasoning. On his view, though, the logic in question is intuitionistic, not classical (see especially Wright 2007, but also his 2001 and 2003). He reaches this conclusion through a close analysis of the paradoxical argument.

As remarked above, the derivation showing that (1), (2), and (3) form an inconsistent trio is intuitionistically correct; Wright duly accepts that derivation. He further accepts that the statements (2) and (3) are true, so that the derivation shows that (1) is false. But he insists that that conclusion is acceptable. We are tempted to assert (1) because we think we know that there is no sharp cut-off between the red members of the sequence and the non-red members. But really, we do not know any such thing. In order to assert (1)—that is, in order to assert that there is no sharp cut-off—we would require an insight into the semantics of the predicate 'red'—that is, an understanding of the way it contributes to the determination of the truth or falsity of statements containing it—that we do not (yet) possess. Of course, we are equally unable to assert that there *is* any sharp cut-off point between the red and the non-red tubes in our sequence, but lack of entitlement to assert should not be confused with entitlement to deny. On a correct view, Wright claims, a predicate's 'vagueness...does not consist in the absence of sharp cut-offs'; rather, it consists in there being 'nothing in our practice with the predicate that grounds the claim that there is a sharp cut-off' (Wright 2007: 439). In particular, where we have a simple statement in which a vague predicate is applied to one of its borderline cases,

the breakdown in convergence of verdicts leaves us without uncontroverted evidence for its truth or for the truth of its negation. Since we lack convincing theoretical grounds to think that one or the other must be true nonetheless—because such grounds would have to regard the determinants of truth and falsity as constituted elsewhere than in our linguistic practice—we are left with no compelling reason to regard either [the statement] or its negation as true. (Wright 2007: 441)

Let us pick a tube from the middle of our sequence—a_{50}, say. That tube, we may suppose, is a borderline case of redness: even given full information about its colour, and full information about how competent English speakers are disposed to use the predicate 'red,' it remains indeterminate whether the predicate correctly applies to the tube. On Wright's view, we should refrain both from asserting and from denying 'Tube a_{50} is red.' Similarly, he thinks, we should refrain both from asserting and from denying 'Some red tube in the sequence is immediately followed by a non-red tube'— that is, we should refrain both from asserting and from denying formula (10). In particular, then, we should refrain from denying (10); equivalently, we should refrain from asserting (1).

What, though, of the second part of the paradoxical argument? As remarked, (1), (2), and (3) are (intuitionistically) inconsistent, and premises (2) and (3) are clearly true. So it seems that (1) is false. But if (1) is false, then its negation is true. So we are entitled to assert—indeed, we are committed to asserting—(9):

(9) $\neg\neg\exists n\,(a_n$ is red $\wedge\neg(a_{n+1}$ is red$))$.

Wright accepts (9). Because (1)—the denial that *red* has a sharp cut-off point— contradicts the evident truths (2) and (3), we must be prepared to deny any denial that there is a sharp cut-off point. Assuming that denying a statement is equivalent to affirming its negation, it follows that we must be prepared to affirm the doubly negated formula (9). However, in order to reach the genuinely unpalatable (10), which asserts that there is a sharp cut-off between the red and the non-red tubes, we would need to eliminate a double negation. And this, Wright argues, we are neither compelled nor entitled to do. Since we cannot assume that statements involving vague expressions are either true or false, 'we should . . . abstain from unrestricted use of the law of double negation elimination' (Wright 2007). Such abstention enables us consistently to deny (1), assert (9), but refrain from asserting or denying (10). This combination of assertions, denials, and abstentions from assertion and denial con- stitutes, Wright proposes, a coherent response both to the *Sorites* proper and to the related Paradox of Sharp Boundaries.[4]

---

[4] The kernel of Wright's position, then, is to deny the denial of (10) while refraining from either asserting or denying (10) itself. This is why I read Wright's essay of 2007 as elaborating the position Putnam took in 1985: 'In my proposal (10) is not denied. What the Read–Wright argument does show, in fact, is that I should deny (10)'s denial—that is, I should accept the *double negation* of (10)' (Putnam 1985: 203, with renumbering of formulae). The double negation of (10) is (9) which, as we have seen, Wright does accept in his 2007 paper.

We have here a novel and interesting argument for intuitionistic logic. *Pace* Frege, there *is* a logic for assessing deductions involving vague predicates, but the logic in question is intuitionistic rather than classical. If it works at all, Wright's argument has very wide application. Many ordinary predicates are such that nothing in our ordinary practice with them grounds the claim that their extensions have sharp cut-offs. For many of those predicates, indeed, it will be possible to construct a *Sorites* sequence. To those predicates, Wright's argument will apply: on pain of contradicting clear truths, we must deny that there is no sharp cut-off point, whilst resisting the unsupported assertion that there is such a cut-off. This position is stable only if there are restrictions on the elimination of double negations; so we have an argument in favour of a logic (like intuitionistic logic) which imposes such restrictions.

What should we make of this argument? In his reply to '"Wang's Paradox,"' Michael Dummett put on record his 'admiration for the beautiful solution of the Sorites advocated by Crispin Wright, clouded by a persistent doubt whether it is correct' (Dummett 2007a: 453). Anyone who appreciates logico-philosophical finesse will share Dummett's admiration for an ingenious proposal, but there are, indeed, two immediate reasons to doubt its correctness.

First, Wright moves too quickly from the observation that we hesitate when pressed to say whether a borderline red-orange object is red, to what is really needed to sustain the distinction the intuitionist wishes to draw between (9) and (10). The observation is correct, and if one further supposes that the only possible grounds for the truth of a vague statement must be overt in our ordinary practices of asserting and denying it, then it cannot be assumed that an atomic statement in which a vague predicate is applied to one of its borderline cases is either true or false. This point, however, is not immediately relevant to the status of the formulae (9) and (10). These formulae are not atoms, and (on intuitionistic principles) we may assert that a complex statement is bivalent even when we cannot assert the bivalence of any of its components.[5] So we need a further argument to show that we cannot assert the bivalence of (9) and (10).

Wright seems to have in mind a traditional Heyting-style semantics for the language of intuitionism on which truth goes with assertibility and bivalence goes with decidability.[6] Under that semantics, the existentially quantified formula (10) may be assumed to be bivalent if (a) the relevant domain of quantification is surveyable, and (b) each matrix instance is decidable. Condition (a) is surely met. In a *Sorites* deduction, the relevant sequence of objects is finite; indeed, as noted, the deduction would go through in essentially the same way if (10) were replaced by a disjunction. So Wright must deny that all the matrix instances of (10) are decidable; in particular, he must deny that we have a general ground for denying any formula of the form 'a_n is red $\land \neg(a_{n+1}$ is red).' For, in a Heyting-style semantics, such a ground would render (10) false, so we would have a proof of (1), as well as of (9), and

---

[5] For example, an intuitionist is always entitled to assert the falsity, and hence the bivalence, of $\neg(A \lor \neg A)$, whatever the status of A.

[6] See especially the discussion on p. 441 of Wright (2007), which concludes: 'We should therefore abstain from unrestricted use of the law of double negation elimination. We do so on the grounds that it is nonconservative of knowledge—exactly the Intuitionistic reservation.'

Wright's way out of paradox would collapse back into contradiction. The problem is that there is a plausible general ground for denying any such conjunction—namely, Timothy Williamson's thesis that knowledge of inexact matters demands margins for error (Williamson 1994: 226–30). According to that thesis, knowledge that a_n is red requires that a_{n+1} may also be red, for all we know; this excludes knowledge that a_{n+1} is not red. Since Wright takes it that an entitlement to assert requires knowledge, Williamson's thesis provides a general ground for rejecting any assertion of the form 'a_n is red $\land \neg(a_{n+1}$ is red).' On Heyting's principles, this would amount to a ground for denying (i.e., affirming the negation of) any such conjunction. This is turn yields a ground for deeming (10) to be false, which is inconsistent with Wright's claim that its negation, (1), is also false. If he wishes to persist with a Heyting-style semantics, Wright badly needs to dispel the considerable attractions of the thesis that inexact knowledge demands margins for error, for that thesis brings his position into ruin.[7]

Second, and more importantly, Wright's suggestion that the logic of vague statements is intuitionistic is not sustained by any detailed semantic analysis. He points to broad analogies between the intuitionists' attitude to mathematical statements and the attitude that he recommends taking towards vague statements: in each case, statements' 'truth and falsity have to be thought of as determined by our very practice, rather than by principles which notionally underlie' that practice.[8] As Dummett has remarked, however, it is most implausible to suppose that the Heyting semantics for the language of intuitionist mathematics correctly describes the contribution that the connectives and quantifiers make to the meaning of vague statements:

> Wright can hardly say that to grasp the meaning of the statement ['This curtain is red,' said of a borderline red-orange curtain] is to be able to recognize a demonstration that the colour of the curtain is red, not orange, or that it is orange, not red. Presumably he must say that it consists in an ability to judge that the colour of the curtain is neither definitely red nor definitely orange, but is definitely either one or the other. If so, however, the theory of meaning for vague statements on which Wright wishes to rest intuitionist logic as applying to them interprets disjunction quite differently from the intuitionist theory of meaning for mathematical statements.   (Dummett 2007: 451–2)

As we shall soon have occasion to note, Heyting's is far from being the only semantic theory that sustains intuitionistic logic, but Dummett's general point remains: 'It is not enough to show that the Sorites paradox can be evaded by the use of intuitionistic logic; what is needed is a theory of meaning, or at least a semantics, for sentences

---

[7] This thesis about knowledge does not entail Williamson's own epistemicism about vagueness—the claim that there is a sharp cut-off point in the sequence of red things, but that we cannot in principle know where it is. Someone who denies, or refrains from accepting, that there is a sharp cut-off can still hold that knowledge that such-and-such an object is red requires a margin for error.

[8] Wright (2007: 441). One of Wright's theses in his essay is that competence with a word like 'red' is not underpinned or informed by adherence to any rule. If that thesis is correct, then a semantic theorist cannot sensibly aim to articulate the rules that govern competence with 'red.' But in calling for a more detailed semantic analysis than Wright provides, I am not asking for that, but only for something that it is reasonable to seek on any plausible view. For the predicates 'is not red,' 'is either red or orange,' etc., are clearly semantically complex, so it must be in order to ask for a statement of the general principles that relate their meanings to the meanings of their components. That is what I mean by a semantic analysis.

containing vague expressions that shows why intuitionistic logic is appropriate for them rather than any other logic' (Dummett 2007: 453).

Wright might, I suppose, contest the need for a semantic theory. At least, he might argue that the construction of such a theory is not needed to vindicate his use of intuitionistic logic. We are dealing, after all, with a paradox: statements involving vague terms which we are disposed to assert yield contradictions given classical logic, but they do not yield absurdities when the operative logic is intuitionistic. One might reasonably think that this fact provides strong reason for restricting some classical laws even in advance of constructing a semantic theory. Indeed, Wright might pray in aid some of Dummett's own writings here. At the end of the chapter in *The Logical Basis of Metaphysics* which presents his reasons for adopting intuitionistic logic, Dummett remarks that 'the meanings of the intuitionistic logical constants can be explained in a very direct way, without any apparatus of semantic theory, in terms of the use made of them in practice' (1991: 299). 'A metalanguage whose underlying logic is intuitionistic,' he goes on, 'can be understood, and its logical laws acknowledged, without appeal to any semantic theory and with only a very general meaning-theoretical background. If that is not *the* right logic, at least it may serve as a medium by means of which to discuss other logics' (1991: 300). The 'very general meaning-theoretical background' to which Dummett refers includes the requirement of 'harmony' between the introduction and elimination rules for a connective. Wright might fairly point out that the requirement of avoiding outright contradiction is more clearly compelling than that of harmony, and that the former requirement is all he really needs to press his case for intuitionistic logic as the 'right logic' when reasoning with vague terms.

For all that, a semantic theory remains desirable: it may help us to understand *why* the distinctively classical rules are invalid when applied to vague terms. As Dummett was wont to say in the rather different context of the foundations of mathematics, to observe that a certain postulate leads to contradiction is to 'wield the big stick' when what is really wanted is some explanation of why the postulate is troublesome. In this chapter, I want to take up Dummett's challenge and construct a semantic theory for vague terms which offers some hope of vindicating intuitionism as the logic of vagueness. I am not sure that the semantic theory I propose is fully satisfactory; indeed, §5 will present some reasons for doubting it. Those reasons, however, are far from conclusive, and it would be interesting to know whether Crispin Wright is disposed to accept it. If he were, it would give him a way of meeting Dummett's challenge head on by showing why intuitionistic logic, rather than any other logic, sets the standards for assessing deductive arguments involving vague terms.

3. A Semantics for Vague Predicates that Validates Intuitionistic Logic

The semantic theory I have in mind is really no more than an application of a theorem of Alfred Tarski's.

Topologists have the notion of an *open* set. The mark of such a set is that, whenever an object belongs to it, there is a 'neighbourhood' around the object, all of whose members also belong to the set. Suppose we have a domain which comprises all the possible coloured objects, not merely those that actually exist. Then it is very natural to postulate that the set of red objects in this domain will be open in a suitably defined topology. For suppose that a possible object x is red. Then it is natural to postulate that there is a neighbourhood of possible objects around x that are also red. An object's perceived colour does not depend only on the wavelength of the light it reflects or emits: it also depends on the ambient illumination and on the colours of nearby objects. All the same, if our red object, x, reflects light of wavelength 700 nm, then it is plausible to claim that all the possible objects which are observed in the same context as x and in circumstances where human colour perception is as it actually is, but which reflect light of wavelengths between 699 and 701 nm, will also be red. In a suitably defined topology, then, these possible objects will constitute a neighbourhood around x, all of whose members are also red. An object that reflects light of 700 nm is a paradigm case of redness, but the argument may be adapted to show that there is a neighbourhood around any red object. Suppose that y is a red object that reflects light of 620 nm, and so is close in colour to some orange objects. Even assuming that colour perception is as it actually is, it may well be that some possible objects that reflect light of wavelengths between 619 and 621 nm are not red. However, so long as y is red, it is plausible to claim that there is *some* neighbourhood around y—for example, those possible objects that are viewed in the same context as y but which reflect light of wavelengths between 619.9999 and 620.0001 nm—all of whose members are red. This smaller neighbourhood is enough to ensure that the set of red objects in the domain is open. It is also natural to postulate that the set of possible objects that are *not* red will be another open set in the same topology. For if z is a possible object that is not red, perhaps because it reflects light of only 550 nm, it is plausible to claim that all the possible objects which are viewed in the same context as z, but which reflect light of wavelengths between 549 and 551 nm, are also not red.

Arguments analogous to that just sketched may be advanced for many vague predicates. Suppose that x is a tall man in the domain of all possible men. Then, whatever precise height x has, it is plausible to claim that there is a neighbourhood around x all of whose members are also tall men. If x is close to the borderline for being tall, the members of the neighbourhood may have a precise height that is very close to x's. All the same, the set of tall possible men will be open in an appropriately defined topology. So will the set of possible men who are not tall.

We cannot assume that the set of *actual* men who are tall is open in the relevant topology. Suppose that every man of height greater than 5'8" dies, except for one man, Adam, whose height is 6'1". Adam is tall, but in the circumstance envisaged there is no neighbourhood of actual tall men around him. All the same, it is entirely natural to take account of merely possible existents when giving a semantic theory for predicates. This is certainly the case when we are concerned with logical relations between sentences and predicates. When a conclusive statement follows logically from some premiss statements, it is logically necessary that it is true whenever they are all true. Similarly, when a conclusive predicate follows logically from some premiss predicates, it is logically necessary that it is true of whatever they are all true of. Given that, it is natural to take the semantic value of a predicate to be a subset of a domain of all logically possible objects, not merely the actual ones. This raises, of course, the vexed question of how we might best conceive of that domain, but there is a way of bracketing that question for present purposes. Rather than taking the semantic value of 'tall' to be a subset of the domain of all possible men, we might take it to be a set of specific heights, determined by a real number (with 'foot' as the unit, say): a man will be tall if his specific height belongs to this set. Since a specific height will exist even if no man has that height, this semantic value of 'tall' will be an open set regardless of men's actual heights. This manoeuvre also enables us to be specific about the topology in question: on this approach, the semantic value of 'tall' is an open set in the standard topology on the real line. Similarly, we may take the semantic value of 'red' to be an open set in the same topology, where the real numbers in this case specify wavelengths of reflected or emitted light. An object may be said to be 'red' if it reflects or emits light of a wavelength belonging to this semantic value. These assignments of semantic values are of course idealized. No determinate set of real numbers, whatever its topological properties, contains the heights (in feet) of all and only tall men; and an object's perceived colour depends on the ambient illumination and the colours of nearby objects as well as on the wavelengths of light it emits or reflects. But all the semantic theories we currently have involve considerable idealization and (as we shall see) the present idealization leads to very different results from the familiar one, which postulates sharp boundaries even for vague predicates such as 'red' and 'tall.'

How might these ruminations about open sets encourage a theorist who recommends using intuitionistic logic when assessing deductive arguments involving vague terms? Where S is any topological space, and L is any propositional language whose connectives are '$\wedge$,' '$\vee$,' '$\rightarrow$,' and '$\neg$,' let us say that a map v is an *interpretation* of L into S if v maps each well-formed formula of L into an open subset of S in such a way that

$$v(A \wedge B) = v(A) \cap v(B)$$
$$v(A \vee B) = v(A) \cup v(B)$$
$$v(A \rightarrow B) = Int\,(v(B) \cup (S - v(A)))\text{ and}$$
$$v(\neg A) = Int(S - v(A)).$$

(Thus $v(\neg A) = v(A \rightarrow \perp)$, where $v(\perp) = \emptyset$.) The interior $Int\,(X)$ of a set X is the largest open set contained in X. Accordingly, the clauses for the conditional and for negation ensure that a conditional or negated formula is mapped into an open set in S. Since the intersection and union of two open sets are also open, v maps every statement in L to an open set in S. Let us then say that an argument '$A_1, \ldots, A_n$; so B' in L is *valid* in the topological space S if and only if

for every interpretation v of L into S, $v(A_1) \cap \ldots \cap v(A_n) \cap S \subseteq v(B)$.

Let us define a consequence relation $A_1, \ldots, A_n \models_{Top} B$ as one which obtains when the argument '$A_1, \ldots, A_n$; so B' is valid in every topological space. A theorem due to Tarski—the Second Principal Theorem of his paper 'Sentential Calculus and Topology' (1938: 448)—then tells us that the intuitionistic propositional calculus is sound and complete with respect to this notion of consequence. That is, where $\vdash_{Int}$ signifies deducibility by the rules of the intuitionistic propositional calculus,

(A) $A_1, ..., A_n \vdash_{Int} B$ if and only if $A_1, ..., A_n \models_{Top} B$.

In fact, this Theorem of Tarski's tells us more. Let us define the *closure* $Cl(X)$ of a set X in a topological space S to be the complement of the interior of the complement of X. That is, $Cl(X) = S - (Int\ (S-X))$. Let us call a topological space S *dense-in-itself* when, for every $x \in S$, $x \in Cl(S - \{x\})$. Let us call S *normal* when, for any two subsets X and Y of S, if $Cl(X)$ is disjoint from $Cl(Y)$ there exist disjoint open subsets U and V of S such that $Cl(X) \subseteq U$ and $Cl(Y) \subseteq V$. Let us say that S has a *countable basis* if there is a countable family $U = \{U_i\}_{i \in \omega}$ of open subsets of S such that every non-empty open subset of S can be represented as the union of a sub-family of U. Finally, let us say that a topological space is *Euclidean* if it is dense-in-itself, normal, and has a countable basis. Then Tarski (*ibid.*) also proved

(B)   Where S is any given Euclidean topological space, $A_1, \ldots, A_n \vdash_{Int} B$ if and only if the argument '$A_1, \ldots, A_n$; so B' is valid in S.

The *n*-dimensional Euclidean space $\mathfrak{R}^n$ with its usual topology is Euclidean in the present sense. Although the usual basis of open balls in $\mathfrak{R}^n$ is not countable, every open set in $\mathfrak{R}^n$ can be represented as the union of a sub-family of the restricted basis comprising those open balls with rational radii and whose centres have rational Cartesian coordinates; this restricted basis is countable.[9] In particular, then, (B) tells us that $A_1, \ldots, A_n \vdash_{Int} B$ if and only if the argument '$A_1, \ldots, A_n$; so B' is valid in the usual topology on the real line $\mathfrak{R}$. This result immediately shows that $\ulcorner A \vee \neg A \urcorner$ is not a theorem of the intuitionistic sentential calculus and that $\neg \neg A \vdash A$ is not a correct sequent in that calculus. For consider a sentence letter P whose value $v(P)$ in $\mathfrak{R}$ is the open half-line $(0, \infty)$. Then $v(\neg P) = Int\ (\mathfrak{R} - v(P)) = Int\ ((-\infty, 0]) = (-\infty, 0)$. Thus $v(P \vee \neg P) = v(P) \cup v(\neg P) = (0, \infty) \cup (-\infty, 0)$, so that $\ulcorner P \vee \neg P \urcorner$ does not meet the necessary condition for validity, viz. $\mathfrak{R} \subseteq v(P \vee \neg P)$. To show the invalidity of $\neg \neg A \vdash A$, consider a sentence letter Q whose value $v(Q)$ is set to be $(-\infty, 0) \cup (0, \infty)$. In that case, $v(\neg Q) = Int\ (\mathfrak{R} - v(Q)) = Int\ (\mathfrak{R} - ((-\infty, 0) \cup (0, \infty))) = Int\ (\{0\}) = \emptyset$. So $v(\neg \neg Q) = Int\ (\mathfrak{R} - v(\neg Q)) = Int\ (\mathfrak{R} - \emptyset) = Int\ (\mathfrak{R}) = \mathfrak{R}$. So the sequent $\neg \neg Q \vdash Q$ does not meet the necessary condition for validity, viz. $v(\neg \neg Q) \subseteq v(Q)$.

In fact, Tarski's proof of this Theorem may be adapted to establish corresponding results about monadic predicates rather than closed formulae. Let L' be a language consisting of a number of such predicates, formed by freely applying the predicate-forming operators 'and,' 'or,' 'if…then,' and 'not' to a stock of atomic monadic predicates. L', then, contains no singular terms or quantifiers. Indeed, it contains no closed formulae: $\ulcorner A$ and $B \urcorner$ is a conjunction of two monadic predicates, meaning (as it might be) 'red and spherical.' That need not stop us from applying deductive rules to the predicates in L': it makes sense to say, for example, that 'red and spherical' entails 'red.' Let us consider in particular the deductive system in which the rules of the intuitionistic propositional calculus are applied to these simple and complex monadic predicates, and let $\vdash_{Int'}$ signify deducibility by the rules of this system. Where S is again any topological space, we say that a map v is an *interpretation*

---

[9] So, for example, in the usual topology on $\mathfrak{R}$, the open set $(0, \sqrt{2})$ is the union of the countable collection of open sets $(0, r)$, where r is a positive rational number such $r^2 < 2$.

of L' into S if v maps each predicate in L' into an open subset of S in such a way that $v(A \text{ and } B) = v(A) \cap v(B)$, $v(A \text{ or } B) = v(A) \cup v(B)$, $v(\text{if } A \text{ then } B) = Int\,(v(B) \cup (S-v(A)))$, and $v(\text{not } A) = Int\,(S-v(A))$. As before, we say that an argument '$A_1, \ldots, A_n$; so B' in L' is *valid* in the topological space S if and only if

for every interpretation v of L' into S, $v(A_1) \cap \ldots \cap v(A_n) \cap S \subseteq v(B)$,

and we define a consequence relation $A_1, \ldots, A_n \models_{Top'} B$ as one which obtains when the L' argument '$A_1, \ldots, A_n$; so B' is valid in every topological space. Then Tarski's Second Principal Theorem also yields

(A')　$A_1, \ldots, A_n \vdash_{Int'} B$ if and only if $A_1, \ldots, A_n \models_{Top'} B$

and

(B')　Where S is any given Euclidean topological space, $A_1, \ldots, A_n \vdash_{Int'} B$ if and only if the L' argument '$A_1, \ldots, A_n$; so B' is valid in S.

Which of these four deliverances of Tarski's Theorem may help to vindicate the suggestion that intuitionistic logic is the logic of vagueness? The ones to look at are (A') and (B'). The intuitive idea we are exploring is that the extension of a vague *predicate* may be expected to be an open set in a suitably defined topology. It is the topological interpretation of L' that directly bears on that idea, not the interpretation of L. We shall eventually have to return to the logic of vague statements. Even if we could show that an intuitionistic system was the right logic for assessing deductions involving vague monadic predicates, this would not vindicate Wright's solution to the *Sorites*, which requires a propositional logic, not merely a logic for monadic predicates. All the same, the intuitive idea invites us to reverse the usual order of explanation and start by considering the logical behaviour of predicates rather than the behaviour of complete statements.

What, then, might (A') tell us about the logic of vague predicates? It tells us that when a monadic predicate B is not intuitionistically deducible from some premiss predicates $A_1, \ldots, A_n$, there is *some* topological space S, and *some* interpretation v from L' into S, for which $v(A_1) \cap \ldots \cap v(A_n) \cap S$ is not a subset of $v(B)$. As things stand, though, we do not know that this particular topological space S is one in which a vague predicate has an open extension. Theorem (A'), then, is unlikely to be effective in showing that intuitionistic logic is the strongest logic that may legitimately be employed in deductions involving vague predicates. I shall therefore explore the possibility suggested by (B')—that the standards for assessing such a deduction are set by some particular Euclidean space, determined by the deduction's context.[10]

Given our discussion so far, certain Euclidean spaces suggest themselves as setting the standard for assessing deductions involving vague predicates. Let us suppose that the atomic predicates of L' are the colour terms 'red' and 'orange.' As we have seen, if we take the semantic values of these terms to be sets of precise wavelengths of light,

---

[10] In discussing Wright's hypothesis in my book *The Boundary Stones of Thought* (Rumfitt 2015: §8.3), I explored only the possibility of vindicating it by way of (A'). Accordingly, the rejection of the hypothesis in that book was premature.

then (a) the semantic value of each predicate in L' (whether simple or complex) will be an open set in the natural topology on $\mathfrak{R}$ and (b) that topology is Euclidean. Accordingly, (B') tells us that the logic of L' will be intuitionistic so long as the following holds:

An argument '$A_1, \ldots, A_n$; so B' in L' is valid *simpliciter* if and only if it is valid in $\mathfrak{R}$.

Given the idealizations already made, this account of validity is *prima facie* plausible. We are supposing that the semantic values of colour terms are open sets in $\mathfrak{R}$. Under that supposition, the conditions on v would appear to be faithful to the meanings of 'and,' 'or,' 'if…then' and 'not' as these expressions connect predicates. These considerations by no means establish the proposed account of validity, but they do make it worth exploring in detail.

Similar considerations apply when other vague terms stand as the atomic predicates in L'. Suppose that the atomic predicates are now 'short' and 'tall' (as applied to men). As before, we take the semantic values of these terms to be sets of real numbers, which measure men's precise heights (in feet, say); a man will be tall if his precise height belongs to the value of 'tall.' In this case, too, (a) the semantic value of each predicate in L' (whether simple or complex) will be an open set in the natural topology on $\mathfrak{R}$ and (b) that topology is Euclidean. Given these idealizing assumptions, it is again plausible to hold:

An argument in L' is valid *simpliciter* if and only if it is valid in $\mathfrak{R}$.

It then follows from (B') that the logic of L' will be intuitionistic.

All this might encourage the following *Conjecture*:

An argument moving from vague predicate premises to a vague predicate conclusion is valid *simpliciter* if and only if it is valid in some particular Euclidean space.

If the Conjecture were correct, then intuitionistic logic would indeed be the right logic to use in assessing the validity of such arguments.

4. The *Sorites* Revisited

Even if the semantic values of vague predicates are always open sets in a Euclidean topology, the Conjecture may still be false. The Conjecture also requires that having its semantic value be an open set in a Euclidean topology is the strongest topological condition that may be imposed on a vague predicate. If the semantic values of such predicates have to satisfy a stronger topological condition, that condition might well sustain a stronger-than-intuitionistic logic. In §5, I shall consider whether the Conjecture is vulnerable to objection along these lines. On any view, however, our analysis is incomplete. If the Conjecture is correct, then intuitionistic logic will set the standards for assessing deductions in which we move from some vague (monadic) predicates as premises to a conclusion which also a monadic predicate. Intuitionistic logic, in other words, regulates such deductions as 'Red and large; so red.' It remains an open question, however, whether that logic also regulates the paradoxical deduction of §1. In that argument, '∧,' '∨,' and '¬' are all used to connect sentences, not predicates, and it is essential that they should be so used. A step in a *Sorites* argument cannot be analysed as an inference from premises saying that a given object has such-and-such a property, to the conclusion that the same object has some further property: the thrust of the *Sorites* is to argue for assigning a given vague property to a newly considered object. So the logic to be employed in evaluating the *Sorites* is moot until we have extended our account of the meanings of 'and,' 'or,' 'if...then,' and 'not' to yield an account of the meanings of '$\land$,' '$\lor$,' '$\to$,' and '$\neg$.' We do not, however, need to extend our theory further and explain the meanings of quantified formulae. Instead we can replace existentially quantified formulae such as Wright's (10) with disjunctions such as (10'). In what follows, I shall write as though the deduction of §1 has been recast in just this way, so that it is the rule of $\lor$-introduction, not $\exists$-introduction, which validates the step from (5') to (6').

How, then, might we extend our semantics for 'and,' etc., so as to yield a semantics for '$\land$,' etc.? There is a natural way of doing this. Consider the formula 'a_1 is red $\land$ $\neg(a_2$ is red)' as an example. That sentence is true if and only if the two-place predicate 'ξ_1 is red and not (ξ_2 is red)' is true of the ordered pair $<a_1, a_2>$. Given that the semantic value V_1 of 'ξ_1 is red' is an open set in $\Re$ with its usual topology, the Cartesian product $V_1 \times \Re$ will be an open set in $\Re^2$ with *its* usual topology. Similarly, given that the semantic value V_2 of 'not (ξ_2 is red)' is an open set in $\Re$ with its usual topology, the Cartesian product $\Re \times V_2$ will be an open set in $\Re^2$ with its usual topology. So $V_1 \cap V_2 = (V_1 \times \Re) \cap (\Re \times V_2)$ will be another open set in $\Re^2$. As noted, 'a_1 is red $\land$ $\neg(a_2$ is red)' is true if and only if $<a_1, a_2> \in V_1 \cap V_2$. Now in addition to the sentential connectives '$\land$,' '$\lor$,' '$\to$,' and '$\neg$,'[11] the language L of our *Sorites* argument consists of a single atomic predicate, 'red,' together with a hundred singular terms: 'tube a_1,'..., 'tube a_{100}.' By extending the method just indicated, we can construct, for any sentence A in L, a one-hundred place predicate P such that A is true if and only if P is true of $<a_1, ..., a_{100}>$, where the semantic value of P is always an open set in $\Re^{100}$ with its usual topology. The predicate P will belong to a language whose atomic predicates ascribe redness to various members of sequences of one hundred objects (thus one atom will say 'First member is red, second member is red, third member is red'), and whose only compositional devices are the predicate connectives 'and,' 'or,' 'if...then,' and 'not,' interpreted to signify intersection, union, interior of relative complement, and interior of complement. The construction is such that a deduction '$A_1, ..., A_n$; so B' in L is valid if and only if the corresponding deduction involving one-hundred place predicates, '$B_1, ..., B_n$; so Q,' is valid in $\Re^{100}$. $\Re^{100}$ with its usual topology is another Euclidean space. So, by (B'), we conclude that intuitionistic logic sets the standard for assessing deductions in L as well as for deductions in L'.

The method used to construct the predicates is of course inspired by Tarski's insight that a closed formula in a language with quantifiers is true if and only if it is satisfied by every infinite sequence of objects in the domain (see Tarski 1935: 195, Definition 23). Because, however, we are dealing with a language with sentential

---

[11] The particular *Sorites* deduction given in §1 happens not to contain '$\to$' but other such deductions will contain the conditional, so it needs to be treated: in intuitionistic logic, '$\to$' cannot be defined in terms of the other connectives.

connectives, not quantifiers, we need consider only finite sequences: any deduction is finite and so will contain only finitely many singular terms.

The analysis above could be applied to many *Sorites* deductions. A *Sorites* argument for 'tall' may be analysed in essentially the same way, given that the semantic value for the monadic predicate 'tall' is another open set in $\mathfrak{R}$. Of course, the analysis rests on the Conjecture. Given that, however, it provides a strong case for Wright's hypothesis that intuitionistic logic regulates deductions involving vague predicates. Given the master assumption that the semantic values of vague predicates are open sets, the semantic principles for 'and,' 'or,' 'if...then,' and 'not' are natural—indeed, they are well-nigh inevitable. Moreover, since $\mathfrak{R}$ and all its finite powers provide Euclidean topologies in which many vague predicates have open sets as their semantic values, the analysis extends to account for the validity of deductions involving different sorts of vague predicate: the analysis accounts for the validity of 'John is either rich or tall; John is not tall; so John is rich' as easily as it accounts for the validity of 'Tube *a* is either red or orange; tube *a* is not orange; so tube *a* is red.' In defending his hypothesis, Wright can say more than that adopting intuitionistic logic staves off contradiction in the Paradox of Sharp Boundaries. The semantic theory just sketched provides what Dummett was missing: an explanation of why intuitionistic logic is the right logic to use in assessing deductions involving vague predicates.

5. Problems for the Proposal

Despite these attractive features, the suggested explanation faces problems. The problems may not be insuperable, but I shall conclude by mentioning the two which seem to me to be the most serious.

The first is that Tarski proved the Theorem on which the semantic analysis rests in a classical metalogic: at various points in his proof, he employs classical rules of inference that are not acceptable to an intuitionist. In one respect, this casts further doubt on the thesis that classical logic provides the standards for assessing deductions involving vague predicates: even if we assume classical principles in the metalogic, the semantic theses we have been considering validate a weaker-than-classical logic for the object language. In the end, though, we shall want to prove that intuitionistic logic is sound and complete with respect to the proposed semantics *in an intuitionistic metalogic*. The thesis, after all, is that intuitionistic logic is the logic of vagueness, and the metalanguage in which the semantic theory is cast is just as vague as the object language. The colour predicate 'red' is vague, but so is the metalinguistic predicate 'satisfies the predicate "red."' I do not know whether Tarski's Theorem—or, more particularly, its consequence (B')—can be proven in an intuitionistic metalogic, but this question is important for the vindication of Wright's hypothesis. The claim that intuitionistic logic is the logic of vagueness would be better supported if the answer were affirmative.

The other major problem the theory faces concerns the Conjecture itself. As I argued at the start of §3, it is highly plausible to claim that the semantic value of a vague predicate will be an open set in a suitably defined topology. By itself, though, this claim does not sustain the Conjecture. We also need it to be the case that the strongest topological condition which may be imposed on a vague predicate is that its semantic value should be an open set in a Euclidean topology. There are reasons to doubt that this holds in general. Indeed, there are reasons to doubt that it holds even in the particular cases we have been examining. For how did we show that Excluded Middle and Double Negation Elimination were not valid in $\mathfrak{R}$? In each case, the proofs involved assigning the subset $(0, \infty) \cup (-\infty, 0)$ of $\mathfrak{R}$ to be the semantic value of either a simple or a complex predicate. That subset is indeed an open subset of $\mathfrak{R}$ under the usual topology. One might reasonably doubt, however, whether a set such as this can really serve as the semantic value of a vague predicate. Can it really be that the wavelengths of red light, for example, are those lying in a certain range of a real numbers *with the sole exception* of a single real number lying within that range? The natural answer is no: our understanding of the term 'red' seems to preclude such a set as an admissible extension of the term. It is, however, vital that such sets should be admissible extensions of predicates if Tarski's proposition (B') is to hold.

Indeed, the case suggests a stronger condition for a set to serve as the semantic value of a vague predicate. Topologists call a set *regular open* if it is the interior of its own closure. Any regular open set is open, but $(0, \infty) \cup (-\infty, 0)$ is an example of a set which is open without being regular. The closure of $(0, \infty) \cup (-\infty, 0)$ is $\mathfrak{R}$ itself, which is also the interior of the closure. Regular open sets, then, have no 'cracks' in them, and one might postulate that any admissible extension of a vague term will be regular open, not merely open. Given appropriate definitions of meet, join, and complement, however, the regular open subsets of any topological space form a Boolean algebra.[12] If the extensions of vague predicates are required to be regular open, and 'and,' 'or,' and 'not' are interpreted using the appropriate definitions of meet, join, and complement, then the semantics validates the full classical logic. The stronger topological requirement, then, refutes Wright's hypothesis that intuitionistic logic is the logic of vagueness and therewith his solution to the *Sorites*.

That stronger requirement, though, opens the way to an alternative solution to the ancient puzzle. As the case of $(0, \infty) \cup (-\infty, 0)$ shows, the union of two regular open sets need not be regular open. Accordingly, the join operation of the Boolean algebra of regular open sets is not $X \cup Y$, but $Int\ Cl\ (X \cup Y)$. If the Boolean algebra is to validate classical logic, it is this join operation that must be assigned to interpret the predicate connective 'or' and hence the sentential connective '$\vee$.' Under this interpretation, '$A \vee B$' may be true even though neither A nor B is true. Now Wright's 'Unpalatable Existential' (10), viz.,

(10) $\exists n\, (a_n \text{ is red} \wedge \neg(a_{n+1} \text{ is red}))$,

is no more than the following long disjunction

(10') $(a_1 \text{ is red} \wedge \neg(a_2 \text{ is red})) \vee ... \vee (a_{99} \text{ is red} \wedge \neg(a_{100} \text{ is red}))$.

---

[12] This result follows directly from the Stone Representation Theorem (Stone 1936). Although Stone was the first to publish the result, Tarski liked to claim credit for it: 'This fact was noticed by me as far back as 1927, and was implicitly stated in Theorem B of [Tarski 1929], where, however, a different terminology was used' (Tarski 1938: 449). The interested reader may like to ponder in what sense the result in 'implicit' in Theorem B of Tarski 1929; see Tarski (1983: 29).

On the alternative semantics now proposed, then, (10′) may be true even though no disjunct 'a_n is red $\wedge \neg(a_{n+1}$ is red)' is true. Wright solves the *Sorites* by revising classical logic so that we do not reach (10) or (10′). On the alternative approach, the inference to (10) or (10′) need not be resisted. For, under the alternative interpretation, Wright's 'Unpalatable Existential' is not really unpalatable: it does not ascribe a sharp boundary to the red members of the *Sorites* sequence.

I developed this solution to the *Sorites* in chapter 8 of *The Boundary Stones of Thought* (see Rumfitt 2015: 250–1). I argued there that the extension of a vague predicate will be regular open whenever the predicate is 'polar' in Mark Sainsbury's sense—that is, when its meaning is given by reference to a system of contrary paradigms or poles (see Sainsbury 1990). I stand by that argument. However, I never pretended to have an argument for the conclusion that every vague predicate is polar. So I have no argument for the thesis that every vague predicate has a regular open extension.

For this reason, it seems to me to be well worth exploring further the semantic theory sketched in §§3–4. It seems likely that the semantic values of *some* families of vague predicates must be regular open, not merely open. When reasoning with these predicates, we would appear to be entitled to apply all the rules of classical logic, so we shall need a solution to the *Sorites* different from Wright's. There is, though, no reason I can discern to hold that *all* vague predicates must have regular open sets as their semantic values. Perhaps having an open set as semantic value is the strongest *general* requirement that any vague predicate must satisfy. If the topological space in which the set is open is Euclidean, then this requirement will sustain intuitionistic logic. That logic, then, may be the strongest logic that we are entitled to use in reasoning with an arbitrary vague predicate. Perhaps we are entitled to use the classical logic only when reasoning with certain families—albeit central and common families—of vague term. As the many occurrences of 'may be' and 'perhaps' in this paragraph show, I do not know the answers to these questions. I hope, though, to have indicated some promising directions for further inquiry.[13]

Note added, October 2019:
I completed this chapter in 2015. Since then, I have returned to the question of how to justify the use of intuitionistic logic when reasoning with vague concepts in joint work with my colleague, Susanne Bobzien. In our paper 'Intuitionism and the Modal Logic of Vagueness' (forthcoming in the *Journal of Philosophical Logic*), we advance a justification which proceeds via a translation into a modal language whose necessity operator is understood to mean 'it is clearly the case that'. Our argument that the modal logic of clarity is the system S4M draws on some of Crispin Wright's

---

[13] I first found myself entertaining these ideas while scribbling on whiteboards during tutorials on vagueness given to finalists and BPhil students at Oxford; I hope that on this occasion the ideas are at least legible. Drafts of the chapter have provided the basis for seminars and lectures at Oxford and London, and for talks at the Universities of Bristol and Leeds and at the Moral Sciences Club in Cambridge. I thank members of all my audiences for helpful questions and remarks. I am particularly indebted to Timothy Williamson for written comments, and to Mark Sainsbury for his advice and encouragement.

ideas about higher-order vagueness, so this new justification of intuitionism may be a more consequent development of his position than that presented in this chapter.

References

Dummett, M. A. E. (1975). 'Wang's Paradox,' *Synthèse* 30, 301–24.

Dummett, M. A. E. (1991). *The Logical Basis of Metaphysics*, London: Duckworth.

Dummett, M. A. E. (2007). 'Reply to Crispin Wright,' in R. E. Auxier and L. E. Hahn (eds.), *The Library of Living Philosophers: Michael Dummett*, Chicago, IL: Open Court, 445–54.

Frege, G. (1896). 'Letter to Peano, 29 September 1896,' in G. Gabriel et al. (eds.), *Wissenschaftlicher Briefwechsel*, Hamburg: Felix Meiner, 1979, 181–6. English translation in *Philosophical and Mathematical Correspondence*, H. Kaal (trans.), Oxford: Blackwell, 1980, 112–18.

Frege, G. (1897–8). 'Begründung meiner strengeren Grundsätze des Definierens,' in H. Hermes et al. (eds.), *Nachgelassene Schriften*, Hamburg: Felix Meiner, 1969, 164–70. English translation in G. Gabriel et al. (eds.), *Wissenschaftlicher Briefwechsel*, Hamburg: Felix Meiner, 1979, pp. 152–6.

Frege, G. (1902). *Grundgesetze der Arithmetik*, volume II, Jena: Hermann Pohle.

Frege, G. (1976). *Wissenschaftlicher Briefwechsel*, G. Gabriel et al. (eds.), Hamburg: Felix Meiner.

Frege, G. (1979). *Posthumous Writings*, P. Long and R. White (trans.), Oxford: Blackwell.

Frege, G. (1980). *Philosophical and Mathematical Correspondence*, H. Kaal (trans.), Oxford: Blackwell.

Putnam, H. W. (1983). 'Vagueness and Alternative Logic,' *Realism and Reason: Philosophical Papers*, volume 3, Cambridge: Cambridge University Press, 271–86.

Putnam, H. W. (1985). 'A Quick Read is a Wrong Wright,' *Analysis* 45, 203.

Read, S. L., and C. J. G. Wright (1985). 'Hairier than Putnam Thought,' *Analysis* 45, 56–8.

Rumfitt, I. (2015). *The Boundary Stones of Thought: An Essay in the Philosophy of Logic*, Oxford: Clarendon Press.

Sainsbury, R. M. (1990). *Concepts without Boundaries*, London: King's College. Reprinted in Keefe and Smith (eds.), *Vagueness: A Reader*, Cambridge, MA: MIT Press, 1996, 251–64.

Stone, M. H. (1936). 'The Theory of Representations for Boolean Algebras,' *Transactions of the American Mathematical Society* 11, 37–111.

Tarski, A. (1929). 'Les Fondements de la Géometrie des Corps,' *Annales de la Société Polonaise de Mathématique*, supplement, 29–33. Page references are to the translation by J. H. Woodger, entitled 'Foundations of the Geometry of Solids,' in J. H. Woodger and J. Corcoran (eds.), *Logic, Semantics, Metamathematics: Papers from 1923 to 1938*, second edition, Indianapolis: Hackett, 1983, 24–9.

Tarski, A. (1935). 'Der Wahrheitsbegriff in den formalisierten Sprachen,' *Studia Philosophica* 1, 261–405. Page references are to the translation by Woodger, entitled 'The Concept of Truth in Formalized Languages,' in J. H. Woodger and J. Corcoran (eds.), *Logic, Semantics, Metamathematics: Papers from 1923 to 1938*, second edition, Indianapolis: Hackett, 1983, 152–278.

Tarski, A. (1938). 'Der Aussagenkalkül und die Topologie,' *Fundamenta Mathematicae* 31, 103–34. Page references are to 'Sentential Calculus and Topology,' in J. H. Woodger and J. Corcoran (eds.), *Logic, Semantics, Metamathematics: Papers from 1923 to 1938*, J. H Woodger (trans.), second edition, Indianapolis: Hackett, 1983, 421–54.

Tarski, A. (1983). *Logic, Semantics, Metamathematics: Papers from 1923 to 1938*, second edition, J. H. Woodger and J. Corcoran (eds.), Indianapolis: Hackett.

Williamson, T. (1994). *Vagueness*. London: Routledge.

Wright, C. J. G. (1975). 'On the Coherence of Vague Predicates,' *Synthèse* 30, 325–65.

Wright, C. J. G. (2001). 'On Being in a Quandary: Relativism, Vagueness, Logical Revisionism,' *Mind* 110, 45–98.

Wright, C. J. G. (2003). 'Rosenkranz on Quandary, Vagueness and Intuitionism,' *Mind* 112, 465–74.

Wright, C. J. G. (2007). '"Wang's Paradox,"' in R. E. Auxier and L. E. Hahn (eds.), *The Library of Living Philosophers: Michael Dummett*, Chicago, IL: Open Court, 415–44.

7

Quandary and Intuitionism
Crispin Wright on Vagueness

Stephen Schiffer

1. The philosophical problems of vagueness are set by the sorites paradox, an instance of which is given by the following inference (SI):

(1) A person with only 2¢ isn't rich.
(2) There isn't a non-negative integer n such that a person with n¢ isn't rich but a person with n + 1¢ is rich.
(3) ∴ A person with 5,000,000,000¢—that is, $50 million—isn't rich.

Put into symbolic abbreviation, this reads:

(1) ¬R(2)
(2) ¬∃n[¬R(n) & R(n + 1)]
(3) ∴ ¬R(5,000,000,000)

SI is a paradox because it presents four appearances that cannot all be veridical: first, it appears to be valid—after all, it's both classically and intuitionistically valid; second, its sorites premiss, (2), seems merely to state the obvious fact that in the sorites march from 2¢ to 5,000,000,000¢ there is no precise point that marks the cutoff between not being rich and being rich; third, premiss (1), which asserts that a person with only 2¢ isn't rich, is surely true; and fourth, the conclusion (3), which asserts that a person with 5,000,000,000¢— that is, $50 million—isn't rich, is surely false.

2. SI is evidently unsound, since it's conclusion is evidently false. So one thing we expect from a resolution of the paradox it presents is a pinpointing of why it's unsound. Perhaps this pinpointing would reveal that one of the four mutually incompatible appearances was determinately spurious; or perhaps it would merely reveal that it was in some way indeterminate which of the appearances was spurious. Either way, the pinpointing would show that one or more of the appearances didn't have the plausibility it might at first have seemed to have. So a second thing we should expect from a resolution of the paradox is an account of why it is that the now-identified problematic appearance, or appearances, ever seemed to have a better standing.

Stephen Schiffer, *Quandary and Intuitionism: Crispin Wright on Vagueness* In: *Logic, Language, and Mathematics: Themes from the Philosophy of Crispin Wright*. Edited by: Alexander Miller, Oxford University Press (2020). © Stephen Schiffer.
DOI: 10.1093/oso/9780199278343.003.0007

Many philosophers, including Crispin Wright, would suppose that the first task is easily accomplished: since SI is obviously valid, (1) is obviously true, and (3) is obviously false, we are simply bound without further ado to regard SI as a reductio of its sorites premiss (2). In other words, SI is determinately unsound because premiss (2) is determinately false. For these philosophers, the only difficult task is the second, which for them means explaining why (2) isn't true, given that it seems merely to be denying the apparent falsehood that there is a precise 1¢ cutoff that separates the non-rich from the rich.

For the classical logician, the negation of (2),

$$[\text{DN}] \; \neg \neg \exists n[\neg R(n) \, \& \, R(n+1)],$$

is equivalent, via double-negation elimination (DNE), to

$$[\exists] \; \exists n[\neg R(n) \, \& \, R(n+1)],$$

which seems to assert that there is the precise 1¢ cutoff in question. The classical logician has the formidable task of explaining how $\exists$ can be true, and supervaluation-ism and epistemicism dominate, if they don't exhaust,[1] the ways she has to do this.

The supervaluationist foregoes the semantic principle Bivalence (every statement is true or false) in order to keep the logical principle LEM (the law of excluded middle: every statement of the form $(S \lor \neg S)$ is necessarily true) and the rest of classical logic. She holds that if a statement is borderline, then it's neither true nor false, notwithstanding that the instance of LEM formed by it and its negation is necessarily true. She can hold this because, for her, a statement is true just in case it's true under every admissible precisification, false just in case it's false under every admissible precisification, and neither true nor false just in case it's true under some admissible precisification and false under another. An admissible precisification of a statement is, roughly speaking, the assignment to it of a precise bivalent interpret-ation under which the statement may be either true or false if it's borderline, but will be true under the interpretation if it's determinately true, and false under the interpretation if it's determinately false. In this way, a statement of the form $(S \lor \neg S)$ is guaranteed to be true even when, being borderline, neither S nor $\neg S$ is true, since in every admissible precisification one of the statements S or $\neg S$ will be true. Likewise, *mutatis mutandis*, for $\exists$: it may be true for the supervaluationist even though there is no particular number that makes it true; for it may be that even though there is no number n such that the statement that $[\neg R(n) \, \& \, R(n+1)]$ is true in every admissible precisification, there is for each admissible precisification a non-negative integer n such that the statement that $[\neg R(n) \, \& \, R(n+1)]$ is true in that precisification, thus making $\exists$ true in every admissible precisification, and thus true absolutely. So the supervaluationist might try to explain why the truth $\exists$ seems implausible by specu-lating that we mistakenly take it to imply that there is a sharp cutoff, that is that there is some non-negative integer n such that the statement that $[\neg R(n) \, \& \, R(n+1)]$ is true.

---

[1] I count Edgington's (1996) degree-theoretic treatment of vague discourse as a notational variant of supervaluationism: see Schiffer (2003: 193–4).

But this isn't a very good way to explain away ∃'s apparent implausibility. There remains the unanswered question of why, if the supervaluationist has given the correct account of ∃'s truth conditions, we supposed in the first place that in order for ∃ to be true there must be something that satisfies its quantified open sentence '[¬R(n) & R(n + 1)].' One is perhaps more likely to think that supervaluationism has it wrong and that in order for a disjunction to be true it really must have a true disjunct, and that in order for an existential generalization to be true, it really must have a witness. The more important problem, however, is that there are telling objections to supervaluationism as an account of vagueness.[2] For example, it may be that the statement that Louise won't go out with Harold because he's bald may be true, even though it's *false* on every admissible precisification (e.g., it's false that Louise won't go out with Harold because he has fewer than 2,152 hairs on his scalp),[3] and, as Wright observes, it also gives the wrong result on 'No one can knowledgeably identify a precise boundary between those who are tall and those who are not.'[4]

Wright has another telling objection to supervaluationism, one that suggests, as we'll see, his own novel account of vagueness. He points out that supervaluationism doesn't cohere with the characteristic mental state of one for whom a proposition presents itself as borderline. Actually, this objection isn't leveled just against supervaluationism, but to all *third-possibility views of indeterminacy*—that is, to every view according to which "indeterminacy consists/results in some kind of status other than truth and falsity—a *lack* of a truth value, perhaps, or the possession of some other truth value."[5] Wright's objection to any third-possibility theory, and thus to supervalutionism, is that:

It is quite unsatisfactory in general to represent *in*determinacy as any kind of *determinate* truth status—any kind of middle situation, contrasting with both the poles (truth and falsity)—since one cannot thereby do justice to the absolutely basic datum that in general borderline cases come across as *hard cases*: as cases where we are baffled to choose between conflicting verdicts about *which polar verdict applies*, rather than as cases which we recognise as enjoying a status inconsistent with both. Sure, sometimes people may non-interactively agree . . . that a shade of colour, say, is indeterminate; but more often—and more basically—the indeterminacy will be initially manifest not in (relatively confident) verdicts of indeterminacy but in (hesitant) differences of opinion (either between subjects at a given time or within a single subject's opinions at different times) about a polar verdict, which we have no idea how to settle—and which, therefore, we do not recognize as wrong.[6]

If the supervaluationist account of borderline propositions were correct, we would expect a person who took a proposition to be borderline to judge that it was neither true nor false. If, for example, you take Harry to be borderline bald, we should, given the supervaluationist account of vagueness, expect you to know that it's not the case that Harry is bald and not the case that he's not bald. Wright's point is that that isn't what we find. You won't judge that it's neither true nor false that Harry is bald but will instead be in a *quandary* about what to think about Harry's being bald. Your state of mind will be one of *ambivalence*, in that you will have some inclination to say

---

that Harry is bald and some inclination to say that he's not bald, and, if you're reflective, you're apt to doubt that anything could possibly be forthcoming to resolve the issue for you. The supervaluationist, or any other third-possibility theorist, must explain why you're in a state of quandary, or ambivalence, rather than in the entirely unambivalent state of knowing that it's neither a fact that Harry is bald nor a fact that he's not bald. What will the supervaluationist say, that you will have to master his theory in order for you to be in the correct mental state when Harry comes across to you as borderline bald?

The epistemicist accepts classical logic and Bivalence. For her, $\exists$ does entail, as it seems to do, that there is a non-negative integer n such that the statement that $[\neg R(n)\ \&\ R(n + 1)]$ is true. Indeed, epistemicism entails that everything we state has absolutely precise truth conditions. If I say that Mara is old, then the epistemicist must hold that there is some precise moment of time t and some number n such that my statement is true iff t is the precise moment of time at which Mara began to exist and n nanoseconds have passed since t. The epistemicist then devotes as much ingenuity as he can muster to explaining why we seem unable to locate these precise boundaries and why we mistakenly doubt their existence.

One problem with epistemicism is that there are good reasons for doubting that there are reference-determining factors that could determine precise truth conditions for our statements.[7] You call me in New York from Los Angeles and ask about the weather. I reply, "It's very cold." There seem not to be extension-determining factors that could make it the case that there is some precise area of space s and some precise number n such that my statement is true iff the temperature in s at the time of my utterance was $n°C$ or less. The epistemicist's commitment to there being such exquisitely fine-tuned extension-determining factors is on a par with Leibniz's commitment to this being the best of all possible worlds.

A second problem with epistemicism is that, like third-possibility theories of vagueness, it doesn't cohere with the characteristic mental state of one for whom a proposition presents itself as borderline—but in a different way from the way in which third-possibility theories don't cohere. You are presented with Harry in circumstances that are epistemically optimal for your judging whether or not he's bald and for judging whether or not he's tall. Suppose you take him to be borderline bald and borderline tall, and that you take there not to be any connection between a man's height and the extent of his hirsuteness. Since you take Harry to be borderline bald, you have some inclination to say that he's bald and some inclination to say that he's not bald, and likewise, *mutatis mutandis*, for his being tall. Suppose that these inclinations are of equal strength. To what degree should you be inclined the say that Harry is bald and tall?

If epistemicism were correct, you should be much less inclined to say that Harry is bald and tall than you are to say either that he's bald or that he's tall. Your inclinations-to-say would be manifestations of, and thus explained by, partial beliefs about propositions you took to be either true or else false, and these partial beliefs would be manifestations of your *ignorance*, and thus *uncertainty*, about facts beyond

---

[7] See, for example, Schiffer (1999) and (2003: 181–7); and Wright (2003b).

your ken—that he was bald, or else not bald, and that he was tall, or else not tall. These partial beliefs would arguably be normatively governed (relative to certain idealizations and qualifications) by the axioms of classical probability theory. So, if you believed to degree 0.5 that Harry was bald and to degree 0.5 that he was tall, and you took these two propositions to be independent, then you should believe the conjunction that Harry was bald and tall to degree 0.25. That is indeed how it should be, given that you think that what is at issue is a combination of facts that is beyond your ken. After all, you think each conjunct can go either way, but it's a much stronger thing to suppose that they will both go the same way.

But I don't think you would, or should, be much less inclined to say that Harry is bald and tall than you are to say either that he's bald or that he's tall. You would, and should, be as inclined to say that he's bald and tall as you are to say that he's bald and that he's tall, given that the degree to which you're inclined to say that he's bald is the same as that to which you're inclined to say that he's tall. Your situation isn't that you have only a small degree of evidence for a conjunction that may or may not be a fact. As far as you're concerned, all the evidence you could possibly have is already in. Your situation isn't one of *uncertainty*; it's one of *ambivalence*. You don't suppose that there are two secret facts—one concerning whether or not Harry is bald, the other whether or not he's tall; all the facts are right there before you, and your ambivalence just is your competing inclinations concerning how to describe the facts you know to be right under your nose. You're not lacking *evidence* for something that is determinately one way or determinately the other; it's that the underived conceptual roles of your concepts of baldness and of tallness are each pulling you in opposing directions, when you know there is no conceptual court of appeals to resolve the matter for you.

Inclinations-to-say regarding what you take to be matters of determinate fact are manifestations of partial beliefs. Is your inclination to say that Harry is bald a manifestation of a partial *belief* that Harry is bald, when Harry presents himself to you as a borderline case of a bald man? The answer may not be obvious.[8] If you do have a partial belief that Harry is bald, when you take him to be borderline bald, then that would show that there are two kinds of partial beliefs, distinguished by the principles that normatively govern them. In *The Things We Mean* I argued that there were these two kinds of partial beliefs, and I called them *standard partial beliefs* (SPBs) and *vagueness-related partial beliefs* (VPBs). SPBs are (under suitable idealization) normatively governed by the axioms of classical probability theory.[9] They are partial beliefs that manifest *uncertainty* about propositions we take to be determinately true or determinately false. If you believe to degree 0.5 that you left your keys in your office, then that is most likely an SPB and it represents your ignorance about where you left your keys, when you suppose that the question of where you left them has a determinate answer. When x comes across to you as borderline F, you will have

---

[8] See Schiffer (2003: ch. 5).

[9] This needs qualification for the interaction of SPBs with VPBs; a more accurate statement would be that SPBs would be normatively governed by the axioms of probability theory if they were the only kind of partial belief.

some inclination to say that x is F, and some inclination to say that x isn't F. If the mental states these inclinations manifest are beliefs, then they are VPBs, partial beliefs that are not normatively governed by the axioms of probability theory, and if in circumstances that are optimal for judging both Harry's baldness and his tallness you v-believe to degree 0.5 both that he's bald and that he's tall, then you should also v-believe to degree 0.5 that he's bald and tall. If, however, epistemicism were correct, the manifested partial beliefs would be SPBs, and you should believe the conjunction to degree 0.25, given that you take the two conjuncts to be independent. For present purposes, we don't need to resolve the issue of whether there are two kinds of partial *beliefs*. It's enough to see that the ambivalent state of mind of one who takes a proposition to be borderline isn't a manifestation of an SPB, the sort of partial belief in a proposition that goes with taking that proposition to be either determinately true or determinately false. But it would be if epistemicism were correct.

So, by my lights, neither supervaluationism (nor any other third-possibility theory of vagueness) nor epistemicism can accommodate the characteristic mental state of one for whom a proposition presents itself as borderline.

3. At this point we have something of an argument for an intuitionistic response to the sorites—at least for one who, like Wright, is confident that SI is a reductio of its sorites premiss. For this theorist accepts DN ($\neg\ \neg\exists n[\neg R(n)\ \&\ R(n + 1)]$), and she takes herself to have eliminated the only ways of defending $\exists$ ($\exists n[\neg R(n)\ \&\ R(n + 1)]$). She must therefore deny the validity of DNE (double-negation elimination), precisely the position of the intuitionist about logic, who is led to deny the validity of DNE by way of refusing to affirm LEM while, at the same time, having a disproof of its negation, $\neg(S \vee \neg S)$.

So far, perhaps, OK, but for the intuitionist response to the sorites to be part of a fully adequate resolution of that paradox, it must be shown to be motivated by an account of what it is to be a borderline case, and this in a way that explains the specious plausibility of the now denied sorites premiss. If the sorites premiss, $\neg\exists n[\neg R(n)\ \&\ R(n + 1)]$, is false, how did it come to seem plausible, and what is it about being a borderline case that allows us to accept DN but precludes our accepting the classically valid inference from it to $\exists$? Wright's answers to these questions, in so far as he has answers, may be gleaned from his account of vagueness.

Wright cannot accept any third-possibility account of vagueness, and he cannot accept the epistemic account of vagueness—for all that matters, any account that accepts Bivalence for borderline propositions and is thereby forced to accept the existence of the sharp cutoffs so much in doubt. But if he can't accept either a semantic or an epistemic account of vagueness, what sort of account is there that is left for him to accept? His implicit answer is that he can accept a *psychological* account of vagueness: an account of vagueness in terms of the characteristic mental state of one for whom a proposition presents itself as borderline, the state he calls *quandary*. If this is right, then among the questions confronting him are these:

(A) What is quandary, the characteristic mental state of one for whom a proposition presents itself as borderline?

(B) How does quandary enter into an account of vagueness?

(C) How does quandary help to motivate an intuitionistic response to the sorites?

Wright's most complete answer to (A) is given by the following conditions:

A proposition P presents a quandary for a thinker T [who is assessing P in circumstances that are optimal for judging P] just when the following conditions are met:

  (i) T does not know whether or not P;
  (ii) T does not know any way of knowing whether or not P;
  (iii) T does not know that there is any way of knowing whether or not P;
  (iv) T does not know that it is (metaphysically) possible to know whether or not P; and
  (v) T does not know that it is (metaphysically) impossible to know whether or not P.[10]

It seems indisputable that one who takes a proposition to be borderline will satisfy conditions (i)–(iii), but what about conditions (iv) and (v)?

Condition (iv) needs to be recast if it's to be correct. It needs to be recast because there is a clear sense in which you *do* know that it's metaphysically possible to know that Harry is bald, even though you take him to be borderline bald: you know that there is a metaphysically possible state of affairs in which Harry is as bald as a billiard ball and you know it. I take it that what is really at issue in both (iv) and (v) is the view Wright calls *Verdict Exclusion*, which he puts by saying that, where P is borderline, no judgment that P is true or that P is false is "knowledgeable,"[11] but which I'll take the liberty of restating as the view that

P's being borderline metaphysically entails that no one knows either P or not P.

Condition (iv) would like to say something to the effect that, while T knows that P is borderline, T does not know that P's being borderline is metaphysically compatible with either P or not P being known. But two things prevent Wright from having (iv) actually say that. First, P might present a quandary for T, and in that way "come across" to her as borderline, even though she lacks an explicit concept of being borderline and is thus incapable of judging that P is borderline. Second, even if that were not a problem, Wright can't allow a restatement of (iv) that explicitly uses the notion of being borderline, since his account of quandary is to be used to say what vagueness is, and (nearly enough for all that presently matters) a proposition's being vague just is the possibility of its being borderline. Evidently, the precisified statement of (iv) would have to find a condition C that did not involve any notion which quandary is earmarked to explain such that (iv) could be rewritten as:

T knows that P satisfies C but does not know that P's satisfying C is metaphysically compatible with P or not P being known.

<hr />

[10] Wright (2003a: 504–5).
[11] Wright (2003b: 92). In (2003a) Wright defined Verdict Exclusion as the view that "a borderline case is something about which we know that a knowledgeable positive or negative verdict is ruled out" n. 47, p. 490, but the (2003b) rendition is the most recent. Wright (personal communication) uses 'knowable' [= capable of being known to be true or to be false] as a predicate of propositions, but uses 'knowledgeable' [= expressive of knowledge] as a predicate of opinions or verdicts.

Since I don't know what C might be, I'll leave the conditions as they are but understand them as though they enjoyed the needed precisification.

A corresponding revision is required for (v); but (v) raises a further issue. Just as (iv) would like to say that T does not know that P's being borderline is metaphysically compatible with either P or not P being known, so (v) would like to say that T does not know that P's being borderline is metaphysically *in*compatible with either P or not P being known. That is to say, (v) aims to express a refusal to accept Verdict Exclusion without explicitly using the notion of being borderline. Now, insofar as (iv) is intended to express the thought that no one knows that it's possible for a proposition to be both borderline and known,[12] it seems clearly to be correct. The same, however, can't be said for (v)'s agnosticism about Verdict Exclusion, as many writers on vagueness, including myself, take Verdict Exclusion to be clearly correct: we take it to be a datum that it's metaphysically *im*possible for a proposition to be both borderline and known. So why does Wright not accept Verdict Exclusion? He proposes two reasons in "Quandary."[13]

His first reason was that he found plausible a certain principle of *Evidential Constraint* (EC) and recognized that EC was incompatible with Verdict Exclusion. EC is a principle Wright takes to be satisfied by the propositions expressed by sentences involving predications of those atomic predicates—'red,' 'bald,' 'heap,' 'tall,' 'child,' etc.—that provide the standard examples of sorites paradoxes:

[A] wide class of vague expressions seem to be compliant with an intuitive version of Evidential Constraint: If someone is tall, or bald, or thin, that they are so should be verifiable in normal epistemic circumstances. Likewise if they are not bald, not tall, or not thin.[14]

Suppose 'F' is a predicate that satisfies EC. Then, following Wright, we may represent the satisfaction of EC by 'F' by the pair of EC conditionals

Fx $\rightarrow$ it is feasible to know that Fx
$\neg$Fx $\rightarrow$ it is feasible to know that $\neg$Fx

Now, the predicates that satisfy EC may have borderline cases, and for these atomic predicates Fx is borderline iff $\neg$Fx is; so if it were not feasible to know a borderline proposition, then the consequents of the two displayed EC conditionals would be false, and we would have the contradiction $\neg$Fx & $\neg$ $\neg$Fx, thus proving that EC is incompatible with Verdict Exclusion. To whatever extent Wright is inclined to accept EC, at least to that same extent he must be inclined to deny Verdict Exclusion.[15]

---

[12] We may re-express the thesis at issue in this abbreviated way since the negation of a borderline proposition is itself borderline.

[13] In a footnote (n. 47, 489–91) Wright suggests that a further argument against the Verdict Exclusion view is that it can't accommodate higher-order vagueness. The argument relies on the questionable premiss that if P is on the borderline between being definitely true and being first-order borderline, then one isn't warranted in believing that one isn't warranted in believing P or in believing not P. But if one knows that P is second-order borderline in that way, then one knows that it's indefinite whether P is definitely true or neither definitely true nor definitely false, and knowing that precludes one from being warranted in believing either P or not P.

[14] Wright (2003b: 96).

[15] To deny Verdict Exclusion is to assert that it's not impossible for a proposition to be both borderline and known, but this doesn't mean that if Wright denies Verdict Exclusion then he must accept that it is

Wright's second reason for denying Verdict Exclusion was that

(a) if we knew that no verdict about a borderline case could be knowledgeable, then we would be "committed to regarding anyone who advanced a verdict, however qualified, as strictly out of order—as making an ungrounded claim and performing less than competently";[16] but

(b) we are not so committed, for it is an "absolutely basic datum"[17] about borderline cases that "the impression of a case as borderline goes along with a [warranted] readiness to tolerate others' taking a positive or negative view—provided, at least, that their view is suitably hesitant and qualified and marked by a respect for one's unwillingness to advance a verdict."[18]

This argument can bear elaboration, but I'll reserve it for when I scrutinize the argument in the critical part of this chapter.

So much for Wright's answer to question (A), his account of *quandary*, which is what he calls the characteristic mental state of one for whom a proposition presents itself as borderline.

4. Wright gives an incomplete answer to question (B) (How does quandary enter into an account of vagueness?). He recognizes that he can't answer (B) by saying that borderline propositions are just those that give rise to quandary:

[S]ome quandaries—Goldbach's conjecture, for instance—feature nothing recognizable as vagueness; and others—that infidelity is alright provided nobody gets hurt, perhaps—may present quandaries for reasons other than any ingredient vagueness. So the task of a more refined taxonomy remains—the notion of quandary is just a first step.[19]

He then offers a tentative proposal to the effect that an account of what it is for x to be borderline φ—which of course is one leading way for the proposition that x is φ to be borderline—would combine three conditions: (i) that quandary is the characteristic mental state of one for whom x presents itself as borderline φ (notwithstanding that quandary is also the characteristic mental state associated with other kinds of indeterminacy); (ii) that it is known that there are, or at least could be, things that were known to be φ and things that were known not to be φ; and (iii) that it isn't known that it isn't known whether or not x is φ.

But even if we accept the entertained three additional conditions, the fact that "quandaries are relative to thinkers... and to states of information"[20] shows that we would still lack a sufficient condition for x's being borderline φ—or indeed for any kind of indeterminacy. For if a proposition is borderline, or in any other way indeterminate, then, while there may be possible worlds in which it isn't indeterminate, it is actually indeterminate at all times in the world in which it is indeterminate, no matter who is attempting to judge it.

It's possible Wright would claim that vagueness can be explained in terms of quandary in a sense that is compatible with there not being any way of *defining*

---

possible for a proposition to be both borderline and known: denying DNE as he does, he has the option of not inferring the possibility of knowledge from the denial of its impossibility.

[16] Wright (2003b: 94).   [17] Wright (2003a: 476).   [18] Wright (2003b: 92).
[19] Wright (2003b: 504).   [20] Wright (2003b: 505).

vagueness in terms of quandary. After all, what in philosophy can be defined? In the end, Wright suggests, not uncryptically, that "the notions of observationality and response-dependence would provide two obvious foci for the search" for a final account of vagueness; but he leaves that search for another day, and concludes his article on quandary on the hope that he "conveyed ... something of the general shape which a stable intuitionistic philosophy of vagueness might assume."[21]

5. This brings us to question (C) (How does quandary help to motivate an intuitionistic response to the sorites?). Wright, we have already noted, would regard it as "really only common sense" that SI—viz.

(1) $\neg R(2)$
(2) $\neg \exists n[\neg R(n) \& R(n + 1)]$
(3) $\therefore \neg R(5,000,000,000)$

—is a *reductio* of its sorites premiss, (2), and on this the majority of philosophers who take on the sorites would agree with him. What distinguishes Wright from those other philosophers is that he would take himself to be warranted in not accepting

$$(\exists) \quad \exists n[R(n) \& \neg R(n - 1)],$$

notwithstanding a disproof of its negation, a position that, of course, requires him to deny the validity of classical logic's DNE, which *would* enforce the inference to $\exists$ from its double negation,

$$(DN) \quad \neg \neg \exists n[\neg R(n) \& R(n + 1)]$$

His warrant, we know, is based on—although not exclusively on—his quandary account of vagueness.

There are three things Wright must do to achieve a fully adequate response to the sorites, and he intends quandary to play a crucial role in all three achievements: (a) He must show that we are justified in not accepting the validity of DNE; (b) he must show why we are justified in refusing to accept $\exists$, notwithstanding a disproof of its negation;[22] and (c) he must explain how SI's sorites premiss, $\neg \exists n[\neg R(n) \& R(n + 1)]$, came by what he takes to be its spurious plausibility.

(a) Wright, qua anti-realist, holds that vague atomic predicates apt to enter into sorites reasoning satisfy evidential constraint (EC). Letting F be such a predicate, we then have

$$(i) \quad (F\alpha \rightarrow FeasK[F\alpha]) \& (\neg F\alpha \rightarrow FeasK[\neg F\alpha])$$

Suppose that α is borderline F. Then, given the quandary account of borderline propositions, we don't know whether or not it's feasible to know either $F\alpha$ or $\neg F\alpha$, that is,

---

[21] Wright (2003b: 508).
[22] Wright needs (b) as well as (a), because merely showing that we should accept a logic in which DNE isn't valid does nothing to show that the proposition doubly negated in a true double negation isn't also true.

$$\text{(ii)} \quad \neg K(\text{FeasK}[F\alpha] \vee \text{FeasK}[\neg F\alpha])$$

A problem arises if we now further assume that LEM holds for borderline propositions, that is that

$$\text{(iii)} \quad F\alpha \vee \neg F\alpha$$

For (i) and (iii) entail

$$\text{(iv)} \quad \text{FeasK}[F\alpha] \vee \text{FeasK}[\neg F\alpha]$$

The problem is that we would then, quite inexplicably, be unable to know both (i) and (iii), for if we could know (i) and (iii), then we could also easily come by deduction to know (iv), which knowledge is ruled out by (ii). It is to this line of reasoning that Wright alludes when he writes:

[I]t cannot stably be supposed that each of EC, LEM and NKD [(ii), for our purposes] is known. Anti-realism supposes EC is known a priori, and NKD seems incontrovertible.... So the anti-realist must suppose that LEM is not known—agnosticism about it is mandated so long as we know that we don't know that it is feasible to decide any significant statement. Since logic has no business containing first principles that are uncertain, classical logic is unacceptable in our present state of information.[23]

Now, while for the anti-realist ($F\alpha \vee \neg F\alpha$) isn't known, it's also the case that she has a disproof of its negation, and thus a proof of its double negation,

$$\neg\neg(F\alpha \vee \neg F\alpha),$$

thereby precluding her from accepting the validity of DNE. In this way, the anti-realist is precluded from accepting that one can validly infer

$$\text{(}\exists\text{)} \quad \exists n[\neg R(n) \,\&\, R(n+1)]$$

from

$$\text{(DN)} \quad \neg\neg\exists n[\neg R(n) \,\&\, R(n+1)]$$

But this line of argument falls short of establishing what Wright needs to establish. It shows that *the anti-realist*—who by definition claims to know EC—cannot accept the validity of DNE; but what Wright owes us is an argument to show that *we*—who may not be anti-realists—cannot be justified in accepting the validity of DNE. Since NKD is indeed incontrovertible, the contest is between EC and LEM. Many philosophers will take this to be an argument for rejecting EC; so where is Wright's *argument* for EC?

He doesn't offer one, but instead acknowledges that

The general thrust of our discussion involves—as one would naturally expect of an advertised intuitionistic treatment—a heavy investment in EC. As I have said, I believe the principle is plausible for the kinds of statement that feature in the classic examples of the Sorites paradox....[24]

---

[23] Wright (2003b: 470).   [24] Wright (2003b: 489–90, n. 47).

Wright does, however, suggest that he will have the argument he needs if he can show that one should regard EC as merely an epistemic possibility. For suppose we accept that quandary, as characterized, is true of borderline cases and that while we don't actually endorse EC, we nevertheless leave open the possibility that it might be right. Then

> it must be that our (presumably a priori) grounds for LEM are *already* inconclusive.... But in that case we should recognize that LEM already lacks the kind of support that a fundamental logical principle should have....[25]

So much—for now—for Wright's case for refusing to accept the inference from DN to ∃.

(b) We also need to see what reason there is for being agnostic about ∃, especially given that we're supposed to be convinced that its negation, the sorites premiss, is false. The argument again turns on EC, and may be reconstructed in the following way.

If ∃, then it has a unique realizer—a unique truth of the form $[\neg R(\alpha) \,\&\, R(\alpha + 1)]$.[26] If—as we're assuming on behalf of the anti-realist—EC holds for R, then it would also be the case that it was feasible to know $\neg R(\alpha)$ and feasible to know $R(\alpha + 1)$, and thus feasible to know $[\neg R(\alpha) \,\&\, R(\alpha + 1)]$. But we know from construction of the sorites series in question that both $\neg R(\alpha)$ and $R(\alpha + 1)$ would be borderline propositions, and, given the quandary account of borderline propositions, it isn't known that it's feasible to know either proposition, and therefore isn't known that it's feasible to know their conjunction. So—since we're taking EC as known—we know that if ∃, then it has a realizer that is feasibly known. But we don't know that it has a feasibly known realizer. Evidently, for any P, Q, if one knows that if P, Q, and one doesn't know Q, then, ceteris paribus, one doesn't know P. We may take cetera to be paria in the present case. Therefore, we don't know ∃. Since we do know that ∃'s negation is false, we see that ∃ is another quandary.[27]

This argument does seem sound to me—given EC.

(c) But how is Wright to explain the seeming plausibility of the sorites premiss, given that he takes it to be false? When I glossed the intuitive case for SI's sorites premiss, I stressed the fact that the proposition it denies, $\exists n[\neg R(n) \,\&\, R(n + n)]$, seems to imply the incredibly implausible claim that there is a precise 1¢ cutoff between the non-rich and the rich. But Wright would give a different—or, at least, an additional—gloss. He would say that we are inclined to suppose that it's a conceptual truth about vagueness that 'rich' would not be vague if there were the sharp cut off entailed by $\exists n[\neg R(n) \,\&\, R(n + n)]$, and that because of this we take the very vagueness of 'rich' to consist, at least in part, in the truth of $\neg \exists n[\neg R(n) \,\&\, R(n + n)]$. But he would then go on to say that we can see that that isn't a good motivation for regarding the sorites premiss as true once we realize that the vagueness of 'rich' consists not in the truth of the sorites premiss, but rather in the *quandary status* of

---

[25] Wright (2003b: 486, n. 44).

[26] A *unique* truth of that form by virtue of its being a conceptual truth about being rich, that if having a given amount of money makes you rich, then having more than that also makes you rich.

[27] The argument this reconstructs is at Wright (2003b: 487).

the "unpalatable existential"[28] $\exists n[\neg R(n) \ \& \ R(n + n)]$, and in the realization that we are warranted in refusing to accept the unpalatable existential even while denying the sorites premiss. Still, I think Wright would allow that he hasn't yet offered a complete explanation of the apparent plausibility of the sorites premiss, since he hasn't yet explained why *common sense* is prone to make the *mistake* of denying that vague terms have precise extensions—and thus precise cutoffs in sorites series— when it is merely entitled to be agnostic on that score.

That concludes my rendering of Wright's quandary-based intuitionistic theory of vagueness. I am very sympathetic to a good deal of Wright's ingenious theory, and I regard my own view on vagueness as a relative of his. But intra-familial disputes are not unknown, and the remaining sections are critical of Wright's theory.

6. Quandary, we may recall, is defined thus:

A proposition P presents a quandary for a thinker T [who is assessing P in circumstances that are optimal for judging P] just when the following conditions are met:
   (i) T does not know whether or not P;
   (ii) T does not know any way of knowing whether or not P;
   (iii) T does not know that there is any way of knowing whether or not P;
   (iv) T does not know that it is (metaphysically) possible to know whether or not P; and
   (v) T does not know that it is (metaphysically) impossible to know whether or not P.

I agree that what is needed to account for vagueness is neither an epistemic nor a third-possibility theory of vagueness, but rather a psychological theory of vagueness which explains vagueness in terms of the characteristic mental state of one for whom a proposition presents itself as borderline. I do not, however, think that quandary, as defined by Wright, is an adequate characterization of that mental state, even if conditions (iv) and (v) are adequately repaired to convey his intended meaning.

One problem with those conditions is that they are too easily satisfied to be of any use, given that 'T' ranges over all thinkers, including the most unreflective and least intelligent of us. For even if it were possible for some of us to know that there is a way to know whether or not borderline Harry was bald, such knowledge might simply be beyond the cognitive reach of dull Clyde; yet that wouldn't preclude the proposition that Harry is bald from presenting a quandary for him. The needed revision cannot be to replace 'does not know' with 'cannot know,' since that would conflict with the intended import of (iv) and (v). Somehow T's not knowing must be circumscribed so that, even if it's over determined, P's being borderline is sufficient to secure it. But we already know that Wright can't use the notion of a borderline proposition in stating the conditions on quandary. I assume the revision must in effect take the form of each condition's beginning with an appropriate completion of the form

P's having features ____ is sufficient for T's not knowing that ... ,

---

[28] This is what Wright calls existentials $\exists$ in (2003a).

although I myself don't know the features that can fill in the blank.

A second problem, to which I have already alluded, is that condition (v) aims to express (without explicitly invoking the notion of being borderline) the claim that we can't know that Verdict Exclusion is true, and that, I submit, is wrong: we *can* know that it's metaphysically impossible for a proposition to be both borderline and known.

And a third problem is that even if the conditions on quandary were revised just enough to entail Verdict Exclusion, those corrected conditions would still fail to get at what is truly distinctive about the state of mind that defines what it is to take a proposition to be borderline.

Let me elaborate these two last problems in turn.

One who accepts Verdict Exclusion, and thus rejects (v), takes herself to have two good reasons for thinking she can know that it's metaphysically impossible for a proposition to be both borderline and known.

The first reason is the simple observation that it's evidently metaphysically necessary that, say, if Louise *knows* that the ball is red, then the ball is definitely red, and thus not borderline. This simple observation certainly gets my immediate assent, and I take my response to be fully justified. Is there any reason to doubt it, apart from Wright's anti-realist driven brief? Well, being borderline is one way of being indeterminate, and Cian Dorr has argued that it's consistent with P's being indeterminate that it's indeterminate whether one knows P.[29] It's clearly possible for a proposition to be borderline, so if Dorr is right, then it should be indeterminate whether one can know a borderline proposition, since all that can be determinately true is that, necessarily, if P is borderline, then no one *determinately* knows P. But I don't think Dorr can be right. I think he overlooks that it's as much of a platitude to say that

If someone knows that S, then it's determinately true that S

as it is to say that

If someone knows that S, then S.

The second good reason for accepting Verdict Exclusion is available by reflection on one's epistemic position when confronted with what one knows to be a borderline proposition. Suppose that you are holding a ball in your hands in circumstances that are as good as you can conceive of them getting for judging the ball's color. You are certain you know what color the ball is, whether or not you have a word for that color in your vocabulary; you know that you have mastery of the concept of red; and you know with certainty that the color you know the ball to have is not one you can now justifiably say either is or is not red. Furthermore, you cannot even *conceive* of how, given all you know, you could come to have warrant for judging either that the ball is red, or that it isn't red. You rightly take yourself to be in the best possible position to verify whether or not the ball is red, and you can't imagine what you could conceivably find out that would give you knowledge that the ball was, or wasn't, red. Given all that, I would think that you would be entirely justified in thinking that

[29] Dorr (2003).

it's impossible, given the obtaining facts, for you, or anyone else, to know whether or not the ball is red. Examples involving any other sorites-prone concept can be used to make the same point.

But what about Wright's reasons for denying Verdict Exclusion?

His first reason was that Verdict Exclusion is incompatible with the satisfaction of EC by the vague terms used to express borderline propositions. There are two problems. First, *every* vague expression must satisfy EC if Wright is to have a fully general account of vagueness and a complete resolution of the sorites; but even if *some* vague expressions, such as 'red,' satisfy EC, it's highly doubtful that they *all* do. 'Brave,' for example, is vague, but it's doubtful that it's evidentially constrained. Second, I should think that whatever case there may be for EC pales by comparison with the case for thinking that it's metaphysically impossible for a proposition to be both borderline and known, and anyway I don't see that there is much of a case for EC. What *does* seem right about EC is that if a thing is *definitely red*, then it may be known to be red, and if a thing is *definitely not red*, then it may be known not to be red.

Wright's second reason for denying Verdict Exclusion was that:

(a) if we knew that no verdict about a borderline case could be knowledgeable, then we would be "committed to regarding anyone who advanced a verdict, however qualified, as strictly out of order—as making an ungrounded claim and performing less than competently";[30] but

(b) we are not so committed, for it is an "absolutely basic datum"[31] about borderline cases that "the impression of a case as borderline goes along with a [warranted] readiness to tolerate others' taking a positive or negative view— provided, at least, that their view is suitably hesitant and qualified and marked by a respect for one's unwillingness to advance a verdict."[32]

My response is that, while there are two ways in which (b) may be true, (a) is false on both those ways, and I am not aware of a true reading of (b) that sustains (a). In other words, the ways in which we must recognize that a person needn't be at fault in taking a "suitably hesitant and qualified" view about a borderline proposition are entirely consistent with our knowing that the proposition can't be known. There are two kinds of case to consider.

The first adverts to my argument (above, on p. 157) to show that epistemicism doesn't cohere with the characteristic mental state of one for whom a proposition presents itself as borderline. Harry comes across as borderline bald both to Al and to Betty. Consequently, both Al and Betty have some positive inclination to say that Harry is bald, and some positive inclination to say that Harry isn't bald. At the same time, we may suppose, there is this difference between Al and Betty: Al's concept of baldness places Harry closer to being bald than to being not bald, so that his inclination to say that Harry is bald is stronger than his inclination to say that Harry isn't bald, whereas just the opposite is true of Betty. We may even suppose that these inclinations are fairly strong, so that, if we pretend they can be measured by

---

[30] Wright (2003b: 94).   [31] Wright (2003a: 476).   [32] Wright (2003b: 92).

real numbers in the interval [0, 1], we may say that Harry is inclined to degree 0.75 to say that Harry is bald, while Betty is inclined to degree 0.75 to say that he is not bald. Such differences in inclinations-to-say are an inescapable feature of vagueness—or, if you will, of the vagaries of the conceptual roles of vague concepts both inter-personally at a given time and intra-personally at different times—and in such a case, neither Al nor Betty may be said to be in any way cognitively at fault. In such a case we may, if we want, say that Al's inclination to say that Harry is bald manifests a "suitably hesitant and qualified" positive stance towards the proposition that Harry is bald, while Betty's inclination to say that Harry isn't bald manifests a "suitably hesitant and qualified" negative stance towards that proposition. But recognizing that we are in no position to find epistemic fault either with Al's stance or with Betty's doesn't preclude our knowing that, given how things are hair-wise for Harry, no one can know either that he is bald or that he is not bald. But how can this be? Surely, a person who is inclined to degree 0.75 to assert a proposition is somewhat confident that the proposition is true, and while it's possible for one to be justified in having that degree of confidence in a proposition that can't be known, that would be because one's epistemic position with respect to the proposition was less than ideal and that, for whatever reason, one was ineluctably limited to incomplete evidence. But that can't be Al's situation; he is in the best possible position for judging whether or not Harry is bald and thus has all the evidence he could possibly have right under his nose. So how can Al's being confident to degree 0.75 be anything other than cognitively misbegotten?

The answer is that Al's degree 0.75 inclination to say that Harry is bald doesn't represent a 0.75 degree of *confidence* that Harry is bald. In fact, it doesn't represent *any* degree of confidence that Harry is bald. It's the borderline-case-induced inclination-to-say that someone could rationally have even if he fully believed that there was no knowable fact of the matter about Harry's baldness. If Al were a philosopher, he might say that he takes Harry to be determinately borderline bald, but that while he places Harry squarely in the penumbra of his, Al's, concept of baldness, he places Harry much nearer to the edge of the penumbra that shares a very fuzzy border with definite baldness than to the edge that shares a very fuzzy border with definite non-baldness. Al's 0.75 degree inclination to say that Harry was bald would call into question Al's epistemic competence only if the inclination manifested what I earlier called an SPB, the kind of partial belief that is normatively governed by the axioms of probability theory. If Al s-believed to degree 0.75 that Harry was bald, then he would be prepared to say that it was quite likely that Harry was bald, and he would take himself not to be in the best circumstances for judging whether or not Harry is bald. Then, subject to a qualification presently to be discussed, we could correctly judge that Al had a belief—an SPB—that was unwarranted. But, by con-struction of the example, Al's inclination to say that Harry is bald doesn't manifest any SPB; although he is inclined to degree 0.75 to say that Harry is bald, he has no inclination to say that it's quite likely that Harry is in fact bald, as he would have if his inclination-to-say manifested an SPB. It either manifests no belief state, properly so called, or else it manifests a VPB, a kind of propositional attitude that isn't norma-tively governed by the axioms of probability theory. Either way, Al's inclination to say that Harry is bald merely manifests the positive side of the two-sided pull that defines the ambivalent, or quandary, state of mind of one who takes a proposition to be borderline, and Al's having it is entirely consistent with his knowing that he is in the best possible circumstances for judging whether or not Harry is bald, and thus that no one can know whether or not Harry is bald.

So much for the first kind of case. In that kind of case, Harry comes across as borderline bald to Al and to Betty, and, therefore, their dispositions to say that Harry is bald are manifestations not of SPBs, but of the states I have called VPBs, and that is why neither can be held to be epistemically at fault by one to whom Harry also comes across as borderline bald. In the correlative instance of the second kind of case, Harry doesn't come across as borderline bald either to Al or to Betty; their dispositions to say that Harry is bald, or not bald, manifest SPBs; they enjoy the same optimal circumstances for judging baldness as those enjoyed by a third person—call him Carl—to whom Harry does come across as borderline bald; and yet Carl, and everyone else, is still correct not to find either Al or Betty to be epistemically at fault. In this instance of the second kind of case, we may even have it that:

- Not only does Harry come across to Carl as borderline bald, but Carl, at ease with his concept of being borderline, explicitly s-believes Harry to be borderline bald.
- Al denies that Harry is borderline bald and s-believes him to be bald.
- Betty denies that Harry is borderline bald and s-believes him not to be bald.
- None of Al, Betty, or Carl is epistemically at fault.

The further facts of the story which vindicate these bulleted points are as follows. Al, Betty, and Carl are educated native speakers of English, none of whom uses 'bald' incorrectly. At the same time, each of their personal concepts of baldness differs somewhat from the others,' even to this extent: that under conditions that are optimal for judging whether 'bald' applies to Harry, Al's use of 'bald' at t induces Al to judge that 'bald' applies to Harry, Betty's use of 'bald' at t induces her to judge that 'bald' doesn't apply to Harry, and Carl's use of 'bald' at t induces him to judge that Harry is a borderline case of a man to whom 'bald' applies. This could happen if Harry's scalp appears as hairless as a billiard ball but it's common knowledge that he shaves his scalp and that no one would judge him to be bald if he stopped shaving it and let his hair grow out. Is Harry bald? Suppose, what I have reason to believe is true, that among native speakers of English none of whom could be said to use 'bald' incorrectly, some would answer yes, some would answer no, and some would indicate in one way or another that the question has no determinate answer. Then we may make good sense of Al's, Betty's, and Carl's utterances as follows. For many speakers of English, including Carl, Harry falls into the penumbral region determined by their use of 'bald'; but Al's use of 'bald' is such that Harry falls outside of the penumbral region determined by Al's use of 'bald' and into, so to say, the positive region the use determines; and for Betty the same is true, except that her use of 'bald' has Harry falling in the negative region determined by her use of 'bald.' I trust it's clear that none of our trio need be epistemically at fault. At the same time, it's not clear what to say about what they believe. Should we say that, owing to their different concepts of baldness, each has a true SPB about Harry, an SPB that Al would express by uttering 'Harry is bald,' Betty by uttering 'Harry isn't bald,' and Carl by uttering 'It's indeterminate whether Harry is bald,' or should we say that they are all addressing the same determinate question—viz. whether or not Harry is bald—and disagreeing about the answer? Our criteria for individuating what is said in an utterance, and thus the criteria we use for deeming two people to have said the same thing, are course-grained, context-dependent, and suffer large areas of indeterminacy, so much so that I doubt that the question just posed has a determinate answer. Its not having a determinate answer is, however, compatible with the further claim that, once apprised of the relevant facts about how English speakers divide on what they are inclined to say about shaved heads, it ought to be clear to all concerned that *as 'bald' is used among English speakers*, it is indeterminate whether it applies to someone who wouldn't be bald if he didn't shave his scalp. In any case, these ways of giving a true reading to Wright's (b) leave him without a true a reading of (a), and thus without a good basis to maintain that we don't know that it's metaphysically impossible for a proposition to be both borderline and known.

My third objection to Wright's account of quandary applies even if his conditions were minimally altered so as to accept Verdict Exclusion. The objection is that the resulting conditions would still leave out what is truly distinctive about the characteristic state of mind of one who takes a proposition to be borderline. This is because even with that correction the conditions wouldn't capture the all-important fact that the characteristic mental state of one for whom a proposition presents itself as borderline is a *state that induces inclinations-to-say but isn't an SPB*.

7. If what I've said so far is correct, Wright's case for an intuitionistic solution to the sorites inference SI is undermined. We can't accept Wright's reason for saying that we should be "agnostic" about what the sorites premiss denies—viz. ($\exists$) $\exists n[\neg R(n)$ & $R(n + 1)]$—while accepting its double-negation, and we have no reason to accept his explanation of why we mistakenly supposed $\exists$ to be false. Nevertheless, I think that an important component of his solution may be right. This is that $\exists$ presents a quandary, and is thus indeterminate. I'll try to explain.

Wright began his approach to the sort of sorites SI presents by taking it to be a datum beyond dispute—"merely common sense"[33]—that the sorites premiss was false, and thus that DN ($\neg \neg \exists n[\neg R(n)$ & $R(n + 1)]$) was true. But then one was confronted by the classically valid inference from DN to $\exists$, and by the fact that the only ways of making sense of $\exists$'s being true seemed unacceptable. There was then no wiggle room left in which to maneuver: if one had to accept DN but couldn't accept $\exists$, one was forced thereby to agree with the intuitionist that DNE wasn't valid, and thus to deny that $\exists$ was entailed by DN. The trouble is that that intuitionistic move needs to be sustained by an explanation that progresses beyond the bind of finding DN but not $\exists$ acceptable, and it's precisely Wright's attempt to provide that explanation that has been called into question. My suggestion isn't that there is some other explanation to sustain the intuitionistic response; it's that one shouldn't accept that the sorites premiss, $\neg \exists n[\neg R(n)$ & $R(n + 1)]$, is false, and thus shouldn't accept its negation, DN. Rather, one should agree with Wright that $\exists$ is indeterminate, but disagree with him about the falsity of its negation—one should hold that $\exists$ is

---

[33] Wright (2003b: 99).

indeterminate in that it's indeterminate whether or not it's true and indeterminate whether or not it's false. (This doesn't, however, mean that one should be "agnostic" about the sorites premiss, as though it might turn out to be true; being agnostic about a proposition means s-believing it to more or less the same degree that you s-believe its negation, and s-believing it to any positive degree is precluded by your taking it to be indeterminate.)

The rub—for me—is that if one takes SI's sorites premiss to be indeterminate, then, in the sense of validity according to which an argument is valid only if it's metaphysically impossible for its premises to be true and its conclusion false, one will also have to say that it's indeterminate whether or not SI is valid, since it's indeterminate whether it has true premises and a false conclusion (I'm assuming that SI's first premiss is determinately true and that its conclusion is determinately false). To make matters worse, one will also have to say that it's indeterminate whether *modus ponens* is valid. For consider the very long sorites inference SI':

$R(5,000,000,000)$
$R(5,000,000,000) \rightarrow R(4,999,999,999)$
$R(4,999,999,999) \rightarrow R(4,999,999,998)$
o
o
o

$\therefore R(2)$

This inference is valid if *modus ponens* is a valid rule of inference, but if I'm to say that SI's sorites premiss is indeterminate, then I shall also have to say that each conditional premiss of SI' is indeterminate if it isn't determinately true, and of course the first premiss—that a person with $50 million is rich—is determinately true, and the conclusion—that a person with only 2¢ is rich—is determinately false. How terrible is this?

In saying that it's indeterminate whether *modus ponens* is valid—in the sense of 'valid' according to which it applies to an inference only if it's metaphysically impossible for it to have true premises and a false conclusion—I'm not saying that I *doubt* that *modus ponens* is valid, as though its turning out to have counterexamples were an epistemic possibility. Doubt and epistemic possibility require SPBs, and they are precluded when one takes a proposition to be indeterminate. It *isn't* epistemically possible that, or an "open question" whether, an instance of *modus ponens* should have true premises and a false conclusion. Nor is my claim that it's indeterminate whether *modus ponens* is valid *in the stipulated sense of 'valid'* inconsistent with its being valid in some other sense. For example, in Hartry Field's recent Kleene-enhanced logic he recognizes three notions of validity, which he calls *validity, strong validity*, and *universal validity*.[34] *Modus ponens* in his logic satisfies all three notions of validity. Nevertheless, Field shares my view that SI's sorites premiss is indeterminate and that the conditional premises in the just-displayed long inference are indeterminate when not determinately true.

* * *

[34] Field (2003).

In this discussion of Wright's theory of vagueness I have been critical of his attempt to resolve the sorites along intuitionistic lines, and I have been critical of his account of quandary. At the same time, I think his theory is an important step forward on this extremely difficult problem. His objections to supervaluationist and epistemicist theories of vagueness are by my lights spot on, and, more importantly, I think he is correct that vagueness needs to be explained in terms of the sort of quandary state one is in when one takes a proposition to be borderline.

References

Dorr, C. (2003). "Vagueness without Ignorance," *Philosophical Perspectives* 17, 83–114.

Edgington, D. (1996). "Vagueness by Degrees," in R. Keefe and P. Smith (eds.), *Vagueness: A Reader*, Cambridge, MA: MIT Press.

Field, H. (2003). "Semantic Paradoxes and Vagueness Paradoxes," in J. C. Beall (ed.), *Liars and Heaps: New Essays on Paradox*, Oxford: Oxford University Press.

Schiffer, S. (1999). "The Epistemic Theory of Vagueness," *Philosophical Perspectives* 13, 481–503.

Schiffer, S. (2003). *The Things We Mean*, Oxford: Oxford University Press.

Wright, C. (2003a). "On Being in a Quandary: Relativism, Vagueness, Logical Revisionism," in *Saving the Differences: Essays on Themes from Truth and Objectivity*, Cambridge, MA: Harvard University Press.

Wright, C. (2003b). "Vagueness: A Fifth Column Approach," in J. C. Beall (ed.), *Liars and Heaps: New Essays on Paradox*, Oxford: Oxford University Press.

PART III
Logical Revisionism

8

Wright and Revisionism

Sanford Shieh

The nature of logic is a perennial focus of philosophical interest. It is all but indisputable that the laws of logic constitute the most fundamental standards of correctness in reasoning. In virtue of what, however, does something qualify as such a norm of reasoning? This question would have no obvious point if there were universal agreement on the laws of logic. Some principles of deductive reasoning have indeed been widely accepted from Aristotle through Frege to the present day. One such principle is the law of excluded middle, the form of inference defining classical logic that has been indispensable to mathematics from Euclid to the present day.[1] Yet Aristotle himself already seems to have questioned this law in the notorious chapter nine of *De Interpretatione*.[2] In recent times, L. E. J. Brouwer advocated abandoning excluded middle in mathematical proof, and was prepared to pay the price by reconstructing mathematics in its absence.[3]

But how can one question or criticize a law of reasoning? Conversely, what could be one's grounds for defending such a law? The puzzle, of course, arises from the fact that criticism and defense generally involve providing arguments, and so must be governed

---

[1] On some accounts excluded middle is necessary but not sufficient for classical logic. A jointly sufficient set of necessary properties must also contain the principle of non-contradiction, the monotonicity and idempotence of implication, the commutativity of conjunction, and De Morgan's laws for dual logical connectives. See Gabbay (1994).

[2] More precisely, as Jan Łukasiewicz argues in Łukasiewicz (1951), in *De Interpretatione* 9 Aristotle clearly expresses doubts about his version of the principle of bivalence—"every affirmation or negation is true or false" (18a34). But Aristotle's view of truth and falsity as presented in *Metaphysica* is that "it is false to say of that which is that it is not or of that which is not that it is, and it is true to say of that which is that it is or of that which is not that it is not" (*Metaphysica* Γ, 7, 1011b26-7). If there are grounds for doubting that not every saying is true or false, then we have grounds for doubting Aristotle's version of the law of excluded middle: "the principle of demonstration . . . that it is necessary in every case either to affirm or to deny" (*Metaphysica* B, 2, 996b26-30). I quote Ackrill's translation of *De Interpretatione* and Ross's of *Metaphysica*.

Łukasiewicz is one of the first logicians to insist on the distinction between the principle of bivalence and the law of excluded middle; see Łukasiewicz (1930: §6 and appendix). Łukasiewicz claims also that Aristotle distinguishes between excluded middle and bivalence in *De Interpretatione* 9, and that in this chapter Aristotle means to reject bivalence while retaining excluded middle (Łukasiewicz 1951: 82).

[3] Brouwer's rejection of the law of excluded middle is first advanced in Brouwer (1908). A major refinement of the argument is given in Brouwer (1927).

Sanford Shieh, *Wright and Revisionism* In: *Logic, Language, and Mathematics: Themes from the Philosophy of Crispin Wright.* Edited by: Alexander Miller, Oxford University Press (2020). © Sanford Shieh.
DOI: 10.1093/oso/9780199278343.003.0008

by norms of reasoning. There is no real problem if the norms governing the arguments are in some way more fundamental than the principle being criticized or defended. But what about an entire system of logical laws, or laws, like excluded middle, without which we would have to forgo many of our intellectual achievements? On what grounds can we adjudicate whether they have authority over our thought?

Frege gives us one answer to this question. He claims that "logic can answer" "[t]he question why and with what right we acknowledge a law of logic . . . only by reducing it to another law of logic" (Frege 1964 [1893]: 15). Unsurprisingly, Brouwer's critique of classical logic presupposes a rejection of Frege's view. His reasoning has three main parts.[4] First, he holds an idealistic view of mathematics: mathematical proof is essentially a mental activity through which mathematicians create mathematical reality. Second, it follows that the limitations of our mental capacities constrain the extent of mathematical reality. Finally, excluded middle fails because of these limitations: whenever we cannot produce a proof or a refutation of a mathematical claim there is a gap in mathematical reality so that neither that claim nor its negation holds. Thus, Brouwer is committed to holding that the normativity of logic is constrained by extra-logical considerations, specifically facts about our cognitive limitations.

Few have been persuaded by Brouwer, mostly because his view of mathematics as created by and hence dependent on mental activity seems obscure and implausible. But Brouwer's case against classical logic has been revived by Michael Dummett. As popularly understood, Dummett differs from Brouwer primarily in rejecting Brouwer's psychologistic view of mathematics. Dummett famously proposed that traditional metaphysical oppositions between realisms and idealisms concerning various subject matters should be reinterpreted as disagreements over whether it is possible for various classes of statements to be true if we cannot, even in principle, know that they are. Realisms insist that this is a genuine possibility, while idealisms, which Dummett calls anti-realisms, reject this possibility. Thus, Dummett replaces Brouwer's view of the nature of mathematics with the thesis that a mathematical statement cannot be true if we cannot know that it is. That thesis, in turn, is supported by a theory of meaning for mathematical statements. The two remaining components of Brouwer's criticism are left more or less intact. To begin with, Dummett, like Brouwer, insists on our cognitive limitations—there are undecidable statements whose truth or falsity we might never discover. While for Brouwer these limitations constrain mathematical reality, for Dummett they constrain the truth-values of statements. Specifically, undecidable statements are neither true nor false, and hence are counter-examples to the principle of bivalence. Finally, as in the case of Brouwer, for Dummett the law of excluded middle founders on our epistemic limitations. A statement has to be false in order for its negation to be true; so, since an undecidable statement is not false, its negation is not true. Thus, for any undecidable statement, neither it nor its negation is true, which yields a counter-example to the law of excluded middle. Since Dummett takes his analysis of meaning to be completely general, applicable to all language, not just to mathematical statements, he also takes this

---

[4] Exactly how Brouwer's argument works is, naturally, controversial, but my simple account in the text is neutral with respect to these controversies. For some recent discussions, see Detlefsen (1990); Hintikka (2001); and Raatikainen (2004).

argument against classical logic to apply to reasoning in any subject whatsoever. Clearly, on this widely held interpretation, Dummett, like Brouwer, also judges the legitimacy of classical logic by extra-logical facts about our epistemic capacities.

Dummett's revisionism, as his case against classical logic is known, has of course generated much discussion. Crispin Wright has made undoubtedly the most signifi-cant contributions to both sides of the debate over anti-realist revisionism. His critical study of Dummett's *Truth and Other Enigmas* (Dummett 1978), "Anti-Realism and Revisionism" (Wright 1981) is *the locus classicus* of arguments *against* Dummett's revisionism. A decade later, in *Truth and Objectivity* (Wright 1992), Wright advances an argument *in favor of* revisionism. As Wright puts it yet a further decade later, in the essay "On Being in a Quandary" (Wright 2001), the pro-revisionism argument of *Truth and Objectivity* was intended to "finesse" the anti-revisionary doubt he raised in the earlier critical study. Moreover, part of the "Quandary" essay is an attempt to fix an "awkward wrinkle" (Wright 2001: 67) in the pro-revisionary argument of *Truth and Objectivity*. This fix provides another pro-revisionism argument different in form from the original.

The overarching topic of this chapter is an investigation of the nature and limitations of anti-realist revisionism, as developed by Dummett and Wright.

Dummett's writings are notoriously difficult, and in the current state of discussion, it remains somewhat unclear how exactly his revisionary argument works. Thus, one of my principal aims is to clarify some main lines of reasoning associated with anti-realist revisionism, in order to situate and explicate Wright's more recent pro-revisionary arguments. This is the task of §§1–7. In §1, I set out the basic tenets of the view of meaning that underlies anti-realism. In §§2–5, I show that the popular understanding of anti-realist revisionism sketched above can be made precise in two lines of argument for the conclusion that excluded middle is invalid, based on the assumption that there exists undecidable statements. I call these the Naïve Revisionary Arguments. In §6, I discuss the problems that undermine these Naïve arguments. For one Naïve argument there is a tension between its two main premises—the existence of undecidable statements and the epistemic constraint on truth; for the other Naïve argument these very two premises turn out to be compat-ible with classical logic. In §7, I argue that one of the main contributions of Wright's later revisionary argument is to provide explicitly a way of sidestepping the tension that undermines one strand of naïve revisionism. Wright stays away from the strong conclusion that some classical forms of reasoning are invalid, and aims for the weaker thesis that we don't know that all classical forms of reasoning are valid. This weaker conclusion does not require the premise that undecidable statements exist; instead, all that's needed is the weaker premise that we don't know whether we can decide every statement. One might sum up Wright's move as dropping denials of various claims in favor of disavowing knowledge of those claims. I call this maneuver "epistemic ascent." Thus, Wright's revisionism can be characterized as an attempt to isolate a subsystem of classical logic that does not go beyond what is warranted by our epistemic limitations.[5]

---

[5] I owe this formulation to a suggestion by Michael Glanzberg.

My other principal aim is to engage philosophically with Wright's revisionism. I will discuss three objections to Wright's revisionary arguments, occasionally drawing on Wright's earlier, anti-revisionary arguments. The first objection is that in order for Wright's revisionary argument to be sound, there cannot be truth-value gaps, statements that are neither true nor false; but Wright's arguments in *Truth and Objectivity* for his minimalist conception of truth do not suffice to rule out the existence of such gaps. This objection is presented in §§8 and 9. The second objection derives from one of Wright's earlier anti-revisionary arguments. Wright had argued that anti-realists could accept the principle of bivalence without being committed to any position that they have to reject. I argue in §10 that a related form of reasoning allows an anti-realist to accept Wright's later revisionary argument as valid without having to reject classical logic. Finally, in §§11 and 12, I argue that Wright's revisionary arguments do not, in the end, escape an epistemic version of the original tension, in one Naïve Argument, between anti-realism's epistemic conception of truth and the existence of undecidable statements.

In view of the difficulties facing the anti-realist revisionary arguments discussed in the first twelve sections, I outline in §13 a proposal about how revisionism might still proceed. The central idea of the proposal is that anti-realist revisionism should proceed without the anti-Fregean assumption that forms of deductive inference should be held accountable to extra-logical facts about the reach of our epistemic capacities. I begin with a different conception of anti-realism, one founded on conceptual rather than epistemological considerations. This style of anti-realism does not generate an epistemic constraint on truth; instead it leads to an identification of conditions of truth with conditions of justification. It follows that the validation of a form of reasoning on the basis that it is truth-preserving becomes a demonstration that that form of reasoning preserves justification from premises to conclusion. On this view, more fundamental forms of inference provide the standard to which a form of deductive inference is held accountable. That is to say, this view comes closer to the Fregean idea that the justification of logic by logical means adverts to nothing extra-logical. Revisionism, on this picture, has to be proof-theoretic: it has to demonstrate that classical reasoning fails to respect more fundamental forms of deductive justification. I finish by noting that there are aspects of Wright's later revisionism that already point in this proof-theoretic direction.

1. Background: Anti-Realism

To begin with, I want to say a word about Dummettian anti-realism. The fundamental thesis about language underlying anti-realism is that linguistic meaning must be communicable. The consensus interpretation of this thesis is that it must be possible for one speaker, A, of a language, L, to *know* what meanings another speaker B of L, associates with the expressions of L that B uses.[6] A further claim is that the essential condition of possibility for attaining such knowledge is for B's uses of

---

[6] The best expression of this consensus is McDowell (1981). There are other ways of construing communicability, in non-epistemological terms. See Shieh (1998) and §13 of this chapter.

expressions to provide evidence for what he or she means by those expressions. This is Dummett's notorious "manifestation" requirement or constraint: "The meaning of a...statement...cannot be...anything which is not manifest in the use made of it, lying solely in the mind of the individual who apprehends that meaning," because "the meaning of a statement consists solely in its rôle as an instrument of communication between individuals" (Dummett 1973: 216).

None of this has anything to do with logic yet. The connection to logic arises from the idea that the meanings of sentences are ultimately explainable in terms of conditions under which those sentences are true. Of course many qualifications and refinements are needed to make this idea plausible,[7] but since our focus is on the revisionary consequences of anti-realism I will not go into these issues. For our purposes it is enough that, in order to satisfy the manifestation requirement, a speaker's use of expressions must have certain features that provide evidence for the truth conditions that she attaches to the statements she makes.

On the basis of this manifestation requirement for truth conditions, anti-realism proceeds to both a negative and a positive thesis. The negative thesis is a rejection of the essentially non-epistemic realist conception of truth, according to which it is possible for a statement to be true (or false) while we cannot, even in principle, come to know that it is. If truth transcends recognition or verification in this way, then it is possible for the truth conditions that a speaker associates with a sentence to be such that she cannot recognize that those conditions obtain when they do, and she cannot recognize that they fail to obtain when they don't. But then how can someone else tell that she has associated those truth conditions with the sentence? If she were incapable of recognizing whether or not those conditions obtain, then she might never be in a position to use the sentence in response to those conditions. Thus, at the very least, it is a question how anyone else can know from her uses of the sentence that she means to express those conditions. That is to say, anti-realism poses to the realist conception of truth a *challenge* to show how it can satisfy the requirement that truth conditions be manifestable. Since it is unclear how the challenge can be met, at a minimum one should reject any claim to *know* that the realist conception of truth is compatible with the manifestation of truth conditions. More strongly, one may go on to claim that, in the absence of a satisfactory account of the manifestation of realist truth conditions, the realist view of truth is incompatible with the manifestation requirement. For the remainder of this chapter, I will assume that the notion of recognition, as used in these debates, is the notion of recognizing that something is the case, and so is just the concept of knowledge.

The positive thesis is that, if truth is conceived of as epistemically constrained, then truth conditions can be manifested. Precisely how truth is epistemically constrained is not a simple issue. Wright gives at least two formulations. In *Truth and Objectivity* he states the following "Epistemic Constraint": "If P is *true*, then evidence is available

---

[7] Two salient complications are the phenomena of indexicality, which requires us to accept that it's the statements that we make using sentences rather than the sentences themselves that are true or false, and the existence of speech-acts, other than statement-making, that are not assessable as true or false, performed using both declarative and non-declarative sentences. See Dummett (1981: 456–7) for an account of the basic form of philosophical explanations of meaning in terms of truth conditions.

that it is so," abbreviated "EC" (Wright 1992: 41; emphasis mine). In "Quandary," EC is sometimes called an "Evidential Constraint," and is the principle that "any *truth* of the discourse in question may feasibly be known," formally represented as "$P \rightarrow \text{FeasK}(P)$" (Wright 2001: 66; emphasis mine). However the constraint is formulated, it is supposed to imply that conditions of truth cannot obtain without the possibility of our knowing or recognizing that they do. But note that while in the prose formulations Wright mentions the *truth* of a statement or proposition, the formal representation does not contain a truth predicate. I will come back below to these distinct formulations to show that they indicate a limitation in Wright's revisionism.

Since most of the arguments that we will examine do not substantially exploit the fact that the constraint on truth has both a modal and an epistemic component, I will formulate the constraint with a single sentential operator *R*. Where φ is any statement,

$$R(\varphi) \leftrightarrow_{df} \text{we can recognize (i.e., know) that (it is the case that) } \varphi.$$

Following Wright, I formulate the Epistemic Constraint as follows:

$$(\forall p)(p \rightarrow R(p)) \tag{EC}$$

Since for Wright the possibility in question is feasibility, in what follows, whenever I use the '*R*' operator to formulate Wright's views, it should be understood as 'FeasK.'

Note that my formulation differs from Wright's in using propositional quantification. The reason is this. Much of the literature on the logic of revisionism fails to make clear whether, in formal representations of general claims about statements figuring in the arguments under discussion, signs such as '*p*' are occurrences of schematic letters or of variables, and if variables whether they are free or tacitly bound by initial universal quantifiers. Indeed, sometimes it appears that conclusions that seem to have to be understood as using quantifiers ranging over statements or propositions are derived from premises that are explicitly characterized as schemata. Purely for the sake of consistency, I will use propositional quantification throughout in my formulations.

Before continuing, I want to emphasize two points. First, the central commitment of anti-realism is the thesis that meaning must be manifestable. It takes further argument to reject the realist conception of truth and to establish that truth must be recognizable. Second, I would like to leave it as an open question whether these two additional theses are independent of each other, that is, whether one could accept that there is a tension between realism and the manifestation of requirement without also accepting that the conceptions of truth embodied by the various epistemic constraints do satisfy the manifestation requirement.

2. Background: Dummett's Revisionary Argument(s)

Logic surely has an essential connection with truth. A truism about logic is that it comprises standards of correct reasoning applicable to all subject matters, and the correctness of these standards consists in the fact that in any inference conforming to them truth is (necessarily) preserved from the premises to the conclusion. Since anti-realism claims that truth cannot be recognition-transcendent and must be epistemically constrained, it is to be expected that anti-realism may have consequences for logic.

But what consequences? Dummett's remarks on this score are not altogether transparent. In "Realism" he writes,

The conflict between realism and anti-realism is a conflict about the kind of meaning possessed by statements of the disputed class. For the anti-realist, an understanding of such a statement consists in knowing what counts as evidence adequate for the assertion of the statement, and the truth of the statement can consist only in the existence of such evidence. For the realist,... we...have a conception of the statement's being true even in the absence of such evidence. For this reason, the dispute can arise only for classes of statements for which it is admitted on both sides that there may not exist evidence either for or against a given statement. It is this, therefore, which makes acceptance of the law of excluded middle for statements of a given class a crucial test for whether or not someone takes a realist view of statements of that class. The anti-realist cannot allow that the law of excluded middle is generally valid: the realist may, and characteristically will.   (Dummett 1963: 155)

This passage expresses two main ideas. The first is that the contrast between the realist and the anti-realist conceptions of truth leads to a disagreement only when the two conceptions are applied to "classes of statements for which it is admitted on both sides that there may not exist evidence either for or against a given statement." Such statements, as I noted above, have come to be called "undecidable" statements. The second idea is that the disagreement is over whether the law of excluded middle applies to undecidable statements: anti-realists cannot accept that this application, while realists "may, and characteristically will."

Such a condensed argument is bound to raise many questions. Two are most salient. First, what exactly is an "undecidable" statement? Second, since Dummett himself is one of those chiefly responsible for bringing to our attention the distinction between logical laws and semantic theses,[8] how did we get, in this argument, from two conceptions of truth to a disagreement over the validity of a logical law?

In *Frege: Philosophy of Language*, we get some clarification on both of these questions. On undecidability Dummett writes,

An undecidable sentence is simply one whose sense is such that, though in certain effectively recognizable situations we acknowledge it as true, in others we acknowledge it as false, and in yet others no decision is possible, we possess no effective means for bringing about a situation which is of one or other of the first two kinds.   (Dummett 1981: 467)

On the consequences of the anti-realist conception of truth for undecidable sentences he writes,

The truth of such a sentence can consist only in the occurrence of the sort of situation in which we have learned to recognize it as true, and its falsity in the occurrence of the sort of situation in which we have learned to recognize it as false: since we have no guarantee either that a situation of one or other kind will occur, or that we can bring about such a situation at will,

---

[8] See, inter alia, Dummett (1978: xix).

only a misleading picture of what we learned when we learned to use sentences of that form can give us the impression that we possess a notion of truth for that sentence relative to which it is determinately either true or false.   (Dummett 1981: 468)

These two passages strongly suggest that for Dummett the connection between conceptions of truth and validity of classical logical laws goes through the question of whether undecidable statements (whatever exactly they are) conform to the semantic principle of bivalence. The realist conception of truth implies that all statements of an area of discourse, even undecidable ones, are determinately either true or false, and therefore satisfy the principle of bivalence. Conformity to the principle of bivalence, in turn, implies the validity of the law of excluded middle for that class of statements. The anti-realist conception of truth, in contrast, implies that undecidable statements do not satisfy the principle of bivalence. Moreover, if a class of statements does not conform to bivalence, then the law of excluded middle is not valid for those statements; hence, classical logic is not valid for any area of discourse containing undecidable statements.

So it is widely assumed that Dummett's case against classical logic has two parts. In one part he argues that anti-realism leads to a view about the truth-values of a type of statement. In the other part he argues that classical logic is not valid for such statements. But, beyond this widespread impression that Dummett advances a two-part argument, there is relatively little agreement or certainty about exactly how the argument works.

Obviously, if we are to make any progress in clarifying the argument, we need some account of decidability and undecidability. Only if we are clear on what an undecidable statement is can we see how there might be some difficulty in taking such a statement to be determinately either true or false. In §3, I make a beginning on an account of the notion of undecidability.

3. (Un)decidability: A First Try

As we have seen, there are two origins of Dummett's revisionism. The first is an attempt to give a non-psychologistic version of L. E. J. Brouwer's rejection of classical logic and justification for intuitionistic reasoning in mathematics. The second results from the realization that, since the bases of this de-psychologization are completely general considerations in the theory of meaning, the arguments against classical reasoning can be applied outside of mathematics.

So, it is plausible that Dummett's notions of decidability and undecidability also have their origin in mathematics. One obvious place to look for a model is, of course, in Gödel's first incompleteness theorem, which states that there are sentences (i.e., closed formulas) in the language of arithmetic such that neither they nor their negations are derivable in systems of arithmetical axioms and rules of inference in which primitive recursion can be formalized. Derivation in a formal system is, of course, intended to make the intuitive idea of mathematical proof precise and gap-free. Thus it is natural to take formally undecidable formulas to be arithmetical sentences for which there is neither proof nor refutation (proof of its negation) relative to the given set of axioms and rules of inference. Now, if these axioms and rules of inference can be taken to constitute all the means of arithmetical proof available to us, then we can take the undecidable sentences to be neither provable nor refutable.[9] Furthermore, proof and refutation are surely our principal, if not sole, means of knowing the truth and falsity of mathematical statements. So now we have arrived at the idea that given a suitable formal system, formally undecidable formulas are formal representations of arithmetical statements that we cannot know to be true and cannot know to be false.

Once we have reached this last formulation, we can abstract from the fact that knowledge in mathematics is achieved primarily by proof. That is to say, we can form general concepts of decidability and undecidability applicable to any statements, even those that we know to be true or false by means other than proof. On this general conception, a decidable statement is one that we can either know to be true or know to be false, and an undecidable statement is one that we cannot know to be true and cannot know to be false.

Do these definitions capture Dummett's notions? They do, at least prima facie, have some connection with Dummett's concerns in "Realism." There, Dummett focuses on those statements for which "there may not exist evidence either for or against." It's plausible that, whenever there is no evidence for a statement, we cannot know that that statement is true; similarly, whenever there is no evidence against a statement we cannot know that it is false. This gibes with our account of undecidability. Unfortunately, what Dummett literally says is not that there *is no* evidence for or against, but that there *may be no* evidence for or against. So there is a doubt that these conceptions of decidability and undecidablity are quite exactly what Dummett uses in his revisionism.

4. Reconstructing Dummett's Revisionism: The Two-Step Naïve Revisionary Argument

Still, using these conceptions of decidability and undecidability, we can offer a fairly straightforward reconstruction of *a* two-part revisionary argument, which corresponds well to the popular picture of Dummett's revisionism mentioned above.

The first part of the argument is supposed to show that, given the anti-realist conception of truth, undecidable statements do not satisfy bivalence. The reasoning goes as follows. Suppose that there is an undecidable statement p that satisfies bivalence. Then p is determinately either true or false. The anti-realist conception of truth is represented by the Epistemic Constraint: if a statement is true, then it is possible for us to know that it is, and similarly if it is false. Hence either it is possible for us to know that p is true, or it is possible for us to know that it is false. But this contradicts the supposition that p is undecidable, that is, that we cannot know p to be true and cannot know it to be false. Hence p does not satisfy bivalence.

Let's make this first sub-argument more explicit with a formal representation. In order to formalize our present account of (un)decidability, we need truth and falsity

---

[9] Of course, many take Gödel's first incompleteness theorem to show that the totality of means of arithmetical proof can never be formally represented.

predicates, or rather, since we're using propositional quantification, truth and falsity sentential operators. Where φ is any statement,

$$T(\varphi) \leftrightarrow_{df} \text{ it is true that } \varphi,$$

$$F(\varphi) \leftrightarrow_{df} \text{ it is false that } \varphi,$$

The definitions then are:

p is decidable$_T \leftrightarrow_{df} R(T(p)) \vee R(F(p))$

p is undecidable$_T \leftrightarrow_{df}$ it is not the case that p is decidable $\leftrightarrow \neg R(T(p)) \wedge \neg R(F(p))$

Note that in the following I will also use the terms 'verify' and 'falsify' in a slightly regimented way to mean, respectively, 'know to be true' and 'know to be false.' In these terms a decidable statement is one that we can either verify or falsify, and an undecidable statement one that we can neither verify nor falsify.

The first sub-argument has three premises. One, of course, is the principle of bivalence, which we formalize thus:

$$(\forall p)(T(p) \vee F(p)) \tag{BV}$$

Another premise is that there exist undecidable statements:

$$(\exists p)(\neg R(T(p)) \wedge \neg R(F(p))) \tag{Und$_T$}$$

The third premise is the Epistemic Constraint. We cannot use Wright's formalization of the Epistemic Constraint here, because, as we saw, it doesn't mention truth or falsity. So we will use a slightly different formalization of the Epistemic Constraint:

$$(\forall p)((T(p) \to R(T(p)) \wedge (F(p) \to R(F(p)))) \tag{EC$_T$}$$

In contrast to Wright's formalization, the present version explicitly mentions truth and falsity. Let's call this version the "Epistemic Constraint on Truth." Perhaps it is not a drastic deviation from Wright's intentions, since, as we have seen, in *Truth and Objectivity* the Epistemic Constraint is "If P is *true*, then evidence is available that it is so." One might indeed take the present version to be a stricter representation of anti-realism, which insists that we must have capacities for recognizing the obtaining of conditions in which (some of) our statements have the properties of truth and falsity.

In order to try to make the reasoning underlying revisionism as explicit as possible, I provide formal deductions in the Appendix. Deduction 1 establishes, using only the rules of inference of minimal logic, that these premises imply a contradiction.

I turn now to the second part of the argument. Here we have to go from the failure of bivalence to the invalidity of classical reasoning. In order to see how this might be accomplished, let's turn to Wright's discussion of Dummett in "Anti-realism and Revisionism," where he proposes and rejects a line of reasoning for precisely this transition. Although Wright rejects this reasoning, we can use a part of it to complete our two-part revisionary argument.

Wright first sketches an argument from bivalence to excluded middle:

If every proposition, P, is determinately either true or false, and if the negation of P is true just in case P is false, then matters have to arrange themselves in such a way that, one way or the other, the disjunction of P with its negation is true ....

[T]his reasoning tacitly appeals to various presuppositions: the orthodox introduction and elimination rules for disjunction, the convertibility of 'P is true' and 'P' ....

(Wright 1981: 433–4)

Here Wright has in mind an argument depending on three premises. The first premise is the principle of bivalence (BV). The second premise is a principle connecting the truth-values of statements with the truth-values of their negations:

$$(\forall p)(T(\neg p) \leftrightarrow F(p)) \qquad \text{(FN)}$$

Let's call this premise the "Falsity-Negation Principle." The final premise is a version, for propositional quantification, of what Wright in *Truth and Objectivity* calls the "Equivalence Schema," which I will call the "Equivalence Principle":

$$(\forall p)(p \leftrightarrow T(p)) \qquad \text{(EP)}$$

The full argument is given as Deduction 2 in the Appendix.

Wright claims that one might be tempted to infer from this argument that if bivalence fails for some statements, then excluded middle doesn't apply to them. This additional step is what Wright rejects. This argument shows that bivalence, the Falsity-Negation Principle, and the Equivalence Principle together provide *one* basis for the validity of excluded middle. But, unless it's the only possible validation of excluded middle compatible with anti-realism, the rejection of bivalence on anti-realist grounds doesn't imply a rejection of excluded middle.

But let's reconsider Wright's deduction of excluded middle from bivalence. The required additional premises, Falsity-Negation and Equivalence, are both generalized biconditionals. Hence one can reverse the direction of Wright's argument. According to excluded middle, the disjunction of each statement with its negation holds. If a statement q holds, then by the Equivalence Principle q is true. If q's negation holds, then that negation is true. By Falsity-Negation, q is false. Hence q is either true or false. This is Deduction 3 in the Appendix; it uses exactly the same rules of inference in the same order as Deduction 1. What this shows is that, in the presence of Falsity-Negation and Equivalence, excluded middle and bivalence are provably equivalent.[10] The direction from excluded middle to bivalence provides a reconstruction of the second part of Dummett's revisionism, for it shows that if bivalence fails, then so does excluded middle. Since, by our first argument, anti-realism implies the failure of bivalence for undecidable statements, by this second argument it also implies the failure of excluded middle for these statements.

Now, of course the second argument assumes the Falsity-Negation Principle and the Equivalence Principle. But these two principles are, as Wright showed, needed to give a realist justification of classical logic from bivalence. So we have an argument

---

[10] See Kneale and Kneale (1962: 47).

ad hominem: a realist who accepts classical logic on the basis of Wright's argument has no grounds to object to an anti-realist's use of these principles. Given anti-realism, the law of excluded middle doesn't hold of undecidable statements.

Let's call this reconstruction of Dummett's revisionism the two-step Naïve Revisionary Argument.

5. The One-Step Naïve Revisionary Argument

In the first step of the two-step Naïve Argument, I used an Epistemic Constraint different from Wright's. The reason was that in that argument we were trying to connect facts about truth-values of statements (given by bivalence) to the possibilities of knowing these truth-values (which fix their decidability), but Wright's Epistemic Constraint doesn't mention truth or falsity at all. Instead, Wright's formalization of the Constraint connects statements directly to the possibilities of knowing what they express.

Now, this feature of Wright's Constraint suggests that we might be able to use it to relate excluded middle directly to decidability, without going through the intermediary of bivalence. According to excluded middle, for any statement p, either p or its negation is the case. It follows, by Wright's Epistemic Constraint, that either it is possible to know that p is the case, or it is possible to know that the negation of p is the case, that is, p is either verifiable or falsifiable. This of course is not the claim that p is decidable, as we have formalized decidability. But it is clearly a generalization of the idea of a mathematical statement's being either provable or refutable, which is the intuitive basis of Dummett's notion of decidability. For, to repeat, a proof of a mathematical statement p is our primary means of knowing that p. So we can formally represent (un)decidability thus:

$$p \text{ is decidable}_P \leftrightarrow_{df} R(p) \vee R(\neg p)$$

$$p \text{ is undecidable}_P \leftrightarrow_{df} \neg R(p) \wedge \neg R(\neg p)$$

Since (un)decidability in this sense is based on the possibilities of knowing what statements express, let's call it "propositional (un)decidability," and the earlier account "truth-value (un)decidability." So excluded middle implies that all statements are propositionally decidable. This conclusion contradicts the claim that undecidable statements exist:

$$(\exists p)(\neg R(p) \wedge \neg R(\neg p)) \tag{Und$_P$}$$

For consistency of terminology, let's call Wright's Epistemic Constraint,

$$(\forall p)(p \rightarrow R(p)) \tag{EC$_P$}$$

the "Epistemic Constraint on Propositions."

We can obtain this direct, bivalence-free, argument against classical logic, if we make three replacement in the first sub-argument of the two-step Naïve argument: the Epistemic Constraint on Truth with the one on Propositions, bivalence with excluded middle, and truth-value undecidability with propositional undecidability. This is Deduction 4 in the Appendix. Call it the one-step Naïve Revisionary

Argument. Perhaps something like this argument is what Dummett had in mind in "Realism" where, without mentioning the principle of bivalence, he asserts that "[t]he anti-realist cannot allow that the law of excluded middle is generally valid."

Before going on, let's think a bit about the relation between the two Naïve Arguments. To begin with, since our focus is on the revisionary consequences of anti-realism, the main issue about these arguments is: what is the relation between the two versions of the Epistemic Constraint that are stand-ins for anti-realism? Is either a better representation of the commitments of anti-realism?

Here is one way to think about the two Constraints. Suppose we restricted the Epistemic Constraint on Propositions to propositions ascribing truth and falsity to (other) propositions. Then we would obtain the Epistemic Constraint on the truth-values of those propositions to which truth and falsity are ascribed. That is, the Constraint on Truth would be a special case of that on Propositions.

There is another way to think about how these Constraints are related. Let's assume, in addition to the Equivalence Principle and the Falsity-Negation Principle, the principle that whenever two statements are equivalent, one is recognizable to be the case if and only if the other is:

$$(\forall p)(\forall q)[(p \leftrightarrow q) \leftrightarrow (R(p) \leftrightarrow R(q))]$$

Then we can establish that $(\forall p)(R(p) \leftrightarrow R(T(p)))$ and that $(\forall p)(R(\neg p) \leftrightarrow R(F(p)))$, which allows us to derive each of the Epistemic Constraints from the other.

6. The Rejection of the Naïve Arguments

Neither of the two Naïve Revisionary Arguments, based on the conception of undecidable statements as ones we cannot know and whose negations we cannot know, is a target of Wright's criticism. I suspect the reason is that there are fairly simple problems with these arguments, which make it implausible to attribute either of them to Dummett.

Let's start with the one-step argument. Many, including Dummett himself, have pointed out the problem with this Naïve argument.[11] This argument rests on the joint incompatibility of the Epistemic Constraint on Propositions, excluded middle, and the existence of propositionally undecidable statements. In essence the argument takes the contradiction to demonstrate that, if any discourse contains undecidable statements, then an anti-realist conception of the truth of these statements implies that excluded middle does not hold for them. But, in fact, there's an incompatibility between the Epistemic Constraint by itself, without the help of excluded middle, and the existence of undecidable statements. The reasoning is this. The Epistemic Constraint requires, for any statement to hold, that its holding be recognizable. Now, if some statement q is undecidable, then we neither can recognize that q is the case nor can recognize that q is not the case. Since we cannot recognize that q is the case, the Epistemic Constraint implies that q is not the case. Using the Epistemic Constraint again, it follows that we can recognize that q is not the case.

---

[11] See Dummett (1977: 17); Tennant (1997: 181); Salerno (2000: 214).

But that contradicts the undecidability of q.[12] Thus, by parity of reasoning with the one-step Naïve Argument, if any discourse contains undecidable statements, then an anti-realist conception of the truth does not hold of these statements. This is Deduction 5 in the Appendix.

A different problem arises for the two-step Naïve Argument. The form of reasoning that generated the problem we have just spelt out does *not* show that the Epistemic Constraint on *Truth* is incompatible with the existence of truth-value undecidable statements. With respect to truth, the Epistemic Constraint requires that, if any statement has a truth-value, its possession of that value be recognizable. If some statement q is truth-value undecidable, then we neither can recognize that q is true nor can recognize that q is false. Since we cannot recognize that q is true, the Epistemic Constraint implies that q is not true. But to the conclusion that q is not true we cannot apply the Epistemic Constraint on Truth, since this conclusion does not claim that q *has* a truth-value. So we cannot derive a contradiction from this conclusion and our inability to recognize that q is false. We *can* conclude, from our inability to recognize that q is false, that q is not false. So we can reach only the conclusion that truth-value undecidable statements are neither true nor false, that is, that they are truth-value gaps. This is Deduction 6 in the Appendix.

So the first sub-argument of the two-step Naïve Argument shows that the existence of truth-value undecidable statements is incompatible with these statements' satisfying bivalence. But clearly statements that are neither true nor false are not either true or false. Hence the conclusion that truth-value undecidable statements are truth-value gaps does not conflict with the first sub-argument. However, this conclusion *does* conflict with the use of the Equivalence Principle required in the second sub-argument. The Equivalence Principle is used twice, to license two steps in that sub-argument:

- If q, then it is true that q
- If it is not the case that q, then it is true that it is not the case that q

If the particular statement q is not true and not false, then the consequent of the first conditional—'it is true that q'—is false. Hence that first conditional has an antecedent that is neither true nor false but a consequent that is false. On most accounts of the truth conditions of conditionals with truth-value gaps as components, this conditional is not true. What of the second conditional? Given the Falsity-Negation Principle, since q is not false, its negation, which is the antecedent of the second conditional, is not true. Hence the consequent of the second conditional—'it is

---

[12] An additional problem is that, if we apply the Brouwer–Heyting–Kolmogorov interpretation of intuitionistic negation to all statements, then there can be no intuitionistic grounds for the undecidability of any particular statement p. On this interpretation, a verification of the claim that p cannot be verified is a verification of the negation of p. Thus, the supposition that there is a verification of the claim that p is undecidable leads to an absurdity. So even if the one-step Naïve Argument succeeds as a rejection of classical logic, it would not support adopting intuitionistic logic.

However, this problem is not a decisive objection to the Naïve Argument. Wright shows that it is not clear how precisely intuitionistic negation can be applied to contingent statements without giving rise to counter-intuitive consequences. In particular, it follows from a natural account that if we have grounds for believing, for example, that all human beings will soon be destroyed, then we will have grounds for asserting the negations of just about every statement that we have not yet decided (Wright 1981: 437–8).

true that it is not the case that q'—is false. Now, presumably a counterpart of Falsity-Negation is the principle that the negation of a statement is false just in case that statement is true. If we adopt this principle, then the negation of q is also not false. Thus, the second conditional also has an antecedent that is neither true nor false but a consequent that is false, and so is not true.

The problem, then, is this. If there are truth-value undecidable statements, then not all instances of the Equivalence Principle are true; in particular, those instances required in the second sub-argument are not true. The second sub-argument establishes an implication from excluded middle, Falsity-Negation and the Equivalence Principle to bivalence. Even if we accept that this argument is valid, we cannot move from the failure of bivalence, which is the conclusion of the first sub-argument, to a rejection of excluded middle, unless Falsity-Negation and the Equivalence Principle are both true. So we no longer have a sound argument against the validity of excluded middle for truth-value undecidable statements.

Moreover, it is well known that there are a number of semantic theories with respect to which excluded middle is valid and truth-value gaps exist. Indeed, both Dummett and Wright have elaborated supervaluational semantic theories that accomplish precisely this (Dummett 1969: 365-7; Wright 1981: 438-41). So it is hard to see that an anti-realist critique of classical logic can be based on the two-step Naïve Revisionary Argument.[13,14]

---

[13] It has been argued that the existence of truth-value gaps is not consistent with intuitionistic logic. If a statement p is neither true nor false, then its negation is also neither true nor false. Hence it's not the case that either p or its negation is true. But then the negation of the last claim *is* true. However, that negation is an instance of the double negation of excluded middle, which is intuitionistically valid. This argument should not be confused with the argument, set out in note 12, for the conclusion that it is intuitionistically inconsistent to claim the undecidability of any particular statement. It may be that S. A. Rasmussen and J. Ravnkilde (1982: 404) present something like this argument, but I am not confident of this attribution because I don't fully understand what they are saying.

Obviously, it would be question-begging to add this consideration to the two-step Naïve Argument in order to provide grounds against classical reasoning. In addition, it's not clear to me how compelling this argument really is, since intuitionistic sentential logic has weakly characteristic valuation systems with countably many truth-values.

[14] Kit Fine (1975), in his classic account of the logic of supervaluations, shows that although supervaluations allow the validation of excluded middle without bivalence, the Deduction Theorem fails for the logic of supervaluations (p. 190), although Fine does not claim that this logic is non-classical. Timothy Williamson shows that *reductio ad absurdum*, contraposition, and argument by cases similarly fail (Williamson 1994: §5.4; see also 2004: 118-20). For further details on the logic of supervaluations see the Introduction of Keefe and Smith (1996: 29-32) and Keefe (2000: 174-81). Williamson argues that these facts show that the logic of supervaluations is not classical.

If we accept this argument, we might take *an* anti-realist critique of classical logic to work as follows. To begin with, the first part of the two-step Naïve Argument shows that the Epistemic Constraint requires truth-value undecidable statements to be neither true nor false. Then, we argue that the best, or at least a plausible, semantics of such statements is supervaluational, and hence the logic governing such statements is non-classical. Revisionism of this form may well be best prosecuted on a discourse-by-discourse basis, since perhaps, for example, supervaluational semantics is more plausible for vague statements or statements of set theory than for statements about the past. This type of revisionism is obviously not that of either Dummett or Wright. For one thing, the logic replacing classical reasoning is not intuitionistic.

It is also not clear to me that Williamson's argument is conclusive; see McGee and McLaughlin (1998 and 2004) for plausible replies.

7. Wright's Basic Revisionary Argument: Epistemic Ascent

The failure of the two Naïve Arguments suggests that an anti-realist case for revisionism can't turn on the *existence* of undecidable statements, at least if we conceive of undecidability as impossibility of verification or falsification. The reason, of course, is that anti-realism's Epistemic Constraint implies either that there are no undecidable statements, or that these statements are truth-value gaps.

A closer look at Dummett's actual words suggests that he never did take the existence of undecidable statements to be the key to revisionism. I have already pointed out that, in "Realism," Dummett focuses on statements for which there *may* not exist evidence for or against. So it appears to be the possibility rather than the existence of undecidable statements that divides realism from anti-realism. Closer scrutiny of the discussion in *Frege* confirms that the role of undecidability in Dummett's revisionism is not as simple as we have made it out to be. Here Dummett describes undecidable statements in two ways. First, we "possess no effective means for bringing about a situation" in which "we acknowledge [such a statement] as true" or a situation in which "we acknowledge it as false" (Dummett 1981: 467). Second, we have "no guarantee" either that a "situation in which we have learned to recognize it as true" or one "in which we have learned to recognize it as false" will occur, or that "we can bring about" one or the other of such situations "at will" (Dummett 1981: 468). I take it that Dummett has in mind the following. If we don't have effective means for bringing about either a situation in which we verify the statement or one in which we falsify it, then we have no guarantee, or cannot guarantee, that we will ever be in a position to verify or to falsify the statement. The significance of *not having a guarantee* is that we *don't know* whether we will ever be in one or the other of these positions; henceforth this is how I will interpret the notion of possessing a guarantee. Thus the modality in the "Realism" formulation is epistemic: the sense in which, according to Dummett, "there may not exist evidence for or against a given statement" is that it is compatible with all we know that there is no evidence for or against that statement. But then we don't know that the statement in question is one that we can come to know to be true or know to be false.

We can make these ideas more precise in one of two ways. On the one hand, we can follow Dummett's lead in *Frege* and define a statement as undecidable just in case we don't know that we can verify it or falsify it. Then we can try to reconstruct Dummett's revisionism argument(s) on the basis of the existence of undecidable statements in *this* sense of 'undecidable.'[15] On the other hand, we can follow his lead in "Realism" and hold that the fact on which revisionism depends is not the *existence* of undecidable statements, but rather our ignorance of whether all statements (of a discourse) are decidable. On this second path, we would of course not change our conception of decidability and undecidability.

The important point for our purposes is that in both his anti-revisionary and his pro-revisionary arguments, Wright seems to conceive of the role of undecidability in

---

[15] I have tried to do this in Shieh (1997).

this second way. The revisionary arguments that he criticizes in "Anti-Realism and Revisionism" concern statements that he describes as "not known to be effectively decidable." Moreover, one key premise in the pro-revisionary argument of *Truth and Objectivity*, which in "Quandary" Wright calls the Basic Revisionary Argument, is

$$\neg K[(\forall p)(R(p) \vee R(\neg p))] \qquad \text{(NKD}_\text{P})$$

(K is obviously a knowledge operator). That is, it is not known that all statements (of an area of discourse) are (propositionally) decidable. Let's call this premise "Ignorance of Decidability."

In the context of my presentation of the background for the revisionism debate, Wright's Basic Revisionary Argument may be seen as the result of reformulating the one-step Naïve Argument in response to the problem elaborated in the last section, using the premise of Ignorance of Decidability rather than the problematic existence of undecidable statements. The strategy of the reformulation is to give up trying to show that, assuming anti-realism, excluded middle is directly incompatible with the existence of undecidable statements. Instead, proceed in the following two stages.

First, show that, given anti-realism, excluded middle implies that every statement is such that either it or its negation is knowable, which of course is the claim that every statement is (propositionally) decidable:

$$(\forall p)(R(p) \vee R(\neg p)) \qquad \text{(Dec}_\text{P})$$

Clearly the claim that (Dec$_\text{P}$) is not known is just Ignorance of Decidability ((NKD$_\text{p}$) = $\neg K$(Dec$_\text{P}$)), so let's call (Dec$_\text{P}$) Decidability. In the deduction that Wright gives for this demonstration, the Epistemic Constraint on Propositions represents anti-realism. The reasoning is this. By the Epistemic Constraint, if any proposition p holds, then it is verifiable. Hence it is either verifiable or falsifiable. Similarly, if its negation holds, then again it is either falsifiable or verifiable. Hence if excluded middle holds, all propositions are decidable. This is Deduction 7 in the Appendix.

Second, we subject the conclusion of the first stage to what I will call *epistemic ascent*. Epistemic ascent consists of applying a basic, if not uncontroversial, principle of reasoning about knowledge: given knowledge that some premises imply a conclusion, knowledge of those premises implies knowledge of that conclusion. Formally:

From knowledge of $p_1, ..., p_n \vdash p$, we may infer $K(p_1), ..., K(p_n) \vdash K(p)$.[16]

In the present case, applying epistemic ascent to the first stage of the argument establishes that knowledge of excluded middle and knowledge of the Epistemic Constraint on Propositions together imply knowledge of Decidability. That is, the first stage establishes that

---

[16] The basis of the principle is this. Knowledge requires at least justified true belief. So knowledge of the implication $p_1, ..., p_n \vdash p$ requires that the implication holds and that we are justified in believing that it holds. If $p_1, ..., p_n$ are known, then $p_1, ..., p_n$ are all true, all believed, and we have justifications for each of them. Since $p_1, ..., p_n$ are all true and together imply p, p is also true. Moreover, combining these justifications with the implication $p_1, ..., p_n \vdash p$, we obtain a justification for p. Finally, since $p_1, ..., p_n$ are all believed to be true, and it is believed that they imply p, p is also, at least potentially, believed. Hence p is known. (Naturally the "potentially believed" part is where the controversy comes in.)

$$(EC_P), (LEM) \vdash (Dec_P).$$

Having established this implication, we can conclude that we know that it holds. Hence we can apply the epistemic principle to obtain

$$K(EC_P), K(LEM) \vdash K(Dec_P).$$

It follows that for a class of statements with respect to which "we have no guarantee of decidability" (Wright 2001: 65), that is, of whose Decidability we are Ignorant, *knowledge* of excluded middle is incompatible with *knowledge* of the Epistemic Constraint, that is, $K(EC_P), K(LEM), (NKD_P) \vdash \bot$. Hence anti-realism, which according to Wright holds the Epistemic Constraint to be "known a priori" (Wright 2001: 66), implies that excluded middle is not known to hold for these statements, $K(EC_P)$, $(NKD_P) \vdash \neg K(LEM)$. Revisionism follows thus: "Since logic has no business containing first principles that are uncertain, classical logic is unacceptable in our present state of information" (Wright 2001: 66).

8. The Two-Step Revisionary Argument

Can we apply the epistemic ascent strategy to the two-step Naïve Revisionary Argument? Here is one attempt to do so. The existence of undecidable statements is used in the first step of this Naïve argument. So, to reformulate this argument, we again give up trying to show that, assuming anti-realism, bivalence is directly incompatible with the existence of undecidable statements. Instead, we begin by arguing that the Epistemic Constraint on Truth, together with bivalence, implies truth-value decidability, $(EC_T), (BV) \vdash (Dec_T)$. Here, of course, we need a different version of the premise of Decidability, one in terms of truth-value decidability:

$$(\forall p)(R(T(p)) \vee R(F(p))) \tag{Dec_T}$$

The reasoning is given in Deduction 8 of the Appendix.

Next, we apply epistemic ascent to this result, in order to obtain the conclusion that knowledge of the Epistemic Constraint and knowledge of bivalence imply knowledge of Decidability, and are thus incompatible with Ignorance of Decidability, which now is:

$$\neg K[(\forall p)(R(T(p)) \vee R(F(p)))] \quad (= \neg K(Dec_T)) \tag{NKD_T}$$

That is, $K(EC_T), K(BV), (NKD_T) \vdash \bot$. Thus, given anti-realism, we don't know that bivalence holds of statements that we don't know to be decidable, $K(EC_T)$, $(NKD_T) \vdash \neg K(BV)$.

We now apply epistemic ascent to the second of the two steps of this Naïve Argument. This yields the claim that knowledge of excluded middle, together with knowledge of Falsity-Negation and Equivalence, implies knowledge of bivalence. That is, from (LEM), (FN), (EP) $\vdash$ (BV) we conclude that $K(LEM), K(FN), K(EP) \vdash K(BV)$.

Putting the two steps together, we reach the conclusion that if we assume we know Falsity-Negation and Equivalence, anti-realism implies that we do not know that excluded middle is valid for discourses of whose Decidability we're ignorant: $K(EC_T)$, $(NKD_T), K(FN), K(EP) \vdash \neg K(LEM)$.

Does this reformulated revisionary argument, call it the "two-step Revisionary Argument," avoid the difficulty for the two-step Naïve Argument? The problem there was that the existence of undecidable statements, used in the first step, implies the existence of truth-value gaps, which is incompatible with the Equivalence Principle used in the second step. The first step of the reformulated argument now yields only the conclusion that we don't know that bivalence holds of statements we don't know to be decidable, which does not obviously imply that such statements are truth-value gaps. So it seems we have evaded the original objection.

Unfortunately, for this very reason, it's not clear that the two steps of this revisionary argument can succeed in working together to generate a cogent criticism of classical logic. The reason is this. We saw earlier that if there are truth-value gaps, then the Equivalence Principle doesn't hold. It surely follows that we can claim to know that the existence of truth-value gaps is incompatible with the Equivalence Principle. Recent epistemology has made it plausible that, in order to count as possessing knowledge, a subject has to pass certain threshold conditions with respect to what is known. What exactly these conditions are is of course highly controversial, but there is some agreement that aspects of the contexts in which the putative knower is placed play a central role.[17] Perhaps there is also some agreement that, in the context of philosophical reflection, the interests of the subjects of potential knowledge are, or ought to be, focused only on the truth of the matter under investigation. If so, then it seems plausible that in the context of our discussion, a claim to knowledge can be sustained only if alternatives known to be incompatible with (knowledge of) what is claimed to be known have been ruled out.[18] So, in the present case, in order to know that the Equivalence Principle applies to a discourse, one would have to rule out the existence of truth-value gaps in that discourse, that is, one would have to know that there are no such gaps.

There is a more direct route to this conclusion if we use a "principle of agnosticism" that Wright discusses in "Quandary": "P should be regarded as unknown just in case there is some possibility Q such that if it obtained, it would ensure not-P, and such that we are (warranted in thinking that we are) in no position to exclude Q" (Wright 2001: 67-8). Since the existence of truth-value gaps ensures that the Equivalence Principle doesn't hold, by the principle of agnosticism, if we are in no position to exclude the possibility of truth-value gaps, then we would be justified in concluding that the Equivalence Principle is not known. Contraposing, we conclude that knowledge of the Equivalence Principle requires being in a position to exclude the possibility of truth-value gaps, that is, requires knowledge that there are no such gaps.[19]

Let's return to the two-step Revisionary Argument. The conclusion of the first step is that bivalence is not known. That is, we don't know that every statement is either

---

[17] Some central accounts of contextualist epistemology are Cohen (1988); DeRose (1995); Lewis (1996); Unger (2002); Hawthorne (2004); and Stanley (2005).

[18] See especially Lewis (1996) for the view expressed in this and the previous sentence.

[19] As we will see in §11 below, Wright rejects this principle of agnosticism. But Wright's reason for this rejection is that on his view there are propositions that we can fail to know even if we *can* rule out every counter-possibility to them. My argument, in contrast, turns on *not being able* to rule out a salient counter-possibility. Hence it's not clear that Wright would have cause for rejecting my use of the principle of agnosticism.

true or false. But then, do we know that there are no truth-value gaps? If we don't know that every statement is either true or false, then it is consistent with what we know that not every statement is either true or false. Now, that not every statement is either true or false implies, by classical reasoning, that there are statements that are neither true nor false. So, if we allow classical logic, then we can conclude from our ignorance of bivalence that it is consistent with what we know that there are truth-value gaps, so we don't know that there are no gaps. But then, classically, ignorance of bivalence implies ignorance of the Equivalence Principle.

Now the problem becomes clear. The second step of the two-step Revisionary Argument can arrive soundly at the conclusion that excluded middle is not known only if the Equivalence Principle is known. But what we have seen is that if, according to the first step of this Argument, bivalence is not known, then the only way in which we can avoid concluding that the Equivalence Principle is not known is by prohibiting classical reasoning. That is to say, the two-step Revisionary Argument is question-begging.

9. A Problem for Wright's Basic Revisionary Argument

The difficulty presented in the last section for the two-step Revisionary Argument amounts in essence to an epistemic version of the objection to the two-step Naïve Argument. The root of the problem is the incompatibility between the Equivalence Principle and truth-value gaps. The reason for this incompatibility is that if a statement p is neither true nor false, then 'p is true' is false, and so the conditional 'if p then p is true' is not true. Now let's ask, what about the conditional, 'if p then it is possible to know that p'? The analysis of the concept of knowledge has of course encountered many complications. Still, it is surely not questionable that truth is a necessary feature of what is known. It follows that if p is not true, it is not known. But is it *knowable*? Is it *possible* for p to be known when it is not true? It has been argued that an actually false but contingent proposition might be true in counterfactual circumstances and so might be known.[20] However, it seems to me that knowability in this sense is not the knowability that anti-realism demands of truth. Anti-realism requires that if conditions of truth obtain (or fail to obtain), then those conditions must be within the reach of human cognitive capacities. Thus, those counterfactual circumstances relevant to assessing the anti-realist knowability of a proposition differ from actual circumstances in the (extent and) exercise of our cognitive capacities, not in the conditions detected by that exercise. Since it is the reach of our actual cognitive capacities that underlies Wright's notion of feasibility, this point holds even more so for feasibility. As Wright puts it,

The modality involved in *feasible knowledge* is to be understood...as constrained by the distribution of truth values in the actual world.... [T]he range of what is feasible for us to know goes no further than what is actually the case: we are talking about those propositions whose actual truth could be recognised by the implementation of some humanly feasible process. (Wright 2001: 59-60, n. 17)[21]

---

[20] See Cogburn (2002: 238).

[21] In recent literature on anti-realism and Fitch's paradox of knowability, the view that what is not true is not knowable is sometimes referred to as the factivity of knowability; see, for instance, Brogaard and Salerno (2002: 145-6). Above I sketched what I take to be a plausible argument in favor of a commitment

So, if p is not true, then it is not possible, not feasible, to recognize the obtaining of conditions in which it is true, and hence 'it is feasible to know that p' is false, or, at least, not true. The upshot is that if an area of discourse contains truth-value gaps, then instances of the Epistemic Constraint for these gaps are not true, and thus the Epistemic Constraint is not true of that discourse. But if so, then surely the Epistemic Constraint is not known. We have seen that knowledge (indeed, *a priori* knowledge) of the Epistemic Constraint is required for the second stage of Wright's Basic Revisionary Argument. It follows that this argument also requires the premise that there are no truth-value gaps. Of course, it may be that anti-realism rules out the existence of truth-value gaps. Be that as it may, it is not clear that Wright can adopt this line. For, as I mentioned above, in "Anti-Realism and Revisionism" Wright sketched a supervaluational semantic theory allowing for the existence of truth-value gaps, and argued that it is consistent with anti-realism (Wright 1981: 438-41).

It should be stressed that the problem just raised concerns the soundness of the Basic Revisionary Argument. As Wright points out, the deduction of stage one of the argument is valid with respect to supervaluational semantics (Wright 2001: 67, n. 23).

Of course, one obvious response that Wright can make to this problem is that it fails to take into account the context of the Basic Revisionary Argument. By the time the argument is presented in *Truth and Objectivity*, Wright had already argued for his minimal conception of truth. On this conception, "when a predicate has been shown to have [certain] features, and to have them for the right reasons, there is then no further question about the propriety of regarding it as a truth predicate" (Wright 1992: 24-5), and one of the features in question is "the Equivalence Schema: it is true that P if and only if P, for any substitution instance of P" (Wright 1992: 24).

Now, one might ask: why is this principle one of the features that any truth predicate has to satisfy? Wright's answer is that "standing just behind the [Disquotational Schema, from which the Equivalence Schema can be inferred,] is

to factivity for anti-realism, but the truth of this conclusion, as opposed to Wright's commitment to it, is not required for my argument. I'm grateful to an anonymous referee for urging me to clarify this point, and for suggesting the following argument against factivity.

Anti-realism claims that a speaker's understanding a sentence consists in her ability to recognize verifications of that sentence if she were presented with them. That is to say, a speaker may be credited with such an understanding even if she is never actually presented with such verifications. Consider an undecided but decidable contingent sentence S. Since S is not decided we don't know that it's true or false, but since S is decidable we are licensed by anti-realism to accept that it may *be* true or false. A speaker who understands S, according to anti-realism, could recognize a verifier for the truth of S, and could also recognize a verifier for the truth the negation of S, if presented with either. This implies that S and its negation are both knowable in the sense relevant to anti-realism. We have seen, however, that according to anti-realism, S may be false, in which case its negation is knowable. But then a false sentence is knowable in the sense relevant to the anti-realist.

If I've correctly understood the referee's argument, it seems to me that it is in fact consistent with my argument for factivity. If S is false, then what one who understands S would be able to recognize is *not* a verification of S, but rather a falsification of it, or a verification of its negation. So if S is false it does *not* follow on anti-realist grounds that someone who understands it can know that S; rather, what does follow is that she can know that it's not the case that S. Formally, factivity is the principle that, for all p, $\Diamond K(p) \rightarrow p$, which is equivalent to $\neg p \rightarrow \neg \Diamond K(p)$. So, the falsity of this principle requires a statement q such that either $\Diamond K(q) \wedge \neg q$ or $\neg q \wedge \neg \neg \Diamond K(q)$. In the present case we do have $\neg S$, but we have only $\Diamond K(\neg S)$, not $\Diamond K(S)$, furthermore, $\Diamond K(\neg S)$ implies $\neg \neg \Diamond K(S)$.

the basic, platitudinous connection of assertion and truth: asserting a proposition—a Fregean thought—is claiming that it is true" (Wright 1992: 23). But is it really such a platitude that, as Wright formulates it, asserting that p is *the same thing* as claiming that p is true? I am not confident of the answer to *this* question; however, I will now argue that Wright, in *Truth and Objectivity*, provides the materials for a negative answer.

One part of Wright's minimalism rests on the view that truth and warranted assertibility coincide "in normative force" but diverge "in extension" (Wright 1992: 18–19). Wright gives a general account of how coincidence in normative force can combine with divergence in extension:

Suppose F and G are so related that, while the only kind of reason we can have for supposing that something is G is that it be F, the reason supplied is a defeasible reason. Then having reason to think that an item is G will involve having reason to think that it is F; and having reason to think it is F will amount, when so far undefeated, to reason to think it is G. Hence, if either predicate is normative with respect to some practice, the two predicates will be normatively coincident with respect to it. But, precisely because an item's being F supplies only a defeasible reason for its being G, space is left for divergence in extension between the two predicates.   (Wright 1992: 19)

This account holds of truth and warranted assertibility in virtue of the fact that any reason for supposing that a proposition p is true is that an assertion of p is warranted. This suggests that one cannot be in a position to assert p without also being in a position to claim that p is true, and vice versa. But, prima facie, one could accept this connection between assertion and truth without being committed to holding that asserting p is the same thing as claiming that p is true. Indeed, if assertoric warrant diverges in extension from truth, then, since assertion is surely governed by the norm of assertoric warrant, while plausibly claiming-true should be governed by the distinct norm of truth, then assertion and claiming-true are distinct actions.

All this goes to show that Wright has not ruled out that the truly platitudinous connection between truth and assertion supports no more than the view that any inference from an assertion of p to a claim that p is true, or *vice versa*, is valid. This view, in turn, supports the thesis that a rule of inference from p to 'p is true' and vice versa is valid for all statements p. But, as Bas van Fraassen argued in the course of introducing the notion of a supervaluation, the adoption of such a pair of rules of inference is consistent with supervaluational semantics, and so with the existence of truth-value gaps.[22]

Obviously the foregoing is not a decisive argument against the Basic Revisionary Argument. I have only provided a prima facie case that Wright's minimalism about truth is not sufficient to ground the soundness of this argument. Hence, at this point it's unclear that Wright can claim to have finessed his earlier anti-revisionary argument unless he shows that supervaluationism is not consistent with his minimal conception of truth.

---

[22] Van Fraassen (1966: 493–5). For further argument that features of the concept of truth that underlie our acceptance of the Equivalence Principle are in fact compatible with the sort of gap between truth and falsity opened by supervaluationism, see McGee (1989) and Keefe (2000: 213–20). Note, though, that McGee's supervaluationism does not allow statements to be neither true nor false, just neither definitely true nor definitely false.

10. Anti-Realism, Knowledge of Bivalence, and Knowledge of Excluded Middle

The objection to the two-step Revisionary Argument presented in §8 turns on a problem about how the two steps fit together. We have not questioned the conclusion of the first step, namely, that anti-realism is incompatible with knowledge of bivalence for statements we don't know to be decidable. However, in one of the densest passages of "Anti-Realism and Revisionism," Wright might be understood as rejecting something like that conclusion. In this section, I begin with an interpretation of this passage in order to make explicit some key modal and epistemic assumptions of this aspect of Wright's earlier anti-revisionism. I will then use these assumptions to raise another problem for Wright's Basic Revisionary Argument.

Here is the relevant text:

[I]f somebody accepts Bivalence for a class of statements for whose truth-values we cannot in every case guarantee means of decision, then he is at least committed to holding that we cannot guarantee that truth everywhere coincides with decidable truth. But, unless he accepts the transition from 'we cannot guarantee that P' to 'it is a possibility that not-P,' he has not thereby committed himself to the possibility of verification-transcendent truth. This transition is intuitionistically suspect: of any mathematical statement which is not effectively decidable it would be intuitionistically correct to say that we cannot guarantee the existence of means of verifying or falsifying it, but it is not, in view of the intuitionists' account of negation, acceptable as an intuitionistic possibility that there should simply *be* no means of verifying or falsifying the statement in question. There is therefore a doubt whether endorsement of Bivalence for other than effectively decidable statements is of itself an admission of the possibility of verification-transcendent truth.   (Wright 1981: 434)

The first sentence in this passage clearly alludes to an implication; let's call it the "First Sentence Argument." It seems to consist of a transition from two claims:

(1) Every member of a class of statements satisfies the principle of bivalence.
(2) It is not the case that for every statement in this class we can guarantee means of deciding its truth-value.

to:

(3) We cannot guarantee that for every statement in this class its truth coincides with decidable truth

I will try to understand this implication by first attempting an interpretation of the content of these three claims in order to see what an inference from (1) and (2) to (3) might be.

Obviously there is no interpretive difficulty over (1). Given the interpretation of the notion of guarantee that I've adopted, (2) means:

It is not the case that for every statement in this class we know that we have means of deciding its truth-value.

It is plausible that to have means of deciding the truth-value of a statement is to be able to know that it is true if it is true and to be able to know that it is false if it is false. So (2) is

It is not the case that for every statement in this class we know that we can know that it is true if it is true or we can know that it is false if it is false.

We can take this in two ways, but one of them is just Ignorance of Decidability, (NKD_T):

$$\neg(\forall p)[K(R(T(p)) \vee R(F(p)))] \qquad\qquad (NKD')$$

$$\neg K[(\forall p)(R(T(p)) \vee R(F(p)))] \qquad\qquad (NKD_T)$$

Finally, what about (3)? It seems to be this

We don't know that for every statement in this class it is true just in case we can know that it is true or it is false just in case we can know that it is false.

That is,

$$\neg K[(\forall p)((T(p) \leftrightarrow R(T(p)) \vee (F(p) \leftrightarrow R(F(p))))].$$

The proposition of which we are ignorant, according to this conclusion, is a stronger claim than the Epistemic Constraint on Truth; it states that knowability of the truth-value of a statement is not just necessary but sufficient for that statement to have that truth-value. The sufficiency condition is perhaps not implausible on any reasonable anti-realist construal of knowability, since as discussed above, it's unclear that falsehoods can be knowable for Wright's version of anti-realism. But in any event, it's also not clear that anti-realism really demands anything more than knowability as a necessary condition. So we may perhaps reasonably interpret the conclusion to be ignorance of the Epistemic Constraint on Truth: $\neg K(EC_T)$.

On the interpretation just sketched, the inference from (1) and (2) to (3) is the argument that acceptance of bivalence for statements of whose Decidability we are Ignorant implies ignorance of the Epistemic Constraint on Truth. Now the question is, what does Wright mean by "acceptance"? If he means "knowledge," then (1) is the claim that bivalence is known to hold. On this reading, the First Sentence Argument is intended to establish the implication: $K(BV), (NKD_T) \vdash \neg K(EC_T)$. If this is right, then the First Sentence Argument establishes an implication equivalent to the one established by the first step of the two-step Basic Revisionary Argument.

In the remainder of the quoted passage Wright argues that the First Sentence Argument fails to show that anti-realism must reject "endorsement of Bivalence for other than effectively decidable statements." It seems clear that Wright has in mind here statements not known to be decidable. Given the reading of the First Sentence Argument just sketched, Wright's reasoning seems to go as follows. To begin with, the conclusion of the First Sentence Argument—the Epistemic Constraint on Truth is not known—does not imply "the possibility of verification-transcendent truth," which defines semantic realism. Anti-realism *is* committed to rejecting semantic realism. But since semantic realism is not implied by the conclusion of the First Sentence Argument, anti-realism is not committed to rejecting that conclusion or either of the premises that imply that conclusion. Hence anti-realism is not commit-ted to rejecting knowledge of bivalence for statements of whose Decidability we are Ignorant. Let's call this interpretation of Wright's line of reasoning the "Argument for Anti-realist Bivalence."

This argument raises three questions. Why does ignorance of the Epistemic Constraint fail to imply the possibility of verification-transcendent truth? Even if knowledge of bivalence for statements not known to be decidable doesn't imply realism, is that knowledge nevertheless compatible with anti-realism? If not, does Wright's objection commit him to accepting the coherence of a position differing both from realism by rejecting possibly verification-transcendent truth and from anti-realism by accepting bivalence for statements we don't know to be decidable?

The quoted passage might suggest that Wright's answer to the first question is that the transition from ignorance of the Epistemic Constraint to the possibility of verification-transcendent truth is not intuitionistically valid.[23] But in a later clarification of this argument Wright explains that the basis for rejecting this transition is not its intuitionistic invalidity, but a distinction between "real" and epistemic modality closely connected to the distinction between metaphysical and epistemic modality made familiar by Saul Kripke's *Naming and Necessity*.[24] Using '$\Diamond$' for the real possibility operator, we can formulate Wright's position as the claim that the form of inference from $\neg K(p)$ to $\Diamond(\neg p)$ is not valid; call this the "ignorance–possibility" form of inference. He writes,

> The semantic realist does hold . . . that the extensions of 'true' and 'ascertainably true' may possibly diverge. And this 'possibly' is not merely epistemic. The realist's view is not that it is consistent with everything we know, or whatever, that truth can outrun evidence; it is that it *really* can outrun evidence—that the availability or otherwise of evidence is no *conceptual* constraint on the capacity of a statement to be true. Now obviously, to lack a guarantee that the extensions of two predicates are the same is not to be obliged to admit the (real) possibility that they diverge: we have, for instance, no guarantee that the extensions of the predicates 'counter-example to Goldbach's Conjecture' and 'prime number between 32 and 36' coincide, since we have no guarantee that there are no counter-examples to Goldbach's Conjecture, that is, that it is true. But to find ourselves in this state of ignorance is, emphatically, not to have to admit that there is a possibility that the extensions of those two predicates diverge; we simply do not know whether that is a possibility or not.
>
> Thus anyone—realist or anti-realist—should demur at the form of transition—from 'There is no guarantee that P' or 'We have no reason to suppose that P' to 'It is a possibility that not-P'—which is involved here.   (Wright 1986: 460; second emphasis mine)

I want to point out that it is critical to this non-intuitionistic argument that Goldbach's Conjecture is a mathematical proposition, and so not contingent. Because it is not contingent, if it is true it is necessarily true, and so it is *not* possible for it to be false; while if it is false obviously it *is* possible for it to be false. More generally, if a proposition is not contingent, then whether its falsity is possible depends on its truth-value. It is for this reason that, whenever we're ignorant of the truth-value of a proposition we know to be non-contingent, "we simply do not know whether [its falsity] is a possibility or not," that is to say, it is compatible with all we know (which includes its non-contingency) that its falsity is not possible. This

---

[23] This is how Rasmussen and Ravnkilde understand Wright's argument, see Rasmussen and Ravnkilde (1982: 390–1).

[24] Kripke (1980: especially 35–7). Kripke argues for this distinction primarily through the example of Goldbach's Conjecture, at 36–7.

doesn't hold for contingent propositions. Whether a contingent proposition is true or false, it is *possible* for it to be false—this possibility follows from its contingency. Hence if we know that a proposition is contingent, then, even if we're ignorant of its truth-value, it is *not* compatible with all we know that its falsity is not possible.

We need one more point about Wright's argument. The conclusion of the ignorance–possibility inference that Wright rejects is:

$$\Diamond[\neg(\forall p)((T(p) \rightarrow R(T(p)) \vee (F(p) \rightarrow R(F(p)))] \qquad (\Diamond\neg EC_T)$$

But, prima facie, this is *not* the claim that verification-transcendent truth is really possible. What exactly is the claim that truth might really transcend verification, or that "the extensions of 'true' and 'ascertainably true' may possibly diverge" or that "truth can *really* outrun evidence"? A natural construal of these formulations is

It is possible that some statement is such that, either it is true and we cannot know that it is true, or it is false and we cannot know that it is false.

That is to say,

$$\Diamond[(\exists p)((T(p) \wedge \neg R(T(p)) \vee (F(p) \wedge \neg R(F(p)))] \qquad (\Diamond VT_T)$$

It seems, then, that Wright is tacitly assuming, at the very least, that $(\Diamond\neg EC)$ implies $(\Diamond VT)$, if not also that these two claims are equivalent.[25]

To sum up Wright's argument for Anti-realist Bivalence: in order to reach semantic realism from knowledge of bivalence, one takes three steps:

1 From knowledge of bivalence to ignorance of the Epistemic Constraint
2 From ignorance of the Epistemic Constraint to the possibility of its negation
3 From the possibility that the Epistemic constraint fails to the possibility of verification-transcendent truth

Wright's objection is that the ignorance–possibility inference in step 2 is not valid.

I turn now to reflect on the consequences of the invalidity of ignorance–possibility inferences for Wright's Basic Revisionary Argument. Once again, recall that that Argument rests on the joint incompatibility of knowledge of excluded middle, knowledge of the Epistemic Constraint, and Ignorance of Decidability: $K(LEM)$, $K(EC_P)$, $NKD_P \vdash \bot$. From this Wright inferred that knowledge of the Epistemic Constraint and Ignorance of Decidability imply that excluded middle is not known, $K(EC_P)$, $NKD_P \vdash \neg K(LEM)$. But, of course, from a purely logical standpoint, the joint incompatibility also allows us to conclude that knowledge of excluded middle and Ignorance of Decidability imply ignorance of the Epistemic Constraint,

---

[25] Once we've distinguished $(\Diamond\neg EC_T)$ from $(\Diamond VT_T)$, it's not altogether clear that Wright needs to argue against ignorance-possibility inferences in order to block the transition from ignorance of the Epistemic Constraint to possibility of verification-transcendent truth. The reason is that, while $(\Diamond VT_T)$ implies $(\Diamond\neg EC_T)$, the converse implication, from $(\Diamond\neg EC_T)$ to $(\Diamond VT_T)$, holds classically but not intuitionistically. Thus, even if ignorance of the Epistemic Constraint *does* imply that it might really be false, contrary to what Wright argues, it's not clear that from the conclusion of this implication one can validly infer the possibility of verification-transcendent truth.

Of course this point also puts in question the very assumption I've just now attributed to Wright.

K(LEM), NKD$_P$ ⊢ ¬K(EC$_P$). Now let's apply the form of reasoning Wright used in the Argument for Anti-realist Bivalence. What's so bad about ignorance of the Epistemic Constraint for an anti-realist? It would be a problem only if it implied something anti-realism is committed to rejecting, namely, the possibility of verification-transcendent truth,[26] that is, if analogues of the three steps just described hold in the present case:

1′ From knowledge of excluded middle to ignorance of the Epistemic Constraint (on Propositions)
2′ From ignorance of the Epistemic Constraint to the possibility of its negation
3′ From the possibility that the Epistemic constraint fails to the possibility of verification-transcendent truth

But, as Wright has argued, the ignorance–possibility inference of step 2′ is invalid. Hence, the assumption that we *know* classical reasoning holds of a set of statements of whose Decidability we are Ignorant does not imply a commitment to semantic realism. It follows that the deduction in Wright's Basic Revisionary Argument does not demonstrate that anti-realism is committed to rejecting classical logic for statements not known to be decidable.

This argument would not present any problems for Wright's Basic Revisionary Argument if anti-realism is committed, in addition to rejecting realism, to knowledge of the Epistemic Constraints. This is clearly Wright's view; as we have seen, he describes his Epistemic Constraint as "known *a priori.*" So far in this discussion of revisionism I have not questioned anti-realism's commitment to claiming knowledge of some version of the Epistemic Constraints. But it's not beyond question. Recall from the account of anti-realism in §1 that we can think of Dummett's semantic anti-realism as based essentially on *only* the manifestation of meaning, so that the knowability of truth requires further argument(s). What are those arguments? On one view, what we can immediately infer from the manifestation requirement is *only* that semantic realism is untenable. That is to say, the supposition that the truth-value of some statement is unknowable is logically inconsistent with the claim that the meaning of that statement can be communicated. Hence if all meaning can be communicated, then this supposition of unknowable truth-value implies a contradiction. Moreover, if the communicability of meaning is a *conceptual* truth about meaning, then not just the existence, but even the possible existence of recognition-transcendent statements would imply a contradiction. So, if we assume the communicability of meaning, then we can rule out the (real) possibility that some statement's truth-value transcends recognition. But, at this point, a somewhat bearish anti-realist can dig in her heels and hold that we have already reached the only conclusion firmly established by this reasoning. That is, she holds that although anti-realism provides grounds for ruling out the possibility of any statement having a truth-value that

---

[26] Since the Basic Revisionary Argument is formulated in terms of propositions rather than truth, we need a different formulation of verification transcendence:

$$◊[(\exists p)(p \wedge \neg R(p))] \hspace{4cm} (◊\text{VT}_P)$$

transcends recognition, and that perhaps this even gives us grounds for ruling out the failure of any Epistemic Constraint, these grounds are not enough to give us knowledge of these Epistemic Constraints. And so, from the perspective of this anti-realist, Wright's Basic Revisionary Argument does not require her to hold that we don't know that excluded middle applies to statements not known to be decidable.

Such an anti-realist can bolster her position by arguing that, provided we accept the invalidity of ignorance–possibility inferences, ignorance of the Epistemic Constraints *has to be* consistent with the impossibility of verification-transcendent truth. The reasoning is as follows. Suppose that the ignorance and the impossibility at issue are not jointly consistent: $\neg K(EC)$, $\neg\Diamond(VT) \vdash \perp$. Then, ignorance of the Epistemic Constraint implies the double negation of the possibility of verification-transcendence, and hence, classically, that possibility: $\neg K(EC) \vdash_{CL} \Diamond(VT)$. But, classically, the possibility of verification-transcendence implies the possibility that the Epistemic Constraint fails: $\Diamond(VT) \vdash_{CL} \Diamond(\neg EC)$. Hence ignorance of the Epistemic Constraint implies its possible falsity, contrary to the assumption that ignorance–possibility inferences are invalid. The fact that this line of argument uses classical logic (both in the object- and the meta-language) is no more question-begging than is any rejoinder to it on the basis that if classical logic is prohibited, then the required demonstration of consistency fails, if this failure is then used to argue against knowledge of excluded middle.

Now, at this point, one might wonder: how, if she is so bearish, could this anti-realist be at the same time so bullish as to hold that she *knows* that classical logic applies to statements of whose Decidability she is Ignorant? But why not? Why can't she think that no one counts as knowing the meanings of the logical constants of negation and disjunction unless he accepts the validity of the law of excluded middle? So knowledge of excluded middle is simply constitutive of understanding the logical constants. This view is closely related to a position that Wright himself considers in "Anti-Realism and Revisionism" (Wright 1981: 449). The position of this mildly bearish anti-realist, who accepts knowledge of classical logic but not knowledge of the Epistemic Constraint, might be an odd blend, but, so far, it is not obviously incoherent.

The position just described rests at bottom on the assumption that Wright's case against ignorance–possibility inferences is that, in general, ignorance of a proposition *does not logically imply* the real possibility of its negation. For, if p does not imply q, then it must be consistent to hold p and not-q; moreover, classically the inconsistency of p and not-q would license the implication from p to q. But another look at Wright's argument suggests that perhaps he was advancing a different objection to ignorance–possibility inferences. Remember that Wright's argument required those propositions for which such inferences fail to be non-contingent. For a non-contingent proposition, whether its negation is possible is determined by its truth-value, so if one is not in a position to know that truth-value, one is in no position to affirm or deny the possibility of its negation. But from this point it does not obviously follows that if one *is in a position to deny* the possibility of the negation of a non-contingent proposition, one might nevertheless not be in a position to know that it holds. Our mildly bearish anti-realist accepts that the possible falsity of any Epistemic

Constraint can be ruled out on the ground of an analysis of the concept of meaning. Does this not give us *a priori* grounds for believing that the Epistemic Constraints hold? But then do we not have grounds for claiming to *know* that the Epistemic Constraints hold?

The argument I've just sketched, however, is not conclusive. The anti-realism to which it is vulnerable holds that the possible existence of verification-transcendent statements is (logically) inconsistent with the manifestation requirement. But anti-realists need not be quite so bullish on the strength of the manifestation requirement. Dummett, for one, in more cautious moments, characterizes anti-realism as claiming not to have refuted, but only to have posed a challenge to, semantic realism. The challenge is to explain how the meanings of possibly recognition-transcendent statements can be communicated. Following this line of thought, an even more strongly bearish anti-realist might hold that her resistance to semantic realism goes no further than the claim that, in the absence of a convincing account of the communicability of verification-transcendent statements, we *do not know* that it is possible for there to be any such statements. Being in this state of ignorance, one obviously is in no position to rule out the failure of any Epistemic Constraints, and so the foregoing argument for anti-realist knowledge of these Constraints does not go through. The strongly bearish anti-realist is then free to claim, consistently, knowledge of excluded middle.

Perhaps there remains some residual unease over an anti-realist's claim to *knowledge* of classical logic. But Wright himself gives us the materials for an ultra-bearish anti-realism that is not vulnerable to this doubt. In "Realism, Bivalence and Classical Logic," Wright formulates an objection to the First Sentence Argument of "Anti-Realism and Revisionism":

Suppose our arithmetical intuitionist visits the oracle at Delphi and inquires of the Pythian Priestess, 'Who, intuitionist or platonist, has the truth of the matter as far as number-theory is concerned?' And suppose that her answer is 'Bivalence holds for all number-theoretical statements.'... [T]he response is in keeping with the best traditions of Delphic ambiguity, since from an intuitionist viewpoint it will equally well bear interpretation as the claim that all number-theoretic problems are resoluble.... [I]f the intuitionist accepts the pronouncement under the latter interpretation, it is evident that he can also continue to accept that we cannot guarantee the decidability of all arithmetical statements, yet still incur no commitment to 'we cannot guarantee that (number-theoretic) truth everywhere coincides with decidable (number-theoretic) truth.'

What would generate commitment to 'we cannot guarantee that (number-theoretic) truth everywhere coincides with decidable (number-theoretic) truth' would be the assumption:

1 that we cannot guarantee that all (number-theoretic) statements are decidable, and
2 that we can guarantee that Bivalence holds for all (number-theoretic) statements.

The realist, of course, believes he has the guarantee described by (2), flowing from the nature of statement-meaning. The anti-realist, in contrast, who accepts Bivalence on the say-so of an oracle, or as an expression of his confidence that human ingenuity can surmount all obstacles, or whatever, will presumably only accept its interior that-clause. (Wright 1986: 462-3)

Wright's point, it seems, is that the difference between realism and (some varieties of) anti-realism over bivalence might rest simply on the fact that realism claims *knowledge* of bivalence while anti-realism claims only *acceptance* of that principle. Now, why can't an ultra-bearish anti-realist make use of precisely this distinction between knowledge and acceptance? She agrees with her merely bearish colleagues that anti-realism does not give her grounds for knowledge of the Epistemic Constraints. But, although this opens up the logical space for her to claim to know that excluded middle applies to not-known-to-be-decidable statements, she is unwilling to commit herself to this knowledge claim. Instead, she thinks that, since it is not logically contradictory to claim to know excluded middle, it certainly can't be contradictory to take up the more cautious stance of merely *accepting* excluded middle for all statements, whether or not known to be decidable.[27]

11. The Awkward Wrinkle

We have now discussed two worries about Wright's Basic Revisionary Argument. The first is based on doubts about the truth of Wright's Epistemic Constraint, and so puts the soundness of the Argument in question. The second is based on doubts about whether ignorance of the Epistemic Constraint is compatible with anti-realism, and so furnishes reservations about whether the Argument in fact uncovers any incompatibility between anti-realism and knowledge of classical logic. In this section and §12, I discuss a third worry. This problem is independent of the issue of the commitments of anti-realism that was central in the last section, so now I revert to Wright's assumption that according to anti-realism the Epistemic Constraints are known.

Recall that Wright's Basic Revisionary Arguments can be seen as the result of addressing the fundamental problem afflicting the one-step Naïve Revisionary Argument. That problem consists of a tension between the Epistemic Constraint and the existence of undecidable statements. Central to the response that generates the Basic Revisionary Argument is the maneuver of epistemic ascent. One might thus say that Wright's Revisionary Argument is an epistemic version of the one-step Naïve Argument. A natural question, then, is whether there is an epistemic version of the problem that led to the downfall of this Naïve Argument.

If there is an epistemic version of this problem it would, by epistemic ascent of the original problem, consist of a tension between *knowledge* of Epistemic Constraint and *Ignorance* of Decidability. In "Quandary," Wright discerns an "awkward wrinkle" in the Basic Revisionary Argument, which is precisely such a tension. The problem, as Wright formulates it, centers on the justification of our Ignorance of Decidability.

---

[27] Jon Cogburn argues that Dummett's anti-realism is committed to denying the (conceptual) possibility of verification-transcendent truth, and that Dummett's criticism of classical logic is in fact an argument against classical truth conditional semantics, on the ground that it implies the possibility of verification-transcendent statements (Cogburn 2002). Apart from the point, discussed above, that classical reasoning arguably does not require validation by classical truth conditional semantics, the current argument shows that there is an anti-realism, which can reasonably be ascribed to Dummett, that denies only knowledge of the possibility of verification-transcendent truth, and that is consistent with acceptance of classical logic.

As we saw at end of §8, Wright accepts a principle of agnosticism, called (AG): "*P* should be regarded as unknown just in case there is some possibility *Q* such that if it obtained, it would ensure not-*P*, and such that we are (warranted in thinking that we are) in no position to exclude *Q*" (Wright 2001: 67-8); this may be formally represented as:

$$(\forall p)(\neg K(p) \leftrightarrow (\exists q)((q \to \neg p) \wedge \neg K(\neg q)) \tag{AG}$$

According to this principle, in order to justify Ignorance of Decidability, one needs to find some possibility incompatible with Decidability and show that we don't know that this possibility fails to obtain. Let *A* describe any such possibility—that is, *A*, (Dec$_P$) $\vdash \bot$. It follows that *A* implies that not every statement is decidable:

$$A \vdash \neg(\forall p)(R(p) \vee R(\neg p)).$$

As we have just reminded ourselves, the objection to the one-step Naïve Argument is that the Epistemic Constraint implies something very close, namely, that there are no undecidable statements:

$$(EC_P) \vdash \neg(\exists p)(\neg R(p) \wedge \neg R(\neg p)).$$

The crucial question then is whether the two claims implied by *A* and by (EC$_P$) are mutually consistent. If the answer is no, then *A* and (EC$_P$) themselves are not mutually compatible. But then, by epistemic ascent, knowledge of the Epistemic Constraint implies knowledge that *A* fails: from (EC), $A \vdash \bot$ we get (EC) $\vdash \neg A$, and then $K(EC) \vdash K(\neg A)$. Since *A* is an arbitrarily chosen possibility incompatible with the Decidability of every statement, it follows that knowledge of the Epistemic Constraint implies knowledge that *every* such possibility fails. Hence knowledge of the Epistemic Constraint deprives us of justification for Ignorance of Decidability.

By classical reasoning, the non-existence of undecidable statements, $\neg(\exists p)$ $(\neg R(p) \wedge \neg R(\neg p))$, implies that all statements are decidable, $(\forall p)(R(p) \vee R(\neg p))$. Thus classical logic yields a negative answer to the crucial question. Hence, it follows by classical reasoning that anti-realism cannot justify any claim that a discourse contains statements not known to be decidable. Now, the inference required to reach this conclusion, from the non-existence of undecidable statements to the decidability of all statements, is not intuitionistically valid. That is to say, knowledge of the Epistemic Constraint does not suffice to rule out every possibility incompatible with Decidability unless classical reasoning is allowed. But, as Wright points out, the Basic Revisionary Argument is supposed to provide grounds for rejecting classical logic (Wright 2001: 68). If this Argument doesn't work unless we prohibit classical reasoning, it is at best a question-begging argument for revisionism.

Wright's response to this difficulty is to argue for a modification of the principle of agnosticism, one that allows ignorance of a proposition even when we can rule out all possibilities incompatible with it. One condition that suffices for ruling out all possibilities incompatible with a proposition *p* is where we have grounds for ruling out the negation of *p*. This is because if any possibility *q* is incompatible with *p*, then *q* implies the negation of *p*; hence our grounds for ruling out the negation of *p* are also grounds for ruling out *q*. So, in searching for a modification of the principle (AG),

Wright considers cases in which we fail to know a proposition at the same time that we can rule out its negation.

The cases on which Wright focuses are predications of vague terms in borderline cases; Wright's example is 'x is red' (Wright 2001: 72). Such predications possess two properties. First, the Epistemic Constraint applies to them because Wright holds that color properties are "transparent" in the sense that they "have essentially to do with how things visually appear and their instantiations [,and] may always in principle be detected by our finding that they do indeed present appropriate visual appearances" (Wright 2001: 72). Second, these predications are examples of what Wright calls "quandaries." Quandaries are statements with respect to which we have four levels of ignorance. For any of these statements, we

1 "do not know" whether it is true or false,
2 "do not know how we might come to know" whether it is true or false,
3 "can produce no reason for thinking that there is any way of coming to know" whether they are true or false, and,
4 have "no basis to think that anything amounting to knowledge" of whether it is true or false "is metaphysically provided for" (Wright 2001: 71, 75).[28]

As I understand Wright's argument, the crucial characteristic of quandaries is that they are not known to be decidable. That is, using the predicate 'Q 'to abbreviate 'is a quandary,' $Q(p) \rightarrow \neg K(R(p) \vee R(\neg p))$. Wright argues that if a proposition p both is a quandary and satisfies the Epistemic Constraint, then the instance of excluded middle for p is a counter-example to the original principle of agnosticism. Specifically, $p \vee \neg p$ is not known, but, at the same time, we can rule out its negation and so every possibility incompatible with it.

The argument for being able to rule out the negation of $p \vee \neg p$ is independent of the assumption that p is a quandary. It is simply an intuitionistic (indeed, minimal) proof of the double negation of any instance of excluded middle, which uses only the rules of disjunction introduction and negation introduction (Wright 2001: 67, n. 26).

The argument against knowledge of $p \vee \neg p$ is not as clear; I take it to work as follows. As we know from the pre-epistemic ascent phase of the Basic Revisionary Argument, excluded middle and the Epistemic Constraint together imply that all statements are either verifiable or falsifiable. Hence, if excluded middle and the Epistemic Constraint both apply to p, then p is either verifiable or falsifiable: $p \vee \neg p, p \rightarrow R(p), \neg p \rightarrow R(\neg p) \vdash R(p) \vee R(\neg p)$. By epistemic ascent, knowledge that p satisfies excluded middle, together with knowledge that either p or its negation is knowable, entails knowledge that p is either verifiable or falsifiable, $K(p \vee \neg p)$, $K(p \rightarrow R(p)), K(\neg p \rightarrow R(\neg p)) \vdash K(R(p) \vee R(\neg p))$. But this consequence contradicts

---

[28] Some analyses of vagueness might support all four of these levels of ignorance, but it's not clear that they would be endorsed by all theories of vagueness. In particular, (2) and (3) would be rejected by most contextualist theories, since these theories hold out the possibility that further investigation of contextually provided standards might enable one to come to know whether a borderline predication is true. Such theories need not be committed to claiming that we will always succeed in this sort of investigation, merely that it offers one open avenue of inquiry. Some main contextualist theories of vagueness are presented in Kamp (1981); Tappenden (1993); Raffman (1994); Soames (1999); and Graff (2000). I'm grateful to Michael Glanzberg for pointing this out to me.

the supposition that p is a quandary, so, $K(p \vee \neg p)$, $K(p \rightarrow R(p))$, $K(\neg p \rightarrow R(\neg p))$, $Q(p) \vdash \bot$. Hence, since we're also supposing that p satisfies the Epistemic Constraint, we do not know that p satisfies excluded middle: $Q(p)$, $K(p \rightarrow R(p))$, $K(\neg p \rightarrow R(\neg p)) \vdash \neg K(p \vee \neg p)$. Let's call this argument the Quandary Argument.

From the Quandary Argument, Wright extracts a replacement for principle (AG):

Consider any compound statement, A, whose truth requires that (some of) its constituents have a specific distribution of truth-values or one of a range of such specific distributions. And let the constituents in question be subject to EC. Then (AG$^+$) A is known only if there is an assurance that a suitably matching distribution of evidence for (or against) its (relevant) constituents may feasibly be acquired.   (Wright 2001: 76)

How does the Quandary Argument support (AG$^+$)? I take it that Wright's idea is this. A quandary instance of excluded middle—$p \vee \neg p$—is a disjunction, so its truth requires the truth of at least one disjunct, that is, either the truth of p or of $\neg p$. Suppose also that the Epistemic Constraint applies to p and to $\neg p$. Then $p \vee \neg p$ satisfies the hypothesis for the application of (AG$^+$). Since p is a quandary, we don't know that we either can recognize the truth of p or can recognize the truth of $\neg p$. That is, if p is a quandary, then we don't have "assurance" that we can recognize p and $\neg p$ to have those truth-values required by the truth of $p \vee \neg p$. The Quandary Argument tells us that in these circumstances we also don't know that $p \vee \neg p$; alternatively, that if we do know that $p \vee \neg p$, then p can't be a quandary.

Unfortunately it is not entirely clear how this new principle (AG$^+$) is to be applied to save the Basic Revisionary Argument. The awkward wrinkle was that, according to (AG), Ignorance of Decidability loses its justification when knowledge of the Epistemic Constraint enables us to rule out all possibilities incompatible with Decidability. In order to apply (AG$^+$) to this case, we would have to show that (Dec$_P$) is a compound statement whose truth requires some distribution(s) of truth-values for its constituents, and we don't know that we can recognize that its constituents have that or those truth-values.

However, it seems that the Quandary Argument can be used on its own to provide a case for revisionism, without going back to the Basic Revisionary Argument. We might take the argument to show that no instance of excluded middle for quandary statements subject to the Epistemic Constraint is known. Alternatively, we might argue as follows. We know, by minimal logic, that double negations of all instances of excluded middle hold as a matter of logic. So, if we accepted double negation elimination as a valid form of inference, we would be able to infer all instances of excluded middle by logic alone, and then conclude that we know all these instances. But since we have independent grounds for thinking that we know there are quandaries subject to the Epistemic Constraint, we have independent grounds for thinking that we don't know that all instances of excluded middle hold. Hence we have reason to reject double negation elimination, and so classical logic, as generally valid.

12. The Awkward Customer

My worry about revisionism based on the Quandary Argument derives from thinking about the notion of a quandary statement. Wright's models for quandaries, as we have seen, are statements predicating vague terms of borderline cases. On Wright's view, our intuitive reaction to such statements is that it is indeterminate whether they are true or false, and this indeterminacy is epistemic. That is, we don't know whether the statement is true or false. The first three levels of ignorance defining a quandary constitute Wright's initial account of the ignorance involved in this indeterminacy. Given these three levels of ignorance with respect to a statement, our position can be described by a *conjunction*: we don't have grounds for thinking that we can know that the statement holds, *and* we don't have grounds for thinking that we can know that the negation of that statement holds: $\neg K(R(p)) \land \neg K(R(\neg p))$. Wright adds the fourth level of ignorance because, with only the first three levels, there is no decisive answer to the following objection advanced by an "awkward customer" (Wright 2001: 73).

[M]ight I not have all those three levels of ignorance and still know that it is the case *either that* P *is knowable or that its negation is*? . . .

[I]t does not strictly follow from the . . . characterization . . . of Quandary [in terms of three levels of ignorance] that if '*x* is red' presents a quandary, then we have no warrant for the disjunction,

FeasK(*x* is red) ∨ FeasK(it is not the case that *x* is red)

All that follows, the awkward customer is pointing out, is that we are, as it were, thrice unwarranted in holding either disjunct. To say that someone does not know whether A or B, is ambiguous. Weakly interpreted, it implies, in a context in which it is assumed that A or B is true, that the subject does not know which. Strongly interpreted, it implies that the subject does not know that the disjunction holds. . . . It is uncontentious that such examples may be quandaries if that is taken merely to involve ignorance construed as an analogue of the weak interpretation of ignorance whether A or B. But to run the Basic Revisionary Argument, a case needs to be made that borderline cases of colour predicates present quandaries in a sense involving ignorance under the strong interpretation. What is that case?   (Wright 2001: 74)

We can understand the awkward customer's objection as resting on a logical point. The conjunctive ignorance resulting from the first three levels of ignorance implies that it's not the case that either we know we can verify the statement or we know we can falsify it:

$$\neg[K(R(p)) \lor K(R(\neg p))].$$ (1)

But, in order for the Quandary Argument to work, quandaries must have the property that we don't know that either we can verify a quandary or falsify it:

$$\neg K[R(p) \lor R(\neg p)]].$$ (2)

What justifies the inference from (1) to (2)? One might argue in this way. Suppose that we know the simplest implications of what we know. Either disjunct of a disjunction of course implies that disjunction. Hence, knowledge of either disjunct implies knowledge of the disjunction: $K(p) \lor K(q) \vdash K(p \lor q)$. By contraposition, ignorance of a disjunction implies knowledge of neither disjunct: $\neg K(p \lor q) \vdash \neg[K(p) \lor K(q)]$. But *this* implication does not license the inference we need. It does not justify concluding that *p* has the property required by the Quandary

Argument, (2), from our possessing the first three levels of ignorance with respect to p, (1). What we need is the converse implication—$\neg[K(p) \vee K(q)] \vdash \neg K(p \vee q)$; but this does not, in general, follow from the implication we do have.

Wright's response is to add metaphysical ignorance to the definition of quandaries:

[I]t is crucial to the conception of indeterminacy being proposed that someone who takes a (presumably tentative) view for or against the characterizability of such a case as 'red' is *not known to be wrong*. But that is consistent with allowing that it is also not known whether knowledge, one way or the other, about the redness of the particular case is even *metaphysically possible*.... I suggest that we should acknowledge that borderline cases do present such a fourth level of ignorance: that, when a difference of opinion about a borderline case occurs, one who feels that she has no basis to take sides should not stop short of acknowledging that she has no basis to think that anything amounting to knowledge about the case is metaphysically provided for.   (Wright 2001: 75)

It seems plausible that the notion of metaphysical possibility that Wright uses here is the same as the notion of "real" possibility of his earlier anti-revisionary arguments, discussed in §10 above. How does the ignorance of metaphysical possibility described in this passage lead to the required property of quandaries? The connection, I take it, comes from the assumption that metaphysical possibility is entailed by the stronger modality of feasibility. That is, if it is feasible to know what a statement expresses, then it is metaphysically possible to know it. Given our construal of verifiability as feasibility of knowledge, the assumption yields a principle:

$$(\forall p)(R(p) \rightarrow \Diamond K(p)) \tag{FP}$$

From this principle it follows that if p is decidable, then either it is metaphysically possible to know that p, or it is metaphysically possible to know its negation:

$$R(p) \vee R(\neg p) \vdash \Diamond K(p) \vee \Diamond K(\neg p).$$

Let's use the phrase "it is metaphysically possible to decide p" to abbreviate "either it is metaphysically possible to know that p, or it is metaphysically possible to know its negation." By epistemic ascent, knowledge of decidability implies knowledge of metaphysical possibility of decidability:

$$K[R(p) \vee R(\neg p)] \vdash K[\Diamond K(p) \vee \Diamond K(\neg p)]$$

Hence

$$\neg K[\Diamond K(p) \vee \Diamond K(\neg p)] \vdash \neg K[R(p) \vee R(\neg p)],$$

that is, ignorance of metaphysical possibility of deciding p implies Ignorance of Decidability, which is precisely the property required by the Quandary Argument. Obviously what my reconstruction of Wright's reasoning assumes is that "knowledge whether p is metaphysically provided for" means "either it is metaphysically possible to know that p or it is metaphysically possible to know that not-p."

I can now state my reservations about the Quandary Argument. Let's ask, assuming anti-realism, and so assuming knowledge of the Epistemic Constraint, can we profess ignorance of the metaphysical possibility of deciding any given statement?

Remember why we gave up the one-step Naïve Revisionary Argument: the Epistemic Constraint is incompatible with the existence of (propositionally) undecidable statements. Clearly the reasoning leading to this conclusion remains valid if, instead of the existence of undecidable statements, we used the premise that some particular statement q is undecidable. That is to say,

$$(EC), \neg R(q) \wedge \neg R(\neg q) \vdash \bot.$$

Principle (FP), which connects verifiability with metaphysical possibility of knowledge, implies that if it is not metaphysically possible to know a statement, then that statement is not verifiable: $(\forall p)(\neg \Diamond K(p) \rightarrow \neg R(p))$. So, if it's not metaphysically possible to decide q, then it is not feasible to decide q:

$$\neg \Diamond K(q) \wedge \neg \Diamond K(\neg q) \vdash \neg R(q) \wedge \neg R(\neg q)$$

Hence the Epistemic Constraint is not merely incompatible with the undecidability of q; it is inconsistent with the claim that deciding q is not "metaphysically provided for":

$$(EC), \neg \Diamond K(q) \wedge \neg \Diamond K(\neg q) \vdash \bot.$$

From this conclusion we can of course infer that the Epistemic Constraint implies that it's not the case that it not metaphysically possible to decide q:

$$(EC) \vdash \neg(\neg \Diamond K(q) \wedge \neg \Diamond K(\neg q))$$

If we allow classical reasoning, then the conclusion of this implication implies that either it's not the case that it's not metaphysically possible to verify q or it's not the case that it's not metaphysically possible to falsify q. Hence,

$$(EC) \vdash_{CL} \neg \neg \Diamond K(q) \vee \neg \neg \Diamond K(\neg q)$$

Since we're using classical logic, we have double negation elimination, so we reach the conclusion that the Epistemic Constraint implies the metaphysical possibility of deciding q:

$$(EC) \vdash_{CL} \Diamond K(q) \vee \Diamond K(\neg q).$$

By epistemic ascent, knowledge of the Epistemic Constraint implies that knowledge whether q *is* metaphysically provided for:

$$K(EC) \vdash_{CL} K[\Diamond K(q) \vee \Diamond K(\neg q)].$$

Ignorance of the metaphysical possibility of deciding a quandary is the fourth level of ignorance needed to establish the ignorance of the decidability of quandaries required to run the Quandary Argument. The argument just presented shows, using classical logic, that knowledge of the Epistemic Constraint rules out this fourth level of ignorance with respect to any statement whatsoever, and so a fortiori with respect to quandaries. But the Quandary Argument rests on the assumption that there are statements known to be subject to the Epistemic Constraint that are also quandaries. Hence our argument shows that the Quandary Argument fails to demonstrate that excluded middle is not known to apply to quandaries. Of course, our argument uses classical logic. But this means that the Quandary Argument does not provide a cogent basis for rejecting classical logic.[29]

Let me stress that this objection does not, and is not meant to, show that "there must be a way of adjudicating all borderline colour predications" (Wright 2001: 76). The point, rather, is that if a reason to reject excluded middle is that it implies all these adjudications, then that is also a reason to reject the Epistemic Constraint.

All this, if correct, shows that the Quandary Argument fails as an argument on its own for revisionism. But what about Wright's view, which seems to be that the Quandary Argument supports the revised principle of agnosticism, (AG$^+$), which then allows a justification of Ignorance of Decidability and so allows us to run the Basic Revisionary Argument? What implications does the preceding argument have for the revised principle of agnosticism (AG$^+$)? Recall that the Quandary Argument supports (AG$^+$) in the following way. The Argument is supposed to have two features. First, we don't know the truth of disjunctions satisfying the following two conditions: they are instances of excluded middle for quandaries, and the Epistemic Constraint applies to them. Second, knowledge of the truth-values of the disjuncts of those disjunctions is not feasible, because it is not metaphysically provided for. These two claims support an instance (AG$^+$), namely, that knowledge of the truth of those disjunctions requires feasibility of acquiring knowledge of the truth-values of their disjuncts. But the preceding argument shows that, unfortunately, since the Epistemic Constraint applies to these disjuncts, knowledge of their truth-values *is* metaphysically provided for, and so we no longer have a basis for claiming that that knowledge is not feasible. So we no longer have support for principle (AG$^+$), or, hence, for any conclusion arrived at using it as a premise.

This objection to the Quandary Argument and the main objection to the one-step Naïve Argument have almost the same structure. The Naïve Argument deduces that the conjunction of the Epistemic Constraint and excluded middle implies that there are no undecidable statements, and concludes that the existence of undecidable statements constitutes a reason to reject excluded middle. The problem is that the Epistemic Constraint by itself implies that there are no undecidable statements. Similarly, the Quandary Argument deduces that knowledge of both the Epistemic Constraint and excluded middle implies knowledge that quandaries are decidable, and concludes that our ignorance of the decidability of quandaries constitutes a reason to reject knowledge of excluded middle. The problem we have just uncovered is that the Epistemic Constraint by itself implies, by classical reasoning, knowledge of the decidability of quandaries.

---

[29] Like the Quandary Argument, the two-step Revisionary Argument is also question-begging because it requires a prohibition on classical reasoning. Similar difficulties have been raised for at least two other revisionary proposals. First, one argument in favor of intuitionistic over classical logic is that a formal statement of Church's Thesis can be shown to be intuitionistically consistent but classically inconsistent with Peano Arithmetic. However, the formal statement in question can only be interpreted as Church's Thesis if the meta-language quantifiers are non-classical (see Dragalin 1980). Second, Neil Tennant argues against classical logic on the grounds that it, together with a version of the Epistemic Constraint, imply the existence of decision procedures for provably undecidable discourses (Tennant 1997: ch. 7). Cogburn objects that Tennant's reasoning requires an intuitionistic understanding of existential quantification (Cogburn 2003). I'm grateful to an anonymous referee for bringing these objections to my attention.

In fact, there is a much more direct route to this last conclusion. Let's simply apply epistemic ascent to the reasoning underlying the objection to the one-step Naïve Argument. That reasoning, as already mentioned, shows that the Epistemic Constraint implies that it is not the case that a particular statement q is undecidable: (EC) $\vdash \neg(\neg R(q) \wedge \neg R(\neg q))$. Classically, that it is not the case that q is undecidable implies that q is decidable. Hence, classically, knowledge of the Epistemic Constraint implies knowledge that q is decidable, for any particular statement q, and so also for any particular quandary.

13. Concluding Remarks

All the revisionary arguments we have examined have a common basic form. In each case the argument has two main components. One is an Epistemic Constraint, a linking principle allowing us to go from logical principles to what they imply about our capacities for knowledge. The other is a description of what we are in fact capable of knowing; depending on the argument, this is either (Und) or (NKD). The criticism in each case works by using the Epistemic Constraint to find that classical logic requires us to be able to know quite a lot, more than what is warranted by the facts about our capacities, as described by (Und) or (NKD). The assumption underlying this form of argument is that a standard to which deductive inference is or should be held accountable consists of facts about what we can know.

A number of the objections I have presented to these revisionary arguments also have a common basic form. In each case, we find that the Epistemic Constraint by itself requires our capacities for knowledge to satisfy certain conditions that conflict with the facts about these capacities. To recapitulate, the initial version of the conflict goes like this. According to (Und), in fact there are statements that we can't decide. But the Epistemic Constraint tells us that there are no undecidable statements. As if in response to this conflict, Wright works with a more cautious account of what we are in fact capable of knowing. As he puts it, "(NKD) seems incontrovertible [. . . ,] for does it not merely acknowledge that, relative to extant means of decision, not all statements are decidable?" (Wright 2001: 66). Now, it remains the case that if classical logic is valid, then by the Epistemic Constraint our cognitive abilities do extend to every statement. Thus Wright argues that knowledge of classical logic commits us to claiming knowledge about our abilities that is not warranted by what we in fact know about those abilities. Unfortunately the Epistemic Constraint by itself still conflicts with Wright's account of our abilities. Wright's account does not affect the implication from the Epistemic Constraint to the claim that there are no undecidable statements. So if we know that the Epistemic Constraint holds, then we must know that there are no undecidable statements. But then, don't we know that every statement is decidable, after all? Well, this last "then" *does* depend on the use of classical logic. In an argument for giving up classical logic, however, one proscribes its use at the cost of begging the question.

This suggests two things. First, it's not clear that revisionism can proceed by using the Epistemic Constraints to measure the legitimacy of logical principles against facts about what we can know. Second, it's not clear that facts about our cognitive capacities constitute the right basis for assessing logical laws.

Can Wright's revisionism do without the Epistemic Constraint? Since knowledge of the Constraint is a premise in the Basic Revisionary Argument, I don't see how it can work without the Constraint. The same holds for the Quandary Argument. But as we saw, it is not clear whether, or how, the Quandary Argument could be a revisionary argument on its own. This suggests that we might look at the roles that the Quandary Argument might play in revisionism, and try to see if those roles can be filled by arguments that don't rely on the Epistemic Constraint.

Specifically, let's look at the revisionary argument that I proposed on the basis of the Quandary Argument. We begin with Wright's point that the double negations of all instances of excluded middle are justified by uncontentious principles of logic and therefore known. It follows that if the rule of double negation elimination were logically valid, then all instances of excluded middle would be justified purely logically and therefore known. It's at this point that the Quandary Argument comes in. This argument provides independent grounds against knowledge of some instances of excluded middle. According to this argument, knowledge of quandary instances of excluded middle is ruled out by ignorance of the decidability of quandaries. Wright goes on to propose that a necessary condition on knowledge of disjunctions is knowledge of the feasibility of knowing at least one disjunct. Clearly, then, the proposed revisionary argument asks only two things of the Quandary Argument: first, to show that such a necessary condition holds; and, second, to show that there are instances of excluded middle that fail this necessary condition. Can these two requirements be met without the Epistemic Constraint?

I am far from certain that they can be. But if they are to be met, we have to begin revisionism from a different conception of anti-realism. The conception of anti-realism that we have been assuming is generated by a problem of evidence: how can we have evidence for a speaker's association of truth conditions with her statements? Anti-realism's solution is based on the assumption that the only evidence we can have must consist of the speaker's verbal behavior: what she says and does. In order for this evidence to reveal truth conditions, however, the behavior must arise in response to the speaker's recognition that these conditions obtain or fail to do so. From this point, some form of Epistemic Constraint seems inevitable, since if truth conditions can obtain without being recognized, then they can obtain without any response, any behavior, from the speaker, in which case there would be no evidence that the speaker has these conditions in mind.

Some of Dummett's writings, however, suggest a different objection to the realist view of truth. For example, in one of the most quoted formulations of the manifest-ation constraint, Dummett writes, "if two individuals agree completely about the use to be made of [a] statement, then they agree about its meaning" (Dummett 1973: 216). If a central component of meaning is truth conditions, then we can read this as claiming that it is not possible for two speakers to agree completely in what they would count as justifications for a statement while attaching different truth conditions to it. Elsewhere I have developed this idea into a non standard version of anti-realism.[30] The main thesis of this anti-realism is that we cannot, coherently and completely, distinguish our

---

[30] Shieh (1997).

conception of a statement's being true from its being justified. Hence being true and being justified go hand in hand. But this doesn't imply that a statement can't be justified unless it's possible for us to recognize that it is. That is to say, this thesis does not lead inexorably to an Epistemic Constraint on truth. It follows that if this anti-realism leads to revisionism, it doesn't do so by the sort of revisionary arguments we have discussed here. One might then wonder whether it can support any revisionism at all.

The path to revisionism begins as follows. If we have no basis for distinguishing being true from being justified, how must we conceive of the justification of a form of reasoning, Π, on the basis of the truth-values of its instances? Our standard conception is that Π is justified if its conclusion has to be true whenever all its premises are true. If we do not distinguish truth from justification, this conception becomes: Π is justified if its conclusion has to be justified whenever all its premises are justified.[31] A well-known way of making sense of this conception is to adopt a hierarchical and compositional account of deductive justification. That is to say, some forms of justification are singled out as fundamental, and the fundamental justification of logically complex statements is based on fundamental justifications of their components. Justifications that rely only on fundamental forms are canonical justifications. A non-canonical form of reasoning is valid just in case whenever the premises of any of its instances are canonically justified, so is the corresponding conclusion of that instance. That is, validity is preservation of canonical warrant. What I have described is of course Dummett's proof-theoretic justification of logical laws, most fully presented in *The Logical Basis of Metaphysics* (Dummett 1991).

Let's go back now to our discussion of revisionism using a substitute for the Quandary Argument. From the perspective of a proof-theoretic justification of logic, we proceed as follows. We begin as before with the minimal proof of the double negations of all instances of excluded middle, but now we think of this as the claim that all these double negations have canonical warrant. The question now is, do all instances of excluded middle have canonical warrant? We can argue, on the basis of three claims, that quandary instances do not have canonical warrant. First, canonical justifications of disjunctions require canonical justifications of at least one disjunct.[32] Second, accepting, with Wright, that color properties may always in principle be detected visually, canonical justifications of color predications are observational. Finally, we can't tell by looking whether the quandary cases of color predications hold or not. Hence, if double negation elimination is not a fundamental form of inference, then it is not justified because it fails to preserve canonical warrant.

At least four assumptions in the line of thought just sketched require defense if it is to amount to an argument for revisionism. First, we need to show that canonical justification of a disjunction requires canonical justification of at least one disjunct. Second, we need an account of why the justification of the double negation of all instances of excluded middle is canonical. Third, we need an argument for why the

---

[31] For a fuller argument, see Shieh (1999).

[32] Note that this assumption is not question-begging, because it does not by itself require a rejection of classical logic. Classical logic doesn't have the disjunction property while intuitionistic logic does; but that property is a property of entire systems of logic, in which a distinction between canonical and non-canonical justification has not been drawn.

transparency of color implies that observational warrant for color predications *and their negations* are fundamental. Finally, we have to show that double negation elimination is a fundamental form of inference.

I have hardly provided a revisionary argument. At best I've sketched the promise of one. Moreover, even if we have the four further required arguments, the proposed revisionary argument would have to confront Wright's Wittgensteinian worries about whether anti-realism is consistent with a conception of objectivity that he sees as underlying proof-theoretic justifications (Wright 1981: 452-7). Still, if the promise were fulfilled, it would arguably constitute a cogent argument conforming, at least in part, to Wright's later conception of revisionism.[33,34]

Appendix: Deductions

The following deductions are carried out in what is essentially the system of Quine's *Methods of Logic* (Quine 1982), refined by Warren Goldfarb (Goldfarb 2003), and adapted to propositional reasoning. I sometimes compress steps where it seems full disclosure does not improve perspicuity. The acronyms of the rules appearing in the justification of each line are explained at the end.

Deduction 1 The First of the Two-Step Naïve Revisionary Argument, (Und_T), (EC_T), $(BV) \vdash \bot$

[1]	(1)	$(\exists p)(\neg R(T(p)) \wedge \neg R(F(p)))$	(Und_T)
[1,2]	(2)	$\neg R(T(q)) \wedge \neg R(F(q))$	(1) EII
[3]	(3)	$(\forall p)((T(p) \rightarrow R(T(p)) \wedge (F(p) \rightarrow R(F(p))))$	(EC_T)
[3]	(4)	$T(q) \rightarrow R(T(q))$	(3) UI, $\wedge$E
[1,2,3]	(5)	$\neg T(q)$	(2),(4) $\wedge$E, MT
[3]	(6)	$F(q) \rightarrow R(F(q))$	(3) UI, $\wedge$E
[1,2,3]	(7)	$\neg F(q)$	(2),(6) $\wedge$E, MT
[8]	(8)	$(\forall p)(T(p) \vee F(p))$	(BV)
[8]	(9)	$T(q) \vee F(q)$	(8) UI
[1,3,8]	(10)	$\bot$	(5),(7),(9) $\bot$I, $\vee$E, [2]EIE

[33] For an extensive treatment of proof-theoretic arguments for revision, see Tennant (1997: ch. 10). Tennant articulates and defends a number of proof-theoretic constraints, two central instances of which are conservative extension and harmony. It is clear, for instance, that the rules of inference governing negation in classical logic do not conservatively extend systems of reasoning applying to negation-free fragments of languages: for example, Putnam's Theorem—$(p \rightarrow q) \vee (q \rightarrow p)$—does not contain negation and yet can only be proven using classical negation rules. But the philosophical significance of such mathematical facts, in particular their implications for the sort of anti-realist revisionism proposed in the text, require much more discussion than I can attempt here.

[34] I would like to thank Mihaela Fistioc, Michael Glanzberg, and Riki Heck for helpful comments and advice, which considerably improved this chapter. I would also like to single out an anonymous referee for providing me with invaluable suggestions; I have marked instances of special indebtedness in footnotes. Finally, I would like to thank Julie Perkins for her usual excellent editorial advice.

Deduction 2 Wright's argument from bivalence to excluded middle, (BV), (FN), (EP) ⊢ (LEM)

[1]	(1)	$(\forall p)(T(p) \vee F(p))$	(BV)
[2]	(2)	$(\forall p)(F(p) \leftrightarrow T(\neg p))$	(FN)
[3]	(3)	$(\forall p)(p \leftrightarrow T(p))$	(EP)
[1]	(4)	$T(q) \vee F(q)$	(1) UI
[2]	(5)	$F(q) \rightarrow T(\neg q)$	(2) UI; ∧E
[3]	(6)	$T(q) \rightarrow q$	(3) UI; ∧E
[3]	(7)	$T(\neg q) \rightarrow \neg q$	(3) UI; ∧E
[]	(8)	$q \rightarrow q \vee \neg q$	∨I; →I
[]	(9)	$\neg q \rightarrow q \vee \neg q$	∨I; →I
[3]	(10)	$T(q) \rightarrow q \vee \neg q$	(6),(8) →I, →E
[2,3]	(11)	$F(q) \rightarrow q \vee \neg q$	(5),(7),(9) →I, →E
[1,2,3]	(12)	$q \vee \neg q$	(4),(10),(11) ∨E
[1,2,3]	(13)	$(\forall p)(p \vee \neg p)$	(12) UG

Deduction 3 The Second of the Two-Step Naïve Revisionary Argument, (LEM), (FN), (EP) ⊢ (BV)

[1]	(1)	$(\forall p)(p \vee \neg p)$	(LEM)
[2]	(2)	$(\forall p)(T(\neg p) \leftrightarrow F(p))$	(FN)
[3]	(3)	$(\forall p)(p \leftrightarrow T(p))$	(EP)
[1]	(4)	$q \vee \neg q$	(1) UI
[2]	(5)	$T(\neg q) \rightarrow F(q)$	(2) UI, ∧E
[3]	(6)	$q \rightarrow T(q)$	(3) UI, ∧E
[3]	(7)	$\neg q \rightarrow T(\neg q)$	(3) UI, ∧E
[]	(8)	$T(q) \rightarrow T(q) \vee F(q)$	∨I, →I
[]	(9)	$F(q) \rightarrow T(q) \vee F(q)$	∨I, →I
[3]	(10)	$q \rightarrow T(q) \vee F(q)$	(6),(8) →I, →E
[2,3]	(11)	$\neg q \rightarrow T(q) \vee F(q)$	(5),(7),(9) →I, →E
[1,2,3]	(12)	$T(q) \vee F(q)$	(4),(10),(11) ∨E
[1,2,3]	(13)	$(\forall p)(T(p) \vee F(p))$	(12) UG

Clearly Deduction 1 and Deduction 3 can be merged into a single deduction to establish that

$$(\text{FN}), (\text{EP}), (\text{Und}_T), (\text{EC}_T), (\text{LEM}) \vdash \bot,$$

by replacing line (8) (i.e., (BV)) in Deduction 1 and its consequence on line (9) with lines (1)–(12) of Deduction 3.

Deduction 4 The One-Step Naïve Revisionary Argument: (Und_P), (EC_P), (LEM) $\vdash \perp$

[1]	(1)	$(\exists p)(\neg R(p) \land \neg R(\neg p))$	(Und_P)
[1,2]	(2)	$\neg R(q) \land \neg R(\neg q)$	(1) EII
[3]	(3)	$(\forall p)(p \to R(p))$	(EC_P)
[3]	(4)	$q \to R(q)$	(3) UI
[1,2,3]	(5)	$\neg q$	(2),(4) ∧E, MT
[3]	(6)	$\neg q \to R(\neg q)$	(3) UI
[1,2,3]	(7)	$\neg\,\neg q$	(2),(6) ∧E, MT
[8]	(8)	$(\forall p)(p \lor \neg p)$	(LEM)
[8]	(9)	$q \lor \neg q$	(8) UI
[1,3,8]	(10)	$\perp$	(5),(7),(9) ⊥I, ∨E, EIE

Deduction 5 Objection to the One-Step Naïve Argument, (Und_P), (EC_P) $\vdash \perp$

[1]	(1)	$(\exists p)(\neg R(p) \land \neg R(\neg p))$	(Und_P)
[1,2]	(2)	$\neg R(q) \land \neg R(\neg q)$	(1) EII
[3]	(3)	$(\forall p)(p \to R(p))$	(EC_P)
[3]	(4)	$q \to R(q)$	(3) UI
[1,2,3]	(5)	$\neg q$	(2),(4) ∧E, MT
[3]	(6)	$\neg q \to R(\neg q)$	(3) UI
[1,2,3]	(7)	$R(\neg p)$	(5),(6) MP
[1,3]	(8)	$\perp$	(2),(7) ⊥I, EIE

Deduction 6 Objection to the Two-Step Naïve Argument, (Und_T), (EC_T) $\vdash$ $(\exists p)(\neg T(p) \land \neg F(p))$

[1]	(1)	$(\exists p)(\neg R(T(p)) \land \neg R(F(p)))$	(Und_T)
[1,2]	(2)	$\neg R(T(q)) \land \neg R(F(q))$	(1) EII
[3]	(3)	$(\forall p)((T(p) \to R(T(p)) \land (F(p) \to R(F(p))))$	(EC_T)
[3]	(4)	$T(q) \to R(T(q))$	(3) UI, ∧E
[1,2,3]	(5)	$\neg T(q)$	(2),(4) ∧E, MT
[3]	(6)	$F(q) \to R(F(q))$	(3) UI, ∧E
[1,2,3]	(7)	$\neg F(q)$	(2),(6) ∧E, MT
[1,3]	(8)	$(\exists p)(\neg T(p) \land \neg F(p))$	(5),(7) ∧I, EG, EIE

Deduction 7 First Stage of Wright's Basic Revisionary Argument, (LEM), (EC$_P$) $\vdash$ (Dec$_P$)

[1]	(1)	$(\forall p)(p \vee \neg p)$	(LEM)
[1]	(2)	$p \vee \neg p$	(1) UI
[3]	(3)	$(\forall p)(p \rightarrow R(p))$	(EC$_P$)
[3]	(4)	$p \rightarrow R(p)$	(3) UI
[]	(5)	$R(p) \rightarrow (R(p) \vee R(\neg p))$	$\vee$I, $\rightarrow$I
[3]	(6)	$p \rightarrow (R(p) \vee R(\neg p))$	(4),(5) $\rightarrow$I, $\rightarrow$E
[3]	(7)	$\neg p \rightarrow R(\neg p)$	(3) UI
[]	(8)	$R(\neg p) \rightarrow (R(p) \vee R(\neg p))$	$\vee$I, $\rightarrow$I
[3]	(9)	$\neg p \rightarrow (R(p) \vee R(\neg p))$	(7),(8) $\rightarrow$I, $\rightarrow$E
[1,3]	(10)	$R(p) \vee R(\neg p)$	(2),(6),(9) $\vee$E
[1,3]	(11)	$(\forall p)(R(p) \vee R(\neg p))$	(10) UG

Deduction 8 First Stage of the Two-Step Revisionary Argument, (BV), (EC$_T$) $\vdash$ (Dec$_T$)

[1]	(1)	$(\forall p)(T(p) \vee F(p))$	(BV)
[2]	(2)	$(\forall p)((T(p) \rightarrow R(T(p)) \wedge (F(p) \rightarrow R(F(p)))$	(EC$_T$)
[3]	(3)	$T(p) \vee F(p)$	(1) UI
[1]	(4)	$T(p) \rightarrow R(T(p))$	(2) UI, $\wedge$E
[2]	(5)	$F(p) \rightarrow R(F(p))$	(3) UI, $\wedge$E
[3]	(6)	$R(T(q)) \vee R(F(q))$	(3),(4),(5) $\vee$E
[3]	(7)	$(\forall p)(R(T(p)) \vee R(F(p)))$	(6) UG

Rules of Deduction

UI	universal instantiation	$\rightarrow$E	conditional elimination
UG	universal generalization	EG	existential generalization
$\vee$E	disjunction elimination	EII	existential instantiation introduction
$\vee$I	disjunction introduction	EIE	existential instantiation elimination
$\wedge$I	conjunction introduction	MP	*modus ponens*
$\wedge$E	conjunction elimination	MT	*modus tollens*
$\rightarrow$I	conditional introduction	$\bot$I	absurdity introduction

References

Brogaard, Berit and Joe Salerno (2002). "Clues to the Paradoxes of Knowability: Reply to Dummett and Tennant," *Analysis* 62, 143–50.

Brouwer, Luitzen Egbertus Jan (1908). "De Onbetrouwbaarheid der Logische Principes," *Tijdschrift voor wijsbegeerte* 2, 152–8.

Brouwer, Luitzen Egbertus Jan (1927). "Intuitionistische Betrachtungen über den Formalismus," *Koninklijke Akademie van Wetenschappen te Amsterdam* 31, 374–9.

Cogburn, Jon (2002). "Logical Revision Re-Revisited: On the Wright/Salerno Case for Intuitionism," *Philosophical Studies* 110, 231–48.

Cogburn, Jon (2003). "Manifest Invalidity: Neil Tennant's New Argument for Intuitionism," *Synthèse* 134, 353–62.

Cohen, Stewart (1988). "How to be a Fallibilist," *Philosophical Perspectives* 2, 581–605.

DeRose, Keith (1995). "Solving the Skeptical Problem," *The Philosophical Review* 104, 1–52.

Detlefsen, Michael (1990). "Brouwerian Intuitionism," *Mind* 99, 501–34.

Dragalin, A. G. (1980). *Mathematical Intuitionism: Introduction to Proof Theory*. Newport, RI: American Mathematical Society.

Dummett, Michael (1963). "Realism," in Dummett (1978), 145–65.

Dummett, Michael (1969). "The Reality of the Past," *Truth and other Enigmas*, Cambridge, MA: Harvard University Press, 1978, 358–74.

Dummett, Michael (1973). "The Philosophical Basis of Intuitionistic Logic," in *Truth and other Enigmas*, Cambridge, MA: Harvard University Press, 1978, 215–47.

Dummett, Michael (1977). *Elements of Intuitionism*, first edition, Oxford: Oxford University Press.

Dummett, Michael (1978). *Truth and other Enigmas*, Cambridge, MA: Harvard University Press.

Dummett, Michael (1981). *Frege: Philosophy of Language*, second edition, London: Duckworth.

Dummett, Michael (1991). *The Logical Basis of Metaphysics*, Cambridge, MA: Harvard University Press.

Frege, Gottlob (1964 [1893]). *The Basic Laws of Arithmetic*, Montgomery Furth (trans.), Berkeley, CA: University of California Press.

Gabbay, Dov (1994). "Classical vs Non-Classical Logic," in D. M. Gabbay, C. J. Hogger, and J. A. Robinson (eds.), *Handbook of Logic in Artificial Intelligence and Logic Programming*, vol. 2, Oxford: Oxford University Press, ch. 2.6.

Goldfarb, Warren (2003). *Deductive Logic*, Indianapolis: Hackett.

Graff, Delia (2000). "Shifting Sands: An Interest-Relative Theory of Vagueness," *Philosophical Topics* 28, 45–81.

Hawthorne, John (2004). *Knowledge and Lotteries*, Oxford: Oxford University Press.

Hintikka, Jaakko (2001). "Intuitionistic Logic as Epistemic Logic," *Synthèse* 127, 7–19.

Fine, Kit (1975). "Vagueness, Truth and Logic," *Synthèse* 30, 265–300.

Kamp, Hans (1981). "The Paradox of the Heap," in U. Mönnich (ed.), *Aspects of Philosophical Logic*, Dordrecht: Reidel, 225–77.

Keefe, Rosanna (2000). *Theories of Vagueness*, Cambridge: Cambridge University Press.

Keefe, Rosanna, and Peter Smith (eds.) (1996). *Vagueness: A Reader*, Cambridge, MA: MIT Press.

Kneale, William, and Martha Kneale (1962). *The Development of Logic*, Oxford: Oxford University Press.

Kripke, Saul (1980). *Naming and Necessity*, Cambridge, MA: Harvard University Press.

Lewis, David (1996). "Elusive Knowledge," *The Australasian Journal of Philosophy* 74, 549–67.

Łukasiewicz, Jan. (1930). "Philosophische Bemerkungen zu mehrwertigen Systemen des Aussagenkalküls," *Comptes Rendus de la Société des Sciences et des Lettres de Varsovie* 3 (23), 51–77.

Łukasiewicz, Jan (1951). *Aristotle's Syllogistic*, Oxford: Oxford University Press.

McDowell, John (1981). "Anti-Realism and the Epistemology of Understanding," in H. Parret and J. Bouveresse (eds.), *Meaning and Understanding*, Berlin: Walter de Gruyter, 225–48.

McGee, Vann (1989). "Applying Kripke's Theory of Truth," *Journal of Philosophy* 86, 530–9

McGee, Vann, and Brian P. McLaughlin (1998). "Review of *Vagueness* by Timothy Williamson," *Linguistics and Philosophy* 21, 221–35.

McGee, Vann, and Brian P. McLaughlin (2004). "Logical Commitment and Semantic Indeterminacy: A Reply to Williamson," *Linguistics and Philosophy* 27, 123–26.

Quine, W. V. (1982). *Methods of Logic*, fourth edition, Cambridge, MA: Harvard University Press.

Raatikainen, Panu (2004). "Conceptions of Truth in Intuitionism," *History and Philosophy of Logic* 25, 131–45

Raffman, Diana (1994). "Vagueness without Paradox," *Philosophical Review* 103, 41–74.

Rasmussen, Stig Alstrup, and Jens Ravnkilde (1982). "Realism and Logic," *Synthese* 52, 379–437.

Salerno, Joseph (2000). "Revising the Logic of Logical Revision," *Philosophical Studies* 99, 211–27.

Shieh, Sanford (1997). "Undecidability, Epistemology and Anti-realist Intuitionism," *Nordic Journal of Philosophical Logic* 2, 55–67.

Shieh, Sanford (1998). "On the Conceptual Foundations of Anti-Realism," *Synthèse* 115, 33–70.

Shieh, Sanford (1999). "What Anti-Realist Intuitionism Could Not Be," *Pacific Philosophical Quarterly* 80 (1), 77–102.

Soames, Scott (1999). *Understanding Truth*, New York: Oxford University Press.

Stanley, Jason (2005). *Knowledge and Practical Interests*, Oxford: Oxford University Press.

Tappenden, Jamie (1993). "The Liar and Sorites Paradoxes: Toward a Unified Treatment," *Journal of Philosophy* 90, 551–77.

Tennant, Neil (1997). *The Taming of the True*, Oxford: Oxford University Press.

Unger, Peter (2002). *Philosophical Relativity*, Oxford: Oxford University Press.

Van Fraassen, Bas C. (1966). "Singular Terms, Truth-value Gaps, and Free Logic," *Journal of Philosophy* 63, 481–95.

Williamson, Timothy (1994). *Vagueness*, London: Routledge.

Williamson, Timothy (2004). "Reply to McGee and McLaughlin," *Linguistics and Philosophy* 27, 113–22.

Wright, Crispin (1981). "Anti-Realism and Revisionism," in *Realism, Meaning and Truth*, second edition, Oxford: Blackwell, 1993, 433–57.

Wright, Crispin (1986). "Realism, Bivalence and Classical Logic," in *Realism, Meaning and Truth*, second edition, Oxford: Blackwell, 1993, 458–78.

Wright, Crispin (1992). *Truth and Objectivity*, Cambridge, MA: Harvard University Press.

Wright, Crispin (2001). "On Being in a Quandary: Relativism, Vagueness, Logical Revisionism," *Mind* 110, 45–98.

9
Inferentialism, Logicism, Harmony, and a Counterpoint

Neil Tennant

1. Introduction

In 'Five Milestones of Empiricism,' Quine identifies three levels of theoretical focus in semantics: terms, sentences, theories. These are levels of *syntactic and/or set-theoretic containment*. Terms are constituents of sentences, and sentences make up theories. Another sequence of levels, however, can be determined from the increasing *complexity of logico-linguistic functioning* involved: terms, sentences, and *inferential transitions*.[1] For the inferential semanticist, it is the levelling by complexity of logico-linguistic functioning that is theoretically most useful.

Post-Quinean semantics is of two broad kinds:

1. sentence-focused, truth-conditional semantics; and
2. sequent-focused, inferential semantics.

1.1 *Sentence-Focused, Truth-Conditional Semantic*

The truth-conditional theorizing of Tarski and Davidson is well-known, and easily the dominant paradigm among contemporary Anglo-American analytical philosophers. Their theorizing is based on the central conceptual link between truth and meaning that is provided by the celebrated *adequacy condition* on one's overall theory of truth and of meaning.[2] The adequacy condition is that the theory should yield every instance of Tarski's famous

Schema T: *s* is *T* if and only if *p*.

---

[1] Inferential transitions are here thought of as the basic steps involved in arguments, or justifications that are based on premises. The logician sometimes formalizes inferential transitions as rules of natural deduction or as *sequent* rules. Of course, it is not being suggested that ordinary speakers would need to know anything about rules of inference in natural deduction—or about sequents and sequent calculi—in order for an inferential semantics for their language to be correct and fruitful.

[2] Note the addition, here, of the words 'and of meaning.' This is in order not to prejudge the issue, about to be explained, of the direction-of-dependence of the notion of truth on that of meaning, or vice versa.

Neil Tennant, *Inferentialism, Logicism, Harmony, and a Counterpoint* In: *Logic, Language, and Mathematics: Themes from the Philosophy of Crispin Wright*. Edited by: Alexander Miller, Oxford University Press (2020). © Neil Tennant.
DOI: 10.1093/oso/9780199278343.003.0009

Each instance of this schema is obtained by replacing '*s*' with a metalinguistic term[3] denoting a sentence of the object-language, and replacing *p* by a translation of that sentence into the metalanguage (here, taken to be English). Note the focus on *sentences*, as the minimal unit of linguistic communication.

It is commonly maintained that Tarski took the notion of translation (i.e., meaning-preserving mapping) as given, and used that notion to ensure that fulfillment of the adequacy condition entailed that the defined predicate *T* was indeed a truth-predicate. For Tarski, therefore, the conceptual route would be from translation (meaning) to truth, courtesy of the T-schema.

It is also commonly maintained that Davidson (1967 and 1973) reversed the direction of conceptual dependency. Davidson sought observational constraints on the postulation of (conjectural) biconditionals of the Tarskian form. In such postulated biconditionals, the predicate on the left *would already be interpreted as the truth-predicate*. Consequently, the empirically (albeit holistically) confirmed fulfillment of the adequacy condition would entail that the right-hand sides of the Tarskian biconditionals could be taken as (giving) the meanings of the object-language sentences referred to on the left-hand sides. For Davidson, the conceptual route would be from truth to meaning, courtesy of the T-schema. The standard wisdom, therefore, appears to be that if one takes either one of the notions of truth or translation (meaning-specification) as given, then Tarski's adequacy condition delivers the other notion.[4]

Interpretative truth-conditional theorizing in the manner of Davidson might provide a picture of finished, fully acquired competence with a language. It is not, however, a promising account of how the learner of a first language can *acquire* such competence. The theoretical apparatus presupposes finished competence on the part of the theorizer. Moreover, as the project of 'radical interpretation' proceeds, a high degree of logical sophistication is demanded of the theorizer. For the theorizer has to undertake theory-revision, in order to accommodate his theoretical conjectures to bits of now-confirming, now-disconfirming, incoming data. Learners of a first language cannot learn that way; for, in order to do so, they would have had to acquire a language in the first place.

1.2 Sequent-Focused, Inferential Semantics

For the epistemologist of linguistic understanding, it is tempting to look elsewhere for a characterization of what it is that one grasps when one understands logically complex sentences of one's language. The temptation is to try to characterize sentence meanings via an account of *how one comes to grasp them*. The idea is to attend closely to the to-and-fro in our use of words: not only the to-and-fro of conversation, but also the to-and-fro of *inference*. One infers *to* conclusions, and one infers *from* premises. The meaning of a sentence can be reconceived as based on what it takes to get *to* it; and on what one can take *from* it in getting elsewhere.

---

[3] Tarski spoke of *structural-descriptive* terms.

[4] The present author challenges both of these 'priority-granting' accounts, and proposes instead an account of truth and meaning as *coeval*, even when conformity with the (neutrally stated) adequacy condition is the central goal. See Tennant (2001).

Semantics in the manner of Tarski and Davidson concerned itself solely with the sentence–world relationship. But perhaps even that relationship can be re-conceived?—as involving, say, *language-entry rules* (for moving from perception to observational reports) and *language-exit* rules (for moving from sentences heard or inferred, to actions).[5] Moreover, once *within* language (whether a publicly spoken one, or a language of thought) there are the moves *among* sentences, beginning at the points of entry and ending at the points of exit. Are not those moves governed by rules sufficiently exigent for us to be able to appraise the moves as right or wrong, as well- or ill-advised?

This is the challenge to which inferentialism, broadly construed, responds. In doing so, one is really returning to source. For it was Frege himself who, in §3 of the *Begriffsschrift*, gave a contextual definition of the *propositional content* of a sentence P as what P has in common with any sentence that features the same way as P does as a premise in valid arguments. Thus, P and Q have the same propositional content just in case:

for every set Δ of sentences, for every sentence R, the argument $\Delta, P : R$ is valid if and only if the argument $\Delta, Q : R$ is valid.

2. Inferentialism

Inferentialist accounts of meaning vary considerably. One can identify at least six dimensions of variation.

1. Inferentialist accounts may differ on the question of how ambitious or extensive a grounding of normativity and content is claimed to be rendered by the inferential basis proposed. *Strong* inferentialism says 'inferences are the be-all and end-all of semantics,' and seeks to explain even the notions of singular reference and (relational) predication in terms of inference, taking the notion of inference to be more fundamental—both for constitution of meanings and for acquisition of grasp of meanings. By contrast, *moderate* inferentialism ventures only so far as to claim that inferences (or *patterns* of inference) determine the meanings of all the logical operators,[6] and perhaps also the meanings of some important mathematical ones. But the moderate inferentialist does not seek to reduce every significant meaning-affecting syntactic distinction to patterns of inference.

2. A second dimension of variation is in response to the question 'How complicated a system of inferences is needed in order to transform mere signalling into fully fledged linguistic practice?' *Holistic* inferentialism will insist on the presence of a very wide set of inferences, or inference-patterns, including ones involving complex premises and conclusions, before being willing to grant linguistic status to the system of signalling, or information-transfer, concerned. By contrast, *molecularist* inferentialism maintains that the meaning-determining patterns

---

[5] The original source for the idea of language-entry and language-exit rules is Sellars (1954).

[6] For good measure we include the identity predicate among the 'logical' notions, so perhaps 'logical operators' should read 'logical operators or predicates.'

of inference (where they apply) are reasonably operator-specific. Operators can be grasped individually, and not necessarily only as part of one big package. So one's theoretical account of meaning-determination should seek to isolate and characterize, for each operator, its central or canonical meaning-determining patterns of inference.

3. A third way in which inferentialist accounts can differ is by taking different stands on the question whether human beings and their linguistic practices form part of a reasonably continuous natural order, or whether there is some 'conceptual Rubicon' that separates us from other animals, making our languages essentially different from other animals' signalling systems. *Naturalist* inferentialism will countenance 'phase transitions' in complexity of psychology and of behavior. But a naturalist account will insist that human linguistic and conceptual abilities must have arisen by natural selection working on various more primitive capacities of our non-linguistic primate ancestors. Our abilities and faculties, remarkable though some of us may take them to be, are still the products of natural, evolutionary processes. By contrast, an anti-naturalist, or *hyper-rationalist* inferentialism insists that human thought and language is unprecedented, unique, and of a totally different order than what obtains in the rest of the animal kingdom; it is an all-or-nothing collection of faculties that cannot be possessed piecemeal, and the likes of which have no plausible evolutionary precursors in lowlier animals.

4. A fourth dimension of variation is whether the inferences to which the inferentialist account in question appeals are (i) those that enable us to evaluate logically complex statements for truth and falsity against a background of atomic facts, or (ii) those that enable us to make deductive transitions from logically complex premises to logically complex conclusions. One might call *evaluative* inferentialism the view that inferences of type (i) suffice for the explanation of meaning and content; while the *global* inferentialist insists on the need to consider inferences of type (ii) as well.

5. A fifth *differentia* of inferentialist accounts is whether classical logic emerges unscathed as the right logic. On *logically quietist* accounts, it does; on *logically reformist* accounts, some *proper subsystem* of classical logic is favoured— usually intuitionistic logic.

6. Finally, inferentialist accounts can differ in how they define formally (deductively) correct inference in terms of materially correct inference, *if* they take materially correct inference as an independently available notion.

Interest in the alternative approach of inferential semantics is growing. Prawitz (1974, 1977) and Dummett (1977, 1978, 1991), provided the initial impetus to approach the theory of meaning from an inferentialist perspective. Their interest in inferentialism was driven by manifestationist concerns in the theory of meaning, in turn leading to an anti-realist outlook on language, mind, and world. Given the particular way they pursued their inferentialism, one important outcome was a *reformist* attitude towards deductive logic. In their view, inferentialism enjoined *intuitionistic* logic as the correct logic for our logical words, once their role in inference is properly characterized. Inferentialists of the Dummett–Prawitz school are skeptical about the graspability of the purported 'classical' meanings of logical operators such as negation, implication, and existential quantification.

More recently, however, Brandom (1985, 1994, 2000) has developed a 'pragmatist' version of inferentialism that is significantly different in its details from the accounts that Prawitz and Dummett would favour, especially in eschewing the route to anti-realism.[7]

In terms of the contrast developed above, Brandom's inferentialism is (1) strong, (2) holistic, (3) anti-naturalist/hyper-rationalist, (4) global, and (5) logically quietist. The inferentialism of Dummett and Prawitz is also (4) global, but, by contrast with Brandom's, is (1) moderate, (2) molecularist, and (5) logically reformist. Issue (3) appears, for Dummett and Prawitz, to be moot.

Brandom shares with Prawitz and Dummett their general methodological conviction: that illumination of semantic matters is best sought by studying patterns of inference among sentences, rather than by attempting to characterize those sentences' truth-conditions. Nevertheless it appears to be the case that, for Brandom, *classical* logic survives intact after his inferentialist re-construal of meaning.

2.1 Brandom's Inferentialism and the Choice of Logic

In his early paper Brandom (1985) setting out the main ideas behind his inferentialism, Brandom developed a notion of logical consequence by appeal to a relation of 'material incompatibility.' He argued (p. 48) that one of the 'representative validations' is Double Negation [Elimination]: $\neg\neg p \to p$.

In Lance and Kremer (1994 and 1996), Mark Lance and Philip Kremer study a proof theory and algebraic semantics for the Brandomian notion 'commitment to A is, in part, commitment to B,' but limit themselves to conjunction and the conditional. At the end of Lance and Kremer (1996: 448) they write that '[t]here are reasons for trepidation' about extending their account to deal with the interactions of the conditional with disjunction and with negation. They also concede that the [intuitionistic] logical truth $(A \to \neg A) \to \neg A$ 'is hard to motivate on our interpretation of "$\to$."'

In Lance (1995), Lance takes double negation elimination as an axiom, in a 'set of uncontroversial principles' (p. 116), without questioning whether an inferentialist semantics would validate it. Lance's central notion of *permissive entailment* is also inherently classical:

...if we let f stand for 'The Bad,' say a disjunction of all the claims whch are untenable in a given context, we see that permissive inference is definable in terms of committive: A permissively entails B (hereafter A ⊨ B) iff A & ∼B committively entails f (A & ∼B ⊢ f).

It is the 'if' part of this biconditional that the intuitionist would not accept, since it is tantamount to classical *reductio*.

In Lance (2001), Lance leaves his reader with no clear indication of whether a Brandomian inferential semantics validates classical logic. He ends (p. 456) with the following rather gnomic passage:

---

[7] The inferentialism of Brandom (2000) is also a departure from Brandom's earlier (1976), in which his main concern was with the contrast between truth-conditional and assertibility-conditional theories of sentence meaning. In the later work Brandom (2000), the notion of inference occupies center stage.

If we were to require...that agents be committed to all theorems, as opposed to having the inferential content of their commitments articulated by those theorems, relevance would collapse, and the system would be unsuitable to Brandomian purposes. But this is not to say that there isn't some privileged status had by classical tautologies such as $A \lor \neg A$....

Those developing the technical implications of Brandom's inferentialism would appear not to have engaged satisfactorily with—let alone settled—the question whether it validates all of classical logic. By contrast, inferentialists in the Dummett–Prawitz tradition identify the introduction and elimination rules as central, emphasize the importance of an accompanying notion of *direct proof,* or *warrant,* and proceed vigorously to a principled preference for intuitionistic logic.

2.2 *Other Noteworthy Points*

The present author finds three other points to be noteworthy when comparing Brandom's inferentialism with that of Dummett and Prawitz.

(i) Despite the current climate of materialist metaphysics and naturalized epistemology, none of these named figures has seriously engaged the problem of how *normative* logical relations might be possible even if one were to opt for the naturalism described under (3) above.

(ii) Dummett and Prawitz, unlike Brandom, are inspired mainly by, and focus mainly on, the language of mathematics, with the consequence that their inferentialism might be regarded as not 'strong enough,' in the sense of 'strong' that is described under (1) above.

(iii) Despite his focus on inference, Brandom's account does not engage with the methods, techniques and results of modern proof theory, which, for Dummett and Prawitz by contrast, supplies the main materials by means of which they formulate their reformist case.

Ad (ii): Those who, like Dummett and Prawitz, are inspired mainly by, and focus mainly on, the language of mathematics, can be forgiven if they inadvertently give the impression that their inferentialism is not 'strong enough,' in the sense of 'strong' that is described under (1) above. There are at least three ways that a more sympathetic reconsideration of their contribution could defend them against this objection.

First, one could hold, on behalf of Dummett and Prawitz, that the impression of 'moderateness' in this regard derives from the relative absence, in the case of mathematical language, of any 'language-entry' and 'language-exit' rules. Thus their approach could be understood as apt for *extension* so as to be able to handle language-fragments for which such rules become genuinely operative.

Secondly, if this first option be regarded as special pleading, one could hold that, to the extent that there *must be* 'language-entry' and 'language-exit' rules for *any* language-fragment, one needs to attend to the role that is played in the Dummett–Prawitz account by the requirement of *recognition that a given construction is a proof.* This is the analogue, in the case of mathematics, of perception of worldly things. That is to say, it is the analogue of that which is governed by language-entry rules. Moreover, the speech act of *assertion backed by proof*—which is the basic move in successful communication in mathematics—is the analogue of action directed at worldly things. That is to say, it is the analogue of that which is governed by language-exit rules.[8]

Thirdly, one could look for the 'language-entry' and 'language-exit' rules in a slightly different location within the overall intellectual landscape. One could hold that the 'language-entry' rules *for the language of mathematics* are those by means of which one expresses scientific hypotheses in the language of mathematics—hypotheses that would otherwise be expressed purely 'synthetically,' without using any mathematical vocabulary. And the 'language-exit' rules (again, *for the language of mathematics*) are those by means of which, after taking advantage of the tremendous deductive compression afforded by the use of mathematics, one traverses back from mathematically expressed consequences of one's scientific hypotheses to those purely synthetic, observational predictions by means of which the hypotheses are empirically tested.[9]

3. The Author's Preferred Version of Inferentialism

In Tennant (1987b and 1997), the present author sought to deepen the inferentialism of Dummett and Prawitz, and to argue for *intuitionistic relevant* logic as the right logic (the present author now calls this system Core Logic; see Tennant 2017). So the inferentialism developed in those works was (1) moderate, (2) molecularist, (4) global, and (5) logically reformist.

But it was also (3) naturalist. It was found that inferentialism provided a particularly congenial setting in which to pursue the systematic (if speculative) *naturalistic* explanation, begun in Tennant (1984, 1987a, and 1988), and revisited in Tennant (1994), as to how a logically structured language could have evolved from a logically unstructured signalling system. Crucially, the harmony (see §7 below) that an inferentialist says must obtain between the introduction and elimination rules for any logical operator becomes a necessary precondition for the possibility of evolutionary emergence, and subsequent stability, of the operator within an evolving, logically structured language.

This naturalist, evolutionary view was extended further—first in Tennant (1987b), and later in Tennant (1999). In Tennant (1987b), an account was given of so-called 'transitional atomic logic,' whereby the connectives could be understood in terms of inference rules and derivations. In Tennant (1999), *material rules of atomic inference* (especially those registering metaphysical contrarieties) were used in order to show how negation (and the resulting notion of the *contradictory* of a proposition) could arise from just a prior grasp of contrarieties. It is those contrarieties that precipitate occurrences of the absurdity symbol, by reference to which one can then frame the introduction rule for negation. And that introduction rule fixes the meaning of

---

[8] Sellars's language-exit rules were of course more mundane, involving intentional action within the world, directed towards external things. But remember that we are here seeking, on Dummett's and Prawitz's behalves, some *analogue* of this for the language of mathematics; and the analogy may well have to be stretched. *Showing* one's interlocutor a token of a proof brings to an end the drawing of inferences on one's own part. It is then the interlocutor's turn to recognize the proof for what it is.

[9] The reader familiar with Hartry Field's (1980) will recognize the setting. It is that within which Field himself was concerned to demonstrate that mathematical theorizing afforded a *conservative extension* of the synthetic scientific theorizing that is expressible without recourse to mathematical vocabulary.

negation. (The usual elimination rule is uniquely determined as the harmoniously balancing companion for the introduction rule.) If that systematic account of the origin of negation within an increasingly complex language holds up, then one cannot escape the essentially *intuitionistic* character of negation.

Brandom, too, has appealed to contrarieties to explain the role of negation. He defines the negation of a proposition as its 'minimal incompatible.' In doing so, he gives essentially the same definition as just described. His definition takes the form of a definite description: $\neg p$ is *the* proposition r such that (i) r is incompatible with p, *and* (ii) r is implied by any proposition incompatible with p (see Brandom 1994: 115). This definition does not, however, secure anything more than the intuitionistic meaning for negation. Yet Brandom appears to assume that negation, thus defined, must be classical. (See Brandom 1994: 115: 'It has been shown...how to represent classical logic...by constraints on incompatibility relations.' See also n. 73 on p. 668.)

Both Dummett and Prawitz confined themselves to logical connectives and quantifiers. These are *sentence*-forming operators. The present author has sought to extend the spirit and methods of their approach so as to deal also with *term*-forming operators. It turns out that such operators, too, can be furnished with carefully crafted introduction and elimination rules. And these rules arguably capture the 'constructive content' of the notions involved—such as 'number of' and 'set.' Foremost among these are the definite description operator, the set-abstraction operator, and the number-abstraction operator. Identifying the right introduction and elimination rules for these operators is an interesting challenge, and one that, when met, affords a clear distinction between the *analytic* part of (say) set theory, and its *synthetic* part. See Tennant (2004) for details.

Those of an analytic or logicist bent might find this general kind of approach attractive, for it furnishes a principled way of distinguishing that part of a theory that can be said to be analytic from the part that should be conceded to be synthetic. (See the discussion in Tennant 1997: ch. 9 for more details.) Roughly, the analytic part is generated by the introduction and elimination rules. Analytic results involve either no existential commitments at all, or commitments only to necessary existents. Ironically (in this context) the best example of an analytic portion of a theory is what Quine, the great opponent of the analytic–synthetic distinction, himself called 'virtual' set theory. This is the part of set theory that is free of any existential commitments. (It is precisely this part that is captured by the introduction and elimination rules for the set-abstraction operator.) Furthermore, the analytic results can always be obtained constructively, since one has no truck with classical rules when applying only introduction and elimination rules.

4. Constructive Logicism and Wright's Neo-Fregean Logicism

In Tennant (1987b) the present author developed the doctrine of 'constructive logicism,' as the anti-realist's inferentialist reworking of Crispin Wright's resuscitation (see Wright 1983) of Frege's treatment of natural numbers as logical objects.

Wright had sought to begin his derivations from further downstream within the *Grundgesetze*, by using Hume's Principle:

$$\#xF(x) = \#xG(x) \Longleftrightarrow \exists R\, F \xleftrightarrow{1-1}{R} G;$$

as his starting point. But all the while Wright used standard, *unfree*, classical logic, despite the presence, in the language, of number-abstraction terms of the form $\#xF(x)$ ('the number of *F*s').[10] By contrast, the present author took the challenge to be that of logicizing arithmetic *within the self-imposed constraints of anti-realist inferentialism*. The aim was to derive the Peano–Dedekind axioms for pure arithmetic using *logically more fundamental* principles than Hume's Principle, and to do so within the system of intuitionistic relevant logic.

Any concern for constructivity was puzzlingly absent from Wright's (1983) book. Puzzling also was the casual-seeming acceptance of the 'universal number' $\#x(x = x)$ (*aka* 'anti-zero') whose existence was entailed by Wright's adoption of an unfree logic for such abstraction terms. In Tennant (1987b: 236) the present author raised an objection against Wright's too-ready acceptance of the universal number.[11] That objection was raised later by Boolos (1997), apparently independently, and has since come to be known as the 'Bad Company' objection. The subsequent arguments that Wright has advanced for the analyticity of Hume's Principle also cause one to ask whether he might not have been better off identifying logically more fundamental meaning-conferring rules than Hume's Principle, in order to advance his case that much arithmetical knowledge is analytic. For, to reiterate a point made above: analytic results can always be obtained constructively, since one has no truck with classical rules when applying only introduction and elimination rules.[12]

The logically more fundamental principles advocated in Tennant (1987b) were *meaning-constituting rules for the introduction and elimination* of the arithmetical expressions $0, s(), \#x(\ldots x \ldots)$ and $N()$. The rules had to be stated within a free logic, since one needed to be able to detect exactly where one's existential commitments to the numbers crept in.[13] Note, however, that addition and multiplication were outside the scope of the treatment, as indeed they had been for Wright himself (and, even if somewhat surprisingly, for Frege himself, as is argued in detail in Tennant 2008).

The basic adequacy constraint that Tennant (1987b) imposed on a logicist theory of arithmetic was the derivability of every instance of

$$\text{Schema N: } \#xF(x) = n \leftrightarrow \exists_n xF(x).^{14}$$

---

[10] Wright used 'N' for '#.'

[11] Happily, Wright has since disavowed the universal number, and indeed sought its 'exorcism'—see Essay 13 in Hale and Wright (2001: 314–15). Philosophy, one is all too often reminded, is a constant battle against the bewitchment of the intellect by language.

[12] For this reason, the present author has more recently pursued a study of Lewisian mereology, in order to show that Lewis's important conceptual argument for his Second Thesis in Lewis (1991) can be constructivized, once one has the right system of introduction and elimination rules for the fusion operator in mereology. See Tennant (2012) for details.

[13] The best statement of the rules of constructive logicism can be found in Tennant (2004).

[14] This Schema was formulated in a talk to the Cambridge Moral Sciences Club, and recorded in the minutes for October 23, 1984.

This is to be read as: 'The number of Fs is identical to $\underline{n}$ if and only if there are exactly n Fs.' For example, for $n = 2$ the relevant instance of Schema N is

$$\#xFx = ss0 \leftrightarrow \exists x \exists y (\neg x = y \wedge Fx \wedge Fy \wedge \forall z(Fz \rightarrow (z = x \vee z = y))).$$

The original version of constructive logicism developed in Tennant (1987b) has since been extended, in Tennant (2008), to an inferentialist account that deals not only with zero, successor, 'the number of…' and '…is a natural number,' but also with the operations of addition and multiplication. These two operations receive infer-entialist treatment by employing the notion of *orderly pairing*. The main virtue of this full-fledged inferentialist treatment of arithmetic is that it meets a stringent Fregean requirement that one explain how numerical terms (including additive and multi-plicative ones) find application in counting finite collections. The employment of the notion of orderly pairing, and its correlative notions of 'left projection' and 'right projection,' brings to light an interesting feature of these (arguably, logical) notions: all three functions—pairing, projecting left, and projecting right—are *coeval*, in the sense that the introduction and elimination rules for any one of them involves the other two notions.

As one pursues an inferentialist treatment of important logical and mathematical notions in terms of introduction and elimination rules, one finds an interesting variety of 'grades of logicality.' The highest grade is that enjoyed by the most readily understandable logical operators, the connectives (the subject matter of so-called 'baby logic'). These are governed by rules that deal with them one at a time, in isolation from all other operators. The second grade of logicality is that occupied by the quantifiers in free logic, and the logical predicate of existence. These are governed by introduction and elimination rules in whose formulation one is allowed to use an earlier logical operator, already independently introduced with *its* own rules. (See Tennant 2007 for details.) The third grade of logicality is occupied by those notions, like orderly pairing and its associated projections, that are coeval, in the sense that they interanimate, as it were, in their introduction and elimination rules.

5. Inferentialist Accounts of the Meanings of Logical Operators

A crude conventionalist proposal, misappropriating the Wittgensteinian dictum that *meaning is use*, has it that (i) the meaning of a connective can be fully characterized by the rules of inference that govern it, *and* (ii) *any* rules we lay down, as a 'convention,' will fix a meaning for the connective that they are supposed to govern. It is (ii) which is in serious error. Arthur Prior wrote his classic (1960) paper in an attempt to show that if (ii) were to be conceded, this crude inferentialism would be semantically impotent to distinguish genuine from phoney or freakish logical con-nectives. There followed the rejoinder by Belnap (1962), famously imposing on the deducibility relation certain structural conditions (such as transitivity, or Cut), conditions which (not unreasonably) would need to be in place, as presumed background, before the inferentialist could claim that his chosen rules succeeded in characterizing a logical operator. It is against the background of that early exchange that more recent inferentialists have pursued more fully developed accounts.

5.1 Introduction and Elimination of Sentence-Forming Operators

The introduction rule for a sentence-forming logical operator λ (such as a connective or a quantifier) tells one how to *infer to* a compound *conclusion* with λ dominant. In doing so, one *introduces* the dominant occurrence of λ in the conclusion. The elimination rule for λ tells one how to *infer away from* a compound *premise* with λ dominant. In doing so, one *eliminates* the displayed dominant occurrence of λ in the premise (called the *major* premise for the elimination).

5.2 Introduction and Elimination of Term-Forming Operators

Now, premises and conclusions of rules of inference for an operator are always sentences. So what happens when the operator in question is not a sentence-forming operator? In that case, it must be a term-forming operator—assuming, with Frege, that *Term* and *Sentence* are the only two basic categories in our categorial grammar. For a term-forming operator α, a salient occurrence will be one that is dominant on one side of a general identity claim of the form '$t = \alpha(\ldots)$.' The operator α may or may not be variable-binding. Variable-binding term-forming operators include

$$\#x\Phi(x)\text{: the number of }\Phi\text{s;}$$

$$\{x \mid \Phi(x)\}\text{: the set of }\Phi\text{s;}$$

$$\iota x\Phi(x)\text{: the }\Phi.$$

These were treated in some detail in Tennant (2004).

5.3 Balance, or Equilibrium, between Introduction and Elimination Rules

It is important that one who hears a logically complex sentence sincerely asserted should be able logically to infer *from* it all the information that the asserter ought to have acquired before inferring *to* it. By the same token, it is important that one who undertakes to assert a logically complex sentence should ensure that what any listener would be able logically to infer from it is indeed the case. Viewed this way, the rule (λ-I) states the obligation on the part of any speaker who wishes to assert a compound sentence with λ dominant to his listeners; while the rule (λ-E) states the entitlements enjoyed by any listener who hears such a compound being asserted by the speaker. These obligations and entitlements need to be in balance, or equilibrium.

The question now arises: how best might one explicate this notion of balance? There are three different proposals in the literature:

1. conservative extension of logical fragments by new operators;
2. reduction procedures (and normalizability of proofs);
3. harmony, by reference to strength of conclusion and weakness of major premise.

It is, in the present author's view, interesting and important unfinished business to show that these three proposals are equivalent. Here, they will be explained, and proposal (3) will be improved upon.

6. Explications of Balance

6.1 Conservative Extension of Logical Fragments by New Operators

Suppose one is contemplating adding a new logical operator λ to one's present language L. On the 'conservative extension' proposal, the new rules governing λ must produce a *conservative extension* of the deducibility relation already established within L by the rules of inference that are already in use.[15] That is, if $\Delta: \phi$ is a sequent in L (hence not involving any occurrence of λ), then $\Delta \vdash_{L+\lambda} \phi$ only if $\Delta \vdash_L \phi$. The extra logical moves newly permitted within $L + \lambda$ should not afford any new inferential transitions among sentences not involving the new logical operator λ.

One should also require *uniqueness* on the part of the logical operator governed by the new rules.[16] That is, if μ were another new logical operator of the same syntactic type as λ, and were uniformly substituted for λ in all the new rules governing λ, then in the resulting 'double extension' $L+(\lambda+\mu)$, we would have λ and μ *synonymous*— that is, they would be intersubstitutable, *salva veritate*, in all statements of deducibility-in-$L+(\lambda+\mu)$.[17]

6.2 Reduction Procedures (and Normalizability of Proofs)

On the second proposal, the balance between introduction and elimination rules is brought out by the reduction procedure for λ.[18] The general shape of such a *reduction procedure* (for a two-place connective λ) is as follows:

Subproofs representing the speaker's obligations $_{(\lambda\text{-}I)}$ $\dfrac{}{A\lambda B}$ $_{(\lambda\text{-}E)}$	$\mapsto$	More direct warrant for the listener's entitlements, not proceeding *via* $A\lambda B$
Listener's entitlements		

The unreduced proof-schema on the left in each case shows $A\lambda B$ standing *both* as the conclusion of $(\lambda$-I$)$ *and* as the major premise of $(\lambda$-E$)$. In other words, the operator λ is introduced, and then immediately eliminated. Such a sentence-occurrence within a proof is called a *maximal sentence-occurrence*. It represents an unnecessary detour, introducing and then immediately eliminating logical complexity that is not needed for the passage of reasoning to be negotiated.

The reducts on the right respectively show that one cannot thereby obtain anything that one did not already possess. The introduction and elimination rules balance each other. Speakers' obligations and listeners' entitlements are in equilibrium. One may with justification say, with Prawitz, that the elimination rule *exactly inverts* the introduction rule.

---

[15] This requirement is due to Belnap (1962).

[16] This requirement was argued for by Kosta Došen and Peter Schroeder-Heister (1985).

[17] This criterion of synonymy of logical operators is due to Timothy Smiley (1962).

[18] This proposal is due to Prawitz (1965).

One can readily illustrate these ideas in the case of conjunctions $A \wedge B$.

(i) One who hears $A \wedge B$ sincerely asserted should be able logically to infer from it both A and B: for this is all the information that the asserter ought to have acquired before inferring to $A \wedge B$.

(ii) One who undertakes to assert $A \wedge B$ should ensure that both A and B are indeed the case: for A and B are what any listener would be able logically to infer from $A \wedge B$.

This balance between speaker's obligations and listener's entitlements is brought out by the following *reduction procedure* for $\wedge$:

$$
\begin{array}{cccc}
\Delta \quad \Gamma & \Delta & \Delta \quad \Gamma & \Gamma \\
\Pi \quad \Sigma \;\; \mapsto \;\; \Pi & & \Pi \quad \Sigma \;\; \mapsto \;\; \Sigma & \\
\underline{A \quad B} & \underline{A} & \underline{A \quad B} & \underline{B} \\
\underline{A \wedge B} & A & \underline{A \wedge B} & \\
A & & A &
\end{array}
$$

The unreduced proof-schema on the left in each case shows $A \wedge B$ standing *both* as the conclusion of ($\wedge$-I) *and* as the major premise of ($\wedge$-E). In other words, the operator $\wedge$ is introduced, and then immediately eliminated. The occurrence of $A \wedge B$ is maximal. The reducts on the right respectively show that one cannot thereby obtain anything that one did not already possess.

Note also that each of the reducts on the right of the arrow $\mapsto$ has either Δ or Γ as its set of undischarged assumptions. Whichever one it is, it could well be a *proper subset* of the overall set $\Delta \cup \Gamma$ of undischarged assumptions of the unreduced proof-complex on the left. So with the reduction procedure for $\wedge$ we learn an important lesson: reducing a proof (i.e., getting rid of a maximal sentence occurrence within it) *can in general lead to a logically stronger result*. This is because when Θ is a *proper subset* of Ξ, the argument $\Theta : \phi$ might be a logically stronger argument than the argument $\Xi : \phi$. It *will* be a logically stronger argument if one of the sentences in $(\Xi \backslash \Theta)$—that is, the set of members of Ξ that are not members of Θ—does not itself follow logically from Θ. To summarize: by dropping premises of an argument, one can produce a logically stronger argument. And reduction can enable one to drop premises in one's proof of an argument. So reduction is a potentially *epistemically gainful* operation to perform on proofs.

The upshot of this discussion of the logical behavior of the conjunction operator is as follows. The operator has an introduction rule, which states the conditions that must be met in order to be entitled to infer to a conjunctive conclusion. And it has a corresponding elimination rule, which states the conditions under which one is entitled to infer certain propositions from a conjunctive major premise. *Every* logical operator enjoys an Introduction rule and a corresponding Elimination rule. These rules can be formulated in a two-dimensional graphic way, as was the case with $\wedge$-I and $\wedge$-E above.

One possible drawback with the proposal that one appeal to reduction procedures for an explication of balance (between introduction and elimination rules) is that it does not readily apply to proof systems in which there is a requirement that all proofs be in normal form. In such systems, the pre-images for the reduction procedure will not be proofs; and, often, the reducts will not be, either. (This is because reductions get rid of maximal sentence-occurrences one at a time, not all at once.)

For such systems, in which normality is necessary for proofhood, one can still, however, appeal to a kind of 'normalizability' requirement that is closely enough related to the proposal involving reduction procedures. The idea is really quite simple. Given (*normal*) proofs inviting one to 'apply CUT' n times so as to reap the benefits of transitivity:

$$
\begin{array}{cc}
\Delta_1 & \Delta_n \\
\Pi_1 & \cdots \quad \Pi_n \\
\psi_1 & \psi_n \\
\underbrace{\Delta_0, \psi_1, \ldots, \psi_n} \\
\Pi_0 \\
\phi
\end{array}
$$

one does not form the obvious (and often abnormal) 'proof' and then try to normalize it. Instead, one simply states that, whenever such normal proofs exist (as just displayed), there will always be a *normal* proof, within the system of rules, of some *subsequent* of the overall sequent $\cup_i \Delta_i : \phi$. If the latter proof were always effectively determinable from the given proofs $\Delta_0, \Delta_1, \ldots, \Delta_n$, then that would be an added bonus. And this, in fact, is what obtains for intuitionistic relevant logic.[19]

6.3 Harmony, by Reference to Strength of Conclusion and Weakness of Major Premise

With this much behind us by way of proof-theoretic background, the rest of this study will focus on problems facing an alternative explication of the notion of *balance* or *matching* between, on the one hand, the inferential conditions to which one is beholden in an introduction rule, and, on the other hand, the conditions of inferential entitlement that are set out in the corresponding elimination rule. In framing a *Principle of Harmony*, one is seeking to capture, in another way, the aforementioned balance that should obtain—and, in the case of the usual logical operators, does obtain—between introduction and elimination rules.

In Tennant (1978), such a Principle of Harmony was formulated, and this formulation was subsequently refined, first in Tennant (1987b) and later in Tennant (1997). The reader will find below summaries of these earlier formulations of the Principle of Harmony. The purpose here is to revisit those formulations in light of an interesting example furnished by Crispin Wright.[20] Wright's tonkish example

---

[19] For the author's defence against John Burgess's claim (see Burgess 2005a,b) that this metalogical result cannot be obtained using *IR* as one's metalogic, see Tennant (2006).

[20] Personal communication, April 2005.

concerns only sentential connectives; but the lesson to be learned should generalize also to rules governing quantifiers, and (with the obvious necessary modifications) to rules governing term-forming operators.

7. Earlier Formulations of Harmony

7.1 The Formulation in Natural Logic

The original formulation of the Principle of Harmony (Tennant 1978: 74) was as follows.

Introduction and elimination rules for a logical operator λ must be formulated so that a sentence with λ dominant expresses the *strongest* proposition which can be inferred from the stated premises when the conditions for λ-introduction are satisfied; while it expresses the *weakest* proposition possible under the conditions described by λ-elimination.

By 'proposition' one can understand 'logical equivalence class of sentences.' Thus when one speaks of the 'proposition' ϕ, where ϕ is a sentence, one means the logical equivalence class to which ϕ belongs.

The *strongest* proposition with property P is that proposition θ with property P such that any proposition σ with property P is deducible from θ; while the weakest proposition with property P is that proposition θ with property P that can be deduced from any proposition σ with property P.

Strictly speaking, one should continue to speak here of logical equivalence classes of sentences, or of propositions, but any occasional laxer formulation involving reference only to sentences is unlikely to cause confusion.

Let us illustrate the foregoing formulation of the Principle of Harmony by reference, once again, to conjunction. In this case, the principle dictates that

a sentence $\phi \wedge \psi$ expresses the *strongest* proposition which can be inferred from the premises ϕ, ψ; while it expresses the *weakest* proposition from which ϕ can be inferred and ψ can be inferred.

How might this be established? First, observe that by ($\wedge$-I), the sentence $\phi \wedge \psi$ can be inferred from the premises ϕ, ψ. In order to establish that $\phi \wedge \psi$ is the strongest proposition that can be so inferred, suppose that θ can also be so inferred. But then by the proof

$$\frac{\dfrac{\phi \wedge \psi}{\phi} \qquad \dfrac{\phi \wedge \psi}{\psi}}{\theta}$$

with its two applications of ($\wedge$-E), we have that $\phi \wedge \psi$ logically implies θ. Thus $\phi \wedge \psi$ is the *strongest* proposition that can be inferred from ϕ, ψ.

Secondly, observe that by ($\wedge$-E), the proposition $\phi \wedge \psi$ is one from which ϕ can be inferred and ψ can be inferred. In order to establish that $\phi \wedge \psi$ is the *weakest* proposition from which ϕ can be inferred and ψ can be inferred, suppose that θ is a proposition from which ϕ can be inferred and ψ can be inferred. But then by the proof

$$\frac{\dfrac{\theta}{\phi} \quad \dfrac{\theta}{\psi}}{\phi \wedge \psi}$$

with its application of ($\wedge$-I), we have that θ logically implies $\phi \wedge \psi$. Thus $\phi \wedge \psi$ is the weakest proposition from which ϕ can be inferred and ψ can be inferred.

Now let us provide a foil for the harmony requirement, to illustrate that it has some teeth. Take Prior's infamous connective 'tonk' (here abbreviated as @) with its rules

$$(\text{@-I}) \quad \frac{\phi}{\phi @ \psi}$$

$$(\text{@-E}) \quad \frac{\phi @ \psi}{\psi}$$

How would one show that $\varphi @ \psi$ is the weakest proposition from which ψ can be inferred? It cannot be done. For, suppose θ is a proposition from which ψ can be inferred. The task now is to show that θ logically implies $\varphi @ \psi$. One's first thought might be to use the introduction rule to show this; but, as inspection quickly reveals, that would require φ, not θ, as a premise.

Of course, one could simply cheat by saying that θ logically implies $\varphi @ \psi$ because of the proof

$$\frac{\dfrac{\theta}{\theta @ (\varphi @ \psi)} \;(\text{@-I})}{\varphi @ \psi} \;(\text{@-E})$$

And this will quickly set one on the path that is further explored in §7.3.

7.2 The Formulation in Anti-Realism and Logic

The first main improvement on the foregoing formulation of harmony was offered, in *Anti-Realism and Logic*, in response to an observation by Peter Schroeder-Heister, to the effect that the formulation in *Natural Logic* could not guarantee the uniqueness of the logical operator concerned. The requirements of *strength of conclusion* (called (S)) and *weakness of major premise* (called (W)) remained as the two halves of the harmony condition, now called $h(\lambda i, \lambda e)$. (Note the lowercase 'h.' This was essentially the condition of harmony in *Natural Logic*.) The condition $h(\lambda i, \lambda e)$ was spelled out (in Tennant 1987b: 96–7) as follows. Note that λ is here assumed, for illustrative purposes, to be a binary connective.[21]

---

[21] Minor changes of spelling and symbols have been made here.

$h(\lambda i, \lambda e)$ holds just in case:

(S) $A\lambda B$ is the strongest proposition that it is possible to infer as a conclusion under the conditions described by λi; and

(W) $A\lambda B$ is the weakest proposition that can feature as the major premise under the conditions described by λe.

It was then noted that

To prove (S) one appeals to the workings of λe; and to prove (W) one appeals to the workings of λi. The rules are thus required to interanimate to meet the requirement of harmony.

The treatment in Tennant (1987b) then involved laying down a further requirement. This was called 'Harmony,' with an uppercase 'H':

Given λE, we determine λI as the strongest introduction rule λi such that $h(\lambda i, \lambda E)$; and given [the rule] λI, we determine [the rule] λE as the strongest elimination rule λe such that $h(\lambda I, \lambda e)$.

By requiring Harmony, the intention was to determine uniquely a simultaneous choice of introduction rule λI and elimination rule λE that would be in mutual harmony (with lowercase 'h'). Harmony ensures a kind of Nash equilibrium between introduction and elimination rules: an ideal solution in the coordinate game of giving and receiving logically complex bundles of information.

7.3 The Formulation in The Taming of the True

The earlier formulation of harmony (lowercase 'h'), which was common to both *Natural Logic* and *Anti-Realism and Logic*, was strengthened in the following re-statement in Tennant (1997: 321). In this formulation, the earlier comment about how the rules would interact in the proofs of strength-of-conclusion and weakness-of-major-premise was built into the statement of harmony as an essential feature:

(S) The conclusion of λ-introduction should be the strongest proposition that can so feature; moreover one need only appeal to λ-elimination to show this; but in . . . showing this, one needs to make use of all the forms of λ-elimination that are provided

(W) The major premiss for λ-elimination should be the weakest proposition that can so feature; moreover one need only appeal to λ-introduction to show this; but in . . . showing this, one needs to make use of all the forms of λ-introduction that are provided

Suppose one wished to show that the usual introduction and elimination rules for $\rightarrow$ are in harmony. There would accordingly be two problems to solve:

1. Assume that σ features in the way required of the conclusion $\varphi \rightarrow \psi$ of $\rightarrow I$:

$$\frac{\qquad}{\varphi}{}^{(i)}$$
$$\vdots$$
$$\frac{\psi}{\sigma}{}^{(i)}$$

Show, by appeal to $\to E$, that $\varphi \to \psi \vdash \sigma$.

2. Assume that σ features in the way required of the major premiss $\varphi \to \psi$ of $\to E$:

$$\frac{\varphi \quad \sigma}{\psi}$$

Problem (1) is solved by the following proof:

$$\frac{\overline{\quad}^{(1)}}{\dfrac{\varphi \quad \varphi \to \psi}{\dfrac{\psi^{(1)}}{\sigma}} \to E}$$

This proof uses only $\to E$ to show that if σ features like the conclusion $\varphi \to \psi$ of $\to I$ then σ can be deduced from $\varphi \to \psi$. Thus $\varphi \to \psi$ is the strongest proposition that can feature as the conclusion of $\to I$.

Problem (2) is solved by the following proof:

$$\frac{\dfrac{\overline{\quad}^{(1)}}{\varphi \qquad \sigma}}{\dfrac{\psi \qquad^{(1)}}{\varphi \to \psi}} \to I$$

This proof uses only $\to I$ to show that if σ features like the major premiss $\varphi \to \psi$ of $\to E$ then $\varphi \to \psi$ can be deduced from σ. Thus $\varphi \to \psi$ is the weakest proposition that can feature as the major premiss of $\to E$.

By way of further example, the following two proof schemata show that the usual introduction and elimination rules for $\vee$ are in harmony:

$$VE \ \dfrac{\varphi \vee \psi \quad \dfrac{\overline{\varphi}^{(1)}}{\sigma} \quad \dfrac{\overline{\psi}^{(1)}}{\sigma}}{\sigma}\,^{(1)} \qquad\qquad \dfrac{\sigma \quad VI\dfrac{\overline{\varphi}^{(1)}}{\varphi \vee \psi} \quad VI\dfrac{\overline{\psi}^{(1)}}{\varphi \vee \psi}}{\varphi \vee \psi}\,^{(1)}$$

Note how in the second proof schema we need to employ both forms of $\vee I$ in order to construct the proof schema; the reader should consider once again the precise statement of the Principle of Harmony given above.

8. Wright's Tonkish Example

Wright set out the following problem:[22]

Let λ be a binary connective associated with rules λI and λE. According to the *AR&L* characterisation, these are harmonious just in case

> Condition (i): any binary connective @ for which the pattern of λI is valid may be shown by λE to be such that $A\lambda B \vDash A@B$ (so $A\lambda B$ is the strongest statement justified by the λI premisses); and

> Condition (ii): any binary connective @ for which the pattern of λE is valid may be shown by λI to be such that $A@B \vDash A\lambda B$ (so, in effect, λE is the strongest E-rule justified by the I-rule.)

OK. Let λI be $\vee I$, and λE be $\wedge E$. So, to establish Condition (i), assume $A \vDash A@B$, and $B \vDash A@B$. We need to show via λE that $A\lambda B \vDash A@B$. Assume $A\lambda B$. Then both A and B follow by λE. Either will then suffice for $A@B$, by the assumption. As for Condition (ii), assume $A@B$. By hypothesis, the pattern of λE is valid for @, so we have both A and B. Either will suffice via λI for the proof of $A\lambda B$. So $A@B \vDash A\lambda B$.

What has gone wrong? Manifestly the λ-rules are disharmonious (in fact they are the rules for tonk, of course).

9. The Revised Version of the Principle of Harmony

Perhaps the best way to explain what has gone wrong is to clarify how Tennant (1997) had already, in effect, put it right. The following emended version of harmony is a minor variation of the theme in Tennant (1997), and was communicated to Wright in response to this interesting problem. In Tennant (1997), the statement of (S) (strength-of-conclusion) involved the condition that in establishing strength 'one need only appeal to λ-elimination' and the statement of (W) (weakness-of-major-premise) involved the condition that in establishing weakness 'one need only appeal to λ-introduction.' These two conditions will be made more emphatic: respectively, 'one may not make any use of λ-introduction' and 'one may not make any use of λ-elimination.'

In order to distinguish the 'final' version of harmony offered below from its predecessors, let us use (S') and (W') as the respective labels for strength-of-conclusion and weakness-of-major-premise. Wright's apparent counterexample can be rendered inadmissible by laying down the following more emphatic version of the requirement for harmony in Tennant (1997). It is in spirit of the original, but now also—one hopes—in appropriately captious letter. As has been the case all along, it is framed by reference to a connective λ. The newly emphasized conditions are in boldface.

> (S') $A\lambda B$ is the strongest conclusion possible under the conditions described by λI. Moreover, in order to show this,
> - (i) one needs to exploit all the conditions described by λI;
> - (ii) one needs to make full use of λE; but
> - (iii) **one may not make any use of λI.**

---

[22] Direct quote from personal correspondence, with minor typos corrected, extra formatting supplied, and some symbols changed in the interests of uniformity of exposition. The example arose in a graduate seminar at NYU, conducted with Hartry Field in the Spring of 2005, on revision of classical logic.

(W') $A\lambda B$ is the weakest major premise possible under the conditions described by λE. Moreover, in order to show this,

(iv) one needs to exploit all the conditions described by λE;

(v) one needs to make full use of λI; but

(vi) **one may not make any use of λE.**

Suppose now that we try to follow Wright's foregoing suggestion. That is, suppose we try to stipulate λI as $\vee$-like and λE as $\wedge$-like:

$$
\begin{array}{ccc}
(\lambda I) & \dfrac{A}{A\lambda B} & \dfrac{B}{A\lambda B} \\[2ex]
(\lambda E) & \dfrac{A\lambda B}{A} & \dfrac{A\lambda B}{B}
\end{array}
$$

It will be shown that this stipulation does not satisfy the joint harmony requirement (S') and (W').

In order to establish (S') we would need to show, inter alia, that given the inferences A/C and B/C one could form a proof Π (say) of C from $A\lambda B$, but would (i) need both those inferences to do so, and (ii) need to make full use of λE.

In order to establish (W') we would need to show, inter alia, that given the inferences C/A and C/B one could form a proof Σ (say) of $A\lambda B$ from C, but would (i) need both those inferences to do so, and (ii) need to make full use of λI.

Candidates for the sought proof Π (for (S')) might be thought to be

$$
\dfrac{\dfrac{A\lambda B}{A}}{C} \qquad \dfrac{\dfrac{A\lambda B}{B}}{C} \quad ;
$$

but neither of these proofs exploits both the inference A/C and the inference B/C. So these candidate proofs violate requirement (i) in (S') to the effect that one needs to exploit all the conditions described by λI. Moreover, each candidate proof uses only 'one half' of λE, not full λE. So these proofs also violate requirement (ii) in (S') to the effect that one must use all of λE.

Candidates for the sought proof Σ (for (W')) might be thought to be

$$
\dfrac{\dfrac{C}{A}}{A\lambda B} \qquad \dfrac{\dfrac{C}{B}}{A\lambda B} \quad ;
$$

but neither of these proofs exploits both the inference C/A and the inference C/B. So these candidate proofs violate requirement (i) in (W') to the effect that one needs to exploit *all* the conditions described by λE. Moreover, each candidate proof uses only 'one half' of λI, not full λI. So these proofs also violate requirement (ii) in (W') to the effect that one must use all of λI.

Suppose rather that one were to stipulate a perverse choice of 'halves' of the preceding I- and E-rules for λ. Thus suppose that the rules λI and λE were now taken to be, respectively,

$$(\lambda I) \qquad \frac{A}{A\lambda B}$$

$$(\lambda E) \qquad \frac{A\lambda B}{B}$$

In order to establish (S') we would need to show, inter alia, that given the inference A/C one could form a proof Π (say) of C from $A\lambda B$, but without (by (iii)) making any use of λI.

In order to establish (W') we would need to show, inter alia, that given the inference C/B one could form a proof Σ (say) of $A\lambda B$ from C, but without (by (vi)) making any use of λE.

It might be thought that a candidate proof for Π would be

$$\frac{\dfrac{A\lambda B}{B}\,(\lambda E)}{\dfrac{B\lambda C}{C}\,(\lambda I)}\,(\lambda E)$$

but this violates requirement (iii), since it uses λI.

It might be thought that a candidate proof for Σ would be

$$\frac{\dfrac{C}{A\lambda B}\,(\lambda I)}{\dfrac{A}{A\lambda B}\,(\lambda E)}\,(\lambda I)$$

but this violates requirement (vi), since it uses λE.

10. On the Requirements of Full Use of Conditions and Rules

Salvatore Florio has produced an interesting case which illustrates the need to insist on 'full use' of conditions and rules when establishing the statements (S') and (W') for harmony.

Consider the obviously non-harmonious rules

$$(\lambda I) \quad \frac{A \quad B}{A\lambda B}$$

$$(\lambda E) \quad \frac{A\lambda B}{C}$$

where the rule (λE) is like the Absurdity Rule, in that it allows the conclusion C to be *any* sentence one pleases.

First we prove (S'), but without heeding fully the requirements that have been laid down on such a proof.

Assume that D features in the way required of the conclusion $A\lambda B$ of λ-introduction:

$$\frac{A \quad B}{D}$$

We are required to show that $A\lambda B \vdash D$. The following proof suffices:

$$\frac{A\lambda B}{D} \, {}_{(\lambda E)}$$

Next we prove (W'), again without heeding fully the requirements that have been laid down on such a proof.

Assume that D features in the way required of $A\lambda B$ as the major premiss for λ-elimination:

$$\frac{D}{C}$$

We are required to show that $D \vdash A\lambda B$. The following proof suffices:

$$\frac{D}{A\lambda B}$$

These swift proofs, however, as already intimated, are unsatisfactory. The respective reasons are as follows.

The proof of (S') does not avail itself at all of the assumption made at the outset (that D features in the way required of the conclusion $A\lambda B$ of λ-introduction). The technical infraction on the part of this attempted proof of (S') is its violation of condition (i): the proof does not 'exploit all the conditions described by λI.' (In fact, one can see from the very statement of (λE) that $A\lambda B$ is the strongest proposition *tout court*, and not just the strongest proposition that can feature as the conclusion of λ-introduction.)

The proof of (W') is defective also. It does not use the rule (λI), thereby violating condition (v).

What has been called the 'final' formulation of harmony, in terms of (S') and (W'), still sits well with the standard connectives. This is revealed by inspection of the obvious demonstrations of maximum strength-of-conclusion and maximum weakness-of-major-premise, when the dominant operator in question is $\neg$, $\wedge$, $\vee$ or $\rightarrow$. The reader will find that conditions (i), (ii), and (iii) (under (S')) and (iv), (v), and (vi) (under (W')) are all satisfied by those demonstrations. So, in ruling out Wright's counterexample, one does not rule out too much. And, the reader will be happy to learn, the original tonk, due to Prior, is ruled out also. All that the final formulation of harmony really does is make absolutely explicit features of the demonstrations in question that usually go unremarked, but which need to be emphasized when confronted with tonkish examples masquerading as genuine connectives.

11. Conclusion to the Discussion of Harmony

Wright's example brings out nicely why it is that stipulations concerning licit methods of proof (of strength-of-conclusion and of weakness-of-major-premise) must be built into the formulation of harmony. The author is moderately confident that the new formulation above, with its emphases on prohibited resources, will withstand any further attempted counterexamples. This new version of harmony (lowercase 'h') should, of course, still be coupled with the uniqueness condition that was called Harmony (upper-case 'H').

In closing, it should be stressed that the formulation of a Principle of Harmony is not just a technical exercise in proof theory of limited (or no) value to the philosophy of logic and language. On the contrary: armed with a satisfactory account of harmony, the naturalizing anti-realist can venture an interesting account of how logical operators could have found their way into an evolving language. Harmony is a transcendental precondition for the very possibility of logically structured communication. A would-be logical operator that does *not* display harmony (such as, for example, Prior's infamous operator 'tonk') could not possibly be retained within an evolving language after making a first debut. Because the 'deductive reasoning' that it would afford would go so haywire, it would have been rapidly selected against. Only those operators would have survived that were governed by harmoniously matched introduction rules (expressing obligations on the part of assertors) and elimination rules (expressing the entitlements of their listeners). For only they could have usefully enriched the medium by means of which social beings can informatively communicate. Only those operators would have been able to make their way through the selective filter for the growing medium.[23, 24]

---

[23] For a more detailed development of these ideas, the reader is referred to Tennant (1984, 1987b: ch. 9, 1994, and 1999).

[24] Discussions with Tadeusz Szubka prompted a more detailed examination of Brandom's inferential-ism, and its points of contrast with Dummett's. Thanks are owed to Salvatore Florio for an extremely

References

Belnap, Nuel (1962). "Tonk, Plonk and Plink," *Analysis* 22, 130–4.

Boolos, George (1997). "Is Hume's Principle Analytic?," in R.K. Heck (ed.), *Language Thought and Logic: Essays in Honour of Michael Dummett*, Oxford: Clarendon Press, 245–61.

Brandom, Robert (1976). "Truth and Assertibility," *Journal of Philosophy* 73 (6), 137–49.

Brandom, Robert (1985). "Varieties of Understanding," in Nicholas Rescher (ed.), *Reason and Rationality in Natural Science*, Lanham, MD: University Press of America, 27–51.

Brandom, Robert B. (1994). *Making It Explicit: Reasoning, Representing, and Discursive Commitment*, Cambridge, MA: Harvard University Press.

Brandom, Robert B. (2000). *Articulating Reasons: An Introduction to Inferentialism*, Cambridge, MA: Harvard University Press.

Burgess, John (2005a). "No Requirement of Relevance," in Stewart Shapiro (ed.), *The Oxford Handbook of Philosophy of Mathematics and Logic*, Oxford: Oxford University Press, 727–50.

Burgess, John (2005b). "Review of Neil Tennant," *The Taming of The True: Philosophia Mathematica* 13, 202–15.

Davidson, Donald (1967). "Truth and Meaning," *Synthese* 17, 304–23.

Davidson, Donald (1973). "Radical Interpretation," *Dialectica* 27, 314–24.

Došen, Kosta, and Peter Schroeder-Heister (1985). "Conservativeness and Uniqueness," *Theoria* 51, 159–73.

Dummett, Michael (1977). *Elements of Intuitionism*, Oxford: Clarendon Press.

Dummett, Michael (1978). *Truth and Other Enigmas*, Duckworth: London.

Dummett, Michael (1991). *The Logical Basis of Metaphysics*, Cambridge, MA: Harvard University Press.

Field, Hartry (1980). *Science without Numbers: A Defence of Nominalism*, Oxford: Basil Blackwell.

Hale, Bob, and Crispin Wright (2001). *The Reason's Proper Study: Essays towards a Neo-Fregean Philosophy of Mathematics*, Oxford: Clarendon Press.

Lance, Mark (1995). "Two Concepts of Entailment," *Journal of Philosophical Research* 20, 113–37.

Lance, Mark (2001). "The Logical Structure of Linguistic Commitment III Brandomian Scorekeeping and Incompatibility," *Journal of Philosophical Logic* 30 (5), 439–64.

Lance, Mark, and Philip Kremer (1994). "The Logical Structure of Linguistic Commitment I: Four Systems of Non-Relevant Commitment Entailment," *Journal of Philosophical Logic* 23 (4), 369–400.

Lance, Mark, and Philip Kremer (1996). "The Logical Structure of Linguistic Commitment II: Systems of Relevant Commitment Entailment," *Journal of Philosophical Logic* 25 (4), 425–49.

Lewis, David (1991). *Parts of Classes*, Oxford: Basil Blackwell.

Prawitz, Dag (1965). *Natural Deduction: A Proof-Theoretical Study*, Stockholm: Almqvist & Wiksell.

Prawitz, Dag (1974). "On the Idea of a General Proof Theory," *Synthese* 27, 63–77.

Prawitz, Dag (1977). "Meaning and Proofs: On the Conflict between Classical and Intuitionistic Logic," *Theoria* 43, 2–40.

careful reading of an earlier draft, which resulted in significant improvements. Robert Kraut was generous with his time and expertise on a later draft. Two referees for Oxford University Press provided helpful comments, which led to considerable expansion. The author is fully responsible for any defects that remain.

Prior, Arthur (1960). "The Runabout Inference Ticket," *Analysis* 21, 38–9.

Sellars, Wilfred (1954). "Some Reflections on Language Games," *Philosophy of Science* 21 (3), 204–28.

Smiley, Timothy (1962). "The Independence of Connectives," *The Journal of Symbolic Logic* 27, 426–36.

Tennant, Neil (1978). *Natural Logic*, Edinburgh: Edinburgh University Press.

Tennant, Neil (1984). "Intentionality, Syntactic Structure and the Evolution of Language," in Christopher Hookway (ed.), *Minds, Machines and Evolution*, Cambridge: Cambridge University Press, 73–103.

Tennant, Neil (1987a). "Conventional Necessity and the Contingency of Convention," *Dialectica* 41, 79–95.

Tennant, Neil (1987b). *Anti-Realism and Logic: Truth as Eternal*, Clarendon Library of Logic and Philosophy, Oxford: Oxford University Press.

Tennant, Neil (1988). "Two Problems for Evolutionary Epistemology: Psychic Reality and the Emergence of Norms," *Ratio* 1 (1), 47–63.

Tennant, Neil (1994). "Logic and Its Place in Nature," in Paolo Parrini (ed.), *Kant and Contemporary Epistemology*, Dordrecht: Kluwer Academic Publishers, 101–13.

Tennant, Neil (1997). *The Taming of the True*. Oxford: Oxford University Press.

Tennant, Neil (1999). "Negation, Absurdity and Contrariety," in Dov Gabbay and Heinrich Wansing (eds.), *What is Negation?*, Dordrecht: Kluwer, 199–222.

Tennant, Neil (2001). "Game Theory and Convention T," *Nordic Journal of Philosophical Logic* 6 (1), 3–20.

Tennant, Neil (2004). "A General Theory of Abstraction Operators," *The Philosophical Quarterly* 54 (214), 105–33.

Tennant, Neil (2006). "Logic, Mathematics, and the Natural Sciences," in Dale Jacquette (ed.), *Philosophy of Logic*, Amsterdam: North-Holland, 1149–66.

Tennant, Neil (2007). "Existence and Identity in Free Logic: A Problem for Inferentialism?," *Mind* 116 (461), 1055–78.

Tennant, Neil (2008). "Natural Logicism via the Logic of Orderly Pairing," in Sten Lindström, Erik Palmgren, Krister Segerberg, and Viggo StoltenbergHansen (eds.), *Logicism, Intuitionism, Formalism: What Has Become of Them?*, Synthese Library 341, Berlin: Springer Verlag, 91–125.

Tennant, Neil (2012). "Parts, Classes, and Parts of Classes: An Anti-Realist Reading of Lewisian Mereology," *Synthese* 190 (4), 709–42.

Tennant, Neil (2017). *Core Logic*, Oxford: Oxford University Press.

Wright, Crispin (1983). *Frege's Conception of Numbers as Objects*. Aberdeen: Aberdeen University Press.

PART IV
Metaphysical Possibility

10

CCCP

Bob Hale

1. Preliminary Remarks

Kripke's well-known argument against the identity of pain with C-fibre[1] firing asserts, first, that if that identity held, it would do so necessarily, so that the psycho-physical identity theorist must claim that the apparent conceivability of pain without C-fibre firing and of C-fibre firing without pain is merely apparent, and gives rise only to an illusion of possibility; and, second, that it is difficult to see how the identity-theorist can sustain that claim. In particular, he cannot explain away what he must view as the illusion of possibility in the way in which we *can* explain away the apparent conceivability of water's being other than H_2O—in terms of our confusing conceiving of something superficially indistinguishable from water turning out not to be composed of H_2O molecules with conceiving of water turning out not to be so constituted. Whereas 'fool's water' is a genuine possibility, there can be no such thing as 'fool's pain,' since whatever *appears as* pain just *is* pain. Conceiving of something superficially indistinguishable from pain not being C-fibres firing *is* conceiving of *pain* not being C-fibres firing.

Two things about this argument are clear—first, that it relies upon a principle to the effect that what is conceivable is possible, and, second, that it is clearly not, as it stands, decisive, since it does nothing to show that the identity-theorist cannot explain away the apparent conceivability of painless C-fibre firing and C-fibreless pain in some other way, along different lines from those Kripke thinks appropriate to explaining away illusions of possibility in the case of his stock examples of *a posteriori* necessities.

Crispin Wright[2] formulates and defends a principle—what he calls the Counter-Conceivability Principle[3]—that is reasonably taken to be the one Kripke's argument needs. But then, contra Kripke, he argues that there are other kinds of case in which

---

[1] I am not concerned with the empirical plausibility of Kripke's example. C-fibres, I gather, are any of the unmyelinated fibres, 0.4 to 1.2 micrometers in diameter, which conduct nerve impulses at a velocity of 0.7 to 2.3 meters per second. They are apparently involved only in dull and chronic pain. So the identification of pain in general with C-fibre firing is independently objectionable. I am indebted to an anonymous referee for this information. See also Hardcastle (1999).

[2] In his paper Wright (2002: 401–40). All quotations of Wright's words are from this paper.

[3] Hereafter, just CCP. The extra 'C' in my title is not a slip. No prizes for guessing what it abbreviates.

Bob Hale, *CCCP*, In: *Logic, Language, and Mathematics: Themes from the Philosophy of Crispin Wright.*
Edited by: Alexander Miller, Oxford University Press (2020). © Bob Hale.
DOI: 10.1093/oso/9780199278343.003.0010

what Kripke would accept as a necessity is apparently counter-conceivable, but in which—just as in the pain/C-fibre case—the illusion of conceivability and hence of possibility resists explanation in Kripke's favoured style, and must therefore be explained away along different lines. Wright proposes an alternative model for dealing with these problematic cases. But since there is no evident reason why the psycho-physical identity theorist should not deploy his alternative model to explain away what he must regard as the illusion of possibility in the pain/C-fibre case, it would seem that if Wright is right, Kripke's argument is not merely not decisive, but fails altogether, because the identity-theorist can actually meet the challenge it poses. In what follows, I raise some doubts about the details of Wright's diagnosis of the flaw in Kripke's argument. My aim is not, however, to challenge either his general strategy or his conclusion, with both of which I am in broad agreement. I hope, rather, to make a small contribution to a broader and less polemical project— the aetiology of modal illusion, which forms, in my view, an essential part of the epistemology of modality.

2. CCCP and Its Range of Application—Some Doubts about Its Applicability to Putative *A Posteriori* Necessities Aired and Answered

The putative necessities with which we are concerned are—in contrast with broadly logical or conceptual necessities—ones of which we can have, at best, only *a posteriori* knowledge. Defenders of the claim that it is not only true that water is H_2O, but necessarily so, are not obliged to hold the necessity to be conceptual; likewise, whilst—if Kripke is right—identity-theorists must view the identity of pains with C-fibre firings, say, as a matter of necessity, they are under no obligation to maintain that the necessity is conceptual. Precisely because this is so, it is a good question why our apparent ability to conceive of water's turning out to be other than H_2O, or of pains turning out to be other than C-fibre firings, should be thought to tell against their necessary identity. The apparent possibility of counter-conception obviously and uncontroversially tells against any claim to *conceptual* necessity, since only what is not ruled out by relations among the concepts involved, and so is conceptually possible, is conceivable (in the relevant sense). But what bearing can conceivability have, in cases where the necessity at issue is not in any case supposed to be conceptual? Of course, only what is conceptually possible can be metaphysically absolutely necessary. But the claim that it is *a posteriori* necessary that water is H_2O does not entail that the opposite is conceptually impossible—so why should the apparent conceivability of water's being other than H_2O be thought to raise any problem for that claim? It is one of the merits of Wright's paper that it takes on and suggests an answer to this question. More precisely, Wright makes it clear that his counter-conceivability principle is intended to apply to *all* absolute necessities:

By the Counter-Conceivability Principle, all putative metaphysical necessities, even a poster-iori ones, thus have to face the tribunal of what we can, as we think, clearly and distinctly conceive; and their defeat may consequently be a priori even if their sole possible form of justification is not.   (p. 408)[4]

He then immediately confronts two related objections which challenge the relevance of counter-conception to putatively non-conceptual necessities. According to the first, the CCP should be seen as applying only to would-be conceptual necessities; to suppose it applicable to '*all* absolute necessities, a posteriori ones included' is

the merest blunder: the retention of an intuitively conceptualist epistemology of modality for a range of cases where modal status originates not in the character of concepts at all but in underlying essences which may go quite unreflected in our concepts of the items whose essences they are.   (p. 409)[5]

The second objection has it that if CCP is applied to all putative absolute necessities, it will force us to deny necessity in precisely the cases Kripke and others have wanted to claim it—precisely because 'it may seem readily conceivable that water might have turned out to have a very different chemical constitution to the one it actually has. So it cannot be necessary that water is H2O, even if it is true that it is' (p. 409), and similarly for all other supposed *a posteriori* necessities.

To the second objection, Wright endorses what is, in effect, Kripke's own response—the objection confuses conceiving of something which has all the surface properties by which we standardly identify water turning out not to be H_2O (which we certainly can do) with conceiving of water itself turning out not to be H_2O (which we can't). With this, I have no quarrel. But I am less sure about his reply to the first objection, which I had better quote in full:

Essentially the same point addresses the first, more general concern. Any a posteriori necessity, N, will be associated with a seemingly intelligible imaginative scenario in which the a posteriori investigations which confirm it turn out to disconfirm it instead. So much is a consequence of those investigations being a posteriori. But it does not follow that N will be *prima facie* counter-conceivable unless it is granted that the imaginative scenario involves an appropriate play with the very concepts configured in N. And that is not granted. If **water** is a natural kind concept, then which concept it is depends on what is the essence of water. The impression that the Counter-Conceivability Principle is all at a sea as soon as necessities originating *in rebus* are countenanced turns on the tacit assumption of a separation between concept and essence: that, as it was expressed above, 'underlying essences ... may go quite unreflected in our concepts of the items whose essences they are'. That assumption simply misunderstands what is being proposed about the character of the relevant concepts that feature in necessities of identity, origin and constitution.   (pp. 409–10)

---

[4] '[W]hat we can, as we think, clearly and distinctly conceive' is, perhaps, ambiguous, between what we *can* (and think we can) clearly and distinctly conceive, or: what we *think we can* ... conceive? But clearly it is only if we *can* counter-conceive a supposed metaphysical necessity, that it will be defeated; if we merely think (erroneously) that we can do so, the supposed necessity remains undefeated. But will the defeat by *a priori*, if actual and not merely seeming counter-conceivability is required? I return to that question later.

[5] '[U]nderlying essences which may go quite unreflected in our concepts'—there is, perhaps, a further ambiguity here. It would be enough for the objection to maintain that the facts about underlying essences are *not fully reflected* in our concepts—it need not be claimed that no facts about essence are built into the relevant concepts. Again, I shall return to this point shortly.

Obviously Wright does not mean to suggest that it is, after all, somehow a matter of conceptual necessity that water is H_2O. But if not, how exactly should we understand his claim that if **water**[6] is a natural kind concept, then which concept it is depends on what is the essence of water? And what are we to make of his implied rejection of the tacit assumption of a separation between concept and essence—is he claiming that the underlying essence of water, say, is after all reflected in our concept of water? If so, how is that consistent with acknowledgement that it is not conceptually necessary that water is H_2O? If not, what is the claim? These are delicate questions. I think there are satisfactory answers to them which Wright might endorse, but if I am right, they call for a more cautious reformulation of his claim about natural kind concepts and essences, and some qualification of his earlier claim that putative *a posteriori* necessities may be subject to *a priori* defeat.

Between the extremes of claiming, on the one hand, that essences are *fully reflected* in our concepts of the things whose essences they are (so that, assuming that water is essentially H_2O, it is somehow part of our concept of water that water is composed of hydrogen and oxygen in a certain proportion) and holding, on the other, that essences are *wholly unreflected* in our concepts, there is an obvious middle way. What sets a natural kind concept, such as **water**, apart from what Wright calls criterially governed concepts, is—just as Wright proposes—that a natural kind concept is contingently linked with a range of properties (what Wright calls indicator properties) in terms of which we explain or fix the reference or extension of the concept, and it is part of the concept that what it applies to has some underlying property or properties by which the extension of the concept is really determined and in terms of which possession of the indicator properties may be explained. So it is not strictly part of the concept, for example **water**, that the stuff to which it applies has this or that particular underlying property (e.g., being H_2O). What is reflected in the concept—what is integral to the identity of the concept, if it is a natural kind concept—is *not* the *specific* underlying property which constitutes the essence of whatever falls under the concept, but the fact (or perhaps better, assumption) that *there is* such a property.

On this account, someone who thinks she can conceive of water's being other than H_2O is not conceptually confused, as she would be if she thought she could conceive of cousins whose parents are all only children. So it remains to be explained what exactly is amiss, when someone thinks she can so conceive. Wright's proposed explanation is that absence of conceptual confusion is necessary but not sufficient for successful counter-conception—if a thinker really is to conceive of water's being other than H_2O, then it must be the concept **water** which they deploy in their attempt to do so, and not some other concept. If **water** is a natural kind concept—so that 'which concept it is depends on what is the essence of water'—then, Wright claims, the impression that the alleged *a posteriori* necessity is prima facie counter-conceivable may be misleading precisely because this second condition is not met.

That claim is not obviously correct. What is built into the concept **water**, or any other natural kind concept, is *only*—or so it might be objected—that there is

---

[6] I follow Wright in signalling reference to a concept by bold-face type.

some underlying property which determines the extension of 'water' and explains possession of the indicator properties. Nothing in the *concept* **water** precludes that underlying property being other than being predominantly composed of H_2O molecules. But if that is so, what good reason is there to insist that a thinker who purports to be able to conceive of water's turning out not to be H_2O must be guilty of failure to deploy the relevant concept?

I think the objection is mistaken, but that to disclose the source of the mistake— and to remove any remaining puzzle about the relevance of conceivability considerations to putatively absolute but non-conceptual necessities—we need an extra piece. Relevant counter-conceiving, in cases of the sort with which we are concerned, needs to be *de re* as opposed to merely *de dicto* conceiving. It has to be, for example, *water* of which one conceives that it might turn out to be other than H_2O. This means that, in attempting to conceive of water's being thus and so, one must acknowledge—on pain of failing to deploy the concept **water**—that *what* one is conceiving of is determined by the underlying property or nature, whatever it is, which (actually) determines the extension of our word 'water.' If that property is *being composed of* H_2O *molecules*, then conceiving of something that isn't so composed is *eo ipso* to fail to conceive of water—and failure to acknowledge as much amounts to a failure to deploy the relevant concept. In this sense, I think Wright is right to refuse to grant that 'the imaginative scenario involves an appropriate play with the very concepts configured in' the putative necessity in question.

It would be good to have a fuller explanation of conceivability *de re*—of what is required for a piece of conceiving to count as a conceiving *of* a certain entity (a particular object, a kind, etc.).[7] But even without that, it seems clear enough that whether or not a piece of conceiving is relevantly *de re* (is, e.g., *de aqua*) is not something we can in general know *a priori*. So I do not think Wright can be strictly correct in claiming that 'defeat [i.e., of putatively *a posteriori* necessities] may consequently be a priori even if their sole possible form of justification is not.' We can agree, perhaps, that prima facie counter-conceivability—our apparent ability to conceive of water being other than H_2O, say—is something we can recognize *a priori*. But defeat requires genuine counter-conception, and that is no more discernible *a*

---

[7] So far as I can see, (surface) grammatical considerations are somewhat fragile, and can't bear much explanatory weight, for various reasons. One might be tempted to look for a contrast between cases in which conceiving takes a propositional object, with 'conceive' governing a 'that-clause, and cases in which conceiving takes a direct non-propositional object. But 'conceive that *p*' is awkward English, and we tend to resort to circumlocutions such as 'conceive of its being the case that *p*,' 'conceive of a situation in which *p*.' However, *de re* conceivings can be cast in this form (e.g., 'conceive of its being the case that Hesperus ≠ Phosphorus'). Further, use of the 'conceive of…' construction is not a reliable indicator that we are dealing with a bit of *de re* conceiving, even when what is conceived of is given by a simple noun or noun-phrase rather than something of the form 'its being the case that *p*' or 'a situation in which *p*.' Some conceivings of… are not *de re*—for example, there is (or need be) nothing *de re* going on when I conceive of a pink and blue polka-dotted zebra (it is not that I have in mind this particular zebra, in Regent's Park Zoo say, and am conceiving of it as polka-dotted pink and blue—I am just imagining (there being) a zebra like this—a merely possible zebra, as we might say). My conceiving of… is, in effect, an existential conceiving; it may, and probably will, involve my having a picture before my mind's eye, but the essential thing is that I am imagining or supposing 'There could be something like that'—my mental picture isn't (and isn't intended to be) a picture of any particular animal.

priori than are the necessities in question. Whether it is (*de re*) conceivable that *x* shouldn't be ϕ will typically, if not always, be an *a posteriori* matter—just as *de re* necessity is—and this leads to the question: can one (ever) settle whether it is *de re* conceivable that *x* shouldn't be ϕ without, in effect, first (or thereby) settling whether it is (*a posteriori*) necessary that *x* is ϕ? If, as I am inclined to think, the answer to that question is 'no,' then, strictly speaking, conceivability *can't* serve as a (fallible or defeasible) 'guide' to (metaphysical) possibility. It does not follow, however, that those who have thought of it as such have been completely off target—for there is *apparent* or *seeming* conceivability, and this surely *is* fallible evidence of possibility. As Wright puts it, the 'operational content' of his counter-conceivability principle is that 'if one has what at least *appears to be* a lucid conception of how it might be that not-P, then that should count as a good, albeit defeasible, ground for its not being necessary that P' (p. 408).

3. Wright's Examples

As I observed at the outset, Wright holds that Kripke's argument against the identity theory fails, because one must, for independent reasons, recognize that there is another way in which an argument from counter-conceivability can be answered— a way which not only differs from the only one Kripke considers, in discussing the identity theory, according to which 'a lucid putative counter-conception falls short ... by failing to engage the distinction between an item and surface counterparts of it' (p. 435), but which, unlike this one, remains open to the identity-theorist.

Wright gives two examples, both designed to persuade us that another model for explaining modal illusions, quite different from anything suggested by Kripke, is needed. The first invokes his apparent ability to conceive of himself not having had the biological parents he actually had; the second appeals to the idea that there could be a sceptical mathematician who claims, post-Wiles, to be able to conceive of finding counter-examples to Fermat's Last Theorem and mistakes in Wiles's proof. I am going to discuss them at some length, so I shall quote Wright's presentation of them in full:

SuperWright

Suppose—Kripke would agree—that I am essentially a human being, and that it is an essential characteristic of human beings to have their actual biological origins. So it is an essential characteristic of mine to be the child of my actual parents. Still I can, it seems, lucidly conceive of my not having had those parents but others, or even—like Superman—of my originating in a different world, of a different race, and having been visited on Earth from afar and brought up as their own by the people whom I take to be my biological father and mother. I can, it seems, lucidly imagine my finding all this out tomorrow.   (p. 435)

The Sceptical Mathematician

We can rest assured, I suppose, that Andrew Wiles really has proved Fermat's "Last Theorem," which therefore holds good as a matter of conceptual necessity. But we can imagine a sceptic about the result who flatters himself that he can still conceive of finding counterexamples to the theorem, and of finding mistakes in Wiles' proof. Of course there will be limits on the detail of these 'conceivings,' or the sceptic would be thought-experimentally finding *real* counterexamples and mistakes. Still, we should not deny that he could be conceiving *something*, and doing so moreover—subject only to the preceding point—in as vivid and detailed a way as could be wished.   (p. 436)

Wright's summary statement of the conclusion he draws from these examples is that

There has to be a category of conceivings that fall short of being counter-conceptions to a given proposition, not because their detail fails to be sensitive to the distinction between items that the proposition is about and 'fool's' equivalents of them, but because it is insensitive to another distinction: that between genuinely conceiving of a scenario in which P fails to obtain and conceiving, rather, of what it would be like if, *per impossibile*, P were (found to be) false.

(p. 437)

Wright thinks his two examples *both* fail to fit the standard template for failed counter-conception because he thinks that there can be no question, in *either* case, of claiming that the failure is due to insensitivity to a distinction between what the conceiving is supposed to be about and some 'fool's' equivalent: thus in the first example:

it cannot be dismissed . . . on the ground that it fails to be sensitive to the distinction between myself and a mere epistemic counterpart, a mere "fool's self," as it were, sharing the surface features by which I identify myself but differing in essence. It cannot be so dismissed because I don't, in the relevant fashion, identify myself by features, surface or otherwise, at all. The point is of a piece with Hume's observation that, in awareness of a psychological state as one's own, one is not presented as an object to oneself. When I conceive some simple counterfactual contingency—say, my being right at this moment in the Grand Canyon—I do not imagine someone's being there who presents themselves, on the surface, as being me. Rather I simply imagine *my* having relevant kinds of experience—imagine, that is to say, the relevant kinds of experience from my first personal point of view. No mode of presentation of the self need feature in the exercise before it can count as presenting a scenario in which I am in the Grand Canyon; a fortiori, no *superficial* mode of presentation, open to instantiation by someone other than myself.   (pp. 435–6)

and in the second:

The last diagnosis we should propose, however, is that his conceivings are insensitive to the distinction between finding counterexamples to Fermat's theorem and finding counterexamples to an *epistemic counterpart* of it!—or to the distinction between finding a mistake in Wiles' reasoning and finding a mistake in an epistemic counterpart of that.   (pp. 436–7)

I find Wright's examples, and some of what he says about them, quite puzzling. I think we can and should separate two questions we might raise about them:

(i) Do they demonstrate a need for a kind of explanation of modal illusions quite different from what we get on Kripke's main 'epistemic counterpart' model?

(ii) Do they—or does either of them—demand explanation in terms of the *per impossibile* model which Wright proposes?

One obvious reason why it is important to separate these questions is that, should it prove that the right answers are 'yes' and 'no,' respectively—so that an altogether different model for explaining away merely apparent counter-conceivings is called for—it is so far possible that it should be one which is no more help to the identity-theorist than is Kripke's main model. Even if that possibility were realized, it would not completely undermine Wright's argument for the conceivability of naturalism. It would, I think, weaken it, because—as I understand it—the examples are supposed not only to convince us that an alternative to Kripke's main model is needed, but to motivate acceptance of the *per impossibile* model. But it would only weaken it, because provided that model is at least viable—even if it is not required to handle SuperWright and the Sceptical Mathematician—the identity-theorist could still appeal to it to explain away the (alleged) illusion that painless C-fibre firing and C-fibreless pain are possible.

4. A Possible Defence of Kripke's Argument Rehearsed and Rejected

Kripke is certainly not committed to the claim which seems to me to be decisively refuted by Wright's examples, that is, the claim that *every* merely apparent case of counter-conceivability can be explained away in terms of his epistemic counterpart model. He puts forward that model specifically in connection with the apparent conceivability of water that isn't H_2O, heat that isn't average kinetic energy of molecules, and other cases in which it might be thought that the opposite of an *a posteriori* necessity is conceivable. He never suggests explaining illusions of possibility in this way, when the countervailing necessity is *a priori* or conceptual. Although he doesn't discuss such cases at length, he does say something which suggests how he would, or at least might, wish to handle them.

First, there's one sense in which things might turn out either way, in which it's clear that that doesn't imply that the way it finally turns out isn't necessary. For example, the four color theorem might turn out to be true and might turn out to be false. It might turn out either way. It still doesn't mean that the way it turns out is not necessary. Obviously, the 'might' here is purely 'epistemic'—it merely expresses our present state of ignorance, or uncertainty.[8]

This at least suggests that one could explain some modal illusions by appeal to the idea that we confuse different kinds of possibility, and more specifically, that we confuse merely epistemic possibility with real, metaphysical, possibility.

The potential bearing of this on Wright's second example should be obvious. The supposedly counter-conceived necessity in that example—Fermat's Last Theorem— is an *a priori* necessity; indeed, in Wright's view,[9] it is a *conceptual* necessity. Given that the mathematician Wright envisages—unlike the rest of us—actually doubts whether FLT is true, there is an obvious alternative explanation of his erroneous belief that he can conceive of finding counter-examples to it, and mistakes in Wiles's proof, namely that he takes it to be an open question whether FLT holds, and so naturally takes it to be (metaphysically and so conceptually) possible that it doesn't. But while the Sceptical Mathematician's modal illusion may be explained by hypothesizing that he conflates epistemic with metaphysical possibility, no such explanation can be given in the pain/C-fibre firing case—at least not if we assume, as I think we

---

[8] Kripke (1980: 103).   [9] And mine, but many would disagree.

may, that the victim of the modal illusion (if that is what it is) does not take it to be an epistemic possibility that pain isn't in fact C-fibre firing.

None of this helps with SuperWright, however, since we may suppose that Wright is in no doubt at all about his actual parents and biological origin. But there is a different strategy that might be tried in this case. One quite standard way in which we may come to believe—rightly or wrongly—that it is possible that something should be thus and so (schematically, that $\Diamond\phi a$) is via believing that it is possible for something else to be that way (that $\Diamond\phi b$), and that there is no relevant difference between the two things (between a and b). Now why shouldn't Wright's belief that he might have had a quite different origin be explained in just this way. He believes that other people are such that they might have had different biological origins, because he seems to himself to be able to conceive of their having done so—really he is just conceiving of epistemic counterparts having different origins; and thinks he can conceive of the same thing at home, because he believes that in this regard he does not relevantly differ from other human beings. But once again, no explanation along these lines looks to be on the cards in the pain/C-fibre example.

It might, then, be argued that there are other, plausible, ways of explaining away the appearances of conceivability in Wright's examples which are not apt for extension to the pain/C-fibre case, and which do not therefore threaten to undermine Kripke's argument against the identity-theorist. There are, however, at least two reasons why we should be dissatisfied with this defence.

(i) The proposed alternative explanation of the merely apparent conceivability of the falsehood of FLT and failure of Wiles's proof can only be deployed when the victim of the illusion actually doubts whether FLT is true and whether Wiles's proof is sound—we can't explain someone's mistakenly thinking that it is metaphysically possible that p by the hypothesis that they take it to be epistemically possible that p and are careless of the difference, unless they do take it to be epistemically possible. So it won't help to explain the illusion in the more radical case of a thinker who accepts that FLT is true and takes Wiles's proof to be sound, but denies there is any necessity about these things—a thinker who is the strict analogue of one who accepts that water is H_2O, but denies that it is necessarily so.[10] As far as SuperWright is concerned, Wright could perfectly well agree that one *could* fall victim to the illusion that one might have had a different origin in the indirect fashion suggested, but insist, plausibly enough, that it doesn't have to be like that— why should it not be that one seems to oneself to be able directly to conceive of oneself as having had different origins?

---

[10] I am not sure why Wright builds it into his second example that the mathematician is sceptical about FLT and Wiles's proof, thereby opening a potentially significant disanalogy with standard examples of supposed counter-conceptions to supposed *a posteriori* necessities, where the facts (e.g., that water is H_2O) are not called into question. It might be thought the analogy would have been better served by invoking, instead, the curious character Wright once dubbed 'the Cautious Man,' who accepts logical and mathematical statements as true as readily as the rest of us, but not that they are necessary (see his 1980: 453 ff.). But Wright is surely right to resist any such temptation; the Cautious Man does *not* present himself as *able to counter-conceive* what we take to be necessary—he readily agrees that he can't imagine things otherwise, but complains that he can't see how that warrants taking anything to be necessary.

(ii) In any case, it is insufficient to get Kripke's argument out of trouble merely to point out that there are kinds of explanation which he could give for SuperWright and the Sceptical Mathematician which can't be used by the identity-theorist. For Kripke's argument, as already noted, is a challenge—to uphold it, it needs to be argued that there isn't *any* kind of explanation that will serve the identity-theorist's turn, and so argued, in particular, that Wright's *per impossibile* model won't serve.

5. Two Doubts about Wright's *Per Impossibile* Model

I think there *are* grounds on which one might be sceptical about the applicability of Wright's *per impossibile* model. According to the model, we fall victim to modal illusion through insensitivity to a distinction 'between genuinely conceiving of a scenario in which P fails to obtain and conceiving, rather, of what it would be like if, *per impossibile*, P were (found to be) false.'

One obvious worry is that the model is a useful one only if there is a distinctive way things would be like, were it the case that *p*, when it isn't possible that *p*. Standard semantical accounts of counterfactuals do nothing to encourage the belief that there is. So much the worse, Wright may claim,[11] for such treatments—'For a large class of impossibilities, there are still determinate ways things would seem if they obtained' (p. 437). Perhaps. But this raises large and difficult issues which I can't pursue here, beyond observing that if Wright's model is to be viable, it should be possible to uncover constraints on counterfactual conditionals which allow us to accept some with impossible antecedents as true, while rejecting (nearly all) others as false. I don't myself have any useful idea what such constraints might look like. And I'm not sure that I share the intuitions which seem to convince Wright that there *must be* some (because there are determinate truths about how things would (seem to) be, were some impossibility to obtain). The only facts or claims in this area about which I feel reasonably confident can all, so far as I can see, be accommodated on the view that counterfactuals with impossible antecedents are all vacuously true, but that there are broadly pragmatic reasons for being more interested in some than in others. I can't show that counterfactuals can't be assigned truth-conditions which make relevantly finer discriminations than the standard accounts.[12] So I shall claim only that the assumption that Wright's model is viable carries a significant but undischarged explanatory debt, and that unless the case is compelling for thinking that there are (uncontroversial) illusions of possibility that cannot be otherwise satisfactorily explained, we are under no great pressure to accept it.

There is a potentially much more serious cause for concern, which we may uncover by asking just what distinction it is that Wright intends, when he contrasts 'genuinely conceiving of a scenario in which P fails to obtain and conceiving, rather,

---

[11] And does—cf. p. 437, n. 22: 'This is one reason why semantical treatments of subjunctive conditionals that hold all with impossible antecedents to be true are unfortunate.'

[12] There are, of course, semantical accounts which do so by invoking impossible worlds. Even if I were convinced that vacuous truth of counterfactuals with impossible antecedents is something to be avoided, I would need some convincing that this cure isn't worse that the supposed disease.

of what it would be like if, *per impossibile*, P were (found to be) false.' What exactly is the intended force of the addition '*per impossibile*'?[13] Is it meant to be within the scope of the 'conceiving,' so that it contributes towards specifying the *content* of what is conceived?—or should '*x* is conceiving of what it would be like if, *per impossibile*, *p*' be taken as equivalent to the simple conjunction of '*x* is conceiving of what it would be like if *p*' with 'it is impossible that *p*'? The ambiguity here is broadly similar to that of 'Jack denied Jill's claim that he had been hitting the bottle again,' which can be heard so that it can be true even if Jack doesn't know that Jill has claimed that he'd been drinking heavily again, but can also be heard so that its truth requires Jack to have said something like 'Jill's claim that . . . is completely untrue,' or (to Jill), 'Your claim is just not true.' But whereas I have no difficulty in understanding the words 'Jill's claim' as contributing to the specification of the content of Jack's denial, I can't see how to construe 'conceiving of (what it would be like if, *per impossibile*, *p*)' so that the content of the conceiving differs from that of the plain 'conceiving of (what it would be like if *p*).' So I can see no option but to construe '*x* is conceiving of what it would be like if, *per impossibile*, *p*' as equivalent to the simple conjunction of '*x* is conceiving of what it would be like if *p*' with 'it is impossible that *p*.' But then we have a distinction to the purpose only if there is one between 'conceiving of a situation in which *p* is false' and 'conceiving of what it would be like, if *p* were false.' I'm not sure that there is. But even if I'm wrong about that, a worry remains. For so long as there is no difference in content between conceiving of what it would be like, if *p* were false and conceiving of what it would be like, if, *per impossibile*, *p* were false, what is to prevent one from defending *any* claim to necessity against a charge of counter-conceivability by claiming that the alleged counter-conception is only a conception of what it would be like if, *per impossibile*, the claimed necessity was false, not a conception of a scenario in which it is false?

I don't claim these doubts are decisive. I do claim they are enough to warrant closer scrutiny of Wright's examples, with a view to uncovering a less problematic kind of explanation of modal illusions of the sort or sorts they involve.

6. The Sceptical Mathematician Dissected, a Fresh Diagnosis, and an Alternative Model for Explaining Some Modal Illusions

Wright's sceptical mathematician is supposed to provide us with a plausible (but of course flawed) example of counter-conceiving—one that demands explaining away, but which resists dissolution in terms of the 'official' Kripkean model, and so requires a different explanation. No doubt we can readily enough imagine someone who has studied Wiles's proof but remains unpersuaded by it, and who actually doubts that

---

[13] There is also some unclarity about the intended force of the parenthetical 'found to be.' Obviously a proposition can't be found to be false unless it is false; but does conceiving of what it would be like if *p* were found to be false involve conceiving what it would be like if *p* were false? Obviously the latter does not require the former. So the question is, in part, whether being found to be false is an essential part of the conceiving, or merely an optional extra.

FLT is even true. And we can imagine that his doubts are substantial enough for him to express himself in terms like these: "I doubt that Fermat's Last (so-called) Theorem is true. I reckon that, given time, someone could come up with a counter-example. And as for Wiles's 'proof,' I can certainly imagine someone coming up with a counter-example/finding a mistake there. I suppose it is equally only a matter of time and effort before someone finds one." But I take it that this is clearly not enough for Wright's purpose. Someone's merely doubting FLT, thinking that there is probably a counter-example waiting to be found and that probably there is some mistake in Wiles's proof is nowhere near enough to provide us with a plausible but flawed case of apparent counter-conception. For that, we need a sceptic who appears—to himself, even if not to us—to have a lucid and coherent conception of a specific counter-example to FLT (or at least, of a recipe for locating one) and/or a lucid and coherent conception of a specific mistake in Wiles's proof (or, once again, at least of definite grounds for thinking there to be one, which are not merely consequential upon whatever grounds he has for doubting that FLT is true). On the other hand, whatever more detailed shape his efforts take on, they must fall short of locating—*per impossibile*, assuming (as we are) that FLT is true and that Wiles's proof is sound—any actual counter-example or any actual error in the proof. So a delicate balance needs to be struck, between—to put it roughly—too little detail and too much (or perhaps better, the wrong sort of detail). If there is too little detail, our sceptical mathematician will not have done enough to qualify as even apparently counter-conceiving FLT or Wiles's proof—as distinct from merely thinking that there must be, or probably is, a counter-example, or that there must be, or probably is, a mistake in the proof. But if there is too much detail—or detail of the wrong sort—he will get into conflict with the actual mathematical and proof-theoretic facts.

Wright acknowledges that his sceptic's musings must not be too detailed: 'Of course there will be limits on the detail of these 'conceivings,' or the sceptic would be thought-experimentally finding *real* counter-examples and mistakes.' But he sees no great difficulty here: 'Still, we should not deny that he could be conceiving *something*, and doing so moreover—subject only to the preceding point—in as vivid and detailed a way as could be wished.' As against this, I shall make two claims: first, that Wright's assessment of what, subject to the limitation indicated, his sceptic can accomplish is unduly sanguine—and in particular, that there is reason to question that he can, when operating under that limitation, be seen as conceiving of something to the purpose in 'as vivid and detailed a way' as we should require; and second, that Wright in effect mislocates the real danger involved in exceeding the limit on detail. Let me explain and attempt to justify these claims.

What we need (to imagine) is a sceptic who claims, in regard to some bit of conceiving he has actually performed, that it is a counter-conception— that is, that he is conceiving of finding a counter-example to FLT, or a mistake in Wiles's proof. Let's suppose that what he claims to be able to conceive of is (finding) a counter-example to FLT.[14] Our sceptic can presumably share his putative counter-conception with

---

[14] It think it is clear that what I go on to say about this case applies, with minor adjustments, to the case where the sceptic claims to be able to conceive of finding a mistake in Wiles's proof. The crucial point is that just as, if FLT is true, it is so necessarily, so, if Wiles's proof is sound, it is so necessarily.

us—at least, if he can't, there's nothing to discuss or explain, and we can just stop right there. So let us suppose him to be more articulate. What might he tell us? It seems to me that here we confront something approximating to a dilemma, which lines up roughly with the two extremes—too much and too little detail—between which, as noted, a course needs to be steered. Consider first what is apt to go wrong if there is too much—or the wrong kind of—detail. Suppose our sceptic thinks he can conceive of finding a counter-example because he believes he has actually located one. 'Do tell us more,' we say—all ears. Now the difficulty is obvious. It is not—as Wright would have us believe, that he would be 'thought-experimentally finding [a] *real* counter-example.' If he says, for some definite x, y, and z and definite n greater than 2, that $x^n + y^n = z^n$, then we can just check it out. Either he's right and Fermat (and Wiles, and a lot of others) wrong, or—much more likely—he's wrong. Assuming he's wrong, he's made an arithmetic mistake—and that's all the explanation we need, why his (as he supposed, lucid and coherent) attempt at counter-conceiving FLT is flawed. So let us suppose—as we must, if we are to keep the example in play—that he avoids this much, or this kind, of detail in elaborating his putative counter-conceiving. If he is more coy about it—stops short of saying enough to identify a definite putative counter-example—then, it seems to me, things will go in one of two ways. *Either*, while he says less, he'll still say enough for us to see that he's made a definite arithmetical mistake—as will presumably be the case if he gives what purports to be a (non-constructive) proof that Fermat's equality holds, say, for some n between 17 and 31, with x, y, and z all greater than 10,000—*or* he won't, but will avoid arithmetical error only by failing to say enough to qualify as expressing a putative counter-conception (i.e., as distinct from merely expressing his conviction that there is (probably) a counter-example to be found).

Wright may protest that my construal of his example is unfair, and simply misunderstands the way in which thoughts of his sceptical mathematician are supposed to constitute a plausible (albeit flawed) case of counter-conceiving. He will agree, presumably, that his sceptic's putative counter-conceiving must avoid the kind of detail which would involve him in arithmetical error. But he will deny that the price of doing so must be to degenerate into something which has no claim to be a piece of relevant counter-conceiving at all. The intended force of the claim that his sceptic does enough to convince us that he is 'conceiving *something*,... in as vivid and detailed a way as could be wished,' is that the details of his sceptic's story could pass muster as a conceiving—not of (finding) a counter-example to FLT, but—of 'what it would be like if, *per impossibile*, FLT were (found to be) false.' The sceptic starts with the hypothesis—in fact necessarily false, although he does not, of course, take it to be so—that there are counter-examples to FLT. He then imagines someone finding one—*without, however, attempting to imagine it in any detail*—and goes on to imagine the impact of this 'discovery' on the mathematical community: the initial reactions of shock and even disbelief which, however, quickly disappear (since simple and apparently flawless calculations show the putative counter-example to be the genuine article), the embarrassment of those (virtually everyone) who had been 'taken in' by Wiles's proof, Wiles's own chagrin and reluctant admission that there must be some flaw in his work, the ensuing hunt for a mistake—perhaps a hitherto unnoticed but false assumption—in that 'proof,' and so on. There is, it seems, no doubt that this is a perfectly conceivable flight of fancy, and that it could be elaborated in as much vivid detail as you like, subject to the proviso that it remains silent on the mathematical detail of the supposed counter-example. So isn't Wright right after all?

I think not. Or more precisely, I think he is partly right, but partly wrong. What is right is that this is a perfectly intelligible piece of conceiving. What is wrong—or at least, in my view, very much open to question—is that the flaw in it, viewed as an attempt to counter-conceive FLT, should be diagnosed as Wright proposes, that is, as a matter of insensitivity to a distinction between conceiving of FLT's being false and conceiving of what it would be like *if, per impossibile*, FLT *were (found to be) false.* Nothing in what the sceptical mathematician actually *imagines*—as distinct from the background doubt as to the truth of FLT which *fuels* his imaginative flight—*demands* its description in those terms, *rather than* in these quite different terms: conceiving what it would be like *if, as is perfectly possible*, FLT *were (erroneously)*[15] *thought* to be counter-exemplified. To be sure, the notion that FLT is false does play a part in the exercise as described—as I said, it fuels (that is, encourages or motivates, and perhaps even partly causes) the sceptical mathematician's imaginings. But it is not *part of what he imagines.* Of course, one might say that he *imagines that* FLT is counter-exemplified and so false, and then goes on to imagine...But that should not be allowed to mislead us. It just means that he *thinks that* FLT is counter-exemplified and so false, not that he imagines a scenario in which it is so. Further, although the belief that FLT is counter-exemplified may in fact play a part—be what inspires the sceptical mathematician's imaginings—it is clearly an inessential part in the proceedings. We, who entertain no sceptical doubts about the truth of FLT or about Wiles's proof of it, can just as easily engage in the same imaginative exercise as the sceptic. Of course, the sceptic will likely express himself in less guarded terms than we—persuaded as we are of the truth of the theorem and the soundness of Wiles's proof—would choose. He will doubtless describe himself of conceiving of what it would be like, if FLT *were found to be* false or if Wiles's proof *were found to involve* a mistake—sceptical as he is, he will see no reason to speak more cautiously of his conceiving of what it would be like, if FLT were *thought* to be false, etc. Indeed, it is essential to the generation of his illusion that he should have no qualms about so describing himself—the illusion, on the present suggestion, comes about precisely because its victim confuses conceiving what it would be like, were *p* believed to be false with conceiving what it would be like, were *p* found to be false. Perhaps no one who conceives of what it would be like, if FLT were thought to be false, but is clear that what she is conceiving is what it would be like, if FLT were merely (or falsely) believed to be false, is at all likely to succumb to the illusion. But this is not the position of Wright's sceptical mathematician, for whom the falsehood of FLT is anyway an *epistemic* possibility. Further, it isn't clear one couldn't fall victim to the confusion even if one has no tendency towards scepticism about FLT—perhaps it is enough that one conceives of what it would be like, were FLT to be believed false,

---

[15] As we, but not Wright's sceptical conceiver, would say. It does not have to be part of the conceiving that the belief *is* erroneous.

without being clear that what one is conceiving is how things would or might be if it were *falsely* believed FLT false.

So it seems to me that there is a perfectly viable alternative diagnosis, quite distinct from the one Wright proposes, of the modal illusion to which his sceptical mathematician falls victim—that he simply confuses conceiving of FLT's being found to be false with conceiving what it would be like *if, as is perfectly possible, FLT were thought to be false.*[16]

It should be clear, but is still worth emphasis, that this alternative diagnosis is quite distinct from what Wright dismisses as 'The last diagnosis we should propose, . . . that [the sceptical mathematician's] conceivings are insensitive to the distinction between finding counter-examples to Fermat's theorem and finding counter-examples to an *epistemic counterpart* of it.' There is evidently no play, in my alternative diagnosis, with the obviously highly questionable notion of an epistemic counterpart of FLT. What might such a thing be? A supposed 'theorem' to the effect that $x^n + y^n = z^n$ has no integer solution for n greater than *one*, perhaps! Wright is clearly right to dismiss the last diagnosis as totally incredible. I think he is also right to conclude that the standard Kripkean diagnosis of modal illusion (in terms of illicit passage from the possibility of our being in an *epistemically* analogous situation, qualitatively indistinguishable from the one we are actually in, to an erroneous conclusion about *metaphysical* possibility) can get no grip on this (kind of) example. At least, that conclusion follows, if we grant that any situation epistemically analogous (in relevant respects) to our own with respect to FLT must involve an epistemic counterpart of FLT.[17] The constraints on the epistemic counterpart relation ensure that there can be, in this case, no such thing. An epistemic counterpart of water (or Hesperus, or this table, etc.) has to be something distinct from water (or Hesperus etc.) itself—since it has to be, in the envisaged epistemically analogous situation, other than H_2O (or other than Phosphorus etc.), as water cannot be—but nevertheless (superficially)[18] qualitatively indistinguishable from water (or Hesperus etc.). In general, an epistemic counterpart of x has to be something distinct, but nevertheless (superficially) qualitatively indistinguishable, from x. But in contrast with substances, heavenly bodies and articles of furniture, there can be nothing distinct from a (mathematical) proposition that is yet superficially indistinguishable from it. Propositions may be more or less difficult to grasp, and distinct propositions may sometimes be easily

---

[16] I am assuming here that conceiving of FLT's being believed to be false is not problematic in the way conceiving of its being false is, and more generally, that while conceivability—as I, following Kripke, understand it here—entails possibility, believability does not. That is, we are able to believe what is impossible, even if we cannot conceive of its being so. I think this assumption is correct, but it certainly raises problems. See note 22 below.

[17] As we shall see later, this concession calls for some qualification, but not of such a kind as to affect the present point.

[18] Superficially, of course, because the Kripkean model does not call for counter-examples to the Identity of Indiscernibles—it calls only for something having, in Wright's terms, all the surface indicator properties of water. It can, and must be allowed to, differ with respect to deep structural—and perhaps essential—properties, such as being H_2O.

confused with one another, but they don't have hidden essences waiting to be discovered.[19]

7. Some Stock-Taking

We have thus far encountered four distinct models for explaining modal illusions—specifically, the illusion that possibly not-p, where in reality, necessarily p. We have, first, the two models suggested by Kripke:

(i) *Kripke's main (Epistemic Counterpart) model*—we think p's falsehood possible because we can conceive of a situation epistemically analogous to our own in relevant respects in which some closely related proposition p', which we fail to distinguish from p, would be false. Typically, in the case of *a posteriori* necessities *de re*, this will involve an 'epistemic counterpart' of the *res* in question—something very closely resembling it in respect of surface indicator properties.

(ii) *The Ep & Met model*—we simply confuse the epistemic possibility that p with its being metaphysically possible that p

We then have:

(iii) *Wright's 'per impossibile' model*—we confuse conceiving of p's being false with conceiving of what it would be like if, *per impossibile*, p were (found to be) false

To these, I have suggested we should add what we may call:

(iv) *The Doxastic Consequences model*—we confuse conceiving of p's being (found to be) false with conceiving of what it would be like if, as is perfectly possible, p were thought to be false

I have agreed with Wright that neither of his examples can be satisfactorily handled on the *Epistemic Counterpart* model, but have expressed some doubts both about whether they call for explanation in terms of his *per impossibile* model, and about the viability of that model. My doubts about the need to invoke Wright's model to deal with his examples do, if well-founded, weaken his case in favour of the model, which is precisely that it is needed to handle those examples. If I am right, the *Sceptical Mathematician* is best handled in a quite different way, using my *Doxastic Consequences* model. But even if I am right about that, the force of Wright's critique of Kripke's argument against the identity theory remains largely undiminished, unless my doubts about the viability of Wright's model are well-founded. So long

---

[19] This, it seems to me, is the one solid point of analogy with Wright's other example of a seeming possibility that resists standard Kripkean treatment—his (counterpartless) conceiving of himself as having had a quite different origin. Just as there is no 'fool's self,' so there are no 'fool's propositions.' But whilst I agree with Wright on this point, I think his examples differ significantly in other respects.

as Wright's model is a viable one, it is at the service of the identity theorist, even if it is not required to deal with either of the examples by which Wright seeks to motivate its acceptance. In fact, Kripke's argument is in trouble anyway, even if my doubts about the viability of Wright's model *are* well founded, so long as my *Doxastic Consequences* model is workable. For if that is so, then—whether or not it is agreed to be the best model for explaining the Sceptical Mathematician's modal illusion(s)—it could, clearly, be invoked by the identity-theorist to explain away the apparent conceivability of painless C-fibre stimulation and C-fibreless pain. We can easily imagine how things might be, if pain were believed to be other than C-fibre stimulation. We imagine, for example, media reports of neurophysiologists discovering that as well as C-fibres, normal human beings have D-fibres. Since these are distributed about our bodies much as C-fibres are, it is readily understandable how it came to be thought that pain is C-fibre stimulation. Analysis reveals, however, that whilst most pain-killing drugs reduce *both* C- and D-fibre activity, some newly developed ones reduce only the latter and are everywhere as effective as the usual drugs in reducing pain. Perhaps we go on to imagine reading newspaper articles reporting massive investment by the big pharmaceutical companies in research programmes aimed at developing newer and even better D-fibre 'dullers,' and less well-funded investigations into the (now obscure) function of C-fibres, etc.

As I see it, then, the central thrust of Wright's objection to Kripke's argument remains intact, even if the best way to implement that line of objection is not quite as Wright originally proposes. So I shall have no more to say on that score. Instead, I want now to look more closely at Wright's other extremely interesting but perplexing example. As with the Sceptical Mathematician, I shall suggest that the illusion of conceivability can be explained without recourse to Wright's preferred model, because there is another at least equally viable alternative.

8. The Complexities of Conceivability *De Se*, and another Route to Modal Illusion

SuperWright is a difficult example in part because it is complicated in respects other than the one Wright emphasizes—that is, its being a piece of conceiving concerning himself.

One complication is that it is—in contrast with the other piece of first-person conceiving Wright mentions (his conceiving of his being at this moment in the Grand Canyon)—a piece of *past tense* conceiving. What is (apparently) conceived is Wright's *having had* a past quite different from his actual past. Conceiving of oneself as having done or undergone such-and-such, or as having been thus-and-so, is in general quite different from conceiving of oneself doing or undergoing such-and-such, or as being thus-and-so. When I conceive of myself travelling by train to London, for example, this will most likely take the shape of my imagining doing various more specific things such as taking a taxi from my house to Glasgow Central Station, queuing up for a ticket, boarding the train on Platform 1, looking out of the window as the train crosses the Clyde... getting off the train at Euston, and perhaps experiencing various things characteristic of train travel, such as the motion of the train as it gathers speed, the irritating background noise of mobile 'phone conversations and personal stereos, periodic announcements of progress (or lack thereof), upcoming stations and delays on the tanoy,' etc.[20] But if what I have set myself to conceive of is my *having travelled* to London by train—so that my having (at some, perhaps unspecific, time in the past) completed the journey is to be part of what I conceive, then it would not obviously be appropriate for me to engage in imaginings of this sort. Indeed, it would arguably be quite inappropriate. Suppose you invite me to imagine/conceive of myself having travelled to London by train, and to tell you in as much detail as I can what I'm imagining. I come out with a description of my imagined specific doings and experiences (boarding the train in Glasgow, drinking awful tea, and eating blotting paper sandwiches at exorbitant prices, etc). "No, no," you protest, "you've misunderstood—I didn't want you to imagine *actually making* the journey, I wanted you to imagine *having made* it." You'd be quite right to complain.

So, what is required, if I am to comply with your request? Well, very roughly speaking, what I need to do—at least in this case of past-tensed conceiving—is to imagine the aftermath in a way that builds in the supposition that I got to London by train. So it won't be enough for me just to imagine being in London, and doing and undergoing various things whilst there—since that is entirely consistent with my having got there by car or plane, or indeed, with my having been there all along. I might imagine myself getting off a train and emerging from Euston Station, being asked by friends who meet me how my journey was, and telling them about the two-hour delay between Rugby and Watford Junction, etc. Something of this sort, at least, is what I need to do.

There is, then, nothing intrinsically or necessarily baffling about past-tensed conceiving—conceiving of (oneself) having done or undergone such-and-such, as opposed to conceiving of (oneself) doing or undergoing such-and-such. But Wright's example involves an additional complication—it essentially involves conceiving (or at least of attempting to conceive) of certain facts about *causation* having been otherwise than they actually were. Of course, there is a straightforward way in which that is involved in any piece of past-tensed counterfactual conceiving—say, conceiving of myself as having travelled to London last Thursday. To conceive of this—at least if one takes the conceiving to show, or at least provide reasonable grounds for thinking, that it is possible that I should have travelled to London last Thursday—involves assuming that the determinants of my actions might have been other than they actually were. But it seems to me that conceiving of this need not be part of the *content* of my conceiving of myself having travelled to London etc.—it is rather just a *presupposition* of my conceiving of that. But Wright's (apparent) conceiving of his having had a different origin is *essentially* a matter of conceiving of his causal history having been other than it actually was—it is (part of) the *content* of his (apparent) conceiving.

---

[20] It is worth noticing that nothing in this need tie my imagined journey to any particular time—my conceiving can be, in this sense, timeless, although I could incorporate elements which fixed the time of my imagined journey.

There is, once again, nothing intrinsically or necessarily difficult in conceiving of causes being different. For many types of event or states of affairs, there are—and we know that there are—different ways in which they may come about. So we can readily conceive of a motorway collision being caused by a certain car's brakes failing, rather than by its driver's falling asleep at the wheel, or another's attempting a reckless piece of overtaking. In the one case, we imagine the driver depressing the brake pedal, but with no effect on the vehicle's speed; in another, we imagine the driver struggling to keep his eyes open and to concentrate on the road, but finally losing normal consciousness and careering into the vehicle in front or swerving across the outside lane into the crash barrier. Suppose, however, we set ourselves to conceive, not of something's *being caused* in a certain way, but of its *having been caused* in one way rather than another—say, of some actual collision's having been caused, not as it actually was, by brake failure, but by its driver having fallen asleep. That is, what is wanted is a piece of past-tensed conceiving—not of the accident's coming about or *being caused* in a certain way, but of its having come about in that way, its *having had* such-and-such a cause. Then—at least if my earlier remarks about past-tensed conceiving are on the right lines—what we shall focus on, in our conceiving, is the aftermath. What we'll imagine is something of this sort—investigators on the scene of the collision, trying to figure out how it happened, interrogating witnesses, including any surviving drivers of vehicles involved. Perhaps, if the driver whose car went out of control is still alive, we imagine him being questioned, claiming that his brakes failed, but police failing to find either any fault with them or any characteristic skid marks on the road surface, and finding, on further investigation, that the driver had been on the road non-stop for many hours without stopping to rest, etc., and concluding that he must have fallen asleep.

This is, of course, only one case. But it suggests a general point about how conceiving of something's having been caused in some way other than it was actually caused may—and perhaps typically will—go: what we imagine is our (or someone's) *discovering* that it was so caused. Or perhaps better—for a reason we'll come to— what we imagine is our having compelling evidence or good reason to believe that it was caused in this other way. We imagine a situation in which it is—as is not the case in fact—an *epistemically open* question what caused the accident, or whatever it was, and in which that question gets a different answer from the one we believe or know to be correct. I don't claim that this is the *only* way in which we may conceive of something's having been caused in some way other than that in which it was actually caused. But I do think it is probably how we most naturally and usually go about it. So it is not at all surprising that Wright describes his (apparent) conceiving of himself as having had a different biological origin in just such terms:

I can, it seems, lucidly conceive of my not having had those parents but others, or even—like Superman—of my originating in a different world, of a different race, and having been visited on Earth from afar and brought up as their own by the people whom I take to be my biological father and mother. *I can, it seems, lucidly imagine my finding all this out tomorrow.*

(p. 435; my emphasis)

We should now come back to the question suggested just now, whether one should describe this kind of conceiving in terms of imagining (our) finding out or discovering that such-and-such, or rather in more guarded terms as imagining (our) having compelling evidence or good reason to believe that such-and-such. If Hume's Principle is true—that is, if conceivability implies possibility—then we should insist upon the more guarded description. For if we really could conceive of finding out that p, then it would be possible that we should find out that p. And since 'find out' is factive—we can only find out that p, if p—it would follow (by the sound modal principle that $\Diamond A$ entails $\Diamond B$, when A entails B) that it is possible that p. It is true enough, or at least widely accepted, that in many—perhaps the vast majority of—cases, it is metaphysically, even if not epistemically, possible that the causes of events and states of affairs should have been other than they actually were. In such cases, we may agree that we can conceive of finding out that the causes were otherwise. But if some of the facts about causation are metaphysically necessary—as they must be, if the doctrine of essentiality of origin is true—then whatever it is that we can conceive in regard to causes being other than they actually were, it cannot be that we can conceive of *discovering* that they were other than they actually were. The closest we can come to that is (something like) conceiving of our having compelling evidence, or good reason to believe, that the causes were otherwise.

Wright's description is conspicuously *not* guarded in the way I am recommending. But it *is* guarded in another way—he does not say roundly that he can imagine finding out that he had different parents, etc., only that he 'can, *it seems*, lucidly imagine finding all this out tomorrow' (my emphasis). I take it that he means the qualification to be taken seriously, and that he would deny that he, or anyone else, can—without qualification—imagine finding out, etc. His claim is, in effect, that this description is a *mis*description. The correct description of what he is doing, in his view, is that he can conceive of what it would be like if, *per impossibile*, he had had a quite different origin from the one he actually had. This, he thinks, he really can conceive; and if he were to succumb to the illusion that he might have had a different origin, his doing so would, in his view, result from confusing this perfectly possible piece of conceiving with conceiving of himself as having had a different origin.

What I have said suggests a quite different diagnosis of this particular modal illusion from the one Wright thinks we should adopt. What he can conceive of, according to me, is a situation in which he (and presumably the rest of us) would be confronted with compelling evidence, or at least good reason to believe, that he was not the child of that pair of human beings widely and plausibly supposed to have been his biological parents, but of some other pair of human beings, or perhaps (more radically) of some Superbeings far away on Krypton—evidence, say, which strongly supports the hypothesis that Baby Wright was inadvertently switched with Baby Bloggs shortly after their births, or perhaps something more exotic, such as his turning out to have quite the wrong sort of DNA profile to have sprung from ordinary human parents. Falling victim to the illusion that he might have had a different biological origin would then consist in confusing this with conceiving of himself as having had a quite different origin.

It should be clear that my alternative explanation of the illusion that one might have had some origin other than one's actual origin contrasts not only with Wright's preferred *per impossibile* explanation, but also with Kripke's *Epistemic Counterpart* model—there is no play, in my explanation, with the highly questionable notion of a 'fool's self,' which is somehow mistaken for the genuine article. I think it is also clear that my explanation conforms neither to my earlier *Doxastic Consequences* model, nor to the *Ep & Met* model—what we conceive of is not the consequences of its being believed that our biological origins were otherwise, but our being confronted with compelling evidence which might induce that belief, and it is not being suggested that it is epistemically possible that one had an origin quite different from what one supposes, and that one simply mistakes epistemic for metaphysical possibility. But it is not so clear that the explanation does not involve postulating a conflation of epistemic and metaphysical possibility of a more subtle kind. To conceive of a situation in which one is confronted with compelling evidence that p (e.g., that one had such-and such an origin, where this differs from what one supposes to be one's actual origin) is, as already remarked, to conceive of a situation in which it is (at least prior to one's being confronted with the hypothesized evidence) an epistemically open question whether p. That is, in effect, to say that in the *conceived* situation (as opposed to one's actual situation), it is epistemically possible that p. Given that conceivability entails (metaphysical) possibility, it follows that it is *metaphysically* possible that it be *epistemically* possible that p. One might then see the proposed explanation as suggesting that victims of the illusion do after all conflate epistemic and metaphysical possibility—albeit in a more subtle way, at one remove, as it were—when they slide into thinking it possible that their origins should have been other-wise. That might be how it goes, in some cases.[21] But it does not seem to me that the explanation *has* to be seen as appealing to a confusion of two sorts of possibility—why shouldn't it be just that conceiving of having compelling evidence for p is confused with conceiving of its being the case that p?

In short, I am inclined to think that the suggested explanation exemplifies a further, distinct model for explaining modal illusion, which we might call:

(v) *The Misleading Evidence model*—we think it possible that p because we can conceive of our having compelling evidence or good reason to believe that p, and we fail to distinguish this from conceiving of its being the case that p.

9. Concluding Remarks

I conclude with some brief remarks about the various models for explaining illusions of possibility discussed here.

First, there is a clear point of contrast between the *Ep & Met* model, on the one side, and the *Epistemic Counterpart, Doxastic Consequences, Misleading Evidence* models, and with them Wright's *per impossibile* model, on the other. Although all suggest ways of explaining away illusions of metaphysical possibility, the first works in a quite different way from the others. *Ep & Met* explains the illusion in terms of confusion of one sort of possibility with another, whereas they explain the illusion of

---

[21] It would be somewhat implausible to suggest that the man on the Clapham Omnibus, even in his most philosophical moments, might succumb to the illusion that he could have had different parents as a result of overlooking the crucial difference between epistemic and metaphysical possibility and (mis)applying the S4 principle! But who knows what might go on up the Ivory Tower?

possibility in terms of our genuinely conceiving something close in some way to the putative possibility, but nevertheless distinct from it—we really can conceive of a situation in which q, but we confuse this with conceiving of a situation in which p, where it is p's metaphysical possibility, not q's that is in question.

This suggests, second, that we might view at least *Epistemic Counterpart, Doxastic Consequences, Misleading Evidence*, and perhaps also Wright's model, if it is viable, as species of a single genus. I think we can usefully do so. In general, when we attempt to envisage how things would be, were it the case that p, we are likely to proceed in one of two broad ways—we may seek to envisage what the likely or certain *consequences* of its being the case that p, on the one hand, or we may, on the other, attempt to envisage how, were it the case that p, this fact would manifest itself to us, or how things would *appear* to us (we may, of course, do both). The different models I've discussed seem to fall comfortably on one side or the other of this broad division.

Thus one of the likely—but of course far from certain—consequences of its being the case that p, at least for many values of p, is that we would believe that p, and our so believing would carry with it a whole host of further consequences. Of course, we might well believe that p, even if it weren't the case that p, and those further consequences would still flow. The point is that there is a significant overlap between how things would be if p (and perhaps more narrowly, the consequences of its being the case that p) and how things would be if we believed that p, perhaps falsely. Hence the possibility—exploited in the Doxastic Consequences model—that we should be liable to confuse conceiving the latter with conceiving the former.

In attempting to envisage how things would appear to us, were it the case that p, we are, in effect, imagining ourselves presented with evidence that p. The envisaged evidence may be more or less strong, and it may more or less directly support the proposition that p. But in general, it will be the case that we could be in possession of the envisaged evidence without its being the case that p. There need be no transparent difference—no difference reflected in the envisaged evidence itself—between envisaging how it might compellingly, but misleadingly, seem to us that p, and envisaging how it would be if we had compelling, and as it happens non-misleading, evidence that p. It can, of course, be part of what we conceive, when we envisage ourselves possessed of compelling evidence that p, that our evidence is misleading. But it need not be, and if it is not, then it is clearly possible that we should fall into thinking that we really are conceiving of a situation in which p, when what we are doing is conceiving rather of our being presented with compelling but (outside the scope of our conceiving) misleading evidence. In short, the Misleading Evidence model is clearly in place on this side of my broad division. The Epistemic Counterpart model, I suggest, is best viewed as a special case. In general, where we are concerned with the putative possibility that something, a, should have a certain property, P, we may think this is possible because we really can envisage a situation in which we would have convincing evidence that a is P. It may be that we are misled because, convincing as our envisaged evidence is, it is consistent with our possessing it that a should not only not be P, but be such that it could not be P. Very often, the illusion could come about in one of two ways—it may be that while a could not be P, something else, a', could, and that our envisaged evidence is insensitive to the difference between a and a'; or it may be that while a could not be P, it could be

P', and that our envisaged evidence is insensitive that that difference. In the former case, a' is an epistemic counterpart of a—a kind of 'fool's a.' But as we know, there are cases in which the apparent conceivability of a's being P cannot be put down to being taken in by a fool's a, because there isn't any such thing. So the Epistemic Counterpart model gets no grip. In terms of this account of the matter, Kripke's error lies, in part, in jumping from there to the incorrect conclusion that the more general Misleading Evidence model can't apply—as it clearly can, in the pain/C-fibre case.

Finally, it is worth drawing attention to some presuppositions of the Misleading Evidence and Doxastic Consequences models. The former clearly presupposes that there being evidence for a proposition p does not entail p's conceivability, and the latter that what cannot conceivably be the case may nevertheless be believed. Both presuppositions are independently plausible—and unavoidable anyway if one takes conceivability to entail possibility and accepts, as one surely must, that one can believe and have evidence for the opposite of a necessity.[22] It is true, or at least very

---

[22] I do not claim that belief in the impossible is entirely unproblematic—only that it is (a) possible and (b) not problematic in the way that conceiving of the impossible is (see note 16). Assumption (b) was challenged by an anonymous reader of the penultimate draft of this chapter, who took the view that claims to believe in metaphysical impossibilities are no less problematic than claims to conceive of them. The same reader also acutely observed that my *Doxastic Consequences* model may be argued to raise the very same difficulties as I suggest afflict the *per impossibile* model. In particular, it seems that if one seeks an account of the semantics of belief in terms of (possible) worlds, one faces a dilemma parallel to the one confronting a (possible) worlds treatment of counterfactuals—just as in the latter case, one must either hold that all counterfactuals with impossible antecedents are vacuously true, or admit so-called 'impossible worlds' as a means of differentiating among the truth-conditions of different such counterfactuals, so in the case of belief, one must either accept that all necessarily false beliefs share the same content, or swallow impossible worlds as a means of differentiating among them. Here I can do no more than indicate briefly the main lines along which I would seek to meet this challenge, which demands a much more detailed and careful response than I have space for here. In my view, the two cases—the semantics of counterfactuals and that of belief—differ in crucial respects. In the former case, a semantics based on possible worlds provides an account of the truth-conditions of counterfactuals which agrees well with our pre-theoretical practice, or dispositions to accept or reject such statements. It is true that such a semantics has the counter-intuitive consequence that counterfactuals with impossible antecedents are all (vacuously) true. However, as suggested in §5, there are non-semantic considerations in terms of which we may explain why we find some such counterfactuals more interesting than others—for example, and very roughly, we are with good reason interested in conditionals we have grounds to assert independently of any knowledge of the truth-values of their component clauses. In brief, a possible worlds semantics of counterfactuals works pretty well, and its counter-intuitive consequences can be accommodated by appeal to non-semantic, broadly pragmatic considerations. In the case of a possible worlds semantics of belief, it seems to me, matters are very different. Whilst such a semantics may provide a plausible account of some such idealization as rational belief under optimal conditions of information, etc., it does very badly as an account of the truth-conditions of ordinary belief-reports. Prior to the introduction of so-called impossible (or, in some accounts, incomplete) worlds, such a semantics gets just about everything wrong. It is not just that all necessarily false beliefs wind up with the same content—everyone believes every necessary truth, and all such beliefs have the same content too; everyone winds up believing all the logical consequences of anything they believe; substitution of co-referring terms in the content-clauses of belief-reports is truth-preserving; and one believes that everything is thus-and-so if and only if one believes of each thing that it is thus-and-so (i.e., analogues of the Barcan and Converse Barcan principles hold, as they plainly should not). It would not be plausible to claim that a possible worlds semantics for belief-reports gets things basically right, and attempt to explain away these manifest shortcomings as essentially merely pragmatic phenomena—they are, rather, precisely the things a good semantics of belief needs to get right. Without impossible worlds, a worlds-based approach has no chance of doing that. But while introducing such

plausible, that belief requires intelligibility—that is, that one cannot believe what one cannot understand. But what we can understand need not conceivably be true—we can, for example, understand inconsistent and even overtly self-contradictory statements, but it is not, in my view at least, conceivable that such statements should be true. In particular, what is, in the sense with which we've been concerned here, inconceivable (e.g., that Hesperus $\neq$ Phosphorus, that water isn't H_2O, or that Fermat's Last Theorem is false) may nevertheless be intelligible.[23]

References

Hale, Bob (2007). "'Into the Abyss' (Critical study of Graham Priest: *Towards Non-Being—The Logic and Metaphysics of Intentionality*, Oxford 2005)," *Philosophia Mathematica* 15 (III), 94–110.

Hardcastle, Valerie (1999). *The Myth of Pain*, Cambridge, MA: MIT Press.

Kripke, Saul (1980). *Naming and Necessity*, Oxford: Blackwell.

Wright, Crispin (1980). *Wittgenstein on the Foundations of Mathematics*, London: Duckworth.

Wright, Crispin (2002). "The Conceivability of Naturalism," in Tamar Szabó Gendler and John Hawthorne (eds.), *Conceivability and Possibility*, Oxford: Clarendon Press, 401–40.

worlds may lead to a formal semantics which technically avoids these shortcomings, it does so—as I've argued elsewhere—at the cost of sacrificing all credibility as an account of the truth-conditions of belief-reports (see Hale 2007: 94–110).

[23] I thank audiences in Barcelona and St Andrews for useful discussion of some of this material, and to the editor of this volume for giving me the opportunity to contribute to it. It should go without saying—but it won't—that I owe a great debt of gratitude to Crispin Wright. Our collaboration, which stretches back for many years, has been, for me, an unfailing source of inspiration, insight, and philosophical excitement—as well, of course, as occasional frustration and a few headaches! I can barely imagine what my philosophical life would have been like without it, but I know it would have been much the poorer. Owing him so much, it is both an honour, and a great pleasure, to contribute to this celebration of his work.

Replies

By *Crispin Wright*

Foreword

The publication of this second collection of specially commissioned papers by so distinguished a group of scholars, centred on issues in which I have invested much of my own research effort, is a source of great pride to me. I am immensely grateful to all the contributors for their insightful and challenging essays, from thinking about which I have learned so much.

It is a special pleasure that the volume also includes a previously unpublished early paper by the late George Boolos, ruminating on the general thesis of logicism and the status of second-order logic, which Riki Heck found among George's papers and kindly edited for this collection and to which a brief postscript of their own has been appended.

Once again the essays naturally divide into four groups. Two are focused on aspects of my general interests in the epistemology of logic and in Michael Dummett's revisionism; five on my attempts, over the last thirty-five years or so, to overturn the once dominant impression of Frege's philosophy of mathematics that it had proved a complete failure; two on my even longer-running attempt to understand something of the nature and significance of the pervasive vagueness of natural language and the roots of the Sorites paradox; and one on the epistemology of modality. As with the first volume I have resisted the temptation to try to respond to every point of interest raised in the essays, so that my 'Replies' mix eclectic reactions to various of their arguments and proposals with more general reflections on the issues. Again, I am hopeful there may be some gains in readability from this approach; and perhaps some advance in the various debates.

The suggestion that this second volume might appear somewhat later than its companion was originally Peter Momtchiloff's. He did not, I'm certain, have it in mind that there might be more than a six-year gap between the two! I must apologise to all concerned for, as it must seem, having dragged my heels, although my own judgement, for what it is worth, is that while some of the chapters have already circulated and attracted a degree of independent discussion, none has 'dated' to any significant extent. My thanks to Peter for his patience throughout the delay and, especially, to Alex Miller for all his hard work in putting the volume together and, as people quaintly say, for 'making it possible'. Special thanks are also due to Yu Guo for invaluable help in compiling the lists of works referred to, editing of footnotes, and proofreading, and to all who have helped me with critical feedback on ideas in the various sections of my Replies, including Susanne Bobzien, Paul Boghossian, John Divers, Jim Edwards, Julien Murzi, Ian Rumfitt, Stephen Schiffer, Stewart Shapiro, and Amia Srinivasan. I am especially grateful to Riki Heck who generously read through early versions of each section of the Replies and gave me detailed comments resulting in many improvements.

Almost all the material incorporated in the Replies was originally composed especially for this purpose, but some of it has found its way into other publications in the interim. Part II very substantially overlaps with my contribution to Sergi Oms's and Elia Zardini's anthology of papers on the Sorites paradox.[1] §§ 2 and 3 of Part I closely follow material published in my contribution to Philip Ebert's and Marcus Rossberg's recent anthology on abstraction.[2] §5 of Part III draws extensively on my contribution to a Book Symposium on Ian Rumfitt's *The Boundary Stones of Thought*.[3] My essay in Part IV is a version of my contribution to the Festschrift for Bob Hale edited by Ivette Fred and Jessica Leech.[4] And some of the discussion in §4 of Part I, on Julius Caesar, assembles lightly edited excerpts of co-authored material[5] that was originally written before this volume was commissioned.

In the time that has elapsed during its preparation, two of the contributors, William Demopoulos and Bob Hale, have passed away, each after a long, very courageous struggle with cancer. Both were outstanding philosophers and wonderfully collegial intellectual companions from whom I learned a great deal over many years. With Bob in particular, I had the privilege of a collaboration that ran for almost four decades and amounted to a kind of sustained intellectual adventure that I am sure is given to very few. Both are very much missed. This book is affectionately dedicated to their memory.

---

[1] Elia Zardini and Sergi Oms (eds.) (2019), *The Sorites Paradox*, Cambridge: Cambridge University Press.

[2] Philip A. Ebert and Marcus Rossberg (eds.) (2017), *Abstractionism*, Oxford: Oxford University Press.

[3] *Philosophical Studies* (2018), 175 (8). (This material is used here under the terms of the Creative Commons Attribution 4.0 International License (http://creativecommons.org/licenses/by/4.0/).)

[4] Ivette Fred and Jessica Leech (eds.) (2018), *Being Necessary: Themes of Ontology and Modality from the Work of Bob Hale*, Oxford: Oxford University Press.

[5] This derives specifically from chapter 14 of *The Reason's Proper Study* and from a Book Symposium on that volume in *Philosophical Books* (2003), 44 (3).

Replies to Part I
Frege and Logicism

Replies to Demopoulos, Rosen and Yablo, Edwards, Boolos, and Heck

In 1979 during the first year of my appointment at St Andrews, Bill Lyons and Andrew Brennan invited me to contribute something to their recently instituted (though, unfortunately, now defunct) Scots Philosophical Monographs series. I felt that, newly installed in a venerable Scottish Chair, something like *noblesse oblige* applied and, though I previously had had no intention of publishing any of its content, it occurred to me that the thesis I had written for my Oxford BPhil ten years previously might provide the basis for a suitable short book. So over the summer of 1980 I set myself the task of writing up *Frege's Conception of Numbers as Objects* (FCNO), alongside a piece on strict finitism promised for a special number of *Synthese* on realism that Esa Saarinen was guest-editing. I recall a summer of unusually warm, dry weather for Eastern Scotland and daily near-deafening noise of warplanes taking off from RAF Leuchars—the cottage I was renting was more or less directly under their flight path. But these distractions notwithstanding, I managed to complete both projects, more or less, by the start of the Martinmas term in October, though I recall that the monograph came much more easily than the effort on strict finitism (which is one of my papers that I suspect almost nobody has read).

I say "complete...more or less" because I did not quite succeed at that time in doing everything I planned for the monograph. Three quarters of what I wrote was concerned to explain what I took to be the conception of 'object' that underlay Frege's apparently platonistic attitude to numbers in *Grundlagen* and to argue for its metaphysical merits. That part of the monograph developed reasonably straightforwardly. But I also wanted to go on to sketch a case for something whose epistemological significance, at least for the arithmetic of finite cardinals, I believed would come close to that of Frege's project in *Grundgesetze*, had he been able to execute it successfully. The core of the case was to be the observation that the principle which Frege presents in *Grundlagen* §63 and the sections following as a *definition* of numerical identity—what in the monograph I called $N^=$ but was later canonically (though questionably)[6] retitled "Hume's Principle" by George Boolos (henceforward

---

[6] Questionably, since although Frege himself cites Hume, and the latter did indeed have the idea, roughly, of one-to-one correspondence, there is no evidence that he had any inkling of number as a second-level concept. For discussion, see Dummett (1998).

HP)—provides, when augmented by suitable definitions of *zero, natural number,* and *successor,* everything needed for a formal deduction in a suitable higher-order logic of the axioms of number-theory. Frege himself had given, in §§70–83, an informal sketch of how the definitions and reasoning might go, including crucially the famous recovery of the principle of arithmetical induction, based on his definition of the ancestral of a relation, and an argument that every number has a successor. But these sections were framed in terms of his identification of numbers with extensions of concepts—a move made in response to the apparent insufficiency of *HP* to resolve the truth-values of 'mixed' identity statements, like "Zero is identical to Julius Caesar"—and it seemed to me that the identification with extensions was actually playing no essential role in the informal reasoning Frege sketched. If the axioms of arithmetic could indeed be deduced, in a plausible second-order logic, just from *HP*, without the calamitous play with extensions, and if *HP* itself could indeed justly be regarded as a definition, there would be the makings of an interesting argument that arithmetic is indeed analytic in the sense characterized at *Grundlagen* §3[7] and a vindication of the core thesis of Frege's philosophy of arithmetic would be in prospect.

I had already persuaded myself that the deduction would go through more than a decade earlier at the time of writing the BPhil thesis, though the statutory word limits had prevented me from including in what I submitted for the degree very much of whatever the pencilled scribblings had been that had persuaded me. But now that I came to try to reconstruct a proof, I hit various snags and what I eventually came up with was complex enough to raise doubts that it might conceal mistakes and that it anyway must be distant from what Frege had in mind. It was several months until I summoned the nerve to turn it in to the publisher anyway, and almost seventeen years before the publication of the patient research by George Boolos and Riki Heck that explains the difficulties I had encountered: namely, Frege *didn't* actually sketch a valid proof of what we now call Frege's Theorem in *Grundlagen.* As Boolos and Heck put it,

§§82–3 offer severe interpretative difficulties. Reluctantly, and hesitantly, we have come to the conclusion that Frege was at least somewhat confused in these two sections and that he cannot be said to have outlined, or even intended, any correct proof there.[8]

Still, there is a valid if somewhat cumbersome proof outlined in the final section of *FCNO;* though better ways of doing it have of course since emerged.[9]

---

[7] "The problem becomes, in fact, that of finding the proof of the proposition, and of following it up right back to the primitive truths. If, in carrying out this process, we come only on general logical laws and on definitions, then the truth is an analytic one..." Frege (1884/1950; Austin's translation, p. 4[e]).

[8] From Boolos and Heck (1998), reprinted in in Boolos (1998: 316); and in Heck (2011: 69).

[9] The result is justly known as Frege's Theorem after Heck's painstaking demonstration (Heck 1993) that Frege's proofs of the axioms of arithmetic in *Grundgesetze* do in all essentials proceed from *HP,* without unavoidable recourse to extensions or his Basic Law V. Both Peter Geach (arguably, see Geach 1955: 569) and Charles Parsons (explicitly in Parsons 1965—see the reprint in Demopoulos 1995: 198)—had asserted the theorem some years before the publication of my book, but neither published a proof of it and there is no knowing whether both or either had seen past the inadequacies of Frege's *Grundlagen* outline to a cogent version (though my bet would be that at least one of them did.) My suspicion, though, is

The publication process for *FCNO* proved a frustrating experience. The publisher for the Scots Philosophical Monograph series was Aberdeen University Press, then a subsidiary of Pergamon Press owned by the late and now infamous Robert Maxwell. In those days you published a book by delivering a paper typescript to the publisher which would then be copyedited in ink before manual typesetting. It will have had little to do with Maxwell that the press employed a copyeditor who had so little sense of what they were reading as to see fit to re-invert my quantifiers throughout, rendering them as proper English upper-case letters, 'A' and 'E.' But it did, I imagine, reflect on his business strategy that the first run of my book went to only 500 copies and that, as I later learned, there was absolutely no intention to produce any more. The result was that *FCNO* rapidly became unobtainable, though it could not have been foreseen that second-hand copies would go on to be priced at the ludicrous levels (at one point, four figure US dollar sums on Amazon, Ebay, etc.) that they have attained in the intervening years.[10] I do not know if anyone has ever paid out such a sum for the book. I hope not. If so, I hope they felt they got value for money. I doubt they did.

*   *   *

1. Frege's Theorem

1.1 *Demopoulos*

So, what should one make, metaphysically, or epistemologically, of Frege's Theorem? My own view from the start has been that its primary significance is epistemological. Initially, I was content to express matters in terms of the notion of analyticity. As noted above, a deduction in pure logic of the axioms of arithmetic from something that could be justly regarded as akin to a legitimate definition (together with the supplementary definitions of *zero, successor*, and *natural number*), should count as a demonstration of analyticity in Frege's sense. But "analyticity" is, after all, a technical term. The epistemological significance of such a deduction would be that it would show how someone whose understanding of the cardinality operator, "The number of...," was encapsulated by *HP* could come to recognize the truth of the axioms of arithmetic *a priori*, purely by reasoning on the basis of that understanding. In short, the primary significance of Frege's Theorem would be that of a demonstration of the *epistemic analyticity* of the axioms of arithmetic, in the sense that Paul Boghossian has more recently usefully contrasted with the notion of *metaphysical analyticity*—truth "purely in virtue of meaning"—that bore the brunt of

---

that Michael Dummett, to whom my intellectual debt on these matters is infinite, and who makes a somewhat scornful reference to my efforts to reconstruct the proof (in Dummett (1991: 123), had not.

[10] This experience followed close on a similar disappointment with Duckworth's publication of my first book, *Wittgenstein on the Foundations of Mathematics*, Wright (1980), where the print-run had been extensive enough but the price had been set at a then exorbitant £35—the equivalent of about £170 at the present time. My experiences with publishers since have been nothing like as unsatisfactory.

As far as *FCNO* is concerned, I did eventually manage to reclaim the copyright for the book, and a second edition by Oxford University Press, with a new postscript, has long been planned.

Quine's influential animadversions.[11] And this, I reckoned, would near enough serve Frege's purposes.

Of course, this way of looking at the matter holds out two hostages. One is to the claim that *HP* is indeed "akin to" a legitimate definition. In a strict sense, a legitimate definition should provide for the elimination of its *definiendum* in any meaningful context in which it occurs. *HP* is certainly not a definition by that standard—at least, not if endowed with its intended generality—since, so endowed, it allows the cardinality operator to bind predicates in which that operator itself occurs, as for instance in Frege's definition of the number 1 as: $Nx:(x=Ny:y\neq y)$.[12] My contention had to be therefore that *HP* does serve, nevertheless, to *fix the meaning* of such occurrences of the cardinality operator as it fails to eliminate. This contention immediately confronts the Caesar problem, of which more below. But in addition, there is a natural concern whether anything one might legitimately regard as akin to an innocent definition could spawn a countably infinite population of objects; and a less immediate but no less troubling concern—what I dubbed the 'Bad Company' problem—later stressed by Riki Heck, George Boolos, Alan Weir, and others,[13] about the general pattern of this kind of "definition"—*abstraction*—prompted by the inconsistency of some examples (*Grundgesetze* Law V and a natural abstraction for the ordinals are two salient cases), at least when embedded in the relevant higher-order logic, and the incompatibility of other self-consistent examples with *HP*.[14]

While those matters have been thoroughly debated in the intervening years, the second hostage has received much less attention: the hostage held out to the claim that the higher-order logic whereby the proof of Frege's Theorem proceeds is legitimate logic. There are two components to that. Whether it is *logic* is, of course, important if what is at issue is whether Frege's Theorem serves the interests something worth calling "logicism." Quine's influence, again, memorably summed up in his remark about "set theory in sheep's clothing,"[15] led to a whole generation of philosophers doubting it, without perhaps being as clear as they should have been exactly what it was that they were doubting. But maybe the title to "logicism" is not terribly important. The claim of epistemic analyticity can be disengaged from the claim of logicism provided the higher-order *reasoning* whereby the proof of Frege's Theorem proceeds is legitimate and independent of arithmetic. The crucial question is whether that reasoning is sound and could in principle be productive of (first-time) knowledge of the axioms. Jim Edwards's and Riki Heck's Chapters 3 and 2 (this volume, respectively) bear directly on the issues arising here.

---

[11] Boghossian (1996). Boghossian is there content to jettison the notion of metaphysical analyticity. For one strategy for rehabilitating it, see Hale and Wright (2015).

[12] The effect is that we have no means to eliminate the occurrence of "$Ny:y\neq y$" in the right-hand side of the expansion, in accordance with *HP*, of the statement that the number of moons of the Earth is One.

[13] Boolos (1990; repr. in Boolos 1998: see esp. 2014–15); Heck (1992); Weir (2003).

[14] For example, Boolos's abstraction for *parities* (Boolos 1990), and mine for *nuisances* (Wright 1997). These principles were originally noted to be incompatible with *HP* on standard model-theoretic assumptions. Sean Ebels-Duggan (forthcoming) has recently shown that they are actually formally inconsistent with *HP*. For state-of-the art resumé and discussion of the Bad Company problem, see Cook (2016) and Cook and Linnebo (2018).

[15] Quine (1970: 66).

Although, however, Frege's own characterizations in *Grundlagen* of the notions of apriority and analyticity are overtly epistemological, there are clear signs of a different aspect to his thinking in that book suggestive of a kind of *metaphysical architecturalism* that has had (under the terminology of 'ground') a significant amount of recent attention from metaphysicians.[16] He writes, for instance, that

The aim of proof is, in fact, not merely to place the truth of the proposition beyond all doubt, but also to afford us insight into the dependence of truths upon one another. After we have convinced ourselves that a boulder is immovable, by trying unsuccessfully to move it, there remains the further question, what is it that supports it so securely?   (§2, p. 2^e)

The crucial notion in this passage is that expressed by the word—*Abhängigkeit*—that Austin translates as "dependence." Frege never shakes off a broadly Euclidean view of what should be accomplished by a good axiomatization of a mathematical theory. On such a view, the axioms should codify the *ground* of the truth of the theorems, the body of fact on which those truths rest. As I read William Demopoulos's characteristically subtle and interesting Chapter 1 (this volume), it too is animated by elements of this way of thinking. Demopoulos writes,

Although Frege's attempted demonstration of the analyticity of arithmetic would indeed show that the basic laws of arithmetic can be justified by those of logic, the principal interest of such derivation is what it would reveal regarding the dependence of arithmetical principles on logical laws.   (Chapter 1, this volume, p. 6)

He goes on:

We should distinguish two roles that sound or truth-preserving derivations are capable of playing in a foundational investigation of the kind Frege is engaged in. Let us call derivations that play the first of these roles *proofs*, and let us distinguish them from derivations that constitute *analyses*. Proofs are derivations which enhance the justification of what they establish by deriving them from more securely established truths in accordance with logically sound principles. Analyses are derivations which are advanced in order to clarify the logical dependency relations among propositions. The derivations involved in analyses do not purport to enhance the warrant of the conclusion drawn, but to display its basis in other truths.

(Chapter 1, this volume, p. 6)

Naturally, any logically sound derivation from premises will succeed in clarifying *a* logical dependency relation between those premises and its conclusion. But not *every* such derivation will succeed in disclosing something worth describing as a *basis*. Demopoulos's notion of an analysis is an investment in the idea that some truths of mathematics may provide the foundations for others, in such a way that derivation of the latter from the former discloses the grounds of their truth.

I completely agree that this idea is Fregean and actively at work in *Grundlagen*. It is, however, strictly independent of any thesis about how the axioms of a mathematical theory may be known, since there is no reason why the basis of a particular truth, in the metaphysically architectural sense gestured at, should also be *epistemically* prior to it—something one can know first and thereby advance to knowledge of

---

[16] Thus, for example, Fine (2001, 2012), Schaffer (2009), Rosen (2009).

the latter. We should distinguish between the claim that X describes the basis for the truth of Y and the claim that Y may be known on the basis of knowledge of X. I think the author of *Grundlagen*, if asked, would say that he wanted to make a claim of both kinds. However, the thesis of the epistemic analyticity of arithmetic, based on Frege's Theorem and the contention that *HP* may justly be assimilated to a definition, is a claim only of the second kind. I am tempted to interpret Demopoulos, notwithstanding the epistemological flavour of much of his language, as wishing, in the end, primarily to advance a claim of the first kind. Let me explain why.

The passage in *Grundlagen* quoted in note 1 above continues as follows:

> If, however, it is impossible to give the proof without making use of truths which are not of a general logical nature, but belong to the sphere of some special science, then the proposition is a synthetic one. For a truth to be a posteriori, it must be impossible to construct a proof of it without including an appeal to facts, i.e., to truths which cannot be proved and are not general, since they contain assertions about particular objects. But if, on the contrary, its proof can be derived exclusively from general laws, which themselves neither need nor admit of proof, then the truth is a priori.

In Demopoulos' view, Frege's Theorem has a significance quite independent of any issue about logicism, or neo-logicism, or the analyticity of arithmetic. Specifically, the proof makes a case, in his view, for the *apriority* of the axioms of arithmetic in the light of the conception of the *a priori* marked in the last two lines of the quoted passage. The standard caption of the *a priori*, as that which can be known "independently of experience," encounters, as is familiar, a raft of difficulties.[17] (How, for these purposes, are we to characterize 'experience' and exactly what mode of 'independence' is germane?) For Frege, by contrast, the hallmark of the *a priori* is *generality*: a proof establishes a proposition *a priori* when its premises are "general laws, which themselves neither need nor admit of proof." The contention that arithmetic is *a priori* in this sense encounters the immediate obstacle that it presents as having a specific, particular subject matter—indeed, one of the axioms apparently concerns a specific object, the number zero. But this obstacle is finessed if arithmetic can be shown to follow from "general laws, which themselves neither need nor admit of proof." If *HP* ranks as such a law, does not Frege's Theorem supply exactly the needed argument?

Well, the contention that Frege's theorem reveals arithmetic to be *a priori* in these terms needs, as noted, to be distinguished from the contention that it displays the ground for the truth of the arithmetical axioms. It is not that these claims are necessarily in tension one with another—not at all. But they are not the same and I think that while Demopoulos's discussion suggests an arresting and indeed plausible case for the second, he outlines only a much weaker case for the first.

Let me try to justify that. In order for Frege's Theorem to make, by the letter of Frege's characterization of the *a priori*, a convincing case for the apriority of arithmetic, three questions need to receive affirmative answers. Does *HP* have the kind of generality germane to Frege's intent in his characterization of the *a priori*? Has it no need of proof? And does it admit of no proof? Let me take them in turn.

---

[17] For elaboration of some of them, see Hawthorne (2007) and Williamson (2013).

HP is certainly *syntactically* general. It has two second-order universal quantifiers in prenex position. But Frege speaks of general *laws*. How exactly is the principle law-like? The natural answer is that its content is *prescriptive*: it lays down the condition under which a pair of concepts may be said to have the same number of instances, and thereby goes to determine what sameness of number consists in. That answer, though, takes us in the direction of the idea that *HP* is "akin to" a definition, and thus may be at service of a demonstration of the epistemic analyticity of the axioms of arithmetic—the idea with which Demopoulos advertises his reconstructive proposal as contrasting, or at least as neutral about.

There is, indeed, a different, very significant kind of generality that *HP* exhibits: it manifests that the notion of cardinal number applies as a measure to any concept (or, if you will property, or set) that lies within the range of its initial higher-order quantifiers—and hence displays the universality of which Frege speaks when he writes of the fundamental laws of arithmetic that

we have only to try denying any one of them, and complete confusion ensues. Even to think at all seems no longer possible. The basis of arithmetic lies deeper, it seems, than that of any of the empirical sciences, and even than that of geometry. The truths of arithmetic govern all that is numerable. This is the widest domain of all; for to it belongs not only the actual, not only the intuitable, but everything thinkable. (*Grundlagen* §14.)

This is a point of great importance both to Frege and to Demopoulos's reconstruction, and I'll come back to it shortly. The immediately relevant consideration, however, is that its manifestation of generality of this kind—the generality of the range of the applicability of the notion of cardinal number—does not seem to invite any conclusion about the status of *HP* as a general law.

Moreover, a critic may continue, we must not overlook the role of the definitions—or what were presented as definitions—of the arithmetical primitives in the proof of Frege's Theorem. One of those is the definition of zero:

$$0 = Nx:x \neq x$$

—zero is the number of non-self-identical things. That looks to "contain an assertion about a particular object," and it plays an indispensable role in the proof. So the proof of Frege's Theorem is not one that relies exclusively on general laws, as it must do if it is to be at the service of a demonstration of apriority in Frege's sense. If it is countered that the above is merely a characterization of the meaning of '0,' the critic will reply that it anyway assumes that 'Nx:x≠x' is a successful referential singular term. *HP* does indeed entail that it is.[18] But is it consistent with the status of something as a general law, in the spirit of Frege's characterization, that it entail the existence of particular objects?

Next, the suggestion, in response to the second question, that *HP* has no need of proof, when advanced in the context of a purported demonstration of the apriority of arithmetic, may ring hollow for "bad company" reasons. Frege himself, confronting the collapse of the authorization for the principle that he had supposed was provided

---

[18] Take both *F* and *G* in *HP* as: x≠x. Appeal to the reflexivity of one-to-one correspondence to infer from *HP*, right-to-left, that Nx:x≠x = Nx:x≠x. Existentially generalize on one of the terms in that identity.

by grounding it in Law V, evidently felt that it retained no authority of its own for his foundational purposes; Frege was no 'neo-Fregean.' But given its apparently close kinship to Law V and other "bad apples,"[19] surely its serviceability for Demopoulos's announced purpose requires at least a proof of consistency. True, we have one—or at least, a proof of consistency relative to second-order arithmetic.[20] But, assuming that is the best we can do, there is now a threat of epistemic circularity hanging over any putative demonstration of the apriority of arithmetic based on *HP*. To wit: confidence that *HP* is indeed a general law, in the spirit of Frege's intent, will require confidence in its truth, which will require confidence in its consistency, which will require confidence in the consistency of second-order arithmetic; and the last, if we have it, will arguably, after Gödel, have to be based on our conviction of the truth of second-order arithmetic under its intended interpretation. But a demonstration of apriority in the sense that Frege is gesturing at when he speaks of "finding the proof of the proposition, and of following it up right back to the primitive truths" and then finding the latter to consist in "general laws that neither need or admit of proof" surely requires that we have a grounded confidence in the latter *independently* of our confidence in the proposition proved—even if we are already perfectly confident of the latter. Or so my imaginary critic will contend.

Finally, is it correct to say that *HP* admits of no proof? What this requires, in context, is that any other principles which service a valid deduction of *HP* will include at least some whose epistemic status is in no way stronger than that of *HP* itself, so that there is no improving one's epistemic relation to *HP* by deducing it from something else. However, even if it is granted that *HP* admits of no proof in that sense, it is clear that what Frege intends for the purposes of a demonstration of apriority are premises which are somehow *indubitable*. Now of course, as Demopoulos notes, Frege is, in all his writings, notoriously reticent on what bestows such a status on a first principle, whether logical or not, and on how we are supposed to recognize it. But that is no cause for congratulation, and in order to make the intended case for the apriority of arithmetic, Demopoulos surely has to do better. A case that Frege's Theorem demonstrates the apriority of arithmetic must substantially consist in argument that the first principle involved—*HP*—has a suitable epistemic status. But Demopoulos doesn't embrace that project in his chapter; his observations about *HP* take a different tack.

That is why I think that, not withstanding his official description of the reconstruction of *Grundlagen* that he proposes as a logicism-unrelated argument for the apriority of arithmetic, the case Demopoulos musters is best interpreted as making a metaphysical point rather than an epistemological one. The key to it is the

---

[19] Thus George Boolos (1990: 293): "It is a weighty assumption of Frege's . . . that the first-level concepts can be mapped into objects in such a way that concepts are mapped onto the same object only if they are equinumerous, and *it is a lucky break that the assumption is even consistent*" (Boolos's italics). A similar sentiment is expressed in Boolos (1987)—see the reprint in his (1998: 175–6).

[20] I conjectured that *HP* is consistent in Wright (1983) and argued that no contradiction is obtainable from it in anything like that manner that afflicts Law V. But I did not attempt there to prove its consistency. Boolos (1987) shows that *HP* is interpretable in second-order arithmetic. Burgess (1984) earlier gave a simple model-theoretic argument for the consistency of *HP*, albeit one based squarely on the presumed coherence of our conception of the structure of the natural number series.

observation that *HP* codifies the *applications* of cardinal number, that it "effects an analysis of the basic laws of pure arithmetic by revealing their basis in the principle which controls the applications we make of numbers in our cardinality judgements."[21] Demopoulos takes this to be a philosophical insight that is made salient by John Bell's set-theoretic proof of (something tantamount to) Frege's theorem.[22] Crudely summarized, Bell proves that any function on a collection of subsets of a set which mimics the applications of a cardinality operator—counts them as equivalent just when they are one-to-one correspondent—will, provided the collection meets certain conditions he specifies, single out an omega-sequence—something with the structure of the natural numbers. The structure of the finite cardinals, as characterized by the axioms of second-order arithmetic, is thus imposed by their proper applications.[23]

Now, I must confess that it is not clear to me that this point is made *more* salient by Bell's set-theoretic way of going about things than by a regular higher-order logical deduction of the axioms of arithmetic from *HP* and the appropriate definitions. Nor am I sure that the point is best served by invoking the idea that *HP controls* "the applications we make of numbers in our cardinality judgements" inviting, as it does, engagement with the question of whether, or to what extent, we should think of the principle as active in our actual cognitive processing. But the important thing is the point itself: the suggestion that Frege's Theorem shows that the truth of the axioms of arithmetic, the existence of an instance of the structure that they characterize, is grounded in a principle that encapsulates the role of zero and its successors as measures of cardinality.

There are, of course, philosophers who are sceptical about any notion of metaphysical ground—what in this context Demopoulos gestures at by "logical basis." And matters are, perhaps, especially thorny when the claim under review is that one (presumed) *necessary* truth provides, in a metaphysical sense, the basis of or ground for others. Is not a proposition's necessity in tension with the very idea of its having any kind of basis or support? Well, perhaps there is a tension with the idea that its truth is somehow propped up, as it were. But that need not be involved. One natural way of getting a feel for a relation of the right kind is via the notion of *explanation*—in a sense of "explanation" akin to that we already apply within mathematics whereby some proofs, but not all, are regarded as explanatory of what they prove. The topic is difficult and controversial,[24] but the phenomenon, of proofs which seem to give an insight into *why* what they establish holds good, is generally acknowledged. The

---

[21] Chapter 1, this volume, p. 11.   [22] Bell (1999).

[23] Less crudely, what Bell shows is the following. Let E be any (non-empty) set, and let v be a function from its subsets to its members. Specifically, let v take as arguments any of a family, G, of finite subsets of E which contains the empty set and is closed under union with singletons of members of E (so that if U is a member of G, and x is a member of E that is not a member of U, then the union of U with {x} is a member of G.) Then if v satisfies the following condition:

$$v(U) = v(V) \text{ iff. } U \approx V.$$

where '$\approx$' denotes bijection, Bell proves that we can define a subset N of E which is the domain of a model of the axioms of arithmetic.

[24] See for instance Steiner (1978); Mancosu (2001); Lange (2016).

construction in Euclid's proof that the interior angles of a triangle sum to 180°
plausibly explains *why* they do so (by demonstrating their respective equivalences,
for an arbitrary triangle, to angles which together compose the angle on a straight
line) and thereby shows how triangularity itself grounds this characteristic. By
contrast the construction in Euclid's clever proof of the infinity of the primes does
not seem explanatory in the same way. What would perhaps be explanatory in the
same way would be a proof based on defining a function that for primes, and only
primes, always generates a larger prime—one that uniformly finds a larger prime for
each prime number in a way that essentially exploits just the fact that the latter is
prime. But what Euclid gives us is only a demonstration that any *number n* must be
smaller than some prime—either $n!+1$ or one of the latter's prime factors. Infinity is
not shown to be rooted in primality as having interior angles summing to 180° is
shown to be rooted in triangularity.

I suggest that Frege's Theorem explains the truth of the axioms of arithmetic in a
way that stands favourable comparison with Euclid's theorem about the interior
angles of a triangle. It shows how it follows from what the numbers *essentially* are—
measures of cardinality—that they satisfy exactly those axioms. It is an *explanatory*
proof, in the sense just gestured at, whose premise concerns *essence*. I suggest that in
so far as the idea comes naturally to us that there can be grounding relations among
necessary truths, it is the possibility of proofs that combine these two features which
underwrites it.

I offer that as a way of framing the core thesis of Demopoulos's chapter. There is,
however, a second element of his discussion which merits separate comment. Early
on[25] he announces a second principal contention:

> The objectivity of arithmetic can be secured by adherence to principles which, although
> unquestionably Fregean, are independent of the thesis that numbers are logical objects, and
> involve only a minimalist interpretation of the thesis that they are objects at all.

We should want to agree with the first part of that. Without fussing too much about
what should be meant by 'objective' in this context, it is reasonable to suppose, first,
that ordinary applied judgements of number are characteristically as objective as the
other concepts involved allow—so that it is, for example, an objective matter that the
Earth has exactly one moon, but perhaps not so much so that three of the second
series of episodes of *Fawlty Towers* were much funnier than the others—and, second,
that this objectivity is a reflection of that of the principle that encodes their correctness conditions. So the axioms of arithmetic should inherit, by Frege's Theorem, that
objectivity.[26] But the second part, as Demopoulos says, marks a point of divergence
from my treatment of the numbers in *FCNO*, presumably because that took the idea

---

[25] Chapter 1, this volume, p. 3.

[26] Compare Demopoulos at p. 22 of Chapter 1, this volume:

> Here then is the bearing of our reconstruction on the *objectivity* of arithmetic. Finite Hume
> is not an arbitrary postulate, but is the principle which controls our ordinary judgements of
> cardinality. But these judgements are objective; hence so also is Finite Hume, since it is
> the principle on which their evaluation depends. The basic laws of arithmetic thus inherit
> *their* objectivity from being founded on the principle that controls our ordinary cardinality
> judgements.

of numbers as objects very seriously. However, since I too would want to describe the notion of object that I aimed to develop in that book as 'minimalist,' some clarification is in order.

The *FCNO* conception of object is logico-syntactically driven. According to it, the notion of a singular term (a Fregean *eigenname*) is conceptually prior to that of an object: the latter is to be explained in terms of the former. An object is anything for which a singular term may stand and reference is, as it were, imposed on a term by its occurrence in referentially transparent position in true (par excellence, atomic) statements. Since terms formed by application of the cardinality operator to concept-expressions do so occur in suitable statements in Frege arithmetic—the system formed by adjoining *HP* to a suitable logic—they do refer and their reference is to objects.

I will not at this point offer anything by way of further defence of this way of looking at the matter.[27] But I need to remark on one thing that is, I think, required by it and one thing that is not. The thing that is required is this: if this conception is to apply to numerical singular terms as introduced by *HP*, then the obtaining of one-to-one correspondences, as described by instances of the right-hand side of *HP*, must be *tout court* sufficient for the successful reference of those terms. No hostage must remain to the possibility of a world in which there are one-to-one correspondences between (finite) concepts but the corresponding numerical terms fail to refer. It must thus be of the essence of any concept—at least, any within the range of the higher-order quantifiers in *HP*—to *have* a number. And this, I suggest, forces us to think of numbers minimalistically, as things whose whole being is to be *of* concepts, as it were, and guaranteed thereby, as straight lines are guaranteed to have a direction, human beings are guaranteed to have an age, and geometrical figures a shape. None of these can be entities which might present themselves to one under some other guise, or which one might encounter armed only with some other, imperfect concept of what they are. Rather they are, as I once put it, shadows cast by the syntax of the relevant parts of our language; albeit, like shadows, none the less real for that.

On this conception, *abstracta*—the entities referred to by singular terms introduced by *HP* and other (acceptable) abstraction principles—are real but 'thin': things whose nature is fully disclosed by their role as the referents of the terms in question. In this respect they are comparable to properties conceived as 'abundant,' so that any well-defined predicate is guaranteed to express a property, and what property it expresses is fully disclosed in the sense—the satisfaction-conditions—of that predicate. So this is one "minimalist interpretation" of the thesis that numbers are objects. Demopoulos's "minimalist interpretation," by contrast, is the view that the terms introduced by *HP* do not need to have any specific singular reference at all. He concludes his chapter thus:

The objectivity the analysis imparts to arithmetic is altogether independent of Frege's conception of numbers as objects except in the very weak sense that they must be arguments to concepts of first level. Such an account of arithmetic's objectivity is entirely compatible with

---

[27] The general conception is the topic specifically of chs 1 and 2 of *FCNO*. It is further elaborated and defended throughout Bob Hale's *Abstract Objects* (Hale 1988) and in chs 1, 2, 6, and 7 of our joint *The Reason's Proper Study* (Hale and Wright 2001b). Hale's and my most recent discussion of it is Hale and Wright (2009). I'll have more to say about it below.

the fact that neither [Frege arithmetic] nor [Peano arithmetic] single out the domain of a particular natural number structure as *the* numbers; and neither the objectivity of arithmetic nor its generality rests on appeal to the idea that numbers are logical objects.[28]

The notion of a "logical object" is somewhat up-for-grabs after the demise of Law V, but I believe that Demopoulos might as well have written "specific" for "logical" in that sentence. As evidence, note his earlier remark that

it is not necessary to single out the numbers uniquely. Any *E* which supports a family—including the smallest such family—of weakly finite subsets satisfying the conditions of Bell's Theorem can serve as the domain of a Frege structure. Frege's original question, "What are the numbers?," has . . . been replaced by another, namely, "What is a natural number type structure?"

That we should not think of what are apparently numerical singular terms as genuinely so—that they are merely, as it were, signposts to places in a structure-type, which absolutely any objects might realize—is of course a venerable, distinguished and, as it has proved, resilient view. And that no departure from it is imposed merely by recognition of the objectivity and generality of arithmetic is true. However, if the explanation of those features is grounded in an inheritance of them from *HP* via Frege's Theorem, then the explanation is consistent with such a structuralist conception of the semantics of the terms only if *HP* is. Since a full understanding of numerical terms, as given by *HP*, may proceed in complete innocence of any grasp of Frege structures, or omega sequences, and the like, that looks to be doubtful; and the impression to the contrary provides a cautionary example of how one can easily be misled into reading into the meaning of a range of expressions what are merely features of their model theory.[29]

1.2 *The Epistemological Status of Hume's Principle*

Let us return to the question of the epistemological significance of Frege's Theorem. If we bracket questions to do with the status of the underlying logic used,[30] it is overwhelmingly natural to think, as in *FCNO* and for a long time subsequently I certainly did think, that the issues here just come down to those concerning the epistemological status of *HP* itself, with any concomitant case for the apriority of arithmetic being dependent on whether *HP* can itself rank as knowable *a priori* even if neither a definition, nor a truth of logic, strictly understood. So the issues may seem straightforward: Do we (can we) know *HP* at all? If so, do we (can we) know it *a priori*? If so, how?[31]

---

[28] Chapter 1, this volume, p. 22.

[29] Much more needs to be said of course, and some more, specifically concerning the structuralist interpretation of numerical singular terms, is said in *FCNO* §xv, pp. 117–29. A further, egregious example of the more general tendency to overplay model-theory, I would contend, is the Quinean idea that second-order logic is crypto- set theory. We'll come to that below.

[30] I'll come to them shortly.

[31] There is, of course, another major question about the significance of Frege's Theorem—and indeed about the significance of any technically successful abstractionist project, even for someone who is content that the possibility of *a priori* knowledge of a branch of pure mathematics would have thereby indeed been demonstrated. It is a question a version of which confronts any project in reconstructive epistemology; in

Well, there is, to be sure, the thought advanced by Demopoulos[32] that, at least when the range of its higher-order variables is restricted to finite concepts, *HP* simultaneously presents both a canonical account of how the content of pure arithmetical statements provides for their applications and a correct analytic digest of our conception of the structure of the natural numbers. Finite *HP*, at least, thus has a strong case to be a correct codification of—analytic *of*—everything essential to pure and applied arithmetical thought.[33] But attractive as this answer may be, it puts the cart before the horse as far as the abstractionist project is concerned. For the purposes of that project, at least as understood in *FCNO*, the justification for *HP* cannot turn on its claim to reflect and encode an antecedent body of arithmetical knowledge. Rather its epistemological merits have to be accessible not just *a priori* but, at least in principle, *in advance of* that body of knowledge. In particular, it has to be possible, at least in principle, to *learn* of the truth of the Dedekind–Peano axioms by the derivation of Frege's Theorem.

So, for abstractionist purposes—and assuming we are not, as to his cost was Frege himself, in the market for its derivation from yet more ultimate and basic principles—it may seem that we need an account of how *HP* might be known *noninferentially yet a priori* by someone so far innocent of the axioms of arithmetic. One immediately salient candidate account, accordingly—and perhaps the only one—is the proposal that *HP*'s *a priori* credentials are those of a successful *implicit definition*: in effect, a stipulation whose effect is so to fix a concept of the meaning of the sole hitherto undefined term in its statement—the cardinality operator—that the truth of the principle comes to be knowable *a priori* just in virtue of our prior understanding of the attendant logical vocabulary, our acquired understanding of the meaning of the operator, and our grasp of the syntax of the statement itself. This was, to all intents and purposes, the way I was thinking of it in *FCNO*, and Bob Hale and I subsequently defended the proposal in some detail.[34] Defending it requires making out a connection between implicit definition and *a priori* knowledge in general. It involves explaining how, in at least some cases, stipulation, or acceptance, of an implicit definition can so constrain the understanding of the very sentence(s) which provide(s) the vehicle of the definition that *a priori* knowledge of the truth of what is

---

this instance, what does it do for the status of our ordinary mathematical beliefs if, irrespective of how we actually arrive at them, some philosopher is able to work out an ideal route—very different to anything we actually do—and a persuasive case that if we were to follow it, we would indeed wind up with (much of) the knowledge that we think we actually have in the relevant region? After all, one may in fact possess only a posteriori knowledge of things that can be known *a priori*. Can considerations be marshalled to make a case that successful abstractionist constructions can somehow "rub off" on the status of, for instance, our actual arithmetical knowledge? I think this question is long overdue much more explicit attention than it has received in the 'neo-Fregean' literature to date, but I shan't say more about it here, except to remark that it already faced Frege. (For forceful but disparaging emphasis on the trend of neglect of this issue, see Heck 2011: 158–60.)

[32] In this volume but also in Demopoulos (1998 and 2000); compare Heck (1997a). Related ideas feature in ch. 1 of Heck (2011: 35–7).

[33] It would of course need further argument that knowledge *a priori* of Finite *HP* may be delivered merely by the reflection that it provides such a codification. As the example of Basic Law V arguably shows, a principle may be analytic *of* a concept and yet false.

[34] Hale and Wright (2000).

expressed is the result; and it then requires defending the more specific claim that *HP*, and other suitable abstraction principles, rank as *pukka* implicit definitions of this kind. However, although I continue to have confidence in the idea that implicit definition can constitute one source of basic—non-inferential—*a priori* warrant, I no longer want to rest on the claim that a complete vindication of the possibility of *a priori* knowledge of *HP* in particular can be accomplished just on that basis. But nor, now, do I think that a complete vindication of the possibility of *a priori* knowledge of *HP* is anyway *required* before knowledge *a priori* of the basic laws of arithmetic, based on Frege's Theorem, might legitimately be claimed. I now reject, in other words, the natural thought noted above that abstractionist foundations for arithmetic, based on *HP*, can have their intended epistemological significance only if *HP* itself is first known *a priori*. Let me try to make good on that, perhaps slightly startling, claim.

I need to invoke at this point a notion of *epistemic entitlement* that I have canvassed in other work; specifically, what I have elsewhere called *entitlement of cognitive project*.[35] The notion needs substantially more elaboration and defence than I propose to try to give it here[36] but the important points about entitlements for present purposes are four:

(1) Entitlement, as I intend the notion, contrasts with knowledge, or justification, or any kind of epistemic warrant that is based on a specific cognitive achievement: an entitlement confers a rational warrant to place trust in a proposition or methodological principle, for the purposes of a certain enquiry or range of enquiries, *without any independent evidence of its truth or soundness*. An entitlement in this sense is, in effect, a warrant to take something for granted.

(2) Entitlements are restricted to *authenticity-conditions*, where something—a proposition or methodological principle—is a authenticity-condition of a given enquiry, or range of enquiries, just if any advance doubt about its truth, or soundness, would rationally commit one to doubting, to the same degree, the efficacy or significance of the enquiry.[37] An authenticity condition, so understood, is in effect what Jim Pryor has called an *anti-underminer*.[38] I shall refer to propositions that express authenticity-conditions for a particular enquiry, or cognitive project, as *presuppositions* of it. But this is not to suggest that in running the enquiry, we need have taken up any particular attitude to any particular such proposition, or even considered it. That a particular authenticity-condition is satisfied is presupposed only in the sense that, *if* it is considered, then taking it to be satisfied is rationally required if the enquiry is to be reckoned to be capable of delivering results in good standing.

(3) That such a presupposition is met may, of course, be something we can independently investigate. An entitlement, though, when we have one, precedes any such investigation, even should one be possible. We are *entitled* to a

[35] Originally in Wright (2004a,b).

[36] See Wright (2004a) for first steps towards a fuller account. The notion is further elaborated and defended against a range of objections that have subsequently emerged in the literature in Wright (2014).

[37] See Wright (2014: 214 f.).   [38] Pryor (2012).

presupposition, for the purposes of a certain cognitive project or projects, just when two conditions are met. First (i) there must be no extant sufficient reason to doubt the presupposition—to think it false or unsound. Should such reason emerge, the entitlement will be defeated. Second (ii) *either* the presupposition must be out of range of any independent justification *or* the project of seeking justification for it must be such as to implicate further presuppositions of no more secure an antecedent status, whose investigation in turn, if possible at all, would implicate yet further presuppositions of no more secure an antecedent status.... The overarching thought, which I believe is a central motivation of many of Wittgenstein's remarks *On Certainty*, is that it is of the essence of any enquiry to rest on presuppositions—things implicitly taken for granted in the significance we propose to attach to the enquiry— which either themselves lie beneath investigation or are such that it is, in context, futile, because regressive, to investigate them in any spirit that seeks to place the enquiry on a stronger, that is, evidentially more robust footing.

(4) Entitlements, so characterized, are *project-relative* and *conditional*. They are licenses to trust, or take for granted, *if* one proposes to undertake a particular project, or any project of a certain kind. There may be epistemic or other reason not to undertake a project. Cognitive projects may be badly conceived—the method of answering may be flawed, or maladapted to the question concerned; or, like any other projects, a cognitive project may be pointless, or dangerous, or wasteful. It is, however a non-negotiable part of our rational nature to undertake enquiry.[39] The alternative is a form of intellectual (and thereby bodily) suicide. That simple consideration, I propose, issues in a *right*: to undertake cognitive projects absent reason to think that they will be ineffective, and absent overriding moral or other practical reason to abstain. Say that a project that meets that condition is *unimpugned*. Then when a cognitive project is unimpugned, and when P is a presupposition of it that meets conditions (i) and (ii) above, it is rationally permissible to undertake the project and rational for one who undertakes the project to trust that P is satisfied.

Let me run through the key ideas one more time. Entitlements of cognitive project are all presuppositions of the kind I outlined—conditions such that a doubt about their obtaining would be rationally sufficient for a proportionate doubt about the competence or significance of the particular cognitive project in hand. What makes such a condition into a rational entitlement for someone undertaking the project, it is proposed, is a combination of three factors: first, that the project is unimpugned— that it is one that the agent has an undefeated right to undertake; second, that no information is possessed which would warrant doubt that the presupposition was met; and third, that any project aimed at verifying that it was met, if feasible at all, would implicate further presuppositions of no more secure an antecedent standing... and so on indefinitely. In such circumstances, to run the original cognitive project, and to take its findings on board, is, to be sure, to run a risk—but an

---

[39] Enoch and Schechter (2008) defend the interesting claim that enquiry is actually a rational *obligation*.

unavoidable risk, a risk of a kind that it is part of being a living rational agent to be prepared to undertake and which goes with the right to enquiry itself.

I have argued elsewhere[40] that entitlements, so conceived, embrace presuppositions of the proper functioning, on an occasion, of our perceptual and intellectual faculties and memories, presuppositions of the conduciveness of the prevailing circumstances to the successful operation of those faculties, and presuppositions of the good standing of the concepts essentially exercised in grasping the cognitive project concerned and carrying it through. But there are strong grounds for admitting a fourth kind of case, viz. that we also enjoy entitlements to rely on the soundness of the *basic inferential machinery*, if any, involved in the execution of a project. Not that, if a rule of inference is challenged, we may not often be able to address the concern. But addressing it is going to involve inference, and, familiarly, very often a seemingly unavoidable reliance upon a principle of inference in a meta-language of the very same pattern as the rule under scrutiny. Since any legitimate concern about the original rule should, manifestly, not be assuaged by meta-theoretic reasoning of the very same pattern, it follows that condition (ii) is met and that, at least in cases where we have no antecedent reasons for misgivings about the rule, reliance upon it may be regarded as an entitlement.

There are subtleties here about which exactly might be the rules of inference to reliance on which we may regard ourselves as entitled in this way, and to what extent a principled demarcation can be made between them and rules the right to use which has to be cognitively earned. But however the discussion of those matters may go in detail, it's plausible to assume that, for instance, *modus ponens*, and maybe some form of conditional proof, will rank as basic entitlements of the intended kind if anything does. Notice, crucially, that in saying this, I am implicitly rejecting one historically quite common conception of the epistemological ground for our acceptance of such basic rules. In classifying the acceptance of such rules as entitlements of cognitive project, we affirm (i) their presuppositional status in a given (very large) range of cognitive projects, together with the considerations (ii) that we are possessed of no reason to call their soundness into question, and (iii) that, were an attempt to justify them to be made, it would necessarily involve reliance on an inferential apparatus of no more secure an antecedent standing—(actually, in the case of the conditional, a reliance on essentially the very same inferential apparatus). But reflect that the last point is simply wrong if there is another, *non-inferential* way whereby the soundness of our basic rules of inference might be recognized. Just that possibility is embraced by the venerable thought that the validity of our most basic rules of inference is given to us by a kind of *rational insight* or *intuition*. I make so bold as to suggest that the venerable thought should be set aside. Let me briefly review some of the relevant issues.

To be clear, I am not suggesting that there is no role for some form of non-inferential *a priori* insight *anywhere* in a satisfactory epistemology of logic and mathematics. But any faculty that enables an agent to recognize truths needs to operate in a context where the truths in question are antecedently understood—you need to understand

---

[40] Wright (2004a).

the proposition that your car is parked on Elm Street, for instance, or that 196 is the square of 14, before you can bring your perceptual, or arithmetical, capacities to bear on their verification. So too, in the present instance, a faculty of intuition apt for the recognition of the validity of basic rules of inference, however exactly conceived, would need to go to work in a context where a thinker *antecedently* fully understood the conditional, for example, but was so far *open-minded*—had no particular view— about the status of, say, *modus ponens*. Is there any such possible state for a rational thinker to occupy?

One kind of historically very influential conception of the meanings of the logical constants says not. On such an *inferentialist* view, it is constitutive of an understand- ing of the conditional to acknowledge the basic rules that govern it in inference— including, presumably, *modus ponens*. So an understanding of the conditional cannot coherently be supposed to be separable from a willingness to acknowledge that *modus ponens* is valid. If it could, there ought to be such a thing as understanding the conditional perfectly yet—because of a failure of one's intuitional faculty rather than of understanding—failing to be arrested by the validity of the rule. But, the inferentialist will contend, there is no such possibility, for no one will *count* as grasping the conditional unless they are willing and able to acknowledge the validity of its basic rules. And that means that here there is no work for a putative faculty of intuition to do—no epistemic space for it to work in.

Call that the *squeezing argument*. It is decisive if the relevant form of inferentialism is accepted. But should it be accepted? Timothy Williamson has argued at length that there are no constitutive connections between understanding and assent of the kind that the squeezing argument relies on.[41] Manifestly, for instance, someone *can* grasp the conditional and yet clear-headedly refuse to acknowledge *modus ponens*. As he observes, Vann McGee is living proof![42] Indeed, Williamson argues that for any pattern of inference, however obvious-seeming and elementary, sense can be made of the possibility that someone fully understand it yet not be disposed to acknowledge its validity.

The example of McGee and *modus ponens* is very apt for Williamson's purposes. But stated in full generality, his claim is, to say the least, an adventurous one. What would someone who stoutly maintained an agnostic attitude to, say, conjunction introduction have to do if they are to satisfy us that they perfectly well understand *and* nevertheless? What kind of behaviour could justify that assessment? Maybe there are no unqualified categorical links between understanding and assent of the kind Williamson is targeting, but the suspicion remains that there are more nuanced connections that continue to squeeze the space in which intuition might work. There is little doubt that a clear-headed and deliberate rejection of a basic, uncontroversial rule of inference will normally be taken as canonical evidence of misunderstanding *unless* at the very least, as in the McGee case, backed by an intuitively arresting counterexample or articulated theoretical reservations. This suggests a relationship between understanding the operator concerned and acknowledgement of certain

---

[41] Williamson (2003), (2006), and (2007: chs. 3 and 4).   [42] McGee (1985).

rules of inference that is still suspiciously intimate if intuition, like vision or memory, is supposed to operate as a discrete, superintendent cognitive faculty. Perhaps, for example, the understanding of a logical operator constitutes a *disposition* to assent to its basic proprietary inferential rules, albeit a disposition that is properly defeasible by suitable theoretical considerations.[43] In that case there will still be no work for intuition to do in mediating the transition from understanding to acknowledgement of the rules.

It is no part of my brief here to defend an inferentialist account of our knowledge of the validity of basic rules of inference. It is one thing to make out some kind of constitutive connection between the understanding of a logical operation and the acceptance of certain appropriate inferential rules, quite another—familiarly, much more difficult—project to attempt to parlay that connection into the conclusion that such understanding-grounded acceptances should rank as *knowledgeable*. Inferentialism about logical content is one thing, inferentialism about logical knowledge quite another. But it is clear in any case that Williamson's objection targets a sub-optimal version of the former, content-inferentialist thesis. The content-inferentialist should not deny that, consistently with understanding an operation, doubts may rationally be entertained about the validity of what are in fact meaning-constituting inferential rules for it; and she should not deny that commonly explicitly accepted inferential rules may prove to be out of kilter with each other,[44] or with those rules that actually constitute the meaning of a targeted logical operation. That a given group of rules of inference are valid is, after all, a *theoretical* claim, and may be called in question by bad though well-motivated theory, or indeed—should they indeed contain some hitherto unnoticed flaw—by good theory. What content-inferentialism cannot accept is that someone might grasp the conditional, for example, and yet have no inclination, even in the absence of any theoretical doubt about their validity or coherence, to practice in accordance with what are *in fact* the rules of inference that individuate that operation.

I suggested above that some suitably qualified form of understanding–assent linkage might yet survive Williamson's critique and so serve to support a version of the squeezing argument. But let us suppose that is wrong and that the most plausible content-inferentialist thesis will make a connection not between understanding and acceptance of any propositions but between understanding and inferential *practice*. Then a gap opens between understanding a targeted logical operation and explicit acceptance of its constitutive rules. There may be a constitutive tie between understanding the conditional, for example, and inferential practice in accordance with *modus ponens*—still allowing that subjection to *modus ponens* is

---

[43] A suggestion of Boghossian (forthcoming).

[44] This, incidentally, is the way the McGee cases actually present themselves: they are not pure paradoxes for *modus ponens* but cases where *modus ponens*—the elimination rule for the conditional—presents as inconsistent with an unspecified *introduction* rule somewhat to the effect that a conditional is assertible only if the addition to one's information of the truth of the antecedent would justify the affirmation of the consequent. It is because they violate the latter that the conditionals in which the McGee arguments conclude impress as unacceptable.

indeed, *pace* McGee, constitutive of the conditional—but there will be no such tie, or at least no immediate such tie, between understanding the conditional and acceptance of the proposition that *modus ponens* is valid. The squeezing argument against an intuitional account of basic propositional logical knowledge will be blocked.

However, this form of inferentialism does create a platform for a different challenge to an intuitional epistemology of validity. Consider a chess player who, for some reason, has never explicitly encountered the rule controlling the movement of the Bishop but has, as we say, picked it up by immersion in the practice of the game. Suppose that now, for the first time, she considers a formulation of the rule—say, "From any square it occupies, the Bishop may move diagonally, backwards or forwards, through any number of consecutively unoccupied squares, and may only so move." Her assent to this proposition is to be expected. Indeed she may find the so-formulated rule *obviously* correct. But what she assents to is a proposition whose normative force concerns correct practice of the game: that this is one aspect of how the game is properly—legally—played. Suppose now that someone reflects on their inferential practice with the conditional and, similarly, comes to assent thereby to a propositional formulation of *modus ponens*. Then there is now a question of how to interpret that assent. An analogy with the case of the Bishop's rule would suggest that what is assented to need be no more than something to the effect, roughly, that this rule governs our use of the conditional, that this is one aspect of how acceptable conditional inference properly goes. That, however, is not the proposition that *modus ponens* is valid. It is merely the articulation of a tacitly accepted norm over inferential practice. An assent to the latter, based on reflection on one's practice, is nothing of special relevance to the epistemology of logic. It is of a piece with the general capacity of knowledge we all have of our intentions. Knowing that this is a rule governing how I, or we, play is not the same thing as knowing that the rule is correct. In the case of the Bishop's rule, there is of course no further, as it were external issue about correctness. But in the case of *modus ponens*, crucially, there is—it is exactly the further issue of validity.

To summarize. When inferentialism drops any claimed connection between understanding and assent, and contents itself with the thesis merely that an understanding of the conditional, for example, is constituted by mastery of its distinctive inferential role, then a gap does indeed open between the understanding and the acceptance of any particular proposition putatively encoding that role. But the question now is why we should conceive of the inclination, on reflection, to accept such a proposition as implicating a recognition of validity, or truth, rather than as comparable to the explicit assent to the Bishop's rule on the part of someone who mastered the moves legally open to the bishop by immersion in the practice of playing chess. The challenge to an intuitional account of the epistemology of basic validities is accordingly to justify describing what happens when, having not reflected on the matter before, we think about an explicit formulation of a basic logical rule of inference and are moved to regard it as good, as implicating an intuition-based recognition of *validity* rather than merely a becoming explicitly conscious of an aspect of the nature of one's own inferential practice. The supporter of intuition needs to provide a sound theoretical motive for describing the matter in the former way, and it is not clear that there is any.

I don't expect these remarks to stop friends of intuition in their tracks (probably nothing will), though I do think they are presented with a very significant challenge.[45] What is of interest here, however, is the dialectical situation *if* we now discard the intuitional view. There was already a strong case, prefigured earlier, for saying that an acceptance of the validity of *modus ponens* is at least an entitlement of cognitive project, one operative indeed wherever conditional inference is part of a cognitive project. But now we have the makings of a case for saying that *modus ponens* is a rule of inference to rely on which we have *only* an entitlement of cognitive project—that no superior form of cognitive achievement is here possible.[46]

There are various possible failings—inconsistency, epistemically irresponsible forms of non-conservativeness, etc.—which, in general, an inferential practice may prove to exhibit. And—at least before we think about Vann McGee's examples, or the Curry paradox—there is, I acknowledge, a strong inclination to say that we *know* that our practice with the conditional is, insofar as we are concerned with features just attributable to the role of the conditional, innocent of such failings. If we get into trouble, we are sure it won't be *modus ponens'* fault. But it's hard to see what possible reason we could give ourselves for thinking so that would not variously rely on conditional reasoning. Yet that kind of boot-strapping justification is available for *any* inferential rules. To be sure, for a theorist who wants to construe knowledge in some brutely externalist way, it can still be true that we do, properly speaking, *know* that our rules of inference for the conditional are valid merely in accepting them and in being, in fact, reliable in tracking validity in our basic logical acceptances. But if the question is our right to *claim* such knowledge, then it's hard to see that we are in position to do so; it is hard to see that we are better placed than to claim an entitlement to take it that they are valid. I do not think that we are.

Actually, I do not think this is a terribly surprising conclusion. That basic rules of inference enjoy this status—that of, so to say, mere entitlements, beneath cognitive achievement—is, I think, something which many would have found antecedently quite plausible. What is crucial, though, is that it is not a conclusion that should

---

[45] I pursue these matters in additional detail in Wright (2018).

[46] Making this case fully requires more than discounting the claims of intuitional knowledge of basic validities. In a number of papers—see, for example, his (2001)—Paul Boghossian has argued that we need to recognize that warrant-productive inference can, and in certain basic cases, must be *blind*: that reasoning can produce warrant for a thinker for conclusions in cases where it is uninformed by any beliefs she has about its validity, indeed in cases where she may have no developed concept of valid inference at all. I regard this observation as correct and important. But granting that is consistent with rejecting the application of the idea that Boghossian makes to underwrite the suggestion that *rule-circular* inference can be warrant-productive—that someone might, for example, use *modus ponens* in a blind but warrant-productive proof of its own validity. One problem with this idea is that it is very doubtful whether the range of cases where blind inference can produce warrant includes inferences to *schematically general* conclusions: the sophistication involved in grasping such a conclusion requires, plausibly, a correspondingly general conception of the validity of the reasoning that leads to it and hence explicit beliefs about its validity. Cf. Dogramaci (2010). But in any case our interest is in our right to *claim* knowledge of the validity of *modus ponens*. And where what is at issue is not just the inferential acquisition of knowledge of the validity of an inference pattern but the justification of a claim to knowledge of it, there has to be a presupposition that one is in a position to claim that the inferential machinery deployed is sound. Whether or not knowledge can, *claims* to knowledge cannot be underwritten by blind inference. For further elaboration, see Wright (2004a,b, and forthcoming).

disturb our right to claim knowledge *on the basis of deductions in accordance with such rules.* That is, even if the consequences of *premises* which are mere entitlements cannot, just on account of their being consequences, enjoy an any more robust form of cognitive status than that, it does not seem that the same limitation should apply to the conclusions of inferences from known premises drawn in accordance with *rules* which we are merely entitled to suppose to be sound. If that were not so, then inference in general would not be a means of extending knowledge—not if all inference must in the end depend on the basic rules, and they are merely entitlements. To be tempted by the thought that inference in accordance with merely entitled rules must correspondingly downgrade the status of its conclusions is to be tempted by a false modesty. If we are entitled to claim that a principle of inference is sound, then we are also entitled to claim knowledge of a statement which we have recognized to follow from known premises by inference in accordance with that principle. We are not restricted to a mere entitlement to such conclusions.

The point is actually quite general. In order to acquire knowledge, we do not need to know that the cognitive apparatus utilized can and does deliver knowledge. It is enough that it can and does do so. This goes for all presumed knowledge-acquisitive faculties: perception, memory, and reasoning of all kinds. To be sure, ascending a level, to *claim* to have acquired knowledge will require the claim that the cognitive apparatus concerned *has* delivered. But this claim can be entered as an entitlement. It does not need to be known in turn (and it cannot always be known in turn; though it is a fine thing when it can).

It is high time to bring all this back to the case of *HP.* If all we had, epistemically, were an entitlement to take it that *HP* is true, deductions from it as a premise would not be capable of generating any superior form of epistemic warrant than that. I have outlined two considerations that may be used to address that worry. First, it is arguable that even where basic but (for the majority) utterly uncontroversial principles of logical inference are concerned, all we have, epistemically, is an entitlement to take it that that they are valid. Second, this admission does not disable them from service in the generation and transmission of knowledge. Rather, being entitled to claim that they are valid, we are thereby entitled to claim that they are knowledge-productive.

If this, however, is how it is for *modus ponens,* it is clearly inappropriate to ask more of *HP.* From a purely proof-theoretic perspective, of course, it makes no difference whether we take *HP* as an axiom in a suitable second-order logic, or whether we take it as a pair of additional rules controlling the introduction and elimination of the cardinality operator. But from the perspective of the epistemology of logic, it makes a big difference. One needs to have a justified claim to know an axiom before derivations from it can justify claims to knowledge of their conclusions. One does not need to have a justified claim to know that a rule of inference is valid before derivations using it can justify claims to knowledge of their conclusions; it is enough, *ceteris paribus,* that one is entitled to take it that the rule is valid. This entitlement is then inherited by those claims.

My suggestion then, in summary, is that the epistemology of good abstraction principles should be assimilated to that of basic principles of logical inference, and that this involves recognizing (i) that their validity is *beneath* knowledge, at least if knowledge in such cases is taken to require some form of reflectively certifiable intellectual processing; (ii) that this limitation is consistent with a rational entitlement to take it that such rules are valid; and (iii) that there is a consequential rational entitlement to take it that they are at the service of knowledge production and extension by inference. If this is accepted, then the proof of Frege's Theorem in a system of second-order logic augmented by rules corresponding to the two halves of *HP* can issue in an entitlement to claim knowledge of the Dedekind–Peano axioms— at least, it can do so provided there is no other well-founded misgiving about the good-standing of the second-order logic required.[47] Moreover since the latter entitlement is generated purely inferentially, deploying only basic rules of deductive inference, the knowledge we are thereby entitled to claim is, if we indeed have it, *a priori* knowledge.

1.3 Justifying the Ontology

The preceding perspective, though, is hostage to a major item of unfinished business. We have shifted from a deployment of the notion of entitlement of cognitive project that aimed to justify the acceptance of *HP* as a legitimate implicit definition to one that aims to justify its acceptance as a pair of complementary basic inferential rules governing the cardinality operator. (On an inferentialist conception of the conceptual and semantic significance of basic inferential rules, the latter proposal is, of course, perfectly consonant with the former.) However, whereas the former ploy was open to the objection that there is in general a potential gap between acceptance of a principle as an effective implicit definition and acceptance of the presuppositions of its truth, the latter ploy is surely no less open to the corresponding objection that there is in general no entitlement to accept new rules of inference as valid if their introductory components implicate novel and untested substantive claims—rules, for example, whose validity carries additional ontological purport not already validated in the language to which they are added. But that is just what an introduction rule corresponding to the right-to-left direction of an abstraction principle will do.

Though the epistemological context—that of the conditions for an entitlement to take it that proposed basic inferential rules are valid—is new, this is, of course, an issue that has divided discussants of the abstractionist programme from the start and has already generated a great deal of debate. I do not expect to be able to change entrenched contrary opinion here, but I will give a brief indication of the way I think the case for the defence should be conducted.

The worry, localized to the case of *HP*, is whether there are, or with what right we take it that there are, any objects to serve as the referents of the numerical singular terms that, exploiting its right-to-left direction, we can enlist *HP* to introduce. In the present dialectical setting, we can set to one side irrelevant general nominalist qualms about admitting abstract objects into one's ontology at all and focus on a doubter whose misgivings specifically concern the apparent ontological presumptuousness of abstraction principles. With what right do we take it that identity contexts of the

---

[47] That issue will occupy us below.

kinds introduced by abstraction principles can soundly sustain quotidian first-order existential generalization?

In response, it is notable that there is, in the case of the conditional rules, no terribly impressive corresponding misgiving to be had about the analogous existential generalizations—that is, about the Ramsey-sentences, if you like, obtained by existential generalization on the places occupied by an expression for the conditional in suitable statements of the validity of the inference patterns licensed by the two rules. Rather, we are, most of us, unshakeably convinced that the acceptance of the soundness of the conditional rules is perfectly warranted; and warrant to take it that the patterns of inference licensed by *modus ponens* and conditional proof are sound is *eo ipso* a warrant to take that there is indeed an appropriate such function validating those rules.[48] The cautious view—that *modus ponens* and conditional proof are good *provided* there is any such function—seems merely neurotic. Is there any basis for such confidence in the case of the conditional rules that goes missing with *HP*, or other good abstraction principles?

We can manufacture a context for a non-neurotic doubt. To suppose that the acceptability of the patterns of inference concerned should await some independent reason to allow that there is such a function—rather as the acceptability of the practices implicitly defining 'Jack the Ripper' might await independent reason to suppose there was indeed a unique perpetrator of the relevant crimes—would be sensible if the context were one where, for some reason, an acceptable validation required finding a candidate from within some preselected domain of functions to discharge the described inferential role. For instance, the context might restrict us to the repertoire of binary truth-functions that feature in the standard semantics for classical sentential logic; then we might propose to identify the conditional introduced by *modus ponens* and conditional roof with the most eligible of these (the material conditional, of course.) But in general one's confidence that there is a function that answers to *modus ponens* and conditional proof does not rest on the possibility of such independent identification of it from within such a preselected, constrained domain. Rather, one would like to say, the rules concerned themselves exhibit the function—they *display* the conditional.

That, roughly, is what I want to say about good abstraction principles and the referents of the singular terms they introduce. They are not to be received in a *reference-fixing* spirit—a spirit whereby *HP*, for instance, is viewed simply as introducing a condition of identity that numbers, if indeed there are any, are required to satisfy.[49] Rather they introduce us to the abstracts concerned in the manner in which—at least on an *abundant* conception of properties—an explanation of the satisfaction-conditions of a predicate introduces us to the property it expresses. There are conceptions of properties—various forms of so-called 'sparse' conception—according to which a doubt can still be entertained whether a predicate with well-defined satisfaction conditions actually latches on to any real property, a

---

[48] Prescinding, again, from irrelevant nominalist concerns about whether there are any functions at all.

[49] Writers who implicitly assume that abstraction principles *are* to be taken in such a reference-fixing sprit include Field (1984), Boolos (1997), and Sullivan and Potter (1997). Fine (2002) is a notable exception.

doubt whether the world cooperates in putting up a real natural distinction that the satisfaction-conditions in question track. But those conceptions contrast with that on which, once a predicate has been well-defined, there is no additional element of risk carried by acknowledging that there is a property—a way of being—which the predicate denotes. On the abundant conception, the specification of the satisfaction-conditions of a predicate does not merely set a condition which any property denoted by the predicate must meet—viz. that it has to be a property whose instantiation by an object is ensured by its meeting the satisfaction-conditions in question. Rather, such a specification *displays* the property concerned. It leaves no space for an intelligent doubt about whether the world cooperates.

What obstructs this kind of way of looking at the matter in the case of abstraction principles is the assumption that their ontological implications need to be redeemed by reference to some *independently given* population of objects. (Compare the kind of artificial context envisaged above for an intelligent doubt about the existence of the conditional.) This, however, is a gratuitous and, when generalized, quite misguided assumption. There is no requirement that the objects in question should yet be available to thought other than under the very concepts of them that, for example, *HP* serves to introduce. In order to recognize that there are indeed such objects, it is not required that we hit on some prior range of things, given to us in some other way and so comporting themselves that they are somehow fitted to qualify as the referents of the new numerical singular terms. Rather the sole means of satisfying oneself that there are indeed such objects can be by verification of statements involving reference to them. And indeed when we contemplate the conditions for justified singular thought in general, that *has* to be, ultimately, the pattern: on pain of regress, there has to be such a thing as justified thought involving a reference-demanding singular mode of presentation where no independent such mode of presentation is deployed in the justification. The thoughts concerning abstracts that abstraction principles introduce us to should be received as *basic* singular thoughts in this sense. The requirement that some independent assurance be given that terms introduced by abstraction principles refer misses this epistemologically fundamental point.

A great deal more needs to be said, of course.[50] My intent here has merely been to outline some steps towards a more explicit account of the kind of epistemological stage setting which I believe abstractionism needs if its philosophical significance, and limitations, are to be properly understood. But one cautionary corollary of the foregoing discussion is perhaps worth emphasis. We need to realize that the traditional conception of the basic *a priori* as a realm of apodictic certainty—a conception in which Frege was immersed up to his ears—is a great mistake. *A priori* knowledge, no less than empirical knowledge, is subject to the ultimate groundlessness that Wittgenstein stressed in his last philosophical writings. Here I have tried to outline how the abstractionist project looks when that point is taken on board—and how indeed the point can help it address certain of the criticisms to which it has been subjected.

*   *   *

---

[50] Some of what needs to be said is broached in Hale and Wright (2009).

2. Rendering unto Caesar . . .

2.1 Prolegomena

The so-called Julius Caesar problem arises in *Grundlagen* as a consequence—some might say: as the nemesis—of Frege's conviction that a satisfactory philosophical account of arithmetic must treat the natural numbers as a species of *object*. In Frege's view, the vindication of an objectual conception of any particular type of thing requires that we specify a *criterion of identity* for things of that type—an account, when *a* and *b* are given to us as things of that sort, of when *a* should be reckoned identical with *b* and when distinct. Famously, Frege proposes (*Grundlagen* §63) *HP* as meeting that requirement. Nor does he ever lodge any objection to *HP* in that capacity. Rather, the objection he sustains is, in effect,[51] that *HP* is inadequate as an explanation of so-called mixed identity contexts wherein only one of the related terms is of the form: the number of *F*s. *HP* has no apparent bearing on what we should reckon to be the truth-value of, for instance:

The number belonging to the concept, *not self-identical*, is Julius Caesar,

and indeed no apparent bearing on any identity context featuring a term of the form, 'The number of *F*s,' on one side and a singular term of a different kind on the other. But that is as much as to say that *HP*, while offering a criterion of identity and distinctness *among* the numbers, tells us nothing about *which* objects the numbers are—or so it at any rate appears.

Since a grasp of the truth-conditions of such mixed identity contexts is in general no condition of an ordinary competence in arithmetic or its applications, this objection is apt to impress on first reading—as it impressed me and, no doubt, has impressed many others—as footling. However, the underlying philosophical point is solid: *if* we are to conceive of numbers as objects—and of course plenty have wished no truck with that idea—then we must be able to explain in general terms which objects given in other terms they are to be distinguished from and with which they may, perhaps, be identical. Grasp of any *sortal concept F*—a concept of a sort of object—has to involve not merely an understanding of what makes for identity and distinctness among the *F*s but also an understanding, to some sufficient extent, of what other sortal concepts may be disjoint from, or overlap with the *F*s. If nothing of that kind has satisfactorily been explained for *number*, then we have as yet attained no adequate conception of the numbers as a distinctive species of object.

And of course, as Rosen and Yablo emphasize in their present contribution (Chapter 5), the problem afflicts abstraction principles in general—at least, it kicks in just as soon as an abstraction principle is viewed as an attempted explanation of a new kind of singular term, and as purporting to fix the truth conditions of what are to be understood as genuine statements of identity configuring pairs of such terms. That abstraction principles may, in good cases,[52] be so regarded is the cardinal thesis of the

---

[51] The reader will recall that Frege's discussion at this point actually proceeds entirely by reference to the putatively structurally analogous account of identity and distinctness of the directions of straight lines.

[52] I prescind at this point from issues to do with "Bad Company."

deflationary style of platonism that Bob Hale and I have defended. No one should dispute that we may come to refer to *abstracta*, and to knowledge of their characteristics, provided we can fix the truth conditions of statements involving reference to them in such a way that the satisfaction of those conditions may be known in relatively unproblematic ways. That is barely more than a platitude. What is not platitudinous is the proposal that we can accomplish that by means of suitable abstraction principles: that, where 'ζ' and 'ξ' range over entities of whatever type, and '$\approx$' denotes an equivalence relation of things of that type, then we may—in good cases—regard instances of an abstraction principle:

$$(\forall\zeta)(\forall\xi)(Abs[\zeta] = Abs[\xi] \leftrightarrow \zeta \approx \xi),$$

not merely as fixing the meaning of the abstraction operator so introduced, and thereby that of the associated singular terms, but as providing cognitive access to the facts depicted by instances of its left-hand side via the knowledge, presumed relatively unproblematic to achieve, of the obtaining of instances of the equivalence relation on the right-hand side, so that—in the case of *HP*, for instance—you can knowingly refer to a number and recognize that it is the number both of the *F*s and of the *G*s by verifying that there are just as many *F*s and *G*s.

More will need to be said, of course, by such a platonist—an *abstractionist*—if the aim is to give content to and construct an epistemology for an interestingly wider range of contexts featuring the abstract terms in question. In the first-order case— where 'ζ' and 'ξ' range over objects—there is the clear prospect of a rigorously eliminative range of additional contextual definitions. Thus, for the example by which Frege actually frames his discussion in these sections of *Grundlagen*—viz. the directions of straight lines—we might envisage augmenting the abstraction:

$$(\forall a)(\forall b)(Dir[a] = Dir[b] \leftrightarrow a /\!/ b)$$

—stipulating that the directions of lines a and b are to be reckoned the same just when a and b are parallel—with contextual definitions of a range of predicates of directions, $\phi_1, \ldots \phi_n, \ldots$, in terms of suitable predicates of lines, $F_1 \ldots F_n \ldots$, on the model:

$$(\forall a)(\phi_n Dir(a) \leftrightarrow F_n a),$$

where it is required that each admissible F_n for this schema express a property for which parallelism is a congruence (so that the ϕ-properties behave as they ought with respect to contexts of identity of direction.) And we could go on, if we felt it advisable, to contextually define quantification into places occupied by *Dir*-terms in the obvious kind of way; thus, for example,

$$(\exists x)\phi_n x \leftrightarrow (\exists y)F_n y.$$

One familiar and traditional philosophical take on the resulting situation would hold that such an apparatus of definitions would in effect *explain away* the appearance of reference to and quantification over directions, and so absolve us of any need to 'countenance' such things. The abstractionist view turns that on its head. Rather, the claim will be, we are thereby provided with the means to explain how reference to and knowledge of directions—a novel species of abstract entity—may be unworryingly achieved. However, there is a point in the vicinity that is very relevant to the Caesar problem and that Frege's concentration in *Grundlagen* on the example of directions may lead a reader to overlook. Namely: there is in *second-order* cases of abstraction—where the equivalence relation concerned is a relation on concepts rather than objects—no evident prospect of such a fully eliminative treatment of terms introduced by abstraction unless the ranges of 'ζ' and 'ξ' are restricted to concepts whose specification requires no use of the new abstract terms. And a very salient example where there can be no such restriction is provided by the numerical terms that are introduced by *HP*—at least if Frege's own strategy for proving the infinity of the number series is to be followed—since it is then essential that there be a license to form numerical terms by applying the number operator to open sentences themselves containing such numerical terms. For that purpose, we need—writing P*xy for: *x ancestrally precedes y*—to be able to form, for each number y, the term:

$$Nx:(P^*(x, y) \lor x = y),$$

and then to prove both that there is such a number and that y immediately precedes it. Where zero, for example, is defined as $Nz:z \neq z$, we need to be able to prove that zero immediately precedes $Nx:(P^*(x, Nz:z \neq z) \lor x = Nz:z \neq z)$. It is obvious—irrespective of how we define ancestral predecession—that *HP* cannot provide for the elimination of all occurrences of the numerical operator in such a construction, since it provides no resource for the elimination of the singleton numerical term from the open sentence, $x = Nz:z \neq z$.

The bearing of this point on the Caesar problem will be clear. If an abstractionist foundation, built on *HP*, is to be given for number-theory, then a case must be made that the meaning of the numerical operator has somehow been fixed for every type of context in which it may legitimately occur, even if *HP* itself leaves us apparently short of the resources required for an eliminative definitional paraphrase of all such contexts. In particular, an explanation has somehow to have been given of the meaning—that is, the satisfaction conditions—of open sentences like the above, not because, or not merely because, an objectual conception of the numbers demands it philosophically but because the construal of such open sentences is a prerequisite if we are intelligibly to bind, by the application of the numerical operator or by quantification, the free variables they contain. Since a general explanation of the satisfaction conditions of open sentences of the form, $x = Ny:Fy$, is effectively nothing other than an explanation of what it takes for something to be a number, the requirement of a solution to the Caesar problem is thus overdetermined, imposed by the internal logistical demands of the abstractionist programme for the foundations of arithmetic as well as by the commitments of an underlying objectual metaphysics.

The Caesar problem thus has all of metaphysical, semantic, and programmatic aspects. That *HP* seems to offer no means to address it, so far from being a "footling" objection, is something that Frege was arguably absolutely right to take with full seriousness. For anyone in sympathy with platonism about the numbers, whether or not deflationary, or with abstractionism generally, the shortcoming must be addressed.[53] But how?

There are two possible directions of response. One would be to canvass some kind of additional principle or principles, perhaps quite general, perhaps specific to number, to supplement the explanation of the numerical operator given by *HP* and to bear directly on questions of sortal inclusion and exclusion between number and other sortals. Good luck to readers who would wish to experiment with that direction. My own thinking about the matter from *FCNO* onwards has been conditioned by the thought that *HP* is effectively all we have to go on—that there *is* no relevant additional, independent, and independently supported principle that we have overlooked—and hence that a solution to the problem, if there is one, must be found by discerning some kind of *latent content* in *HP*, and in good abstractions generally, above and beyond that of a (necessitated) universally quantified biconditional.

Gratifyingly, this perspective is shared by Rosen and Yablo.[54] The sequel will review three suggestions of this kind. The first, presented by Hale and myself in our long joint essay in *The Reason's Proper Study*,[55] has been independently criticized by Rosen; and the second, about which Rosen and Yablo express some pessimism in their present contribution (Chapter 5), has received criticism from William Stirton. I shall here take the opportunity to respond to these criticisms. The third is the present proposal of Rosen and Yablo themselves.

A word of caution. None of the three proposals to be reviewed should be received as an attempt to prove the *existence* of abstracts. Rather they concern how someone who is inclined to accept the legitimacy of abstraction as a means of introducing new ranges of genuine, successfully referential singular terms, may attain a sufficient understanding of what kind of thing it is to which means of reference is thereby achieved. To solve the Caesar problem is not to show that terms introduced by good abstractions refer. But it is—or so I have suggested—to meet a necessary condition for showing that.

2.2 Three Proposals about the Latent Content of Good Abstractions

2.2.1 CROSS-WORLD IMPORT

This proposal was originally framed[56] not as a general solution to the Caesar problem but as a response to a particular kind of nominalism: the suggestion of Michael Dummett[57] that a nominalist can take abstraction principles in stride, since they may always be interpreted in such a way that the novel terms refer back into the domain of their abstractive relation—so that Direction terms, for instance, introduced via the Direction Equivalence:

$$(\forall a)(\forall b)(Dir[a] = Dir[b] \leftrightarrow a /\!/ b)$$

---

[53] For resistance nevertheless to this line of thought, see Heck (1997b).
[54] See Chapter 5, this volume, p. 117. "Latent content" is indeed their phrase.
[55] Chapter 14, "To Bury Caesar...," in Hale and Wright (2001a).   [56] In Hale and Wright (2003).
[57] Dummett (1991: 126).

may always be construed as referring to lines—where, as good nominalists, we now think of the latter as very fine concrete physical inscriptions. (The corresponding claim about the numerical terms introduced via *HP* would be that they may harmlessly be construed as referring back into a domain of concrete objects by which the concepts ranged over by the higher-order variables on its right-hand side are instantiated.)

The nerve of the cross-world import argument is that, once the intended but—in the usual formulations—suppressed *modal generality* of abstraction principles is made explicit, it will not be consistent with their requirements to identify their proper abstracts with any contingently existing concrete particulars. Suppose for instance that the direction of line *a* is identified with some object—as it might be, Tiberius Caesar—that might not exist in circumstances where *a* still existed and remained unaltered in its orientational characteristics. A properly modal abstraction principle for directions should say that in those circumstances, *a*'s direction *would then be the same as it actually is*—notwithstanding the fact that the object, Tiberius, that was supposed to *be* its direction would not be around. So much the worse, therefore, for that supposition. If, on the other hand, the nominalist were to try saying that *a*'s direction should be identified with the line *a* itself—which, as we said, *will* still be around—then the problem will be that *a* might not exist in circumstances where another line, *b*, which actually has the same direction as *a*, retained that same direction.

Let us make this argument more explicit. Let parallelism relate lines both *within* and *across* possible worlds, and take 'the direction of' as an operator on <line, world> pairs, refashioning the Direction Equivalence as:

$$Dir<a,w> = Dir<b,w^*> \text{ iff } a,w /\!/ b,w^*$$

where w, w^* range over possible words, including the actual world, and *a* and *b* are any actual or possible lines. Intuitively the gist of the generalized principle is thus that the direction of *a* as it is/would be in w is the same as the direction of *b* as it is/would be in w^* just in case *a* as it is/would be in w is parallel to *b* as it is/would be in w^*. Using this fully modalized abstraction, we may then capture the gist of the cross-world import argument as follows:

Let **a** be the actual world. Let *a* be any (concrete) line and let q be any contingent existent distinct from *a* whose existence is not contingent on that of *a* nor vice versa. Suppose for *reductio* that:

(i) $Dir<a,\mathbf{a}> = q$

Assume

(ii) $<a,\mathbf{a}> /\!/ <a,w>$—that is, that *a* in w would be parallel to *a* as actually oriented.

Then

(iii) $Dir<a,\mathbf{a}> = Dir<a,w>$,

by the modalized Direction Equivalence—that is, *a* would have the same direction in w as it actually has.

Suppose, however, that—as, on our hypotheses, we may—we select w in such a way that

(iv) q does not exist in w

Then

(v) $Dir<a,\mathbf{a}>$ does not exist in w, by (i) and (iv)

But this, apparently, contradicts (iii).

To defeat, on the other hand, the supposition that $Dir<a,\mathbf{a}>$ is a itself, we may suppose a line b, distinct from a, such that a and b are actually parallel, and run the argument again substituting 'a' for 'q' and 'b' for 'a.'

This looks pretty good, as far as it goes. But Rosen makes[58] what may seem a telling objection. In brief, it is that there is no contradiction between lines (iii) and (v) unless we make an additional and, as he contends, unjustified assumption: the assumption that any term of the form, '$Dir<a,w>$,' if it denotes at all, must denote an entity that exists in w. Without this assumption, the truth of (iii) does not require $Dir<a,\mathbf{a}>$ to exist in w—it is enough that it, and $Dir<a,w>$, *actually* exist. And without any requirement that $Dir<a,\mathbf{a}>$ exist in w, q's identity with $Dir<a,\mathbf{a}>$ but non-existence in w is no problem.

In effect, and informally, leaving aside the play with worlds, Rosen's thought comes to this: that the truth of the identity,

The direction a actually has is the same as the direction it would have in circumstances w,

requires a's actual direction to exist in the envisaged hypothetical circumstances w only on the assumption that to speak of the direction a line *would have* in certain circumstances must involve reference among the things that *would exist* in those circumstances: that the referent, if any, of what we might call a *subjunctive descriptive term*—like 'the man who would have been Provost if the College had voted five days sooner'—must both be actual and feature *within* the invoked subjunctive ontology, so to speak. Provided we make this assumption, an identity like that above refers both to an actual object—on its left-hand side—and a 'subjunctive object'—on its right-hand side and thus demands the existence of the actual object in the subjunctive scenario, just as the cross-world import argument says. But—Rosen's point is—with what right do we presuppose that reading of subjunctive descriptive terms in the relevant contexts?

The question is certainly pertinent. It may be elaborated like this. To speak of the direction that a line a would have in counterfactual circumstances is—when talk of the directions of lines generally is introduced by abstraction as outlined above—to speak indirectly of the orientational characteristics that a would have in those circumstances. But that does not require that the apparatus of directions also exist therein but only, since we actually have that apparatus, that we may use it to type the orientational characteristics that a would then have. (Compare how we may use the

---

[58] Rosen (2003).

metric system of lengths to describe the dimensions objects would have in extreme hypothetical circumstances in which the very institution of measurement and the conceptual apparatus involved would be infeasible.)

Rosen's objection is that no proof has been offered of the needed assumption of reference within the subjunctive ontology. I agree: Hale and I said nothing to prove that assumption. It may be granted that, in general, subjunctive descriptive terms allow of a *wide scope construal*, so that we may normally read 'the *x* that would be *G* in circumstances w' as, roughly,

Among actual objects, that one which would be *G* (and thereby exist) in w.

Still, nothing apparently enforces that style of interpretation of terms introduced by abstraction, even when the relevant abstraction principles are endowed with the modal generality proposed.

But why cannot *we*—abstractionists—enforce it? In writing as though there was something here that a proponent of the cross-world import argument should *prove*, Rosen arguably mistakes the dialectical situation. Rather, what his discussion brings out is merely that if a fully modalized abstraction principle is to be at the service of this partial solution to the Caesar problem, it must be received as part of its *latent explanatory content* that the subjunctive descriptive terms it governs are to be read on the model indicated. After all, this is at least a possible reading.[59] The Direction Equivalence is to be understood exactly as a stipulative explanation of the truth-conditions of identity contexts featuring direction terms interpreted in this latter way, so that '*Dir<a,w>*' always refers, if at all, then both actually and within the ontology of w. When abstraction principles in general are so construed, the particular kind of nominalist response envisaged by Dummett—that of allowing the legitimacy of abstraction in principle but insisting on construing the new terms as referring within the domain of actual contingent existents—will not be an option.

Ordinary subjunctive thought makes free use of singular terms read in this way. It is therefore entirely natural that subjunctive thought of abstracts should be no exception and that their governing abstraction principles should be interpreted accordingly. When they are, the cross-world import argument shows that there is no option but to construe the abstracts in question as distinct from *any* (concrete) contingent existents. It is arguably on exactly this point that the conception of the abstracts both as necessarily existing and indeed as *abstract* is grounded. Unless, therefore, there is some further objection to our right to lay it down that abstraction principles should be so construed, it should be granted that the cross-word import argument does make significant anti-nominalist inroads into the Caesar problem, just as originally advertised.

2.2.2 SORTALS AND CATEGORIES

The inroads, however, even if significant, are only partial. Perhaps the foregoing resolves the epitomical issue about Zero and Julius Caesar. But nothing has been

---

[59] I do not mean to deny that such a reading is never excluded. Consider 'If there were no numbers, then the number of numbers that there would be would be zero.' But it is rare for content, as opposed to context, to force a wide, or a narrow reading at the expense of the other.

accomplished to address the identity or distinctness of abstracts given via different equivalence relations, or indeed to address their relationship to any other kind of necessary existents. And surely this remaining problem—specifically, that of the differentiation of what impress as different kinds of abstract—though narrower in scope, is philosophically no less serious than the original. If we have achieved no determinate conception of when the abstracts introduced by different abstraction principles do or do not (partially) coincide, how can we pretend to have any understanding of them as specific kinds of object at all?

In our (2001b)[60] Bob Hale and I proposed a framework designed to remedy the shortfall: to provide specifically—though the proposal naturally extends to abstract terms as a species—a general necessary condition for when a singular term, assumed to denote an already familiar kind of thing, may be regarded as so much as eligible to generate a truth when introduced into the argument place of a predicate of the form, "$Nx{:}Fx = \xi$." Our proposal was conceived as offering a different and better solution to the Caesar problem than the suggestion I had made in $FCNO$,[61] albeit one which draws on the same underlying idea, namely that materials to draw the distinction between numbers, or abstracts of any kind, and things of other sorts, may be extracted from what HP, or the relevant abstraction principle, itself tells about the identity-conditions of the objects it concerns.

The argument we proposed draws on a general meta-ontological apparatus featuring notions of *sortal concept*, *category*, and *criterion of identity* as key components. Sortal concepts, for the benefit of any reader unfamiliar with the notion, are—no surprise—concepts of sorts of object, and are thought of as individuated by *criteria of application*, distinguishing things of the sort in question from things of any other sort and *criteria of identity*, thought of as supplying the ground for the identity, or distinctness, of any objects of the sort in question. (It is the latter characteristic that underwrites the sometime description of sortals as "count nouns" for naturally, there cannot be a fact of the matter, how many Fs there are, unless there are facts of the matter, for any x and y that are both F, whether x the same as y.) Sortal concepts may be *pure* or *impure*. An impure sortal is alienable: it is consistent with an object's satisfaction of an impure sortal that it—the very same object—should not have done so or should cease to do so. (So *person-with-a-walking-stick* and *number-of-moons-of-Jupiter* are both alienable sortals.) Its satisfaction of any pure sortal, however, is of the essence of an object: where F is a pure sortal, being F is essential to anything which is F. (One cannot coherently suppose that a person ceases to be a person yet continues to exist.) *Categories*, finally, are maximally inclusive sortal concepts under a given criterion of identity: C is a category if it is a sortal concept such that any sortal concept that shares its criterion of identity with C is a sub-sortal of C.

This apparatus is put to work as follows. Let us suppose that every object, of whatever sort, belongs to one and only one category. Then sortals F and G are apt to overlap—that is, have an instance or instances in common—only if they are sub-sortals of the same category. It follows from the definition of a category that F and G

---

[60] Pp. 390–6.   [61] Wright (1983: 116–17).

must therefore share their criteria of identity if they are, even partially, to overlap. Hence if *F* and *G* do not share their criteria of identity, their respective extensions must be mutually exclusive.

Take it then that the latent content of *HP* is that it is to be understood as giving the criterion of identity for a pure sortal—*number*—under which the referents of the terms it introduces are to fall. Then the concepts *number* and, say, *direction* cannot overlap because if they did, it would have to be a necessary truth that the results of applying their respective criteria of identity must coincide, in other words, that, as a matter of necessity, for any *x* and *y*, *x* is accounted identical with *y* by the criterion of identity for numbers if and only if it counts as so by the criterion of identity for directions. Since there is manifestly no such necessity, we may conclude that *number* and *direction* are sub-sortals of different categories. On the assumption flagged above, to which I will return, that no object may belong to more than one category, it follows not merely that the direction of the Earth's axis ≠ zero but, generalizing, that no abstract can be identical with any object falling under a pure sortal whose subsuming category is associated with a criterion of identity distinct from that associated with the sort of abstract concerned. So numbers cannot be Roman emperors, nor shapes, nor lengths. Indeed—and strikingly—since whether a pair of concepts sustain a bijection and whether they are coextensive are in general independent questions, at least one way round, there is motive to conclude that *number* and *extension* too are associated with distinct criteria of identity and hence likewise do not overlap.

The last thought, however, puts some pressure on the question: when exactly, if we are going to employ this apparatus, should we regard *criteria of identity* as the same or distinct? For after all *some* extensions—for instance, those with which Frege himself identified the numbers—may be correctly identified and distinguished by exactly the criterion embodied in *HP*. It may be said that even for these extensions, one-to-one correspondence does not embody the *basic ground* of their sameness and distinctness. That is given, rather, as for all extensions, by sameness of membership. It is just that in these cases the two criteria necessarily give the same results. But this suggestion has the 'nose' of a stipulation: how and why was it determined that criteria of identity are hyper-intensional things, that may differ even though necessarily coincident in their verdicts? A sympathetic reader who thinks that Hale and I were on to a way of looking at the Caesar problem that does actually resonate with something at the basis of our intuitive conviction that numbers are—of course!— not Roman emperors nor, for that matter, shapes or directions, is not likely to have any preconceived opinion about their relation to suitable instances of *extension, set*, or *class*.

For this reason, Hale and I did not regard it as a drawback of our proposal that not every instance of a Caesar problem is promised a clear-cut solution by its means. We took it that, provided that what is offered is a condition that will fail in just the cases where ordinary thought would anyway want the *F*s and *G*s to be distinct, it would be progress enough to provide a *necessary* condition for when criteria of identity associated with a pair of sortals, *F* and *G*, should be regarded as the same. However, it was in offering one specific formulation of this necessary condition that we made the move on which William Stirton's objection pounces.

To explain. At one point in our discussion,[62] Hale and I took note of a possible

fast track to the conclusion that no sortal concept of Fregean abstracts can share its criterion of identity with any other kind of sortal,

namely that

The criterion of identity for Fregean abstracts is always an equivalence relation on other kinds of thing.

—that is, the criterion of identity for its proper abstracts given by an abstraction principle is, as I shall say, an *other-sorted* relation. For instance, the criterion of identity for numbers, a kind of object, provided by *HP* proceeds in terms of an equivalence relation on (first-level) *concepts*, and the criterion of identity for directions provided by the Direction Equivalence proceeds in terms of an equivalence relation on lines—*pace* Dummett, a different sort of object. Whereas

the criterion of identity for non-abstracts is always an equivalence relation on those very objects,

—that is, a *same-sorted* relation. For instance, Davidson's account of events gives the criterion of their identity in terms of a relation—sameness of causes and effects—on events themselves, and salient candidates for the criteria of identity of, for example, material objects, or persons, likewise proceed in terms of relations—spatiotemporal continuity, co-consciousness, etc.—on the very particulars concerned.

Hale and I did not dispute that the distinction as formulated gestures at something metaphysically important. But we dismissed its promise of a fast-track pay-off for the Caesar problem on the ground that it is always possible to reformulate an other-sorted criterion as same-sorted, remarking that whenever

$S(a) = S(b) \leftrightarrow Eq(a,b)$ is an abstraction, then the condition for the identity of Ss can always be expressed as a relation whose domain is those very things as follows:

$$x = y \text{ iff } \forall a \forall b (x = S(a) \ \& \ y = S(b) \rightarrow Eq(a,b))$$

Here the right-hand side is now an equivalence relation on the abstracts concerned, albeit one defined in terms of the original abstractive equivalence, $Eq(a,b)$.

Having thus drawn attention to the possibility of reformulating any other-sorted criterion of identity for abstracts by a same-sorted criterion on the above model, we then proceeded to offer a necessary condition for identity of criteria of identity in what, once only same-sorted criteria are under consideration, is the most salient possible way, namely that a pair of sortal concepts share a criterion of identity only if the same-sorted equivalence relations that are respectively definitive of their criteria of identity are *necessarily coextensive*. Formally, where a is the same F as b just if they stand in the relation, $Eq_F(a,b)$, and a is the same G as b just if they stand in the relation, $Eq_G(a,b)$, F and G share a criterion of identity, we suggested, only if it holds of necessity that

[62] Hale and Wright (2001b: 390).

$$(\forall x)(\forall y)\big(\mathrm{Eq}_F(a,b) \leftrightarrow \mathrm{Eq}_G(a,b)\big).$$

Stirton's objection now surfaces.[63] It is that the proposed solution to the Caesar problem is spoiled by an epistemic circularity. For consider how if hitherto innocent of *number*, and knowing only that *HP* is meant to provide you with a criterion of identity for this new sort of thing, you might try to use Hale's and my proposal to decide, for example that "$Nx:x\neq x$" does not denote a person. Letting '$\mathrm{Eq}_{Person}(a, b)$' express that a and b satisfy whatever we think may be a satisfactory criterion of personal identity, what you need to decide is whether the following holds as a matter of necessity:

$(\forall x)(\forall y)(\mathrm{Eq}_{Person}(x,y) \leftrightarrow x$ and y stand in that same-sorted relation which is derivable from *HP* on the model given above)

Partially expanding the right-hand side of that, we arrive at the following condition on x and y:

$(\forall F)(\forall G)(x=Nu:Fu \,\&\, y=Nu:Gu \rightarrow$ there is a bijection between F and G),

and now the problem is evident. I have already argued that a solution to the Caesar problem is needed before we can intelligently parse open sentences of the form, $x=Ny:Fy$, let alone construe the results of binding their free variables. Yet now it appears that, on Hale's and my proposal, you will have to parse the above clause and assess whether the relation it depicts is necessarily co-extensive with $\mathrm{Eq}_{Person}(x,y)$ *before* you can solve the particular version of the Caesar problem concerned—so that in order to achieve that solution, you must *already* understand the embedded bound occurrences of the predicates, "$x=Nu:Fu$" and "$y=Nu:Gu$." In effect— Stirton's point is—only someone who is already apprised of a solution to the Caesar problem can be expected to understand the condition that Hale and I proposed as necessary if the instances of any previously understood sortal are to be eligible to coincide with those of the sortal putatively introduced by *HP*—or indeed, generalizing the point, with those of any sortal introduced by abstraction. The Hale–Wright proposal is therefore either *de trop* or unusable.

In sum: Let it be agreed that a solution to the Caesar problem is a crucial part of a defence of the claim that *HP* may be used successfully to introduce a sortal concept of natural number. Then it is indeed the case, just as Stirton insists, that no solution to the problem can presuppose prior understanding of predicates of the form $y=Nx:Fx$. So if an application of the proposal of "To Bury Caesar" does unavoidably require prior understanding of such expressions, it has to be vulnerable to the charge of an objectionable circularity that Stirton brings against it.

Perhaps the reader will anticipate my reply. It is that application of the proposal of "To Bury Caesar" does not *unavoidably* require prior understanding of such expressions—that a prior understanding of expressions of that ilk became entangled

---

[63] As presented in Stirton (2016: 30 and 42–5). The objection is as descendant of one of a number developed in Stirton (2003) but he advertises the most recent version of it as "a more refined circularity objection, one which displays (hopefully) a greater awareness of the possible counter-objections at the disposal of the neo-Fregean" (p. 37).

in the application of what Hale and I proposed only because we allowed ourselves to frame our suggestion in terms of same-sorted relations. We had no need to do that.[64]

To clarify. The same-sorted framework provides for an agreeably straightforward account of when criteria of identity for sortals F and G should be reckoned distinct, in terms of the failure of their associated equivalence relations to be necessarily co-extensive. But the issue of sameness, or distinctness of the criteria of identity for a pair of sortals, need not be addressed in terms of relations on those sortals' instances. We can instead ask directly about the metaphysical relationship between the facts that ground identity under the one sortal and the facts that ground identity under the other. For if F and G overlap on an object a, then—on the assumption of the exclusiveness of categories—a must belong to a category which subsumes both F and G, which must therefore share their criteria of identity. And that will require that what grounds identity under F must likewise ground identity under G. So we can ask directly whether that is plausibly the case: whether the kind of fact that makes for identity under F and the kind of fact which makes for identity under G are the same kind of fact. And while the notion of *same kind of fact* is, of course, wide open for philosophical refinement, it is clear up front that facts about continuity of consciousness, or spatio-temporal continuity, or parallelism of lines, or one-to-one correspondence of concepts, or sameness of causes and effects, and so on, ought all to rank as mutually independent kinds of fact in any satisfactory metaphysics of facts. That the respective species of objects whose identity and distinctness they ground should rank as distinct species in consequence is the basic idea of Hale's and my proposal.

Coda—a proof that an object belongs to at most one category?[65]
We make just two assumptions:
   First, categories are *extensional*, that is:

$$C = C^* \leftrightarrow \forall x(x\varepsilon C \leftrightarrow x\varepsilon C^*)$$

---

[64] Indeed, as a criterion of identity of numbers, what we suggested is actually flawed for a familiar kind of reason. Let a be some object which is definitely not a number, for example, the turnip in the vegetable basket, and let b denote any other object, such as the Eiffel Tower. Then $a = \mathrm{N}u{:}Fu$ is false for all F, so that

$$(\forall F)(\forall G)(a = \mathrm{N}u{:}Fu \ \& \ b = \mathrm{N}u{:}Gu \ \to \ \text{there is a bijection between } F \text{ and } G)$$

is vacuously true, since its antecedent, $a = \mathrm{N}u{:}Fu \ \& \ b = \mathrm{N}u{:}Gu$, is always false. Our criterion thus delivers the result that the turnip is the Eiffel Tower.

Since the wrinkle stems from the unrestricted generality of the proposed one-sorted criterion which, as it stands, purports to apply to objects of *any* kind, we could smooth it by explicitly restricting it to numbers, that is, by replacing it by:

$$(\forall x)(\forall y)(number(x) \ \& \ number(y) \ \to \ (x=y \leftrightarrow (\forall F)(\forall G)(a = \mathrm{N}u{:}Fu \ \& \ b = \mathrm{N}u{:}Gu \ \to$$
$$\text{there is a bijection between } F \text{ and } G)))$$

whereby, granted that the turnip is no number, nothing follows about its identity. But this revised one-sorted criterion obviously cannot be of service in explaining which things are numbers. So in a project where the goal is to elicit something of useful bearing on that question from the criterion of identity for numbers, it is of no use to us.

[65] This is a reconstruction of an argument in an email that Bob Hale sent me around the time that he and I were discussing Stirton's (2003). Naturally, I would have wished to run this version past Bob before publishing it here but, sadly, missed the opportunity to do so. As it is, I naturally bear responsibility for any shortcomings in it.

Second, the *co-membership* axiom: that if x and y belong to the same category, C, then they are identical if and only if determined to be so by the criterion of identity for C. That is, writing the latter as *aeqCb*, we have:

$$(\forall x)(\forall y)((x\varepsilon C \,\&\, y\varepsilon C) \leftrightarrow (x=y \leftrightarrow x eqCy))$$

We may then proceed as follows:

(1)	$a\varepsilon C \,\&\, a\varepsilon C^*$	Assumption
(2)	$(\forall b)((a=b \leftrightarrow aeqCb) \,\&\, (a=b \leftrightarrow aeqC^*b))$	1, co-membership axiom
(3)	$a\varepsilon C \,\&\, b\varepsilon C \leftrightarrow (a=b \leftrightarrow aeqCb)$	co-membership axiom
(4)	$a\varepsilon C^* \,\&\, b\varepsilon C^* \leftrightarrow (a=b \leftrightarrow aeqC^*b)$	co-membership axiom
(5)	$(\forall b)(aeqCb \leftrightarrow aeqC^*b)$	from 2
(6)	$(a=b \leftrightarrow aeqC\,b) \leftrightarrow (a=b \leftrightarrow aeqC^*b)$	from 5
(7)	$a\varepsilon C \,\&\, b\varepsilon C \leftrightarrow a\varepsilon C^* \,\&\, b\varepsilon C^*$	from 3,4,6
(8)	$b\varepsilon C \leftrightarrow b\varepsilon C^*$	from 1,7
(9)	$C=C^*$	*b* arbitrary; extensionality

2.2.3 REAL DEFINITION

Rosen and Yablo

Readers must make up their own minds whether the foregoing proposals impress as satisfactory, or as at least potentially satisfactory. Rosen and Yablo, for their part, are unconvinced, writing:

Maybe this approach

—the play with sortals, categories and criteria of identity—

can be made to work and maybe not; hopes of a simple, straightforward solution along these lines are now faded, however, and so we propose to try something different.[66]

Well, my own hopes spring eternal! But new and, perhaps, better ideas are always welcome. The different thing Yablo and Rosen propose to try is to exploit the resources provided by the idea of a *real definition*: a definition whose purpose is not to explain a meaning or to introduce a new concept but to characterize the essential nature of the thing defined.[67]

To illustrate, Yablo and Rosen focus on the following definition of a familiar style of Goodmanian property:

(GR)   For all *x*, *x* is *grue* iff *x* is green and observed or blue and unobserved.

When this is understood in the ordinary way as a definition, rather than as a mere characterization of actual extension, it carries at least the modal import of

(GR+)   Necessarily, for all *x*, *x* is grue iff *x* is green and observed or blue and unobserved.

---

[66] Chapter 5, this volume, p. 119.
[67] This is actually very much in keeping with the way that Bob Hale had latterly come to think about the Caesar problem.

However, there is more than that, Yablo and Rosen observe, that an ordinary subject will be expected to understand from the definition of "grue." For imagine someone, they write, who responds,

I see that a thing counts as grue (in a world) whenever it is either green and observed or blue and unobserved. But I am still not sure *what it is* to be grue. Perhaps for a thing to be grue just is for it to be green and observed or blue and unobserved. But it might also be that for a thing to be grue is for it to be *known by God* to be green and observed or blue and unobserved. Or perhaps to be grue is to be green and observed *and such that* $e^{i\pi} + 1 = 0$ or blue and unobserved and such that $e^{i\pi} + 1 = 0$.[68]

Such a response, they suggest, would betray a failure to appreciate the latent content of the definition. That content is:

(GR++)   For a thing to be grue *just is* for it to be green and observed or blue and unobserved.

Being grue is nothing other than being green and observed or blue and unobserved. That is the whole truth about which property grue is. So it is not any property which merely actually happens to apply to all and only things which are green and observed or blue and unobserved. Nor is it even one which necessarily applies to all and only things which are green and observed or blue and unobserved but of which a complete account must mention other aspects—like the value of $e^{i\pi}$ or the knowledge of God.

A correct real definition in the sense Rosen and Yablo intend thus has both an essentialist and a limitative aspect: it gives a *complete* account of what the *definiendum* essentially is. There's nothing else to know about what the *definiendum* is save what is explicitly stated by the definition.[69] Interpretations of the statement of the definition which—even if they save its (necessary) correctness—carry the implication that there is more to the individuation of the *definiendum* than the definition makes explicit are thereby excluded.

Rosen and Yablo propose to try to marshal this basic idea against two kinds of sceptic about the reference of the numerical singular terms introduced by *HP*. One—whom they quaintly label the *libertine*—holds that, for all *HP* has to say about the matter, the reference of the terms it introduces could be to absolutely any objects whatever. The other—the *pervert*—allows that the reference of such terms may be determinate, but insists that we know nothing whatever about which objects they actually refer to. How may the notion of real definition help to address these sceptical characters?

Suppose it agreed that we should regard *HP*—and good abstractions generally—as, in the first instance, real definitions of the functions that they introduce; in the case of *HP*, we are thus given a real definition of the function expressed by the numerical operator, 'N....' We should acknowledge immediately that merely to propose that *HP* should be taken in this way does nothing to address a third, perhaps the most common form of scepticism about abstraction: that which queries our right to suppose, without further work, that there is any such thing as the function an abstraction postulates, let alone the range of entities that are to provide its values.

---

[68] Chapter 5, this volume, p. 118.   [69] Or, they allow, by trivial consequences of it.

This form of scepticism is prominent in, among others, George Boolos's writings on abstraction,[70] and Hale and I have attempted to address it throughout our neo-Fregean excursions. Rosen and Yablo, however, quickly affirm that it is not on their agenda. Their concern is with the resources provided against libertinism and perversity by the assumption that *HP* is a *successful* real definition; they thereby presuppose that the existential commitments carried by instances of its left-hand side are indeed in good order.

The argument Rosen and Yablo offer is quite intricate and draws on a number of assumptions. Note, to begin with, that once it is granted that *HP* is a successful real definition of 'N...' in the sense gestured at, we immediately dispose of the familiar worry, pressed by Harold Hodes[71] and others, that *HP* leaves just too much latitude in the interpretation of 'N...'— that there are just too many functions on concepts whose values coincide when those concepts are bijective, and hence that 'N...' is left massively indeterminate by *HP* and not properly defined thereby at all. The rejoinder to this will follow exactly the pattern illustrated with *grue* above. Maybe we can *interpret* the occurrences of 'N...' in *HP* by means of any of no end of hyper-intensionally distinct but extensionally coincident functions, but only that function— as Rosen and Yablo term it, the *essential numerator*—whose whole nature is exactly as characterized by *HP* has title to be the *definiendum*. The others are excluded for the same reason that "*known by God* to be green and observed or blue and unobserved" and "green and observed *and such that* $e^{i\pi} + 1 = o$ or blue and unobserved and such that $e^{i\pi} + 1 = o$" are excluded as admissible interpretations of *grue* by the real definition given above.

Provided, then, that it may legitimately be stipulated that *HP* is to be received as a real definition of the numerical operator, libertinism about the function it expresses is precluded. And now since determinacy of the function denoted by 'N...' immediately gets us uniqueness of reference for the numerical terms—it is, after all, a *function* we are defining—libertinism about their referents too is precluded. The numbers *cannot*—metaphysically—just be anything you like.[72]

But what about perversity? What about a sceptic who grants that the N-terms introduced by *HP*, when it is taken as a real definition as explained, must be conceived as determinate in reference, but nevertheless sticks to it that we can have no clue about what their reference is reference to? To be sure, we can propose in addition a real definition of the predicate, x is a *Number*, along the familiar lines:

$$(\forall x)x \text{ is a Number iff for some } F, x = Nx{:}Fx.$$

When this is understood as a real definition along the lines sketched, then anyone who understands it comes to know that "Number" picks out a property such that all there is to having the property is being the value of the essential numerator for some concept *F*. That can *sound* like something helpful—that numbers have no nature except to be the values of the essential numerator for suitable concepts. But of

---

[70] See especially Boolos (1997).   [71] Hodes (1984).

[72] Rosen and Yablo acknowledge that there is a possible question whether distinct items might have the same real definition, citing the example of the complex numbers, *i* and –*i*, whose whole essence, arguably, is that the each squares to −1. Their response is to suggest that that is a possibility only for objects.

course there is an ambiguity. It is of the *property* that we have just given a real definition. All there is to *its* nature is to apply to things that are the values of the function. But nothing yet follows about the essential nature of those values. All there is to the nature of the property *red*—or so someone might hold—is to apply to things that look a certain way under certain canonical conditions. But nothing limitative follows about the nature of red things, save that they have to be capable of so looking under such conditions.

How, then, do Rosen and Yablo propose to advance to a conclusion not about the function, or the property *number*, but about the *objects* to which the latter applies? For it is such a conclusion that we need in order to address perversity. Their attempt is an argument with the following simple overall shape:

(P1)   Facts of the form [Nx:Fx=a] cannot be *brute* facts. When they obtain there must be an account of why they obtain, of what grounds them.

(P2)   On the perverse hypothesis some facts of that form would be brute and unaccountable.
So, the perverse hypothesis is false.

If the argument works, then we are barred from supposing that any particular number could be identical with any entity a such that, were that so, the fact of the identity would be brute. The only true identity statements configuring a term introduced by *HP*—or, generalizing, by any abstraction—will have to be, one and all, accountable.

Rosen and Yablo argue in some detail for P2. They have little difficulty in making a plausible case that nothing we can glean from HP, viewed as a real definition of the essential numerator, pooled with whatever we might come to know about Julius Caesar, would provide the wherewithal to explain the identity of the latter with the referent of any numerical term introduced via *HP*. But their overall argument, remember, is intended as metaphysics rather than epistemology. There is a distinction between the thesis that nothing that we can know or reasonably argue for can provide an explanation of a certain claim and the thesis that the claim itself is *metaphysically* surd: that, supposing it true, there would be nothing to explain its obtaining. What would turn the trick as far as P2 is concerned would be a compelling argument that the obtaining of a fact like [Nx:$x\neq x$=Julius Caesar], would have to admit of explanation in terms of knowledge we can in principle obtain concerning the essential numerator and the famous Roman *imperator* respectively. But I cannot find that Rosen and Yablo offer us any such argument. The perverse sceptic is thus left free to accept P1 but deny that any compelling argument has been provided for P2; the most that has been shown is merely that *we have no idea* how to give the explanations which, by P1, must be possible of any 'mixed' identities that may obtain.[73]

---

[73] Their conclusion about P2 is actually pretty qualified:

We cannot say with confidence that every constitutive explanation—every truth of the form *P obtains in virtue of* Q—must be mediated either implicitly or explicitly by a claim about the definitions or essences of the items that figure in *P*. So we shall only say this. It is hard to think what such an explanation would look like in the present case. If this is right, then it is reasonable to accept (P2). Facts of the form [Nx:Fx=a] would be brute if the perverse hypothesis were true.

Suppose P2 accepted, though. What in any case is there to be said for P1? Why should not some such identities simply *be* surd? Presumably some facts have to be, after all. Against this, Rosen and Yablo bring what appears at first sight to be an attractive idea. They introduce a notion of a *non-basic entity*, where an entity is non-basic just if it has a *reductive real definition*: for instance

a reductively definable *relation R* will be one whose real definition takes the form

$$\maltese_R(\forall x_1, ...x_n)(R(x_1, ...x_n) \leftrightarrow \phi[x_1, ...x_n]),$$

where R is totally absent from ϕ.[74] A reductively definable object a might be one whose real definition takes the form

$$\maltese_a(\forall x)(x = a \leftrightarrow \phi[x]),$$

where a is totally absent from ϕ, etc.[75]

They then propose the following principle:

(NB)   If a is a non-basic entity, then facts involving a are non-basic facts.

We are to think of facts that are non-basic as having to be explicable in terms of other kinds of more basic facts. So if NB is accepted, there can be no surd facts concerning entities that have reductive real definitions.

This suggestion surely does resonate with one aspect of our intuitive inclination to think of mixed identities involving abstract terms as generally false—or better: as lacking any ground for truth. An abstraction, after all, is or purports to be a means of explaining the truth-conditions of statements concerning the kind of entity it introduces in terms of the characteristics of things of a previously familiar, more basic kind. It thus goes with the meaning, one would like to say, that we thereby give to such statements that, whenever true, their truth is grounded in facts concerning the entities dealt with in the discourse that supplies the base language for the abstraction concerned. And if that is accepted, then the very consideration that sets the Caesar problem in the first place—the fact that the connections which an abstraction principle, along perhaps with other supplementary definitions, effects between talk of its proper abstracts and talk of previously familiar entities stop short of anything that settles the truth-value of mixed identities—is also sufficient to solve the problem, since we are now entitled to conclude that the problematic identity statements, exactly because they have not been grounded, are hence untrue.

I fear, though, that we are still in the thick of the woods. Suppose it granted that the essential numerator is a non-basic entity in the sense Rosen and Yablo propose. Still, in order to obtain the result they want, do we not need something additional, namely that its *values*—that is, the numbers—are non-basic entities too? And now we

---

[74] Absent, Rosen and Yablo explain, in the strong sense that R figures neither in ϕ nor in the definition of any of the constituents of ϕ, nor in the definitions of any of *their* constituents, etc.

[75] Chapter 5, this volume, p. 127. '$\maltese_x$' is the marker Rosen and Yablo use to indicate that what follows is to be received as a real definition of x.

confront again the problem noted earlier: *HP* is not a definition, real or otherwise, of the *values* of the function it serves, as a real definition, to define.

It may be replied that if we accept NB, then we must allow that *all* facts about the essential numerator, including what kinds of thing are comprised in its value-range, must be non-basic and hence accountable facts. But one can envisage a perverse sceptic replying that to try to exploit NB in that way is to give a question-begging interpretation of it. *Grue*, for example, when given the real definition above, is a non-basic quality. So, by NB, all facts about it, including what it does and does not apply to, have to be accountable. But it is clear that, in explaining any such facts, an *explanans* will have to draw on more than what the right-hand side of its real definition tells us about *grue*. It will have to draw in addition on facts about the entities to which *grue* does or does not apply, facts about their compliance or otherwise with the condition expressed on the right-hand side of the definition. There is thus no well-motivated requirement that absolutely *all* facts about *grue* be explicable *purely* in terms of the materials supplied by its real definition. So likewise, before we can conclude that the supposition of the identity of Julius Caesar with $Nx:x{\neq}x$ would offend against NB, we have to allow that, in attempting to explain the identity, it is allowable to adduce not just facts about the essential numerator as characterized by *HP* but facts about Caesar too. To be sure, we don't know and indeed cannot foresee any facts of the latter kind that look promising for the project. But that, the perverse sceptic will say, for all we are entitled to think to the contrary, is just a point about us—a point about our limitations.

Might we do better to try to apply the rubric for a reductively definable object suggested in the quotation above directly to the objects in the value-range of the essential numerator? Well, let's try that for o. The result, presumably, will be something like this:

$$(\forall x)(x=0 \leftrightarrow (\forall F)({\sim}(\exists y)Fy \rightarrow x=Nx:Fx))$$

—something is 0 just if it is the number of any empty concept. That, though, is not a *reductive* real definition in the sense that Rosen and Yablo intend. True, the symbol, '0,' disappears on the right-hand side. But their requirement was that the defined entity be "totally absent" from the right-hand side—and that way of expressing matters, though hardly pellucid, presumably involves that the right hand-side should do something other than simply place an identifying condition on the *definiendum*, that it should be *about* something else, a different subject matter in terms of which the reduction advertised by the phrase "reductive definition" is to be given.

Rosen's and Yablo's discussion expresses many attractive ideas. In particular, I wholeheartedly agree with their overarching thought that a satisfying solution to the Caesar problem, if there is one to be had, will draw on a developed metaphysics of abstracts. But I do not think they have shown us how to obtain such a solution simply by milking the resources given by the idea of real definition. The basic problem with what they propose, it seems to me, is that even when an abstraction principle is so viewed, the immediate pay-off is only for our conception of the abstractive function it introduces. Extra materials are needed if we are to draw sound conclusions about the nature of the objects that constitute its value-range. One attraction, it seems to me, of

Hale's and my proposed metaphysics of sortal concepts, categories, and criteria of identity is that it promises exactly such extra materials.

But enough. Let me take this opportunity to leave the reader with a sense of just how open the question, how best to lay Julius Caesar finally to rest, remains.

* * *

3. (Neo-)Logicism and Higher-Order Logic

3.1 Edwards

What is the bearing, if any, of Gödel's incompleteness theorems on logicism and neo-logicism? The classical logicist thesis for number-theory is that the arithmetical primitives—*natural number, zero,* and *successor*—allow of correct definition in purely logical terms, and all the truths of arithmetic are, when reformulated under these definitions, truths of logic. It has not been uncommon for philosophers to take it that this thesis is straightforwardly refuted by Gödel's results.[76] That assessment will be correct, certainly, if it is assumed that logical truth should be effectively (recursively) axiomatizable. But why should we assume that? In FCNO I wrote,

There is, a priori, no better reason to expect logical truth to be completely recursively axiomatisable than there is to expect the same of number-theoretic truth; if number-theoretic logicism [is] correct, Gödel...proved the incompletability of logic. As it is, we can still perfectly sensibly raise the question whether the *fundamental* truths of number theory, that is, the Peano Axioms, admit of proof in the way Frege believed.[77]

Indeed, provided its status as logic is granted, Gödel *did* anyway prove the essential incompleteness of the higher-order logic that undergirds Frege's project—though one consideration that has moved some sceptics about the claim of higher-order logic to *be* logic, properly so regarded, is exactly the background thought that logic, properly so regarded, *ought* to be completely axiomatizable. And one reason for that background thought might be provided by an underlying inferentialism according to which logical truth is wholly determined by logical form and the meanings of the logical constants as fixed by their respectively characteristic rules of inference. For how, consistently with such a view, could something be a truth of logic that could not be established by reasoning in accordance with such rules?[78] (There is, to be sure, a lacuna in that train of thought which I shall come to below.)

There are, however, as the quotation from FCNO indicates, two separate questions here. One is whether a sustainable logicism, or neo-logicism, can be developed for the fundamental laws of number-theory—in effect, the Dedekind-Peano axioms. Call the claim that there is such a sustainable (neo-)logicism the *Core logicist thesis*. The other is whether, if the Core thesis holds, something worth regarding as logicism, or neo-logicism, can be maintained for *all* the—non-recursively axiomatizable—truths of

---

[76] For instance Henkin (1962) and Musgrave (1977).   [77] FCNO, p. 131.
[78] Cf. Shapiro (1998).

number-theory. Call the claim that that too is so, the *Supplementary logicist thesis.* The incompleteness theorems have no bearing on the Core thesis. But do they somehow scupper the Supplementary thesis?

It is on the latter question that Jim Edwards' elaborate and challenging chapter bears. One might well expect that a (neo-)logicist, properly so described, will want to maintain that whatever it is of metaphysical or epistemological significance that she takes the logicist project to establish for the fundamental laws of number-theory thereby extends to all the arithmetical truths, whether or not they are deductively accessible in a particular recursive axiomatization in which the fundamental laws feature—that versions of both the Core and Supplementary theses hold good. If the contention is to be that this inheritance is bestowed via the relation of *logical consequence*, then the (neo-)logicist needs to propose a suitable notion of logical consequence: one whereby *all* arithmetical truths may legitimately be regarded as logical consequences of the Dedekind–Peano axioms and, in addition, whereby all such consequences inherit the relevant metaphysical or epistemological status.

Edwards' argument is, in effect, that no such notion of consequence is in prospect. First-order logical consequence, since completely axiomatizable, cannot be of service here. Indeed, Edwards takes it, no purely deductive (proof-theoretic) notion of consequence can support the Supplementary thesis since the annexure of any such notion to the Dedekind–Peano axioms will generate only an incomplete system of arithmetic. So the (neo-)logicist who wants to maintain that *all* the truths of arithmetic are logical consequences either of truths of logic and correct definitions, or of *HP* and correct definitions, is presumed to be compelled to frame the project under the aegis of some *semantic*—model-theoretic—notion of *second-order* logical consequence.

The salient approaches are accordingly two. Both treat the second-order variables as ranging over sets of the elements of any specified objectual domain, D. But whereas classical, or *full*, semantics for second-order logic treats the second-order variables as ranging over the full power set of D, regarding it as fully determinate just what range of sets is comprised thereby, general, or *Henkin*, semantics treats the range of second-order variables as *open to interpretation*, even after D is fixed, depending on what choices are made for predicate and function comprehension axioms (though usually including the full classical power set as one legitimate case). Henkin semantics accordingly allows for a plurality of admissible but differing interpretations of second-order "All...," even after the objectual domain is fixed.

Since which inferences are valid will naturally vary when the range of the second-order quantifiers is allowed so to vary, one might expect the notion of second-order consequence to fragment in such a treatment. Someone who favoured the general approach could pre-empt such fragmentation by going supervaluational, taking a valid second-order entailment to be one that holds good for any selection of objects, D, for the objectual domain and *any*—however eclectic—interpretation of "All subsets of D." The effect of this move, however, would be a notion of second-order consequence that was purified of any substantial Comprehension assumptions. The more natural direction is accordingly a restricted form of supervaluation. The specific notion of consequence which Henkin proposed and which interests Edwards is that determined by the class of interpretations of the second-order domain which are "faithful,"[79] that is, interpretations which provide—from a classical perspective—for sufficient subsets of D to validate every instance of the full impredicative Comprehension scheme.[80] When Edwards speaks of Henkin consequence, he has in mind truth-preservation in all such *faithful* interpretations of the second-order domain.

It is probably fair to say that the majority of logical theorists who are friendly towards higher-order logics tend to regard Henkin-semantics, so understood, as at best something of a mere model-theoretic *jeu d'esprit*, with no claim to recover the intuitive notion of second-order consequence that they reckon they possess.[81] "All," should mean *all*, after all! Edwards, for the reasons developed in his Chapter 3 (this volume), takes a different view, arguing that principles that merit a place in any plausible methodology of interpretation render it extremely doubtful whether an interpreter of our second-order inferential practice could justifiably prefer the ascription to us of the notion of full consequence, rather than that of faithful Henkin consequence. I will not attempt to take the measure here of the complex argument that he develops, though I do have some sympathy for the thought that a notion of consequence defined in terms of Henkin models has a certain pre-theoretic attraction. When first-order validity is understood as truth-preservation in any non-empty objectual domain, it seems merely in keeping, and of a piece with the essential generality of logic, that second-order validity should likewise be robust under *some* degree of variation in the range of predicates and functions that are taken to superintend a given domain of objects.

However that may be, though, the important point for Edwards's purpose is that second-order Henkin validity, understood as outlined, is in any case completely axiomatizable. There is again, therefore, by Gödel's theorem no option of maintaining that every arithmetical truth is a Henkin-consequence, so understood, of the Dedekind–Peano axioms.

So, it may appear at this point that nothing other than full consequence can serve the purposes of any logicist who embraces the Supplementary thesis. Full consequence seems to be the only game in town. It is an old observation that the full consequences of the second-order Dedekind–Peano axioms are exactly the arithmetical truths. The former are the statements that are validated under every full interpretation of the axioms. The arithmetical truths are those statements that are true under the intended interpretation of the axioms. But the categoricity of the axioms, originally proved by Dedekind, ensures that they have only one, viz. the intended, interpretation. So any full consequence of them, *qua* true under every interpretation of them, is true under the intended interpretation, and is thereby an arithmetical truth; and every arithmetical truth, *qua* true under the intended interpretation, is thereby true under every full interpretation of the axioms, and is thereby a full consequence of the axioms. Thus if all arithmetical truths—and specifically all such truths accessible by a recursive process of augmentation of the Dedekind–Peano axioms by successive

---

[79] Shapiro (1991: 89).
[80] They are also interpretations that provide for every function demanded by the Axiom of Choice if that is included among the second-order logical axioms.
[81] Boolos, Chapter 4, this volume.

Gödel sentences[82]—are to be regarded as logical consequences of the Dedekind–Peano axioms as the Supplementary thesis requires, then the contention has to be, it seems, that full consequence is properly regarded as *logical* consequence.

It merits remark that a (neo-)logicist who proposes to defend the Supplementary thesis in this way has of course surrendered any vestige of the *epistemological* project that Frege had. Frege's logicism was a project in ideal epistemology: it was to supply a basis whereby, for any particular arithmetical truth, a route could be constructed, starting in acceptable definitions and involving only logically valid inferences, that could lead an ideal thinker to recognition of that truth. But that an arithmetical sentence is a full consequence of the Dedekind–Peano axioms is no guarantee that there is any such route to recognition of its truth. Full consequence, even if we grant it the title of logical consequence, is not fitted to sustain the kind of epistemic inheritance relationships that are integral to (neo-)logicism, understood as an epistemic thesis.[83]

Edwards' central claim is that full consequence should not be regarded as logical consequence at all. His argument is framed by the following intuitive characterization of logical consequence (PC). A sentence φ is a logical consequence of a set of sentences Γ iff

It is impossible for φ to be false and all the members of Γ true, where the meaning of the logical constants in φ and in the members of Γ, and otherwise their logical forms, are sufficient in context to generate this impossibility.[84]

If (PC) is acceptable, then in order for full consequence to rank as a case of logical consequence, it has to be the case that for any sentence φ and set of sentences Γ in which '∀,' '→,' and '¬' have the meaning assigned by full semantics, if φ is a full consequence of Γ then φ is a logical consequence of Γ by the lights of (PC); that is, it should not merely be impossible for φ to be false while every member of Γ is true, but this impossibility should be grounded in the logical forms of the theses involved and the meanings assigned by full semantics to the logical constants. The problem Edwards finds with this, as I interpret him, is that all we know about the meaning of higher-order '∀' in full semantics is that, for any domain of objects, D, its range is to embrace all the subsets of D that there are. The truth-conditions of clauses in which that quantifier is the principal operator are thus hostage to the behaviour of the classical power set operation, which for all we can say to the contrary, may vary in unsurveyable ways as we advance out into the Cantorian transfinite. In particular, for any given D, the claim that $(\forall F)\varphi$ is a full consequence of Γ, may be protected from

---

[82] I am of course assuming—with Edwards—the orthodox view that the Gödel-sentences do indeed express arithmetical truths.

[83] That full consequence has this limitation would, to be sure, still be consistent with the attempt to annexe it to a thesis about the *metaphysical ground* of arithmetical truth, of the kind that I attributed above to Demopoulos and which Frege also apparently held: the thesis would be that all full consequences of the Dedekind–Peano axioms inherit the ground of the truth of the latter, which are in turn, for a logicist, grounded in logic and definitions. I will not explore that idea further here, except to remark that it is not at all a trivial question whether full second-order consequence, with its intrinsic reliance on the classical notion of power set, is a suitable conduit for metaphysical ground.

[84] Chapter 3, this volume, page 63.

counterexamples due to the local behaviour of the power set operation on D even though there may be, for example, an enlargement of D, D'—one which exists if, for example, certain large cardinal axioms are true—such that the power set operation on D' gives rise to a counterexample to the same pattern of inference.[85]

A possible reaction to this would be to claim that the meaning of the second-order quantifier in $(\forall F)\varphi$ will shift with change in the objectual domain and hence that such potential variations in the status of the claim that $(\forall F)\varphi$ is a full consequence of Γ are still consistent with the claim of full consequence to implement logical consequence as characterized by (PC). However, that suggestion would save the letter of (PC) at the cost of surrendering the generality of logic. Logically valid inference is meant to offer a guarantee of truth-preservation irrespective of choice of a domain. If that guarantee is to be issued in a manner that accords with (PC), the meanings of the quantifiers cannot be allowed to vary with the range of their bound variables.

So the Supplementary thesis may appear at this point to be unsustainable. Henkin semantics offers no notion of consequence suitable for it, since no completely axiomatizable notion of logical consequence can support the thesis. And we are in no position to take full consequence to be even extensionally coincident with logical consequence, at least not if the latter is as pre-theoretically characterized by (PC) and the train of thought just outlined is cogent. However, I think Edwards overlooks another possibility. Let me explain.

Earlier we moved very quickly past the idea that any deductive notion of logical consequence might serve the needs of the Supplementary thesis. The thought was that Gödel has shown that no recursively axiomatizable system of logic could serve, and—we tacitly assumed—any deductive (proof-theoretic) notion of logical consequence would have to be recursively axiomatizable. But the tacit assumption is open to doubt.

The doubt emerges if we digress a little to review a Gödel-inspired challenge to a broadly inferentialist approach to the meanings of the second-order quantifiers. Such an approach will regard neither full semantics nor Henkin semantics as anything more than alternative model-theoretic representations of meaning. The real determiners of the meanings of the higher-order quantifiers are, rather, the inference rules that primitively govern them. Gödel's incompleteness results pose an evident difficulty for such a view. Consider a standard axiom set, 2PA, for second-order arithmetic, and let G be the Gödel sentence for the system resulting from adjoining 2PA to a formalisation of full classical impredicative second-order logic. Let 2G* be the conditional, 2PA $\rightarrow$ G. Clearly this cannot be proved in the system described (if consistent). And the usual informal reasons for regarding G as a truth of arithmetic weigh in favour of regarding this conditional as a truth of arithmetic. But they are not reasons for regarding it as a truth of logic—it is shot through with non-logical vocabulary. However, consider its universal closure $\forall 2G*$. This is expressed purely

---

[85] As Riki Heck has noted (personal correspondence), there certainly are such cases. For example, there is a pure second-order formula A that is satisfiable in all and only domains D whose cardinality is not measurable. If there is no measurable cardinal, then A is satisfied in all domains and so is a validity of second-order logic will full semantics. If there is a measurable cardinal, on the other hand, then A has 'far out' counterexamples.

second-order logically but it too cannot be provable in second-order logic, for if it were, we would be able to instantiate on it to prove the conditional $2G^*$, and thence to prove G in the relevant second-order arithmetic. So if there is reason to regard $\forall 2G^*$ as a logical truth/validity, that will be a reason to regard second-order validity as underdetermined by the second-order inference rules, including those for the quantifiers—a body-blow, seemingly, against our assumed inferentialist conception of their meaning.

There is indeed such a reason. *Third-order* logic gives us the means to define a truth-predicate for second-order arithmetic and thereby to mimic rigorously in a formal third-order deduction the informal reasoning that justifies the conclusion that G holds good of any population of objects that satisfy 2PA. So $2G^*$ can be proved using just third-order logical resources, without special assumption about the arithmetical primitives. There is therefore also a legitimate generalization to $\forall 2G^*$. So now we have a sentence that can be expressed by means purely of second-order logical vocabulary, and established as a theorem of third-order logic but cannot be derived in second-order logic.[86] If third-order logic is logic, this sentence is a logical truth that configures no logical concepts higher-than-second-order yet cannot be established by the inference rules of second-order logic.

The situation generalizes. From a non-model-theoretic perspective, the incompleteness entailed by Gödel's results for each nth order of higher-order logics consists in the deductive non-conservativeness of its hierarchical superiors over it—each nth-order logic provides means for the deduction of theorems expressible using just the conceptual resources of its predecessor but which cannot be derived using just the deductive resources of its predecessor. But then at no nth order is it open to us to regard the meanings of the quantifiers at that order as fully characterized by the nth-order rules—otherwise, what could possibly explain the existence of truths configuring just those meanings but underivable by means of the rules in question?

There are various possible responses. The would-be inferentialist might attempt a case that third-order logic is *not* logic, so that some of its theorems, even those expressed in purely second-order logical vocabulary, need not on that account be rated as logical truths.[87] But the prospects for this direction look pretty forlorn—after all, the details of an inferentialist conception of quantification at third order, as displayed in the relevant introduction and elimination rules, will be a very close match for those at second-order, differing only in the level of the generalization involved. How could the one kind of operation be 'logical' and the other not?

I think the correct inferentialist response to the problem is different, and—the point I have been steering towards—that it also points to a conception of logical consequence that both accords with (PC) and can enable a supporter of the Supplementary thesis to finesse Edwards's critique. The key thought is as follows. Epistemologically, it would be a mistake for the inferentialist to think of higher-order quantifiers as coming in conceptually independent layers, with the meanings of the

---

[86] For details, see §§3.7 and 4.1 of Leivant (1994).

[87] By way of a parallel, the at least countable infinity of the universe may be expressed in purely first-order logical vocabulary. Still, it may coherently be regarded as a necessary *mathematical* truth, rather than a logical one.

second-order quantifiers fixed by the second-order rules, the meanings of the third-order quantifiers fixed by the third-order rules, and so on. Rather, she should maintain that it is the *entire open series* of pairs of higher- and higher-order quantifier rules which collectively fix the meaning of quantification at each order: there are *single* concepts of higher-order universal and existential generalization, embracing all the orders, of which it is possible only to give a schematic, order-neutral statement.[88] Higher-order quantification is a *uniform operation*, open to a single schematic inferentialist characterization, and there is no barrier to regarding each and every truth of higher-order logic as grounded in the operation of rules that are sanctioned by that characterization.

The upshot of this conception of the matter is that while higher-order logic, in full generality, is not axiomatizable—its validities comprise an indefinitely extensible totality in exactly the sense in which Gödel shows that the truths of arithmetic comprise an indefinitely extensible totality[89]—it remains open to a supporter of the Supplementary thesis to maintain that, for all Gödel's results have to say to the contrary, every validity of higher-order logic is provable using the deductive resources available at some (possibly transfinite) nth order. No semantic or model-theoretic conception of validity is needed. The same claim may then be advanced for Gödel sentences as a class: each such sentence, for a particular axiomatization of arithmetic, may be established by ascending to a higher-order logic in which an appropriate truth predicate for that axiomatization may be defined so that an induction may be executed culminating in a proof of the Gödel sentence in question. Of course the resulting system will have its own Gödel sentence, but a deductive proof of that will be available in turn by ascending to a yet higher-order higher-order logic in which an appropriate truth-predicate can be defined.

If this is defensible, then Edwards's critique of logicism rests on a false dilemma. The correct conception of logical consequence for a proponent of the Supplementary thesis is neither full consequence, nor Henkin consequence, nor any semantic conception of consequence, but *deductive* consequence in the indefinitely extensible hierarchy of higher-order quantificational logics.

All that said, it still has to be acknowledged, of course, that the proposal speaks only to the problem posed for a proponent of the Supplementary thesis by Gödel's incompleteness theorems. There is, so far as I am aware, no reason to think that for every arithmetical truth, φ, there is some nth-order quantificational logic such that φ is a deductive consequence of the Dedekind–Peano axioms in that logic. Wiles's proof of Fermat's theorem, for example, has essential recourse to principles of algebraic geometry. To suppose that some recursively axiomatized higher-order logic offers the means to deduce the theorem from no assumptions other than the Dedekind–Peano axioms would be, in the present state of our understanding, an article of faith.

---

[88] Cf. Wright (2007).

[89] Readers familiar with Shapiro and Wright (2006) will appreciate that this is a limited form of indefinite extensibility—it is an instance, as we there expressed it, of up-to-l extensibility, where l is a countable but non-recursive ordinal—rather than proper indefinite extensibility as there characterized. See Shapiro and Wright (2006: 264–9).

3.2   Misgivings about the Logicality of Higher-order Logic

As we remarked, the significance, metaphysical or epistemological, which one will attribute to Frege's Theorem is going to vary not just with one's view of the status, metaphysical or epistemological, of *HP* but also with one's opinion of the inferential machinery—traditionally, classical impredicative second-order logic with unrestricted comprehension—whereby the result is proved.

Classical impredicative second-order logic with unrestricted comprehension was, in effect, Frege's invention,[90] and he seems never to have entertained any doubt that his brainchild was fit for his logicist purpose. It's notable that in *Begriffschrift* he makes nothing of the distinction between quantifying into positions occupied by what he called *Eigennamen*—singular terms—in a sentence and quantification into predicate position or, more generally, quantification into open sentences—into what remains of a sentence when one or more occurrences of singular terms are removed. Not only does he seem to have conceived of both alike as perfectly legitimate forms of generalization, each properly belonging to logic. More: he seems to have conceived of quantification *as such* as an operation of pure logic and, throughout his career, to have recognized no important distinction between first-order, second-order, and higher-order quantification in general.

This way of thinking is actually quite natural. Why did it become controversial? How did it come to be so widely believed that in his treatment of quantification Frege overlooked a major distinction?

The short answer is: Quine. Writing in a philosophical milieu when 'ontic parsimony' seemed a more urgent *desideratum* than perhaps it does now, Quine was preoccupied with possibilities of masked ontological commitments and, conversely, of merely apparent ontological commitments. His proposal, famously, was: regiment a theory using the syntax of individuation, predication, and quantification, and then see what entities you need to regard as lying in the range of the bound variables of the theory if it is to rank as true. You are committed, as a theorist, to what you need to quantify over in so formulating your theory. (As to the question of how we are supposed to recognize the *adequacy*—or inadequacy—of such a regimentation, Quine had, of course, relatively little to say.)

Quine's cardinal thought about quantification was thus that it is intimately tied to—indeed, is the canonical expression of—ontological commitment. This is not *as such* at odds with the Fregean idea of quantification as a level-neutral uniform logical operation. But once one falls in with it, first- and second-order quantification do suddenly emerge as standing on a very different footing. First-order quantification quantifies over objects. No-one seriously doubts the existence of objects. By contrast,

---

[90]   As Riki Heck observes (Chapter 2, this volume, p. 35), this way of characterizing Frege's achievement is not quite right. There are no comprehension axioms in the systems of Begriffschrift and Grundgesetze. Rather, Frege has a rule of substitution: where F is a predicate of whatever degree and [...F...] is a theorem, one may—subject to standard restrictions—infer the corresponding [...φ...] resulting from replacing all occurrences of F in [...F...] by any open sentence, φ, of the same degree. As Heck notes, the substitution rule is equivalent to full impredicative comprehension. The proof of Frege's Theorem in FCNO proceeds via the substitution rule (not there explicitly formulated.)

second-order quantification is naturally taken to demand a domain of universals, or properties, or concepts. And of such entities Quine canvassed an influential mistrust: a mistrust based, initially, on their mere abstractness—though he himself later, under pressure of the perceived needs of science, swallowed something of his antipathy to the abstract—but also on the ground that they seem to lack clear criteria of identity: a clear basis on which they may be identified and distinguished among themselves. It was the latter consideration which first led Quine to propose that the range of the variables in higher-order logic might as well be taken to be *sets*—abstract entities no doubt, but ones with a clear criterion of identity given by the axiom of extensionality—and then eventually to slide into a view in which "second-order logic" became, in effect, a misnomer, unless, at any rate, one regards set theory as logic. By 1970 he had come to his well-known view:

Followers of Hilbert have continued to quantify predicate letters, obtaining what they call higher-order predicate calculus. The values of these variables are in effect sets; and this way of presenting set theory gives it a deceptive resemblance to logic... set theory's staggering existential assumptions are cunningly hidden now in the tacit shift from schematic predicate letters to quantifiable set variables.[91]

Those remarks occur in Quine's chapter, "The Scope of Logic" in the subsection famously entitled: "Set Theory in Sheep's Clothing!" By the end of that chapter, Quine has persuaded himself, and probably most of his readers too, that Frege and others such as Russell and Hilbert who followed him in allowing higher-order quantification at all, simply overlooked the implicit transition from logic properly so regarded—the theory of the valid patterns of inference sustained by the formal characteristics of thoughts expressible using singular reference, predication, quantification, and identity—to set theory which, to the contrary, in Quine's view, is properly regarded as a specialized, though highly general and fruitful, branch of mathematics.

3.3 *Boolos*

Quine's negativity about higher-order logic has of course met with some distinguished resistance over the years, notably—to cite three very different examples—in Stewart Shapiro's masterly *Foundations without Foundationalism*,[92] in various of Bob Hale's writings about the issues,[93] and of course in those of George Boolos's papers that develop the idea that at least as far as monadic higher-order generalization is concerned, quantification can be treated as plural quantification over individuals.[94] Boolos's present Chapter 4 (this volume), however, predates that proposal and

---

[91] Quine (1970: 68).   [92] Shapiro (1991).   [93] Hale (2012, 2013, and 2015).

[94] See especially the articles, "To Be is to be the Value of a Variable (or to be Some Values of Some Variables)" and "Nominalist Platonism" reprinted as chs 4 and 5 in part 1 of Boolos (1998). Boolos was, of course, well aware that there is no obvious way of extending this suggestion to the interpretation of quantification into the positions occupied by predicates of higher degree. His goal was not to give a thoroughgoing non-set-theoretic interpretation of higher-order quantification but rather to provide a means for interpreting certain generalizations about 'all sets' that strike us as intuitive and true. A heroic effort at polyadic extension of Boolos's idea is nevertheless made by Rayo and Yablo (2002). For pertinent criticisms, see Rossberg (2015).

is not concerned with misgivings about the logicality of higher-order logic that flow from the question, how best to understand the range of its higher-order variables. The focus is rather on challenges based on the disparities between key findings in the meta-theory of second-order logic and their counterparts for the first-order case. Such challenges arise for Boolos, interestingly, in the context of a sneaking sympathy for the idea that, in some interesting sense, arithmetic might be reducible to second-order logic. He writes that,

> I think that a fair judgement of the accomplishments of Frege and Dedekind is impossible without a correct understanding of second-order logic, for it does seem to me that their work can be said, with some justice, to have effected a reduction of arithmetic to logic. "Logic" here means standard (or "full") classical, non-axiomatic second-order logic, ...[95]

Let us first pause briefly over the reduction Boolos has in mind.

The reduction is sourced in the observation, already marked above, that the Dedekind–Peano axioms (augmented with suitable recursive clauses for addition and multiplication) allow of categorical formulation at second-order. Since any true arithmetical sentence, S, is true of the unique structure that the axioms characterize, it follows that any such sentence is true in every interpretation in which the axioms are true, and hence that the conditional consisting of the conjunction of the axioms as antecedent and S as consequent is likewise true in every such interpretation and hence—prescinding from our discussion above of Edwards—a (semantic) logical truth. Goldbach's conjecture (G_{old}), for instance, is an arithmetical truth just if the second-order conditional, $2PA \rightarrow G_{old}$, is a second-order logical truth.

Is that relationship worth calling a 'reduction'? Obviously no mustard is cut unless we take it—again, *pace* Edwards—that second-order (full) semantic consequence is indeed a species of logical consequence. Otherwise no case is made for thinking that conditionals of the illustrated ilk are truths of logic. But suppose we waive any doubt about that and grant that, for example, $2PA \rightarrow G_{old}$ deserves, if true, to be regarded as a truth of logic. Then again: do we have here the makings of a case for a genuine reduction?

Frege would certainly have scorned the idea. For Frege, a genuine reduction of arithmetic to logic must be required to show how logical resources alone provide means whereby we can apprehend the finite cardinals as objects and recognize that they comply with the axioms of number-theory. Neo-logicism, in its turn inherits exactly that project and proposes to implement it by the use of abstraction principles. But conditionals of the kind illustrated offer no such means—precisely because they carry no unconditional ontological import. The Goldbach conditional, even if indeed worthy of the title, 'truth of logic,' is consistent with there being no numbers and hence with 0 and its suite being fictions.

The 'reduction' Boolos moots thus gets the truth-conditions of arithmetical statements wrong—*unless* they wear a misleading face and their content is indeed no more than implicitly hypothetical. Obviously there is a debate to be had about that,

---

[95] Chapter 4, this volume, p. 103.

and indeed about the "if-thenist" tendency in general.[96] But I shall not pursue that debate here—except to remark that Boolos's later tendency to regard it as a decisive objection to logicism that nothing worthy of the title, 'logic,' could carry the existential commitments of number-theory, and that no theory with those commitments could legitimately rank as 'analytic,' marks a change of mind in this respect, if indeed he was ever a settled "if-thenist."

It bears re-emphasis in any case that, even for an if-thenist, the prospect of the mooted reduction is of no bearing on logicism unless second-order logic is indeed *logic*. The challenge Boolos's Chapter 4 (this volume) is concerned to rebut draws on the reflection that while first-order logic is compact, complete with respect to the standard semantics and subject to the Löwenheim–Skolem theorems, second-order logic has none of these properties. If first-order logic is the paradigm of logic—logic par excellence, it were—then what title does second-order logic have to be called "logic" in view of these marked differences from the paradigm?

It is easy to miss the point of the challenge. A bluff response would be, "Well, but who said that these features of first-order quantification theory are paradigmatic of logicality as such?" But Boolos is not content with that, aptly observing that these points of difference between first- and second-order logic also mark points of *similarity* between the latter and set theory. If second-order logic walks like set theory, quacks like set theory . . . .

Boolos's discussion of the issues of compactness and Löwenheim–Skolem, however, though only a little less brisk than the bluff response, is, as it seems to me, compelling. He offers the excellent observation that compactness is actually a *limitation*. The inconsistency of, for instance, the infinite set of sentences,

{"It is not the case that there are infinitely many stars," "It is not the case that are no stars," "It is not the case that there is exactly one star," "It is not the case that there are exactly two stars" . . . }

is intuitively of kinship with, and no less 'logical,' than that—capturable, of course, in first-order logic—of the finite set of sentences

{"It is not the case that there are at least three stars," "It is not the case that are no stars," "It is not the case that there is exactly one star," "It is not the case that there are exactly two stars"}[97]

Since any finite subset of the former is consistent, any theory adequate to capture the inconsistency of the set had better be non-compact; and since the inconsistency is intuitively logical in both cases, there is a case for saying that it is first-order logic, on this point, that actually comes short as a logic.

As for Löwenheim–Skolem, that any set of first-order expressible sentences have a countable interpretation if they have any infinite interpretation reflects only that first-order logical resources do not suffice to characterize the distinctions between countable infinity and the higher infinities. To regard that as a *desirable* limitation for

---

[96] For a recent, characteristically subtle discussion of the issues, see Yablo (2017). Yablo's is the lead article in a symposium, with comments by Stuart Brock and Richard Joyce, Mark Colyvan, Seahwa Kim, Gideon Rosen, and Amie L. Thomasson, and further replies by Yablo.

[97] I have tweaked the examples slightly to emphasize the symmetry.

a logic, properly so regarded, is just to assume something that needs to be argued, viz. that second-order logic's expressive richness takes it beyond the bounds of logic, properly so termed.

Completeness, though, in Boolos's view, is another matter. For whereas compactness and the Löwenheim–Skolem property are not obviously—for the reasons given—desirable features of a logic *qua* logic, completeness surely is: it is surely a desideratum that all the validities of a logical system be generatable by its proprietary proof procedures. Boolos's response is that, if that is a desideratum, so also, surely, is decidability. And in this respect first-order logic too falls down. There is no effective decision procedure for polyadic first-order quantification theory. There is one for the monadic fragment. But so there is for the monadic fragment of second-order logic. If possession of a complete proof procedure is regarded as a necessary condition for the status of a theory as logic, why is decidability not also? Or are we seriously in the market for the verdict that the first-order theory of relations is not logic?

There is, though, I think, a significant difference. As noted earlier, the incompleteness of a deductive system presents a special problem for its status as logic if we conceive that logical validity should everywhere be grounded in logical form and the meanings bestowed on the logical operators by their associated rules of proof. For how then can it happen that some valid schemata may be unrecognizable by iterated application of those rules? By contrast, there does not seem to be the same pressure to say—as decidability would require—that *in*validity should be so recognizable; or not, at least, if the system is such that not every invalid schema has a valid negation.

However, if what I offered above in commentary on Edwards is sound, we can augment Boolos's argument here with the reflection that something less than completeness may anyway be sufficient to scratch the itch. Incompleteness does not force us to allow that there is a bona fide conception of *logical* validity operative in higher-order logic that transcends any essential connection with deductive discoverability by methods of proof that characterize the relevant logical operations. The concern can be assuaged by something less than complete axiomatizability. It should be enough that *any* higher-order logical validity can be deductively demonstrated by logical means, even if there is no complete axiomatization of means sufficient for a demonstration of *all* such validities. (To be sure, there is no conclusive argument to hand that this is indeed the situation—only an argument that in so far as the grounds for doubting that it is so are generated by Gödel's incompleteness theorems, the doubt may be addressed.)

Boolos adds one further consideration into the mix: the continuity in the ideology deployed by the standard interpretations of first- and second-order logic and the associated standard semantic notion of validity, writing that

> there is an account of the conditions under which a first-order sentence is true in an interpretation which may be extended to second-order sentences in a perfectly obvious and straightforward way, without change in the notion of an interpretation or the definition of validity as truth in all interpretations.[98]

---

[98] Chapter 4, this volume, p. 112.

This, however, impresses me as a dangerous point for his purposes. The continuity in the standard model-theory for first- and second-order logics is good news for the friend of the latter only if "truth in all interpretations" can be explicated without reliance on the notion of set. Otherwise—and granting, as Boolos would seem to be implicitly assuming, that the ideology and ontology of one's meta-theoretic account of validity for the theses of a putative logic bears directly on its status as logic or not— the consideration is two-edged: someone who thinks that set is not a logical notion, but that recourse to it is essential in explaining what *first-order* validity is, should be concerned that even first-order logic is not pure logic.

True, one needs to remember here that Boolos is implicitly directing his argument against 1970s Quine, and that Quine would certainly not have been in the market for any conclusion of that sort. So *ad hominem* the point is fair and forceful: the Quinean needs to make out a worry about second-order quantification that impugns the status of second-order logic as logic but that does not impact on the status of first-order logic once it is acknowledged that the standard semantics for first-order logic requires a *domain*—a set—of individuals and interprets predicate-letters by assigning *subsets*, or *sequences* of elements of that domain to them. If this style of interpretation is viewed as giving the *meaning* of a first-order formula, then second-order logic crosses no interesting boundary, it may be charged, but merely involves explicit generalization over a range of entities to which first-order logic already implicitly involved commitment.[99]

---

[99] I am, in these remarks, going past a passage in Boolos's Chapter 4, this volume, which seems to be intended to make a related but stronger point but has puzzled me somewhat. I'll quote it in full. He writes (p. 104) that

> In first-order logic, the definition of truth in an interpretation I is usually given by means of the recursively defined auxiliary notion of satisfaction in I by a sequence of elements of the domain of I. Mates and others have noticed that a recursive definition of truth in I (for variable I) can be given which does not mention satisfaction: the trick, if it is one, is in the assumption that interpretations may assign denotations to individual constants, which are of the same type as individual variables. Interpretations will, of course, typically assign denotations to some predicate letters, thereby enabling us to extend Mates's definition of truth of a first-order sentence in an interpretation in a perfectly straightforward way to second-order sentences containing predicate variables. Notice that we have not needed to introduce a separate non-empty domain for each sort of non-individual variable in order to give a satisfactory, "objectual," definition of truth in I which agrees with the usual one that mentions satisfaction. It seems to me, therefore, that there is at least one sense in which second-order logic need not be regarded as "quantifying over sets" (or relations or functions). We do not need to assume that an interpretation associates a domain or a range with each kind of second-order variable in order to state when a second-order sentence is true in an interpretation.

Boolos's thought appears to be that since the style of definition of truth in an interpretation for a first-order sentence noted by Mates can be extended to second-order sentences without explicitly associating a domain of entitles of some kind—sets or sequences of individuals—with the second-order variables, we do not need, in order to explain second-order validity, to confront the question: what is the second-order domain—what do the second-order quantifiers quantify over? But—at the risk of betraying misunderstanding—that seems to me a mere technicality. When V is an n-adic second-order variable, surely the relevant range of possible assignments to it will just comprise all n-tuples of elements of the objectual domain—and haven't we then effectively, even if not in so many words, fixed these as the domain of the second-order quantifiers?

Boolos's own summary of the thrust of his discussion is characteristically reserved—merely that the meta-theoretic differences between first- and second-order logic do not provide for "a sufficient ground for the essentialist claim about logic[100] that Quine wishes to make." I would suggest that rather undersells the significance of his arguments. It would of course be more satisfying, dialectically, if one could do better than argue that certain mooted necessary conditions for the status of logic which second-order logic fails to meet are not really necessary. It would be more satisfying if there were some clear and agreed criterion in the offing satisfaction of which would *suffice* for logicality and which second-order logic could be argued, positively, to meet. But Boolos understandably had no clear proposal to offer on that score, and I don't either. Lacking such a criterion, the most natural direction in which to prosecute the issue is indeed that illustrated by his thought about uniformity of interpretation; namely, to argue that, whether in the way he suggests or some other, there is a perspective from which second-order logic emerges as essentially *continuous* with first-order logic, rather than as involving some kind of radical epistemic or ontic jump. I myself once made a proposal in that direction, which I shall take the opportunity to reappraise in the next section. Perhaps, though, this is a good place to emphasize that the key question for neo-logicism is not how far we can justifiably distribute the honorific term, "logic," but whether the proof-theory needed for an abstractionist recovery of the fundamental laws of arithmetic and analysis is, as it is sometimes said, "epistemically innocent": whether, that is, it imports no epistemological presuppositions at odds either with the generality—the subject-neutrality—of logic that Frege rightly emphasized or with its capacity to serve as a medium for the generation and transmission of knowledge that, at least for those in sympathy with the spirit of the vexed Kantian category, deserves to be captioned as analytic *a priori*.

3.4 Neutralism

The later Boolos who canvassed the treatment of second-order monadic predicate variables as ranging over pluralities of individuals, Hale who interpreted them as ranging over definable properties, and Shapiro who unashamedly regards them as ranging over sets, all retain the Quinean assumption that quantification, of whatever order, is essentially domain-involving, that it essentially involves a *range of entities* about which it gives us the resources to generalize in various ways, and to which its use commits us. This view of the matter has long been explanatory orthodoxy.[101]

---

[100] Viz. "that the [first-order] logic of quantification as classically bounded is a solid and significant unity" (Quine 1970: 91).

[101] For instance in the entry under "Quantifier" in Blackburn (1994) his *Oxford Dictionary of Philosophy*, Simon Blackburn writes,

> Informally, a quantifier is an expression that reports a quantity of times that a predicate is satisfied in some class of *things*, that is, in a 'domain.'   (my emphasis, p. 313)

while the corresponding entry at p. 338 in Antony Flew's and Stephen Priest's *Dictionary of Philosophy* Flew and Priest (2002) observes that

> The truth or falsity of a quantified statement...cannot be assessed unless one knows what totality of objects is under discussion, or where the values of the variables may come from.   (p. 338)

It is, however, not obvious that it is compulsory. In a paper published just over a decade ago,[102] I argued that the conception of quantification as *essentially range-of-entities-involving* is at best optional and restrictive and at worst a serious misunderstanding of what quantification fundamentally is. That paper proposed by contrast a *neutralist* conception of higher-order quantification—and indeed of quantification more generally—as expressed by the following principle:

(*Neutrality*):   Quantification into the position occupied by a particular type of syntactic constituent in a statement of a particular form cannot generate an ontological commitment not already associated with the occurrence of that type of constituent in a true statement of that form.

According to Neutrality, the ontological commitments of quantified statements go no further than those of their instances, whatever the commitments of the latter may be. So the view is open to someone who thinks that simple predications of "red," for example, *do* semantically entrain a universal of redness, or a set, or to some other kind of entity distinctively associated with that predicate as its semantic value. If we take the view that the semantic values of the predicates in a particular language are sets, neutrality is consistent with regarding the standard semantics for a higher-order logic for that language as part of the theory of its meaning. But Quine's attitude is pre-empted. For in Quine's view, while mere predication *is* free of any distinctive ontological commitment, quantification into predicate position must commit a thinker to an *extra* species of entity—in the best case, sets; in the worst, attributes or universals. This, according to neutrality, is an incoherent stance.[103]

What, though, if we *don't* take the view that a semantics of predication should associate a distinctive kind of entity with each meaningful predicate as its semantic value? The basic idea I wanted to set against the range-of-entities-involving conception is that quantification through expressions of any syntactic kind—that is, any expressions associated with a specific type of semantic contribution—should be viewed as essentially a device for generalizing exactly that: a device for *generalization of semantic contribution*. Given any syntactic category of which an instance, *s*, can significantly occur in a context of the form, [...*s*...], quantification through the place occupied by '*s*' is to be thought of as a function which takes us from [...*s*...], conceived purely as a content, to another content whose truth conditions are given as satisfied just by a certain kind (and quantity) of distribution of truth values among contents of the original kind. A quantifier is a function which takes us from a statement of a particular logical form to another statement which is true just in case some range of (possible) statements—a range whose extent is fixed by the quantifier in question—which share that same logical form are true.

According to neutralism, then, what if any entities are associated with higher-order quantification depends on one's semantics of predication. The commitment one undertakes in affirming, for example, a second-order universal generalization is to the truth of all of a certain range of thoughts sharing a certain logical form—

---

[103] Cf. Rayo and Yablo (2001). The objection is prefigured in Wright (1983: 133).

perhaps to all possible such thoughts, perhaps to all such thoughts expressible in a given, open-ended language, perhaps to some more narrowly contextually determined range. If, with the nominalist, we take the view that no thought of that structure makes any reference to, or otherwise draws on, an entity such as a property, or universal, or set, the result is a nominalism-friendly conception of higher-order generality; if, by contrast, we take it that the essence of predication is the ascription of a property, intensionally conceived, then neutrality will deliver the conception of higher-order generality that Quine reacted against. But note in that case—as Boolos in effect points out—that first-order logic too, insofar as it is a logic of schematic predication, will incur the same ontological commitments. For neutralism, there is no major watershed in this regard between first- and higher-order logics, and no distinction in logical character between the different levels of quantification. Second-order logic and indeed nth-order logic have, when their quantifiers are conceived along neutralist lines, exactly the same claim to be logic as first-order logic.

Above I cited Bob Hale as among the critics of Quine's views about second-order logic who nevertheless retain the Quinean assumption of quantification as essentially domain-of-entities-involving. In fact, though, the interpretation of second-order logic that Hale favours is exactly what neutralism provides on what is now standardly termed an *abundant* conception of properties. According to abundance, any well-defined predicate determines a property, that is, a way things can be. For any well-defined predicate is associated with a satisfaction condition, meeting which is one way for a thing, or things, to be. But conversely, also, Hale holds that properties extend no further then the possibilities for well-defined predication: a property just is a way which some possible well-defined predicate might be used to say how a thing, or things, are. In Hale's view, save where contextual restrictions apply, the range of the quantifiers in second-order logic should encompass all properties so conceived. Since any well-defined predicate must be finitely defined, it would seem to follow that the domain of Hale-properties will represent a considerable contraction by comparison with the domain of the predicate variables in second-order logic under its standard set-theoretic semantic interpretation, which encompasses the full power set of the first-order domain and hence, when the latter is infinite, includes an uncountable population of entities that, at least in any particular countable symbolism, allow of no finite specification. Still, it remains the case, so far as I can see, that there is, in one way, very little difference in spirit between Hale's view, so motivated, and the interpretation of second-order logic that neutralism issues on a nominalist view of predication, whereon even the metaphysically lightweight properties involved in the abundant conception are disdained. On both views, the commitment one undertakes in affirming a higher-order quantified statement, and hence what has to be established in order to justify one in doing so, will be that there exists a suitable range of intelligible true statements to bear witness to the truth of the former, *modulo* the extent of the generality involved in the quantification concerned.

3.5 Heck

Concern about the nature of the entities putatively quantified over—call this the *range problem*—is, however, merely one of several misgivings that classical second-order logic with full impredicative comprehension has tended to provoke. In their present chapter, Riki Heck characterizes a second as follows:

In the case of second-order logic, however, there is another and more fundamental problem with which we must contend: we must explain the second-order quantifiers. Absent such an explanation, we do not so much as understand second-order languages... Now, to understand the second-order universal quantifier, one must understand what it means to say that *all* concepts are thus and so. But to understand that sort of claim, so it is often argued, one must have a conception of what the second-order domain comprises. One must, in particular, have a conception of (something essentially equivalent to) the full power set of the first-order domain, and many arguments have been offered that purport to show we simply do not have a definite conception of [that].[104]

Heck wants to claim that their Arché logic, of which more shortly, finesses this second concern—call it the domain problem—by dispensing with second-order quantification altogether. As they realize, however, the point cannot be that quick, since Arché logic has its own resources for the expression of higher-order generality and there has therefore to be a substantive question whether understanding the generality it does express presupposes the allegedly problematic conception of the full power set of the first-order domain. Naturally, that conception is not required on Hale's interpretation of the range of the second-order quantifiers. But it might be contended that an analogue of the problem carries over, only now the problematic conception is that of all possible intelligible n-adic predicates (open sentences). True, the foreseeable grounds for suspicion of this notion are likely to differ from those that some (myself included)[105] have found with the idea of the classical power set of an infinite domain. The latter have to do with the implicit idea of an arbitrary infinite sequence of operations: the unspecifiability of the putative results of such an operation in uncountably many cases, and the implicit reliance on the idea of the completed infinite involved in the very idea that the could be such a thing as 'the result.' Concern about all possible intelligible n-adic predicates, by contrast, is more likely to focus on misgivings whether there can be any determinate such totality—since one would expect to be able to "diagonalize out" of any well-specified collection of such expressions and it is therefore questionable whether it is possible to possess any definite, stable conception of the domain of second-order logic's proprietary quantifiers as Hale conceives of it, at least when the quantification concerned is intended to be unrestricted.

Heck has, as it seems to me, a strong response to a critic who claims that Arché logic sweeps the domain problem under the carpet rather than serving to address it. The example under discussion is the antecedent of their proposed introduction rule for the ancestral; specifically, writing ϕ^*ab for 'a bears the ancestral of the relation ϕ to b,' the rule:

$$(\text{*-intro}) \quad \frac{[\forall x(\phi ax \to Fx) \ \& \ \forall x \forall y(Fx \ \& \ \phi xy \to Fy)] \to Fb}{\phi^*ab}$$

---

[104] Chapter 2, this volume, pp. 46–7.   [105] Wright (1985).

In Arché logic, this is a first-order schema; there is no binding 'F.' Nevertheless, Heck imagines the critic contending, our understanding of it

essentially involves just such a conception [of the full power set of the domain]. How else are we to understand... the premise of the rule... except as involving a tacit initial second-order quantifier? Does it not say, explicit quantifier or no, that all concepts F are thus-and-so? Does understanding that claim not require the disputed conception of the power set?

And they reply:

No, it does not. A better reading would be: a concept that is thus-and-so is so-and-thus. What understanding this claim requires is not a capacity to conceive of *all* concepts but simply the capacity to conceive of *a* concept: to conceive of an arbitrary concept, if you like. The contrast here is entirely parallel to that between arithmetical claims like $x + y = y + x$ involving only free variables and claims involving explicit quantification over all natural numbers. Hilbert famously argued that our understanding of claims of the former sort involves no conception of the totality of all natural numbers, whereas claims of the latter sort do, and that there is therefore a significant conceptual and epistemological difference between these cases. I am making a similar point about claims involving only free second-order variables as opposed to claims that quantify over concepts.[106]

Actually, however, Heck is here making two points at once: first, that there is a significant conceptual and epistemological distinction between having a notion of *an arbitrary* thing of a certain kind and having a notion of *all* things of that kind, in a sense involving some kind of conceptual representation of their totality; and, second, that this distinction aligns with that between grasp of the meanings of statements, like the premise for *-intro, involving free variables, and grasp of the meanings of the statements that result from them by binding those variables with quantifiers. My sympathies are entirely with Heck and Hilbert on the first point (though more needs to be said about it). But the second seems merely legislative. Granting the distinction in question, why insist that its second component is implicated in quantification per se—especially if that component is regarded as problematic in the crucial second-order case? Why should not both free-variable statements *and* universally quantified statements serve to express just schematic generality? Hale can say that that is precisely how quantification over abundant properties is to be understood. And on the neutralist understanding, a schematic interpretation of quantification is provided for from the start.

That is all I want to say here about the domain and range problems. So far, so good. But now the horizon darkens, unfortunately. The key question is whether we can assume that an interpretation of second-order logic of the kind sanctioned by nominalist-neutralism, or as suggested by Hale, has no implications for the practice of the system, in particular, whether such an interpretation poses no barriers to the derivation of Frege's Theorem and the reconstruction of our judgements of finite cardinal number in general. In Wright (2007)[107] I already noted that neutralism will struggle to underwrite the classical second-order comprehension axioms in any case where the domain of individuals is taken to be uncountable or more generally, for

---

[106] Chapter 2, this volume, p. 48.   [107] Wright (2007: 165–6).

whatever reason, to include elements which do not allow of any individuative specification. Classical comprehension has, for example, this instance:

$$(\forall x)(\exists X)(\forall y)(Xy \leftrightarrow y=x)$$

—intuitively, that for anything, there is a property which anything has just in case it is identical with that thing. This chimes with the neutralist connection between a second-order quantification and the intelligibility of the thoughts that constitute its instances only if each object in the first-order domain in question—in the range of 'x'—is itself a possible object of intelligible singular thought. With uncountably infinite domains that is exactly what is doubtful. Only an infinite notation could provide the means canonically to distinguish each classical real number, for example, from every other in the way that the standard decimal notation provides the means canonically to distinguish among the naturals. So, on plausible assumptions, no finite mind can think individuative thoughts of each and every real. Yet the instance of comprehension cited implies that to each of them corresponds a distinctive property.

More generally, consider *Grundgesetze* §118, which in the notation of *FCNO* is

$$Ny:(y=a)=Ny:(y=b)$$

Here it is essential to understanding the point of the theorem to take 'a' and 'b' as ranging schematically over any objects whatever, and not merely over objects for which we are able to grasp some individuative specification. Any restriction of the range of parameters like the displayed occurrences of 'a' and 'b' to objects available to intelligible singular thought will not merely undercut the generality of theorem 118, but will pose impossible obstacles to the construction the theory of the real numbers, at least as classically conceived.[108]

In a recent critical discussion in another place[109] of issues arising around nominalist-neutralism and Hale's construal of higher-order logic, Heck suggests

---

[108] As Heck has pointed out (personal communication), to the construction of complex analysis also. Consider, for example, the proof that $Nx:(x^2=-1)=2$. Since we necessarily lack any mutually individuative conceptions of the two complex square roots of -1, neutralism has, it appears, no satisfactory account to offer of the meaning of, for example, $(\exists x)(x^2=-1)$, nor therefore of $(\exists x)(\exists y)[(x^2=-1 \ \& \ y^2=-1 \ \& \ x\neq y) \ \& \ (\forall z)(z^2=-1 \rightarrow z=x \ \text{or} \ z=y)]$.

[109] Heck (2018: 146–52). Heck's discussion there of nominalist-neutralism does, it seems to me, highlight a number of important difficulties for the proposals about the interpretation of higher-order generality made in my earlier paper. The most serious of them will preoccupy us shortly. Three others to which I should draw the reader's attention but will not take space to discuss here are these. First, if we suppose that thoughts are hyper-intensionally individuated, then the many possible ways of thinking of a single object may result in misconstrual of the truth conditions of numerically definitely quantified statements when they are interpreted in accordance with the nominalist-neutralist suggestion that, as I put it above,

> the commitment one undertakes in affirming a . . . quantified statement, and hence what has to be established in order to justify one in doing so, will be that there exists a suitable range of intelligible true statements to bear witness to the truth of the former, *modulo* the extent of the generality involved in the quantification concerned.

There may be, for example, exactly one thinkable object that is F but many thinkable singular thoughts of that object that predicate F of it. Second, the paper arguably does insufficient to connect and integrate this conception of the commitments of quantification with the suggestion that the basic determinant of the meanings of the higher-order quantifiers is their inferential role. Third, an inferential role semantics seems

that this difficulty for nominalist-neutralism is, in effect, simply a product of classical comprehension applied in cases where the clause on the right-hand side of the biconditional contains *parameters*, and hence that the problem will kick in also in cases, like

$$(\forall x)(\exists X)(\forall y)(Xy \leftrightarrow Gy \lor y = x)$$

where one or more second-order parameters are involved on the right-hand side. Such cases—comprehension applied to open sentences containing parameters of second level—feature conspicuously in the derivation of arithmetic from *HP*; in particular, as Heck points out, they are needed for the proof of the form of induction for the weak ancestral:

$$Q^{*=}ab \rightarrow (\forall F)(Fa \ \& \ (\forall x)(\forall y)(Q^{*=}ax \ \& \ Fx \ \& \ Qxy \rightarrow Fy) \rightarrow Fb),$$

that is in turn required for the proofs that every number has a successor and of arithmetical induction. However, it may seem questionable, at least on the assumptions of neutralism, that such applications of comprehension have to be problematical. The first-order case presents a problem only and precisely in cases where the range of the individual variables includes elements that *have* no individuative specification. But for nominalist-neutralism—and indeed for Hale—everything occurring within the range of 'F,' 'G,' etc., when they feature parametrically does allow of individuative specification, since only finitely intelligible predicates are allowed.

To be sure, that does not answer the point that the tie effected by neutralism between first-order quantification and intelligible singular thought looks like a roadblock in the way of any neo-logicist ambition to recover not just arithmetic but real and complex analysis. But, as Heck rightly emphasizes, there is anyway a deeper worry nearby: a worry whether *any* significant part of the neo-logicist enterprise, including the derivation of Frege's Theorem, can be accomplished if the underlying logic is fully compliant with nominalist-neutralist constraints. The supporter of second-order logic has, after all, *two* distinctive obligations associated with second-order generalization. The first is to explain the *meanings* of the quantifiers—and it is to this task that both the nominalist-neutralist proposal and Hale's abundant-property suggestion are intended to contribute. But the second is to explain the *scope* of the quantifiers—to explain what range of open sentences in one or more free objectual variables are to be eligible for second-order binding. Frege's answer, in keeping with his comprehensive conception of the scope of objectual quantification, was: "All of them"—*any* well-formed such open sentence is so eligible. But the availability of that answer depends on how one responds to the first obligation. The salient worry, of course, is *impredicativity*: specifically, it is whether

anyway unapt for the treatment of quantifiers like "many" and "most"—and that we should, presumably be looking for a uniform treatment of quantifiers in general, not just of those selected for special attention in traditional logic.

These particular concerns do, I believe, admit of satisfactory responses. I hope to return to the issues in other work.

nominalist-neutralism, or Hale for that matter, can make any sense of impredicative higher-order quantification.

Wright (2007) anticipated the concern but I now regard the tone of the discussion there as much, much too sanguine. Reflection on harmless cases like "Federer has all the qualities of a great champion"—where that very quality is, naturally, one that someone who has *all* the qualities of a great champion had better have!—masks the difficulties which nominalist-neutralism and, I would contend, Hale's proposal too here encounter. The vexed and controversial character of the debates about impredicative specification ever since Poincaré and Russell first expressed concerns is evidence that it is challenging to frame the question of its legitimacy precisely. In the present context, though, it is important to keep in focus that we are not concerned with the *general* question whether, or under what conditions, one may determinately single out an entity of some kind by quantification over the elements of a set or other kind of grouping of which it is a member but with the specific question of what it takes to endow a higher-order predicate with a determinate *sense*.

As Heck remarks, following Kripke, the looming problem is naturally conceived as essentially one of *grounding*. Let us pause to think it through. A higher-order predicate is relevantly semantically grounded only if its satisfaction conditions can be finitely, and non-circularly explained by reference to the satisfaction-conditions of the predicates in its range of quantification. Assume that such a predicate has determinate satisfaction-conditions—a determinate sense—only if so grounded and consider accordingly any base class of well-understood first-level[110] (for simplicity's sake, monadic) predicates containing no higher-order quantifiers. Let "Fy" be one of these *basic* predicates. Now consider any first-level predicate "$QX(Xy)$" formed by some mode of higher-order quantification, Q, into "Fy," and let it be understood that the quantification is impredicative, so that along with all the basic predicates "$QX(Xy)$" itself falls within its own range. Any predicate in good order must possess determinate, finitely intelligible satisfaction-conditions. So, what is the condition on an object, a, if it is to satisfy "$QX(Xy)$"? Well, obviously, it is that a should satisfy Q-many of the predicates[111] in the range of the quantifier. But is that a determinate requirement? There are two cases.

Suppose (Case1) that the truth-value of any instance, "$QX(Xa)$," is always determined by a's satisfaction, or violation, of some number of basic predicates. Then the impredicativity of the quantifier is semantically idle. We could generate something with the same truth-conditions as "$QX(Xy)$" by means of a predicative quantification instead. This is how it is with "Federer has all the qualities of a great champion": while the Swiss tennis star had naturally better have that quality too if the predication is to be true of him, he can presumably have it just by satisfying other predicates which are free of higher-order quantifiers. (In effect, an axiom of reducibility holds in Case 1.)

---

[110] Lest there be any confusion: "first-level" here indicates a predicate of objects. Contrast "first-order," which indicates, as is usual, a predicate free of higher-order quantifiers.

[111] To avoid complication, I am here bracketing questions about possible dependence-relationships among the predicates in question. Doing so does not affect the relevant point to follow.

Thus if the impredicativity of "$QX(Xx)$" is not to be idle, but is to make some material difference to its satisfaction-conditions—if it is to be, as we may say, *actively impredicative*—then (Case 2) what makes "$QX(Xa)$" true must turn on a's satisfaction of Q-many predicates in a range that includes some non-basic predicates, involving higher-order quantifiers. But obviously we make no progress in explaining how "$QX(Xx)$" can indeed have determinate satisfaction-conditions if we simply say that it is itself included in its range and that this makes a material difference to whether any particular object satisfies it. So it seems that a model of how an actively impredicative "$QX(Xx)$" can have determinate satisfaction-conditions must assign a range to its quantifier that both goes beyond the basic predates and includes predicates *distinct* from "$QX(Xx)$." Let's call these *secondary* predicates for "$QX(Xx)$."

Now the problem is evident. These secondary predicates must also involve actively impredicative higher-order quantification. Otherwise, the real contribution made by a's satisfaction of a secondary predicate to its satisfaction of "$QX(Xx)$" will be delivered by its satisfaction of appropriately many *basic* predicates instead, so that the situation will revert to Case 1 by proxy, as it were. So a model that explains how and actively impredicative "$QX(Xx)$" can have determinate satisfaction conditions must, it seems, attempt to do so by including in the range of its quantifier some *other* first-level predicate that is actively impredicative but nevertheless possesses determinate satisfaction conditions. But it was of exactly that possibility of which an explanatory model was demanded in the first place. We are spinning our wheels.

In summary: any higher-order predicate "$QX(Xx)$" has determinate satisfaction-conditions only to the extent that it is a determinate matter whether an individual satisfies Q-many of the predicates in its range, and hence only if *they* have determinate satisfaction-conditions. The challenge is therefore to provide a model of how, if "$QX(Xx)$" is actively impredicative—so that its satisfaction-conditions cannot be explained in terms purely of the satisfaction of Q-many basic predicates—a larger class of predicates might be identified over which it ranges and which may be independently credited with determinate satisfaction-conditions. Since such a larger class will, for the reasons given, be constrained to include further actively impredicative non-basic predicates, it will deliver what is wanted only if the problem of semantic grounding—the problem with which we started—has already been solved for *them*.

I hesitate to present this reasoning as a proof that actively impredicative first-level predicates *cannot* be semantically grounded and hence cannot have determinate satisfaction conditions. What it proves, rather—if it proves anything—is that there is no clear prospect of a model that satisfactorily explains how they can. But if the pessimistic conclusion which that may suggest is right, then that kills off any chance that a second-order logic sufficient for the proof of Frege's Theorem, let alone for an abstractionist recovery of analysis, can be vindicated by nominalist-neutralism, or by Hale's approach. Heck's conclusion, "that Wright's 'neutralist' account of quantification will not serve the needs of the neo-Fregean program,"[112] will be

---

[112] Heck (2018: 152).

sustained. And a similarly negative verdict will apply to Hale's attempted reconstruction of the ontology of second-order logic in terms of abundant properties.

To be sure, to draw that conclusion is not to conclude that neither nominalist-neutralism, nor Hale, can respectively offer a satisfactory interpretation—even maybe a uniquely correct interpretation—of higher-order quantification in general. That is a further question. What is in doubt, rather, is, specifically, whether Hale's or my earlier proposal can underwrite the higher-order logic of *Begriffschrift* and *Grundgesetze* and, indeed, whether either can underwrite any logic strong enough to subserve the proof of (an interestingly strong version of)[113] Frege's Theorem.

Which brings us back to Arché logic. I first encountered the central ideas of Heck's present chapter in a presentation they gave at an Arché workshop in 2005. I recall being pretty much 'blown away,' not just by the technical acumen displayed but by a sense that we could at last see clearly just what were the essential demands that Frege's Theorem makes on a suitable underlying logic—and that they were both different from and, crucially, epistemically much more modest than the resources provided by second-order logic as usually formulated. I still think Heck proves at least that much. The question is whether what their argument accomplishes is sufficient to sustain the thought that, as I put it above, the logic minimally required for Frege's Theorem is fit to "serve as a medium for the generation and transmission of knowledge that, at least for those in sympathy with the spirit of that vexed Kantian category, deserves to be captioned as analytic a priori." Back then I thought the answer was clearly, "Yes," that a key question for neo-logicism had been settled positively, and that the crucial remaining issue was now whether essentially the same logical and definitional resources could support an abstractionist recovery of Analysis. I am not now so confident.

Arché logic is essentially a natural generalization of *ancestral logic*: the system one gets by augmenting first-order logic with an operator, *, that when applied to any expression for a dyadic relation, yields an expression denoting the ancestral of that relation. In Heck's formulation, the ancestral operator is controlled—Heck suggests, "defined"—by a pair of rules closely following Frege's original explicit definition. We have already noted the introduction rule:

$$(\text{*-intro}) \quad \frac{[\forall x(\phi ax \rightarrow Fx) \,\&\, \forall x \forall y(Fx \,\&\, \phi xy \rightarrow Fy)] \rightarrow Fb^{114}}{\phi^* ab}$$

The elimination rule is, naturally, simply the converse:

$$(\text{*-elim}) \quad \frac{\phi^* ab}{[\forall x(\phi ax \rightarrow Fx) \,\&\, \forall x \forall y(Fx \,\&\, \phi xy \rightarrow Fy)] \rightarrow Fb}$$

---

[113] The qualification is necessary because, naturally, the strength of the induction axiom will vary with the range of the higher-order variables it contains. I will omit it in what follows.

[114] Where F must not occur unbound in any premise on which the thesis above the line depends.

Heck notes that these two rules illustrate a general pattern of *schematic definition*. In regular second-order logic, wherever $\phi x(Fx,y)$ is a formula with no free variables save F and y, we can define a new predicate, $A_\phi y$ by universally binding F; thus

$$A_\phi y = \forall F\phi x(Fx,y)$$

What the rules for the ancestral illustrate is how we can always get the effect of that resource, without resort to second-order quantification, by laying down a pair of rules, thus:

$$\text{(A-intro)} \quad \frac{\phi x(Fx,y)}{A_\phi y} \quad ^{115}$$

and

$$\text{(A-elim)} \quad \frac{A_\phi y}{\phi x(Fx,y)}$$

The generality involved in rules of this pattern is nothing more exotic more than the schematic generality already involved in theorems of first-order logic like Fx → Fx, ∼(Fx & ∼Fx), and so on. Arché logic is first-order logic augmented with this definitional resource. And it suffices, Heck shows, for the proof of Frege's Theorem.

At least, it does so provided we allow ourselves a rule of substitution analogous to that which Frege assumed for the system of *Grundgesetze*: in effect, that in applying any elimination rule of the illustrated kind, we may substitute for ϕ any formula of the appropriate form whose formation the language of the system allows, that is, any formula with a single free individual variable—including formulae which contain occurrences of the relevant defined predicate, $A_\phi y$, itself. Augmented with this rule of substitution, Arché logic mimics the proof-theoretic power of a system of second-order logic with just Π_1^1—comprehension and no second-order existential quantifier. But that is enough, as Heck shows, for the derivation of Frege's Theorem. Without the substitution rule, on the other hand, the derivation is blocked.

So, is Arché logic "epistemically innocent"? One train of thought that might occasion doubt is that, in conjunction with a first-order schematic version of *Grundgesetze* Law V, it suffices for Russell's paradox. It is very often supposed that the paradox showed that Law V is inconsistent—the 'Bad Company' problem for abstraction principles is often presented in ways that assume so. But as I have stressed in other work, all that is initially clear is that the paradox is the product of the interaction between Frege's axiom and his underlying logic.[116] It is a philosophical question which, if either, deserves the Lion's Share of the blame.[117] One relevant fact is that when Law V is adjoined to a second-order logic admitting only predicative comprehension, no paradox ensues.[118] Another is that a schematic version of Law V is perfectly consistent in the setting of classical first-order logic. Not so, however,

---

[115] Cf. n. 110.   [116] Wright (forthcoming).
[117] Something of a classic exchange on the question is provided by Boolos (1993) and Dummett (1994).
[118] This was first noted by Heck (1996).

when the underlying logic is Arché logic. Here is one way of obtaining a version of Russell's antinomy. First assume Law V in schematic form:

$$(V) \quad \{x{:}Fx\} = \{x{:}Gx\} \leftrightarrow (\forall x)(Fx \leftrightarrow Gx)$$

Substitution of F for G, *modus ponens*, on the right-to-left ingredient conditional and existential generalization gives us *schematic naive comprehension*:

$$(\exists x)(x = \{x{:}Fx\})$$

Next we apply the schematic rules A-intro and A-elim above to the open sentence,

$$y = \{x{:}Fx\} \rightarrow {\sim}Fy$$

Thus:

$$R\text{-intro} \quad \frac{y = \{x{:}Fx\} \rightarrow {\sim}Fy^{\,119}}{Ry}$$

$$R\text{-elim} \quad \frac{Ry}{y = \{x{:}Fx\} \rightarrow {\sim}Fy}$$

By naïve comprehension, we have that R is associated with an extension:

$$(\exists x)(x = \{x{:}Rx\})$$

—call this object r. Suppose Rr. Then by R-elim

$$r = \{x{:}Rx\} \rightarrow {\sim}Rr$$

Since we have the antecedent of the latter, we obtain ${\sim}Rr$—contradiction. So discharging the assumption of Rr by reductio, we obtain ${\sim}Rr$ depending just on schematic V as premise. That suffices for the premise for an R-intro step meeting the necessary restrictions. Whence Rr.

So, the critic may contend, its relatively spare expressive and proof-theoretic resources notwithstanding, Arché logic still suffices for paradox when annexed to what is merely a pale shadow of *Grundgesetze* Law V—a principle which in regular first-order logic is perfectly harmless. The source of the problem is evident: it is the substitution rule which allows an arbitrary predicate introduced by means of the A-schemata to go on to occur within the scope of 'F' in the associated rules. Syntactically viewed, this is the sparest, most modest form of impredicativity. But the manoeuvre it allows *suffices* to generate paradox when aligned with—so the critic may suggest—an otherwise innocent principle, as we have just seen—and, crucially, is *necessary* for the proof of arithmetical induction by Arché-logical means which Heck provides. How then can it be trusted in the latter context?

Though they do not address it to exactly this question, Heck makes an observation *ob iter* which may be thought to answer it. It is that if we add '∈'—set-membership—as

---

a primitive to a regular first-order language with schematic V as an axiom, and set it as controlled by the principle they call ∈-lite:

$$(y)(y \in \{x:Fx\} \leftrightarrow Fy)$$

then, taking Fx as x∉x and y as {x:x∉x}—as naïve comprehension permits—we immediately have the paradox again using only regular first-order logic. So—one might argue—since ∈-lite is absolutely intuitive as a characterization of set-membership[120]—there is cause to think that schematic V is a bad principle anyway, and hence that its generation of paradox in Arché logic need not reflect adversely on the latter. However, finger-pointing at the sources of paradox is a slippery business. What this last paradox shows is that the intuitive notion of set that combines schematic V, hence naïve comprehension, with the conception of set-membership embodied in ∈-lite is per se an incoherent notion. But then we have once again, not a single suspect but two: to allow that every open sentence in one objectual variable determines a set is not yet a commitment to the idea that any satisfier of that open sentence is a member of that set. The paradox in Arché logic does not exploit any particular view on that, or on how membership should be characterized in general, but only the admissibility of the general idea of a set of F-elements not itself being F. Heck's observation shows that if we are committed to ∈-lite, we have no reason to look askance at Arché logic as a source of the latter paradox. But it is also consistent with the stance that schematic V is innocent, but vulnerable to mishap both when set in Arché logic and when conjoined with ∈-lite.

As mentioned, I argue elsewhere[121] that the original paradox in the system of *Grundgesetze* should, at least in the first instance, be viewed not as a *reductio* of law V but as a demonstration of the incoherence of Frege's unthinking combination of the notion of logical object incorporated in his axiom—whereby any well-formed open sentence has an associated object, an extension or more generally a value-range—with the unrestricted predicate comprehension resources made available by his substitution rule which allows any open sentence, of whatever degree of quanti-ficational complexity, to be a legitimate instance of higher-order quantification. What we have with schematic V and Arché logic is a pared-down tension of essentially the same species. Heck's substitution rule, whereby any predicate intro-duced by instances of the A-schemata may be taken in turn as an admissible substituend for their contained occurrences of predicate variables will not marry with the concept of logical object encoded in schematic V. Indeed there is a case for saying that, rather than presenting a system of logic that undercuts critical concerns about the epistemic innocence of the proof-theory needed for Frege's Theorem, Heck has isolated, rather, the most basic, purest form of the tension, freed from the noise associated with the much greater expressive resources and limitless levels of impre-dicativity involved in the logic of *Grundgesetze*. Until that case is rebutted, the suggestion survives that Arché logic, no less than Frege's logic, is tainted by its partnership in paradox.

---

[120] "The definition of membership one would really like to give" (Heck, Chapter 2, this volume, p. 28).
[121] In Wright (forthcoming).

We can set that to one side, though. For I think it is clear that Arché logic with the unrestricted substitution rule must inherit the concerns about determinacy of sense and impredicativity that Heck themself presses against my nominalist-neutralism and Hale's definable property proposal. Instances of the A-schemata are supposed to fix the meaning, that is, the satisfaction-conditions of the predicates for whose introduction and elimination they provide. There is no reason to doubt that they can do so when the range of F in their respective premises and conclusions is restricted to predicates whose satisfaction-conditions are independently determinate. But, as the reader will likely by now be well-tired of being reminded, such a restriction will block the derivation of Frege's Theorem—and specifically the deriv-ation of arithmetical induction via the schematic rules for the ancestral, where we need to be able introduce predicates themselves containing the ancestral operator for the occurrences of 'F' in the schematic rules. Heck's response to this consideration[122] is to insist that, if we understand schematic generality as *full* schematic generality, then the rules must apply to *any* predicate in good standing, including the very predicate they define. There can be no restrictions. And that is surely quite right— *provided that* a candidate predicate is indeed in good standing, that is, has a determinate sense in the first place. But the concern raised about impredicativity— at least the concern I tried to articulate earlier in this section—is of course a concern about exactly that. Suppose you are familiar with o and predecession and then are told that, for any object y, o stands in the ancestral of the predecession relation to y just in case y satisfies every predicate that o satisfies and which is transmitted across predecession. That will probably seem to make good sense—up until the point when it is explained that, in order for this relationship to obtain, *that very predicate* must be included in the scope of "every predicate" in its own formulation, so that in order for o to ancestrally precede y, it is not enough if y satisfies every *other* predicate which o has and which is transmitted across predecession; no, in addition it is required that y satisfies *this one* too. At that point, you may feel—and arguably ought to feel—you simply haven't been told what the satisfaction condition of "P^*oy" is.

Kurt Gödel famously claimed that there is no *general* barrier to the good-standing of the procedure of specifying an entity of a certain kind by means of a formulation that quantifies over a range of entities that purportedly includes it. Whether such a procedure is in good order turns, he suggested, on whether the relevant range of entities are thought of as pre-existing and 'out there'—cf. Frank Ramsey's example, "the tallest man in the room"—or whether one conceives the specification in question in constructive terms, in which case there is indeed a difficulty.[123] By this measure it may seem that if we are comfortable with realism about the classical notion of power set, and with thinking of the semantic values of predicates as sets of objects, or n-tuples of objects, as so conceived, then there should be no objection to full classical impredicative comprehension, *a fortiori* none to the modest form impredicative definition licensed by the A-schemata under unrestricted substitution. But though Gödel's thought strikes me as sound in certain kinds of case, it seems quite misguided in the present context. In order for an impredicatively quantified first-level predicate

---

[122] Chapter 2, this volume, p.45 and following.   [123] Gödel (1944).

to pick out a determinate subset of the objectual domain, it has first to express a *determinate condition* by satisfaction of which membership of the relevant subset is determined. If there is a standing doubt about whether that is so, it cannot be addressed simply by invoking the classical realist notion of power set. The sets may indeed be 'out there' alright. But it cannot be assumed in general that predicates of the kind that are causing concern will pick out any particular subset unless the worry that impredicative higher-order quantification spawns indeterminacy of satisfaction-conditions has first been addressed.

Heck's relatively relaxed attitude to impredicativity does not seem to be fuelled by this Gödelian thought. But in several places, they write in a way that suggests that they wish to sever the tie that Hale wanted between intelligible predication and qualification for membership of the second-order. In the following passage, for instance, noting what they see as a point of agreement with Hale, they write,

I think Quine's central worry is that our understanding of second-order logic is parasitic on our understanding of set theory: Even if second-order variables do not range over 'classes,' ... our appreciation of what they do range over depends upon our understanding that, for every subclass of the first-order domain, there is some element of the second-order domain that corresponds to it. That is what I am trying to fight off. And Hale and I are largely in agreement how to do it: Our understanding of second-order variables only requires us "to know what general condition something must satisfy, if it is to be a possible value...." (Hale, 2013, p. 154). And we are largely agreed, too, what that condition is: being the sort of thing for which a predicate can stand. Where we differ, again, is that Hale wishes to impose an additional condition: that a Concept exists only if there could be some predicate that expressed it. I, by contrast, find that claim unhelpful because unclear, and I simply see no reason to deny, with Hale, that every subset of the first-order domain corresponds to something in the range of the second-order variables.[124]

This passage, admittedly, can stand some interpretation. But I take it that Heck is affirming Hale's idea that the entities in the range of the predicate variables in second-order logic are the kind of thing for which predicates can stand—meaning that they are Concepts, rather than sets—and denying that our understanding of what a Concept is parasitic on our understanding of what a set is. Where, it seems, they are departing from Hale is in refusing to make it a point of the essence of a concept that it exists only insofar as it can be intelligibly predicated, that is, that every concept is such that some possible significant predicate can be understood (by finite minds) as predicating *it*. I do not think it is obvious that there is any fully coherent such view. At least, we surely cannot *explain* Concepts as the kind of things for which predicates can stand while at the same time allowing that some of them cannot intelligibly be predicated—there has to be some other characterization of what Concepts essentially are which provides for that limitation, and Heck nowhere, to the best of my knowledge, ventures such a characterization. At any rate, there is no ducking questions about what these entities are, how they are given to us, and how they could possibly enter into the subject matter of something properly described as

---

[124] Heck (2018: 165).

"logic"—or at any rate as "epistemically innocent." These are questions that, whatever their other shortcomings, Hale's proposal, and nominalist-neutralism have the resources either to answer or to undercut.

But these are skirmishes. The crucial point remains that Heck's Concepts do no better than the elements of the classical power set of an infinite domain when it comes to supplying a second-order domain that evades the central difficulty with impredicativity that I have been concerned to stress. This is because our basic problem, to repeat, is not with second-order entities for which we may have no means of intelligible individuative specification, problematic though that idea may well be felt to be, but with the good-standing—determinate significance—precisely of what are supposed to be certain individuative specifications of such entities. It doesn't matter whether we take the semantic value of a first-level predicate to be a set, or a Concept, or whatever else, or what we take to be the relationship of such entities to possibilities of intelligible predication. An impredicatively quantified first-level predicate, like any predicate, has a semantic value at all only if it expresses a well-defined satisfaction condition on the elements of the objectual domain, and our concern has been—if I may be forgiven for harping on it—whether, when actively impredicative second-order quantification is essentially involved, any determinate such satisfaction condition is indeed expressed. If that worry is soundly conceived, it is an utterly irrelevant response to invoke the idea that the elements of the second-order domain may be 'out there' yet transcend finitely intelligible expression. The worry has to do with whether predicates formed by actively impredicative quantification possess a determinate sense; it is only *after* that has been addressed that questions about the kind of reference they may have, whether the kinds of entity concerned may transcend possible predication, etc., can so much as arise.

So, I find myself in the end with a reaction to Heck's Chapter 2 (this volume) reminiscent of Dummett's reaction to work of mine on vagueness:[125] "admiration for a beautiful solution" to the problem of vindicating proof-theoretic resources that are minimally sufficient for the derivation of an interestingly strong version of Frege's Theorem, "clouded by a persistent doubt" whether the solution does everything needed. It is a presupposition of the significance of the neo-logicist project that there should be some well-grounded alternative way of thinking about the higher-order logic required than the Quinean way—some way consonant with core idea of logicism, that logical knowledge and at least basic mathematical knowledge are, in some important sense, of a single epistemological kind. I think Heck shows that at least some of the problems associated with the Quinean conception of higher-order logic can be finessed in the present context. But if no fully predicative second-order logic can serve the needs of the proof of Frege's theorem and the construction of a complete ordered field, then the neo-logicist project for arithmetic and analysis remains potentially stalled on the misgivings I have here tried to articulate—misgivings not about whether the proof-theory needed should properly count as 'logic,' but about whether it makes good sense at crucial points in the key constructions.

---

[125] Dummett (2007: 453).

References

Bell, J. L. (1999). "Frege's Theorem in a Constructive Setting," *The Journal of Symbolic Logic* 64 (2), 486–8.

Blackburn, S. (1994). *The Oxford Dictionary of Philosophy*. Oxford: Oxford University Press.

Boghossian, P. (1996). "Analyticity Reconsidered," *Noûs* 30 (3), 360–91.

Boghossian, P. (2001). "Inference and Insight," *Philosophy and Phenomenological Research* 63 (3), 633–40.

Boghossian, P. (2003). "Blind Reasoning," *Aristotelian Society Supplementary Volume* 77 (1), 225–48.

Boghossian, P. (forthcoming). "Intuition, Understanding and the A Priori," in P. Boghossian and T. Williamson (eds.), *Debating the A Priori*. Oxford: Oxford University Press.

Boolos, G. (1987). "The Consistency of Frege's Foundations of Arithmetic," in J. Thomson (ed.), *On Being and Saying: Essays in Honor of Richard Cartwright*, Cambridge, MA: MIT Press, 3–20.

Boolos, G. (1990). "The Standard of Equality of Numbers," in G. Boolos (ed.), *Meaning and Method: Essays in Honor of Hilary Putnam*, Cambridge: Cambridge University Press, 261–77.

Boolos, G. (1993) "Whence the Contradiction?" *Aristotelian Society Supplementary Volume* 67, 211–33.

Boolos, G. (1997). "Is Hume's Principle Analytic?," in R. K. Heck (ed.), *Language, Thought, and Logic: Essays in Honor of Michael Dummett*, Oxford: Oxford University Press, 245–61.

Boolos, G. (1998). *Logic, Logic, and Logic*. Cambridge, MA: Harvard University Press.

Boolos, G. and Heck, R. K. (1998). "Die Grundlagen der Arithmetik, 82–3," in M. Schirn (ed.), *The Philosophy of Mathematics Today*, Oxford: Clarendon Press, 407–28.

Burgess, J. P. (1984). "Review of Wright (1983)," *Philosophical Review* 93, 638–40.

Cook, R. T. (2016). "Abstraction and Four Kinds of Invariance," *Philosophia Mathematica* 25 (1), 3–25.

Cook, R. T. and Linnebo, Øystein (2018). "Cardinality and Acceptable Abstraction," *Notre Dame Journal of Formal Logic* 59 (1), 61–74.

Demopoulos, W. (1998). "The Philosophical Basis of Our Knowledge of Number," *Noûs* 32 (4), 481–503.

Demopoulos, W. (2000). "On the Origin and Status of our Conception of Number," *Notre Dame Journal of Formal Logic* 41 (3), 210–26.

Demopoulos, W. (ed.) (1995). *Frege's Philosophy of Mathematics*. Cambridge, MA: Harvard University Press.

Dogramaci, S. (2010). "Knowledge of Validity," *Noûs* 44 (3), 403–32.

Dummett, M. (1981). *The Interpretation of Frege's Philosophy*. Cambridge, MA: Harvard University Press.

Dummett, M. (1991). *Frege: Philosophy of Mathematics*. Cambridge, MA: Harvard University Press.

Dummett, M. (1994). "Chairman's Address: Basic Law V," *Proceedings of the Aristotelian Society New Series* 94, 243–51.

Dummett, M. (1998). "Appendix on the term 'Hume's Principle'," in M. Schirn (ed.), *The Philosophy of Mathematics Today*, Oxford: Clarendon Press, 386–7.

Dummett, M. (2007). "Reply to Crispin Wright," in R. E. Auxier and L. E. Hahn (eds.), *The Philosophy of Michael Dummett*, The Library of Living Philosophers, New York: Open Court, vol. XXXI, pp. 445–54.

Ebels-Duggan, S. (forthcoming). "Deductive Cardinality Results and Nuisance-like Principles," *The Review of Symbolic Logic*.

Enoch, D., and Schechter, J. (2008). "How Are Basic Belief-Forming Methods Justified?," *Philosophy and Phenomenological Research* 76 (3), 547–79.

Field, H. (1984). "Review of Wright 91983," *Canadian Journal of Philosophy* 14 (4), 637–62.

Fine, K. (2001). "The Question of Realism," *Philosopher's Imprint* 1 (2).

Fine, K. (2002). *The Limits of Abstraction*, Oxford: The Clarendon Press.

Fine, K. (2012). "Guide to Ground," in F. Correia and B. Schnieder (eds.), *Metaphysical Grounding*, Cambridge: Cambridge University Press, 37–80.

Flew, A. and Priest, S. (2002). *A Dictionary of Philosophy*. Aylesbury: Pan.

Frege, G. (1884/1950). *The Foundations of Arithmetic: A Logico-Mathematical Enquiry into the Concept of Number*, J. L. Austin (trans.), Oxford: Blackwell.

Geach, P. (1955). "Class and Concept," *The Philosophical Review* 64 (4), 561–70.

Gödel, K. (1944). "Russell's Mathematical Logic," in P. A. Schilpp (ed.), *The Philosophy of Bertrand Russell*, New York: Tudor, 123–53. Reprinted in H. Putnam and P. Benacerraf (eds.), *Philosophy of Mathematics: Selected Readings*, second edition, Cambridge: Cambridge University Press, 1983, 447–69.

Hale, B. (1988). *Abstract Objects*. Oxford: Blackwell.

Hale, B. (2012). "Properties and the Interpretation of Second-Order Logic," *Philosophia Mathematica* 21 (2), 133–56.

Hale, B. (2013). *Necessary Beings: An Essay on Ontology, Modality, and the Relations between Them*, Oxford: Oxford University Press.

Hale, B. (2015). "Second-order Logic: Properties, Semantics, and Existential Commitments," *Synthese* 1–27.

Hale, B. and Wright, C. (2000). "Implicit Definition and the A Priori," in P. Boghossian and C. Peacocke (eds.), *New Essays on the A Priori*, Oxford: Oxford University Press, 286–319.

Hale, B. and Wright, C. (2001a). "To Bury Caesar..." *The Reason's Proper Study: Essays Towards a Neo-Fregean Philosophy of Mathematics*, Oxford: Oxford University Press.

Hale, B. and Wright, C. (2001b). *The Reason's Proper Study: Essays Towards a Neo-Fregean Philosophy of Mathematics*, Oxford: Oxford University Press.

Hale, B. and Wright, C. (2003). "Responses to Commentators—Book Symposium on *The Reason's Proper Study*," *Philosophical Books* 44 (3), 245–63.

Hale, B. and Wright, C. (2009). "The Metaontology of Abstraction," in D. J. Chalmers, D. Manley, and R. Wasserman (eds.), *Metametaphysics: New Essays on the Foundations of Ontology*, Oxford: Oxford University Press, 178–212.

Hale, B. and Wright, C. (2015). "Bolzano's Definition of Analytic Propositions," *Grazer Philosophische Studien* 91 (1), 323–64.

Hawthorne, J. (2007). "A Priority and Externalism," in Sanford Goldberg (ed.), *Internalism and Externalism in Semantics and Epistemology*, Oxford: Oxford University Press, 201–18.

Heck, R. K. (1992). "On the Consistency of Second-order Contextual Definitions," *Noûs* 26, 491–4.

Heck, R. K. (1993). "The Development of Arithmetic in Frege's *Grundgesetze der Arithmetik*," *The Journal of Symbolic Logic* 58 (2), 579–601.

Heck, R. K. (1996). "The Consistency of Predicative Fragments of Frege's *Grundgesetze der Arithmetik*," *History and Philosophy of Logic* 17 (1–2), 209–20.

Heck, R. K. (1997a). "Finitude and Hume's Principle," *Journal of Philosophical Logic* 26 (6), 589–617.

Heck, R. K. (1997b). "The Julius Caesar Objection," in R. K. Heck (ed.), *Language, Thought, and Logic: Essays in Honour of Michael Dummett*, Oxford: Oxford University Press, 273–308.

Heck, R. K. (2011). *Frege's Theorem*, Oxford: Oxford University Press.

Heck, R. K. (2018). "Logicism, Ontology, and the Epistemology of Second-Order Logic," in I. Fred and J. Leech (eds.), *Being Necessary: Themes of Ontology and Modality from the Work of Bob Hale*, Oxford: Oxford University Press.

Henkin, L. (1962). "Are Logic and Mathematics Identical?," *Science* 138 (3542), 788–94.

Hodes, H. T. (1984). "Logicism and the Ontological Commitments of Arithmetic," *The Journal of Philosophy* 81 (3), 123–49.

Lange, M. (2016). *Because without Cause: Non-Causal Explanations in Science and Mathematics*. Oxford: Oxford University Press.

Leivant, D. (1994). "Higher Order Logic," in D. M. Gabbay, C. J. Hogger, C. J. Robinson, and J. Siekmann (eds.), *A Handbook of Logic in Artificial Intelligence and Logic Programming, Volume 2: Deduction Methodologies*, Oxford: Clarendon Press, 230–321.

Mancosu, P. (2001). "Mathematical Explanation: Problems and Prospects," *Topoi* 20, 97–117.

McGee, V. (1985). "A Counterexample to Modus Ponens," *Journal of Philosophy* 82 (9), 462–71.

Musgrave, A. (1977). "Logicism Revisited," *The British Journal for the Philosophy of Science* 28 (2), 99–127.

Parsons, C. (1965). "Frege's Theory of Numbers," in M. Black (ed.), *Philosophy in America*, New York: Cornell University Press, 180–203.

Pryor, J. (2012). "When Warrant Transmits," in A. Coliva (ed.), *Mind, Meaning, and Knowledge: Themes From the Philosophy of Crispin Wright*, Oxford: Oxford University Press, 269–303.

Quine, W. V. (1970). *Philosophy of Logic*. Cambridge, MA: Harvard University Press.

Rayo, A. and Yablo, S. (2001). "Nominalism Through De-Nominalization," *Noûs* 35 (1), 74–92.

Rosen, G. (2003). "Platonism, Semiplatonism and the Caesar Problem," *Philosophical Books* 44 (3), 229–44.

Rosen, G. (2009). "Metaphysical Dependence: Grounding and Reduction," in B. Hale and A. Hoffmann (eds.), *Modality: Metaphysics, Logic, and Epistemology*, Oxford: Oxford University Press, 109–36.

Rossberg, M. (2015). "Somehow Things Do Not Relate: On the Interpretation of Polyadic Second-Order Logic," *Journal of Philosophical Logic* 44 (3), 341–50.

Schaffer, J. (2009). "On What Grounds What," in D. Manley, D. J. Chalmers, and R. Wasserman (eds.), *Metametaphysics: New Essays on the Foundations of Ontology*, Oxford: Oxford University Press, 347–83.

Shapiro, S. (1991). *Foundations Without Foundationalism: A Case for Second-Order Logic*. Oxford: Oxford University Press.

Shapiro, S. (1998). "Induction and Indefinite Extensibility: The Gödel Sentence is True, but Did Someone Change the Subject?," *Mind* 107 (427), 597–624.

Shapiro, S. and Wright, C. (2006). "All Things Indefinitely Extensible," in A. Rayo and G. Uzquiano (eds.), *Absolute Generality*, Oxford: Oxford University Press, 255–304.

Skolem, T. (1922). "Some Remarks on Axiomatized Set Theory," reprinted in J. van Heijenoort (ed.), *From Frege to Gödel*, Cambridge, MA: Harvard University Press, 1967, 290–301.

Steiner, M. (1978). "Mathematical Explanation," *Philosophical Studies* 34 (2), 135–51.

Stirton, W. (2003). "Caesar Invictus," *Philosophia Mathematica* 11 (3), 285–304.

Stirton, W. (2016). "Caesar and Circularity," in P. A. Ebert and M. Rossberg (eds.), *Abstractionism*, Oxford: Oxford University Press, 37–49.

Sullivan, P. and Potter, M. (1997). "Hale on Caesar," *Philosophia Mathematica* 5 (2), 135–52.

Weir, A. (2003). "Neo-Fregeanism: An Embarrassment of Riches," *Notre Dame Journal of Formal Logic* 44 (1), 13–48.

Williamson, T. (2003). "Understanding and Inference," *Aristotelian Society Supplementary Volume* 77 (1), 249–93.

Williamson, T. (2006). "Conceptual Truth," *Aristotelian Society Supplementary Volume* 80 (1), 1–41.

Williamson, T. (2007). *The Philosophy of Philosophy.* Oxford: Wiley-Blackwell.

Williamson, T. (2013). "How Deep is the Distinction Between A Priori and A Posteriori Knowledge?," in A. Casullo and J. C. Thurow (eds.), *The A Priori in Philosophy*, Oxford: Oxford University Press, 291–312.

Wright, C. (1980). *Wittgenstein on the Foundations of Mathematics*, London: Duckworth.

Wright, C. (1983). *Frege's Conception of Numbers as Objects*, Aberdeen: Aberdeen University Press.

Wright, C. (1985). "Skolem and the Skeptic," *Proceedings of the Aristotelian Society* 59, 117–37

Wright, C. (1997). "On the Philosophical Significance of Frege's Theorem," in R. K. Heck (ed.), *Language, Thought, and Logic: Essays in Honour of Michael Dummett*, Oxford: Oxford University Press, 201–44.

Wright, C. (2004a). "Intuition, Entitlement and the Epistemology of Logical Laws," *Dialectica* 58 (1), 155–75.

Wright, C. (2004b). "Warrant for Nothing (and Foundations for Free)?," *Aristotelian Society Supplementary Volume* 78 (1), 167–212.

Wright, C. (2007). "On Quantifying into Predicate Position: Steps towards a New(tralist) Perspective," in M. Leng, A. Paseau, and M. D. Potter (eds.), *Mathematical Knowledge*, Oxford: Oxford University Press, 150–74.

Wright, C. (2014). "On Epistemic Entitlement (II): Welfare State Epistemology," *Scepticism and Perceptual Justification*, Oxford: Oxford University Press.

Wright, C. (2018). "Logical Non-Cognitivism," *Philosophical Issues* 28 (1), 425–50.

Wright, C. (forthcoming). "How Did the Serpent of Inconsistency Enter Frege's Paradise?," in P. Ebert and M. Rossberg (eds.), *Essays on Frege's Basic Laws of Arithmetic*, Oxford: Oxford University Press.

Yablo, S. (2017). "If-Thenism," *Australasian Philosophical Review* 1 (2), 115–32.

Replies to Part II

Intuitionism and the Sorites

Replies to Rumfitt and Schiffer

I have been wrestling on and off with the philosophical challenges presented by vagueness since the early 1970s. At that time, I think it fair to say, almost nothing of real significance had been written on the topic since the contributions of Eubulides. Philosophers of language from Frege on had been for the most part content to theorize in ways that ignored vagueness, or to focus on idealized languages in which there was none. And no one writing before 1970 seemed fully to have taken the measure of the awkwardness of the Sorites paradox, or the depth of its roots in our intuitive thinking about what kind of ability mastery of a language is. My own interest in the topic was originally piqued by conversations with the mathematician Aidan Sudbury and with Michael Dummett, then my senior colleague at All Souls, who around that time was working on the lecture that soon after he published as "Wang's Paradox."[126] I was drawn to the thought that the apparent open-endedness of the extension of a vague predicate might provide a fruitful model for the manner in which a finitist should think about the extension of *natural number* and that a correct logic of vagueness might accordingly be appropriate for a finitist number-theory. My subsequent paper, "Strict Finitism"[127] was the upshot of my reflections in that direction. But while thinking about finitism I became preoccupied with the Sorites paradox itself. Dummett's paper argued, inter alia, that vague expressions affect natural language with inconsistency—that is, that the paradox shows that the rules governing our use of vague expressions are actually inconsistent. That struck me as an incredible conclusion, but one that was nevertheless forced by a certain conception—what, in the context of philosophy of language in Oxford in the mid-1970s, and the then reverberating "Davidsonic boom," I punningly dubbed "the Governing View"—of the sense in which linguistic competence is, through and through, a rule-generated ability. My first two papers on the subject[128] elaborated

---

[126] Dummett (1975).   [127] Wright (1982).
[128] "On the Coherence of Vague Predicates" was published in the same volume of *Synthese* as Dummett's (1975). The volume also included Kit Fine's "Vagueness, Truth and Logic," and proved to be the launch-pad of the intense discussion of vagueness and the Sorites, now into its fifth decade, that has followed since. "Language-Mastery and the Sorites Paradox" (1976) was published in Gareth Evans's and John McDowell's influential edited anthology, *Truth and Meaning*, exploring the issues raised by Davidson's proposal.

that thought. But they proposed no specific solution to the paradox other than to undercut one kind of motivation for (one form of) its major premise.

It was more than a decade before I felt that I had anything further to say on the issues. By then Hilary Putnam had suggested that a resort to intuitionist rather than classical logic might contribute to a solution.[129] After some skirmishing,[130] I became convinced that there might be something to this. The other development in my thinking at that time was the realization that we need to distinguish a variety of Sorites paradoxes, differing in the form taken by their major premises, the various motivations for those premises, and even—in recognition of the so-called Forced March Sorites—in whether they involve explicit inference from premises at all.[131] I was, however, still thinking of vagueness as essentially a phenomenon of semantics—as some kind of deficiency in content—and it was only after trying to come to terms with Timothy Williamson's remarkable book[132] in the mid-1990s that a different way of thinking about the matter began to dawn on me. In essentials, Williamson's epistemicism grafts together two thoughts: a classical, bivalent metaphysics of indicative content coupled with a view of the vagueness of a predicate as essentially a phenomenon of difficulty of judging its application in the area close to the sharp "cut-off" required by the first thought. It now occurred to me that dispensing with the first while developing the second (shorn, of course, of the presupposition of sharp cut-offs) might provide the motivation for a thoroughgoing intuitionist treatment of the topic.

My first explicit foray in this direction was "On Being in a Quandary."[133] The paper was initially rejected by *Mind* with a comment by the referee that it "contained no discernible line of argument." I protested to the Editor at the time, Mark Sainsbury, who solicited other opinions and, gratifyingly, saw fit to publish. The project underwent further motivation and development in a paper I wrote for the *Liars and Heaps* conference held at the University of Connecticut in 2002,[134] but the most complete statement of it I have offered to date is in my contribution to the Library of Living Philosophers volume for Michael Dummett that was published in 2007.

Stephen Schiffer's Chapter 7 in the present volume is principally focused on the 2001 paper, while Ian Rumfitt (Chapter 6, this volume) mainly addresses the account in the later paper I wrote for Dummett. Each marshals what seem to me important arguments and criticisms. Rumfitt objects that I have neglected to provide any semantic foundation for the intuitionistic revisions to classical logic that the 2007 paper proposes,[135] and that the stability of the position taken there is in any case put under serious threat by Timothy Williamson's much discussed "anti-luminosity" argument. However, the latter half of his chapter is given to the development of a semantics that, with some reservations, he offers as a remedy for his first complaint. Schiffer's Chapter 7 offers an exemplarily careful reconstruction of the main argument of the 2001 paper and endorses my principal reasons for discounting

---

[129] Putnam (1983).   [130] Read and Wright (1985); Putnam (1985).   [131] Wright (1987).
[132] Williamson (1994).   [133] Wright (2001; also published in *The Philosopher's Annual* 24).
[134] Wright (2003b).
[135] Here he echoes a complaint of Dummett himself who in his comment on the paper in the Schilp volume records his 'admiration for the beautiful solution of the Sorites advocated by Crispin Wright, clouded by a persistent doubt whether it is correct' (Dummett 2007: 453).

supervaluationist and epistemicist treatments of vagueness. But he rightly stresses the paramount importance, in any satisfactory solution that treats the Sorites reasoning as a *reductio* of its major premise, of explaining away the apparent plausibility of that premise and finds the account offered in terms of my proposed technical notion of *quandary* wanting on that score. Above all, he baulks at my rejection of the idea that knowledge is impossible in borderline cases—the principle I called Verdict Exclusion.

Thinking about the conjoint import of the two chapters, I have found myself wanting to be concessive on some of the points of criticism they develop but unrelenting on others. But rather than offer a listiform set of responses accordingly, it seemed to me that it might be of most interest to a reader who wants to think about intuitionism in this context if I were to present a new, relatively self-contained presentation of the way I now see the case for it, explaining *passim* how an intuitionist view may handle some of what I regard as the more important of the criticisms that Rumfitt's and Schiffer's searching discussions advance. That is what follows. I'll comment briefly on Rumfitt's semantic proposals at the end.

1. The Basic Analogy

Mathematical platonism may be characterized as the conviction that in pure mathematics we explore an objective, abstract realm that confers determinate truth-values on the statements of mathematical theory irrespective of human (finite) capacities of proof or refutation. This conviction crystallises in the belief that classical logic, based on the semantic principle of Bivalence, is the appropriate logical medium for pure mathematical inference even when, as of course obtains in all areas of significant mathematical interest from number-theory upwards, we have no guarantee of the decidability by proof of every problem. In this respect—the conviction that the truth-values, *true* and *false*, are distributed exhaustively and exclusively across a targeted range of statements irrespective of our cognitive limitations—an epistemicist conception of vagueness[136] bears an analogy to the platonist philosophy of mathematics. Let us characterize a vague predicate as *basic* just if it is semantically unstructured and is characteristically applied and withheld non-inferentially, on the basis of (casual) observation. The usual suspects in the Sorites literature—*bald, yellow, tall, heap*, and so on—are all of this character and will provide our implicit focus in what follows. Epistemicism postulates a realm of distinctions drawn by such basic vague concepts that underwrite absolutely sharp "cut-offs" in suitable soritical series,[137] irrespective of our capacity to locate them. For the epistemicist, the principle of Bivalence remains good for vague languages—or if it does not, it is not vagueness that compromises it—and classical logic remains the appropriate medium of inference among vague statements. An indiscernible difference between two colour patches in a soritical series for *yellow* may thus mark an abrupt transition from yellow to orange;

---

[136] As supported by, among others, Cargile (1969); Sorensen (1988); Horwich (1990); and, in its most thoroughgoing development, Williamson (1994).

[137] We understand a "suitable" sorites series for a predicate *F* to be *monotonic*, that is, one in which any *F*-relevant changes involved in the move from one element to its immediate successor are never such that the latter has a stronger claim to be *F* than the former.

the impression of the *indeterminacy* of that distinction is merely a reflection of our misunderstanding of our ignorance of where the cut-off falls. For the epistemicist, the Sorites paradox is accordingly easily resolved. It is scotched by the simple reflection that its major premise will always be subject to counterexample in any particular soritical series. If the initial element is yellow and the final element is orange, then there must be an adjacent pair of elements one of which is yellow while the next is orange. It is just that we, in our ignorance, are in no position to identify the critical such pair.

On one understanding of it, the *ur*-thought of intuitionism as a philosophy of mathematics is a rejection of the idea of a potentially proof-transcendent mathematical reality as a *superstition*: something that there is, simply, no good reason to believe in. For the intuitionist, the mathematical facts are justifiably regarded as determinate only insofar as they are determinable by proof, and the relevant notion of proof needs accordingly to be disciplined in such a way as to avoid any implicit reliance on the platonist metaphysics. So, in any area of mathematics where we lack any guarantee of decidability, the logic deployed in proof construction cannot rely on the principle of Bivalence and hence—according to the intuitionist—cannot justifiably be classical. In particular the validity of the law of excluded middle, which intuitionism understands as depending on the soundness of Bivalence, can no longer be taken for granted. There is evident scope for a similar reaction to the epistemicist conception of vagueness. The latter is a commitment to a transcendent semantics for vague expressions which construes them as somehow 'glomming onto' semantic values—properties in the case of vague of predicates—that are possessed of absolutely sharply demarcated extensions, potentially beyond our ken. The conception that vague expressions work like that may likewise impress as the merest superstition. Perhaps a little more kindly, it may impress as merely *ad hoc*, for there is not the slightest reason that speaks in favour of it except its convenience in the context of addressing the Sorites.[138] Say that an object, *o*, is *F-surveyable* just in case *o* is available and open to as careful an inspection as is necessary to justify the application of *F* to it whenever it can be justified. For the epistemicist, reality is such that the application of any meaningful basic vague predicate, *F*, to an *F*-surveyable object must result in a statement of determinate truth-value, true or false. For an intuitionistic conception of vagueness—one conceived on the model of mathematical intuitionism—a satisfactory semantics and logic for basic vague predicates must eschew commitment to any such claim.

The avoidance of such a commitment is of course, common ground with any instance of the long tradition of theories about vagueness that construe borderline cases as examples of *semantic indeterminacy*: as cases where the rules of the language leave us in the lurch, so to speak, by issuing no instruction for any particular verdict. Here, though, the intuitionist credits the epistemicist with a crucial insight: that vagueness is indeed a *cognitive*, rather than a semantic phenomenon. Our inability to

---

[138] I have sometimes encountered in discussion the impression that the motivational shortfall here is addressed by Williamson's argument that knowledge everywhere requires a margin of error. Not so. What that argument establishes, if anything, is that *if* there is a sharp cut-off in a sorites series, we will not be able to locate it. The argument provides no reason to suppose that the antecedent of that conditional is true. I'll say more about the Williamsonian argument below.

apply the concepts on either side of a vague distinction with consistent mutual precision is not a *consequence* of some kind of indeterminacy, or incompleteness in the semantics of vague expressions but is constitutive of the phenomenon.

Consider this example. Suppose we are to review a line of one hundred soldiers arranged in order of decreasing height and to judge of each whether he or she is at least 5'10" tall—but to judge by eye rather than by using any means of exact measurement. Let the soldiers' heights range from 6'6" to 5'6". This provides a toy model of a sorites series as conceived by the epistemicist. For while there is indeed a sharp cut-off—there must be a first soldier in the line who is less than 5'10" tall—our judgements about the individual cases will expectably divide between an initial range of confident positive verdicts and a later range of confident negative verdicts between which there will be a region of uncertainty, where we return hesitant, sometimes mutually conflicting verdicts and sometimes struggle to return a verdict at all. In this example, of course, we have a conception of canonical grounds for determining whether a hard case has the property expressed by the predicate at issue, so that is a point of contrast with our situation when we face a sorites series for a vague predicate as conceived by epistemicism. But putting that disanalogy to one side, it remains that in the soldiers scenario, our patterns of judgement will have exactly the physiognomy that the epistemicist regards as the hallmarks of vagueness. Hence, in her view, there is nothing in our practice with vague concepts that distinguishes it from judgements concerning sharply bounded properties about whose specific nature we are ignorant.

The intuitionist agrees with epistemicism that such a physiognomy of practical judgement is characteristic of vagueness. But intuitionism drops both the assumption that the judgements concerned are answerable to the extension of a sharply bounded property *and* the notion that a different kind of explanation of these characteristic judgemental patterns is called for, in terms, roughly, of shortcomings in—our lack of guidance by—the semantic rules that fix the meanings of the expressions concerned. For the intuitionist, the vagueness of a predicate *consists in* these distinctive patterns in our use of it. They are the whole story. The intuitionist conception of vagueness is thus a *deflationary* conception: it holds that there is no more to the phenomenon than meets the eye, so to speak—that it is unnecessary, is indeed a mistake, to look to some underlying feature of the semantics of vague expressions to explain our characteristic patterns of judgement in the borderline area. (That is not to say that one should not look for an explanation of a different kind.) It is the view of the intuitionist that both epistemicism and indeterminism commit versions of this mistake.

The justification for this charge when the canvassed alternative is semantic indeterminism rests on a complex variety of considerations whose details, for reasons of space, I cannot rehearse here.[139] However, there is one such consideration—one relevant aspect of our practice with vague concepts—which is particularly important for the grounding of the most distinctive aspect of the intuitionist approach. Semantic indeterminism interprets borderline cases of a distinction as cases where

---

[139] For elaboration see Wright (2003b, 2007, and 2010). For a further set of objections, see Williamson (1994: chs 5 and 7).

there is no mandate to apply either of the expressions concerned—where the rules for their use prescribe no verdict. That suggestion does a poor job of predicting one salient aspect of our judgemental practices with vague concepts, namely our uncritical attitude to polar—positive or negative—judgements concerning items in their borderline regions. Provided a verdict is suitably qualified and evinces an awareness that the case is a marginal one, it is not treated as a mark of incompetence, or mistake, to have a view, positive or negative, about any single borderline case.[140] Suppose X struggles to have an opinion whether some shade from the mid-region of the yellow-orange sorites is yellow enough to count as yellow but Y is of the opinion that it is—just about—yellow. Our sense is that such divergences are just what is to be expected, and that each reaction can be as good as the other. X need not be regarded as coming short; Y need not be regarded as overreaching. Each reaction is quite consistent with full mastery of *yellow* and due attention to the hue concerned.

According to semantic indeterminism, this *laissez faire* attitude should be regarded as cavalier, for X and Y cannot both be operating as the relevant semantic rules require; the rules cannot both be silent on the relevant hue and mandate Y's qualified verdict of "yellow."[141] Yet our ordinary practice reflects no sense of that. We are characteristically open to—as I have elsewhere expressed it, *liberal* about—polar verdicts about borderline cases. To be sure, the indeterminist might be tempted to interpret this liberality as reflecting a sense of respect for our ignorance about in just which cases the rules do in fact fall silent—which are the true borderline cases. But since, if so, there is no evident means of remedying that ignorance, that again would be a step in the direction of objectionably transcendentalizing the semantics of vagueness. For the intuitionist, in contrast, there really need be no sense in which one who returns a (qualified) polar verdict about a borderline cases does worse than one who fails to reach a verdict.

The point may seem slight. But it is crucial. For respecting this aspect of our practice as in good standing requires that, in contrast to the view of semantic indeterminism, we should not regard borderline cases as presenting *truth-value gaps*.[142] If borderline cases are truth-value gaps, then someone who returns a polar verdict about such a case actually makes a mistake. And that is just what, according to

---

[140] For the purposes of this claim, we may take a borderline case to be any that tends to elicit the judgemental physiognomy characterized earlier among a significant number of competent judges.

[141] To be sure, there is another possibility: we might try to think of the rules as, in the borderline area, issuing *permissions*. Then both a tentative verdict and a failure, or unwillingness, to reach a verdict, may be viewed as rule-compliant. But it is very doubtful that any satisfactory proposal lies in this direction. Presumably among the clear cases the rules must *mandate* specific verdicts rather than merely permit them. So we need to ask about the character of the transition from cases where a positive verdict about F is mandated to cases where it is merely permitted. If this is a sharp boundary, then since there is again no possibility of knowing where it falls, the proposal will have 'transcendentalized' the semantics of F in a manner different from but no less inherently objectionable than epistemicism. But if the transition is accomplished by a spread of further borderline cases—cases that are borderline for the distinction between 'mandatorily judged as F' and 'permissibly judged as F'—then the question arises how, in point of mandate or permission, cases in this category are to be described. For argument that contradiction ensues, see Wright (2003b).

[142] Or indeed as having any kind of "Third Possibility" status inconsistent with each of truth and falsity *simpliciter*. For further discussion, see Wright (2001).

liberalism, we have no right to think. It follows that we have no right to regard borderline cases as *counterexamples* to the principle of Bivalence, and hence that vagueness, as now understood, provides no motive to *reject* Bivalence. Since, by rejecting epistemicism, and recognizing that we cannot in general settle questions in the borderline region either, we have also undercut all motive to *endorse* the principle, the resulting position is exactly analogous to the attitude of the mathematical intuitionist to Bivalence in mathematics: that it is a principle towards which we should take an agnostic stance.

With these preliminaries in place, let us turn to review how an intuitionistic treatment of the sorites may be developed in more detail.

2. The Tolerance and 'No Sharp Boundaries' Paradoxes

The classic deductive[143] sorites paradoxes vary in two respects: first, in the formal character of the major premise involved, and, second—where the major premise is shared—in the manner in which that premise is made to seem plausible. And of course different forms of major premise will call for correspondingly different deductive sub-routines in the derivation of paradox. Perhaps the most familiar form of the deductive sorites is what we may call the Tolerance Paradox. As normally formulated, it uses a universally quantified conditional major premise:

$$(TP) \quad (\forall x)(Fx \to Fx'),$$

and proceeds on the assumption of one polar premise, $F1$, and n-1 successive steps of universal instantiation and *modus ponens* to contradict the other polar premise, $\sim Fn$.

As for motivation, the key thought is, as the title I have given to the paradox suggests, that, such is its meaning, the application of F, and/or the justification for applying it, *tolerates* whatever small changes may be involved in the transition from one element of the series to the next: for instance, that if a colour patch is (justifiably described as) red, a pairwise indiscriminable (or even just barely noticeable) change in shade must leave it (justifiably described as) red; that if a man is bald, the addition of a single hair won't relevantly change matters, and so on. For the examples with which we are concerned, claims of this ilk can seem thoroughly intuitive; and they can be supported by a variety of serious-seeming theoretical considerations.[144] In some cases, indeed, the claim of tolerance may seem absolutely unassailable: how could "looks red' for example, fail to apply to both, if to either, of any pair of items that look exactly the same? Unfortunate, then, that "looks exactly the same" is not a transitive relation.

The Tolerance sorites, however, impressive as it may be in particular cases, is not, or at least not obviously, a paradox of vagueness per se. Vagueness is not, or at least not obviously, the same thing as tolerance. Precision must imply non-tolerance, of

---

[143] As distinguished from the so-called 'Forced March' sorites. For brief discussion of that version of the Sorites, see below.

[144] For elaboration of some such, see Wright (1975).

course, but the converse is intuitively less clear. Ought there not somehow to be some distance between a predicate's possession of borderline cases and its being tolerant of some degree of marginal change? While the claim may indeed seem intuitive, it requires—in the presence of paradox—argument to suggest that *yellow, heap, bald,* etc., are tolerant. But no argument is required to suggest that they are vague. That these predicates are vague is a *datum.*

The No Sharp Boundaries paradox, by contrast, impresses as a paradox of vagueness par excellence. It works with a negative existential major premise,

$$\text{(NSB)} \quad \sim(\exists x)(Fx \ \& \ \sim Fx')$$

that may very plausibly seem simply to give expression to what it is for F to be vague in the series of objects in question. For vagueness, surely, just is the complement of precision, and the sentence of which that negative existential is the negation, viz. what I have elsewhere called the *unpalatable existential*

$$\text{(U)} \quad (\exists x)(Fx \ \& \ \sim Fx')$$

surely just *states* that F is precise in the series in question: that there is a sharp boundary between the Fs and the non-Fs, and so no borderline cases. If, then, F is in fact vague, the negative existential seems imposed just by that fact, indeed to be a statement of exactly that fact. And now contradiction follows by iteration of a different but no less basic and cogent-seeming deductive sub-routine, involving conjunction introduction, existential generalisation, and *reductio ad absurdum* as a negation introduction rule.[145]

With both paradoxes, there is the option of letting the reasoning stand as a *reductio* of the major premise. If we take that option with the Tolerance paradox, we treat it as a schematic proof that none of the usual suspects *is* genuinely tolerant of the marginal differences characteristic of the transitions in a soritical series for it. Tolerance, in that case, is simply an illusion. And that is a conclusion we might very well essay to live with, provided we can provide a satisfactory explanation of why and how the illusion tends to take us in, and of what is wrong with the "serious-seeming theoretical considerations" apparently enforcing tolerance that I alluded to above.

But not so fast: even if those obligations can be discharged, the proposed response, in the presence of classical logic, is not yet stable. For (allowing its ingredient conditional to be material) the negation of TP, now regarded as established by the paradoxical reasoning, is a classical equivalent of the unpalatable existential. So if our logic is classical, non-tolerance does after all collapse into precision, and to the extent that one feels there should, as remarked above, be daylight between them, that should impress as a strike against classical logic in this context. Moreover that impression is only reinforced when one considers the option of letting the No Sharp Boundaries paradox stand as a refutation of NSB. For then all that stands between that result and affirmation of the unpalatable existential is a double negation elimination step. And now, once constrained by classical logic to allow the inference to U, we seem to be on

---

[145] That is, the intuitionistically valid half of classical *reductio*, where the latter also allows *reductio ad absurdum* inferences that serve to eliminate negations.

the verge of admitting that *vagueness itself* is an illusion. That, surely, isn't anything we can live with.

Intuitionism, by contrast, aims at winning through to a position where we can accept each of the Tolerance and No Sharp Boundaries paradoxes as a *reductio* of its major premise but refuse in a principled way the inference onwards to the unpalatable existential. We also aim to retain the ordinary conception of an existential statement as requiring a determinate witness for its truth, and thus to avoid any form of the implausible semantic story that construes the statement, "There is in this series a last *F* element followed immediately by a first non-*F* one" as neutral on the question of the existence of a sharp cut-off as intuitively understood. Our path will be to explore, in the light of the general, deflationary conception of the nature of vagueness outlined earlier, what motivation it may be possible to give for broadly intuitionist restrictions on the logic of inferences among vagueness-involving statements. In this we follow the suggestion first briefly floated at the end of Hilary Putnam's (1983).[146] If, in particular, we can justify a rejection of double negation elimination for molecular vague statements in general, then it may be possible comfortably to acknowledge that both the Tolerance and the No Sharp Boundaries paradoxes do indeed disprove their respective major premises without any consequent commitment to the unpalatable existential, nor consequent obligation either (with the epistemicist) to believe it or (with the supervaluationist) to somehow reinterpret it in such a way that it doesn't mean what it (seemingly) says.

3. Constraints on an Intuitionistic Solution

Following Schiffer, I propose that we set the following three constraints on the project. First and most obvious (*constraint 1*), we need to *motivate* the required restrictions on classical logic in general and, in particular to explain how a valid reductio of TP, or NSB, can fail to justify the unpalatable existential.

Second (*constraint 2*), as with all attempts to solve, rather than merely block a paradox, we must offer a convincing explanation of why the premises that spawn aporia impress us as plausible in the first place, of what mistaken assumptions we have implicitly fallen into that give them their spurious credibility. So in the present instances, we must contrive to explain away the continuing powerful temptation to regard the major premises for the Tolerance and No Sharp Boundaries paradoxes as true. I have said much elsewhere to attempt to defuse the attractions of tolerance premises.[147] Here we will focus on the challenge to explain why NSB is *not* a satisfactory characterization of *F*'s vagueness in the series in question. (As the alert reader will appreciate, we have already implicitly shown our hand on this.)

Finally (*constraint 3*), I think it reasonable to require, although I grant it is not wholly clear in advance exactly what the requirement comes to, that constraints 1 and

---

[146] Early discussions of Putnam's proposal, besides my own work from (2001) onwards, include Read and Wright (1985); Putnam himself (1985, 1991); Schwartz (1987); Rea (1989); Schwartz and Throop (1991); Mott (1994); Williamson (1996); and Chambers (1998).

[147] Such an attempt must perforce be somewhat ramified, in order to match the diverse sources of such attraction. My own diagnostic forays run from Wright (1975) through (1987) to (2007).

2 should, so far as we can manage it, be satisfied in a way that draws on an overarching account of what the relevant kind of vagueness consists in (i.e., of the nature of the relevant kind of borderline cases.) We are proposing restrictions on what, from a classical point of view, are entrenched, tried and tested patterns of inference. If such restrictions are justified, it may be, to be sure, that their justification is global, applying within discourses of every kind. That is the character, for example, of the metasemantic considerations about acquisition and manifestation of understanding originally offered by Michael Dummett half a century ago in support of a global repudiation of the principle of Bivalence except in areas where decidability is guaranteed. Whatever one's estimate of such arguments, what constraint 3 is seeking is a justification for relevant restrictions on classical logic that is specifically driven by aspects of the nature of vague discourse. In the present context, that will require putting to work the deflationary conception of vagueness sketched above in §1.

4. Addressing Constraint 1: "The Basic Revisionary Argument"

In the mathematical case, as remarked, the intuitionistic attitude flows from a rejection of the principle of Bivalence, based on a repudiation of platonist metaphysics and insistence that truth in pure mathematics can only consist in the availability of proof. In the case of vague statements, many would be pre-theoretically willing to grant that Bivalence is generally unacceptable anyway. Certainly, the metaphysics of meaning implicit in epistemicism has none of the intuitive appeal of, say, arithmetical platonism. But even if it is granted that Bivalence is *principium non gratum* where vagueness is concerned, repudiating the principle is one thing and motivating revision of classical logic a further thing. Classical logic need not necessarily fail if Bivalence is dropped. How should the intuitionist argue specifically that the *logic* of vague discourse should not be classical?

What I once called the "basic revisionary argument" is designed to accomplish that result.[148] It runs for any range of statements that are not guaranteed to be decidable but are subject to a pair of principles of *evidential constraint*. That is, for each such statement P, each of these conditionals is to hold:

EC :   P → it is feasible to know P

   Not P → it is feasible to know not-P

Now, it is plausible—but with caveats, to be considered in a moment—that each of the usual suspects—*yellow, bald, tall, heap,* etc.—generates atomic predications that do exhibit this form of evidential constraint; that is, intuitively, if something is, in the sense characterized earlier, surveyable for *yellow*, for example (that is, it is available for inspection in decent conditions, etc.), and it *is* yellow, then we'll be able to tell that it is; and if it *isn't* yellow, we'll be able to tell that. Intuitively, what colour something is *cannot hide* if and when conditions present themselves in which it is possible to

---

[148] This argument is central in Wright (1992), (2001), and (2007).

have a proper look at it.[149] And analogously for baldness, tallness, 'heaphood' and the rest. The basic revisionary argument is then the observation that, if the law of excluded middle is retained for all such predications, P, then reasoning by cases across the EC conditionals will disclose a commitment to the disjunction:

> D: It is feasible to know P ∨ It is feasible to know not-P

—in effect, the thesis that P is decidable. But of that, if the relevant predicate is associated with borderline cases, we have no guarantee. Accordingly we have no guarantee of the validity of the law of excluded middle in application to such statements and therefore have no business reckoning it among the logical laws.

Simply expressed, the thrust of the argument is that a range of statements may be such as both to *lack* any general guarantee of decidability in an arbitrary instance and to *have* a guarantee that if any of them is true, it will be recognizably true and, if false, recognizably false. Imposition of the law of excluded middle onto such statements will then enforce the conclusion that each of them is decidably true or false— contrary to hypothesis. It will amount to the pretence of a guarantee that we do not actually have.

Arguably a very large class of statements are in this position, including not merely vague predications but, for instance, evaluations of a wide variety of kinds, including expressions of personal taste, humour and, perhaps, (some aspects of) morality. And of course the argument will run for any region of discourse where we reject the idea that truth can outrun all possibility of recognition but have to acknowledge that we lack the means to decide an arbitrary question—exactly the combination credited by the intuitionists for number-theory and analysis.

Suppose then that we disdain the law of excluded middle on this (or some or other) basis. The soritical series we are considering involve a *monotonic* direction of change: that is, any F element is preceded only by F elements, and any non-F element is succeeded only by non-F elements. The reader will observe, accordingly, that once the law of excluded middle is rejected, the sought-for distinction between the unpalatable existential and its double negation is enforced (albeit not yet explained.) For the latter is surely established by the inconsistency of NSB with the truths expressed by the polar assumptions. But, given monotonicity, the unpalatable existential is equivalent to the law of excluded middle over the range of atomic predications of F on the series of elements in question.

5. One Objection to the Basic Revisionary Argument

So far, so good. But now for the caveats. The EC-conditionals are challengeable on a number of serious-looking counts. First, they are in direct tension with the upshot of Timothy Williamson's recently influential 'anti-luminosity' argument.[150] Familiarly, Williamson makes a case that if knowledge generally is to be subject to a certain form of (putatively) plausible safety constraint, then it must be controlled by a margin of

---

[149] The claim that EC holds good for these cases is thus not subject to 'killer yellow' issues.
[150] Williamson (2000: ch. 4).

error: in particular, if a subject knows that F holds of an object a, it cannot be that F fails to hold of any object that she could not easily distinguish, using the same methods, from a. The effect, Williamson takes it, is thus that, for elements, x, in a suitable soritical series for F, the following conditional is good:

$$(\text{It is feasible to know that } Fx) \rightarrow Fx',$$

which, paired with the first of the EC conditionals, immediately provides the means to show that F applies throughout the soritical series.

Here is not the place for a detailed engagement with Williamson's thesis. But there are several fairly immediate misgivings about it that deserve notice. One is whether the notion of safety that it utilizes is indeed a well-motivated constraint on knowledge everywhere, whatever the subject matter and methods employed. Williamson's intuitive thought, if I may venture a précis,[151] is that if a subject comes to the judgement that Fa by some method—for instance, simple observation—by which a' is pairwise indistinguishable from a, then the subject must be significantly likely also to judge that Fa'—and now, if the latter is false, she is therefore very likely to make a false judgement using the very methods she used in judging Fa. So those methods are not generally reliable, in which case the judgement that Fa, based upon them, ought not to count as knowledgeable in the first place. Yet if that does capture the intuitive thought, then one question that may be salient to the reader is why we should require that, in order to count as a reliable means for settling a question about one item, a method must also be reliable about *others* that, however similar, differ from it. Why could not a machine—a speedometer, for example—that issues a varying digital signal in response to a varying stimulus, have an absolutely sharp threshold of reliability, so that its responsive signals are reliable up to and including some specific value, k, in its inputs but then go haywire for inputs of any greater value. In that case, its signals may be regarded as 'knowledgeable' for any input value i, less than but as close to k as you like. If it is not a priori ruled out that our judgements, for some particular pairings of subject matter and methods, are like that, then it is not a priori guaranteed that Williamsonian safety is everywhere a necessary condition of knowledge.

I envisage the likely rejoinder that as a matter of anthropological fact, *we* are not in any area of our cognitive activities comparable to such a machine. Maybe so. Still, even if so, it is a contingency. And it seems incredible that such a contingent fact about us could somehow enforce the thought that there are yellows, and instances of baldness, etc., that lie beyond our powers of recognition even in the best of circumstances. Something is going wrong with the argument.

What? A possible diagnosis emerges if we follow through on the following reflection. The EC-conditionals express the luminosity both of (whatever is substituted for) P *and* of its negation. But the assumptions which Williamson feeds into his *reductio* include not merely the assumption of luminosity and of a safety constraint on knowledge but also a premise about the likely unreliability of our verdicts about cases that lie extremely close together and where P holds of one of them but not of the

---

[151] Compare Williamson (2000: 97). The passage there does leave latitude for interpretation.

other. In short, he requires a premise—not liable to discharge—to the effect that if we judge correctly that P holds of a case which, in the kind of series we are concerned with, lies next to one where it does not, then we are—as a matter of anthropological fact—very likely to misjudge the latter as a case where P still holds. It is—or ought to be—immediately notable that this anthropological premise is inconsistent with the assumption of the EC-conditional for not-P. For if not-P is luminous, then we will exactly *not* be prone to misjudge any particular case where it holds. Thus Williamson's assumptions beg the question! No surprise, it may be said, that he can make trouble for a proponent of the EC-conditionals if he helps himself to question-begging premises.

This is apt to seem like a quibble. For is the fact not simply that, for the kind of substituends for 'P' that concern us, we will not, for any pair of sufficiently close cases, actually be able reliably to discriminate one as a case where P holds and its neighbour as a case where it does not? Indeed so. But then the proper conclusion is that a supporter of luminosity, and of the EC-conditionals in particular, ought to resist the suggestion *that there can be any such pair of cases.* That is something for which there is in any case ample independent motivation, since the kinds of series with which we are concerned all exhibit the characteristic of *seamless transition,* whereby we start with cases where P holds and wind up with cases where not-P holds but do so precisely *without* there being any last P case directly succeeded by a case where P fails.

This is, to be sure, a maddeningly difficult notion to characterize satisfactorily.[152] Certainly, there is no hope of doing so in a classical framework, where the reasoning of the Least Number Principle will enforce the thought that if cases where not-P holds are reached eventually, there has to be a determinate last P case. The point remains that seamless transition is a real phenomenon that is active in all the cases of interest to us, and Williamson's *reductio* of luminosity simply fails to reckon with it. Even if we have no qualms in general about the form of safety constraint on knowledge that Williamson utilises, his argument needs a transition from

> If we know P in a case, the next (ever so marginally different) case cannot be a case of not-P,

to

> If we know P in a case, the next (ever so marginally different) case must also be a case of P.

But seamless transition excludes *both* sharp P/not-P cut-offs *and* unrestricted preservation of P's truth across proximal cases. That we are not astride a cut-off will not, when seamlessness is in play, entail that both our feet are resting on cases for both of which either P, or not-P, holds good.

Obviously there is much more to say about this, and I hope to return to it in future work. Here, though, let me flag one further independent misgiving about

---

[152] For further, albeit inconclusive discussion, see Wright (2010).

Williamson's, as I suspect, undeservedly influential argument. Let's continue to grant that the general requirement of safety proposed—again: the proposal that in order to know that P in circumstances C, my methods must be such that they could not easily lead me astray in circumstances sufficiently similar to C—grant that this is well-motivated everywhere. It is striking that, in the way Williamson puts the proposal to work, no account is taken of the possibility of *response-dependence*: the idea that some kinds of judgement—and here the critic is likely to be thinking of exactly the kinds of judgement, about sensations and other aspects of one's occurrent mental life, that Williamson means to target in directing his argument against the traditional idea of our 'cognitive home'—are not purely *discriminatory* of matters constituted independently but are such, rather, that the subject's own judgemental dispositions are somehow themselves implicated in the facts being judged. For any area of judgement where this idea has traction, the supposition that in perfectly good conditions of judgement we might easily respond to what is in fact a not-P case in the way we do to a P-case that is very similar to it, is in jeopardy of incoherence. Simply, if P-ness and not-P-ness are response-dependent matters, then it cannot legitimately be assumed that, purely on the basis of their similarity in a particular case, we will be at risk of responding to a not-P-case in the way we do to a P-case.

To be sure, the heyday of the recent discussion of response-dependence has passed, and rigorous but still dialectically useful formulations of it proved hard to come by when the debates were at their height. Still, many may feel that there is an elusive truth in it, with qualities instantiated in one's phenomenal mental life and Lockean secondary qualities of external objects generally providing two examples of domains to which philosophical justice can be done only by keeping a place for the idea of response-dependence on our philosophical agenda.[153]

All that said, then, I do not think that, in our present state of understanding, Williamson's argument comes anywhere near to establishing that the basic revisionary argument is hobbled by its reliance on the EC-conditionals. But of course much more remains to be said.

6. Two Further Objections

There are, however, two less theoretically loaded reservations about the role of the EC-conditionals in the basic revisionary argument that should be tabled when what is envisaged is its application specifically to vague expressions. First, no connection has actually been explained to link the EC-conditionals with vagueness as such. All that has been offered is the suggestion that the conditionals are plausible for some examples of vague predicates—for the 'usual suspects.' A general theoretical connection is wanted before there can be any firm prospect of a solution by this route to the Sorites paradox in general. One senses that a development may be possible of a general connection between vague judgement and response-dependent judgement,

---

[153] Concerns of this character, although he does not mention the notion of response- (or judgement-) dependence by name nor relate his discussion to the literature about it, are nicely elaborated in Berker (2008). A sophisticated attempt to develop Williamson's argument in a manner sensitive to possibilities of response-dependence is Srinivasan (2015).

grounded in the thought that the status of something as a borderline case is a response-dependent matter. That suggestion, though, once again in the present state of our understanding, is merely speculative.

Second, and perhaps more threatening to this particular strategy for underwriting an intuitionistic treatment of the sorites, is the conflict between the EC-conditionals and the principle that in earlier work I called *Verdict Exclusion*:

VE: Knowledge is not feasible about borderline cases.

EC and VE are pairwise inconsistent (since, as the reader will speedily see, they combine to enforce contradictory descriptions of borderline cases). So someone who accepts EC must deny VE. But VE impresses Schiffer—indeed has impressed a number of expert commentators[154]—as a datum. In any case, the principle may seem to have powerful intuitive support from the very deflationary conception of vagueness which, I have proposed, should be seen as the mainspring of an intuitionistic treatment. On that conception, borderline cases are constitutively cases whereby subjects characteristically fall into weak, inconstant and mutually conflicting opinions. Any opinion a subject holds about such a case is one that she might very easily, using just the same belief-forming methods, not have held. Surely on any reasonable interpretation of a safety, or reliability, constraint on knowledge, that must count as inconsistent with such an opinion's being knowledgeable.

Elsewhere, I have suggested that an endorsement of VE proves, on closer inspection, to be in tension with the liberalism about verdicts in the borderline area that I laid stress on in §1.[155] Let me here make a different point. Once it is given that something is a borderline case, I think the line of argument just outlined for VE is likely to prove compelling. But the crucial consideration is that, of any particular element in a sorites series, it is *not* a given—except as a contingent point about the sociology of a particular group of judges—that it is a borderline case. Being a borderline case is judge-relative—x may be such as to induce the characteristic judgemental difficulties and variability in some but not other competent judges. Let the proposition that x is yellow elicit those characteristic responses in some of us but suppose that Steady Freddy consistently judges x yellow (though acknowledges that it is near the borderline). Must we deny that Freddy's verdict is knowledgeable? After all, it is, we may suppose, the verdict of someone who gives every indication otherwise of a normal competence in the concept, has normal vision, and is judging in good conditions—and judging in a way consistent with his judgement of the same shade on other occasions. It is harsh to say he doesn't know.[156] And if it is at least indeterminate whether Freddy knows, then we—theorists—do not know VE.

On the other hand, if we take it that the EC-conditionals *are* known to hold good for surveyable predications of F, must we not also accept the strange claim that VE is known to be false for such cases and hence that each element in a soritical series for F allows in principle of a knowledgeable verdict about its F-ness? It's not clear. There is

---

[154] Schiffer (2016), Rosenkranz (2003), Williamson (1996).   [155] Wright (2003a).

[156] Some will no doubt say that Freddy has a different concept. But that seems merely *ad hoc*. What does the difference consist in? Why not say instead that he is steadier than we are in his judgements involving a concept we share?

a double negation elimination step in the drawing of that conclusion whose legitimacy might be viewed as *sub judice* in the present dialectical context. Rather than take a stand on the matter, it may seem that prudence dictates, *pro tem.*, that we reserve judgement on both VE and EC, committing to neither.

Prudence, unfortunately, comes at a cost. Unfortunately, that is, for the would-be intuitionist, for the agnostic attitude requires that we must also be agnostic about the basic revisionary argument. If, in our current state of philosophical information, the strongest relevant claim we can justifiably make about the EC-conditionals is that it is epistemically possible that they hold good for surveyable predications of "yellow," "bald," etc., then, supposing we accept the validity of the law of excluded middle, we can validly reason only to the epistemic possibility that D above holds good, that is, that it is *epistemically possible* that, for each P in the relevant class of statements, it is feasible to know P or it is feasible to know not-P. But that double-modalised conclusion doesn't look uncomfortable—or anyway, not uncomfortable enough to put pressure on the acceptance of excluded middle. In particular, if it is epistemically possible that Steady Freddy indeed knows, then for each P in the relevant range, there epistemically possibly could be a steady subject who knowledgeably judges that it is true (or that it is false).

The revisionary import of the basic revisionary argument requires more than that the EC-conditionals are epistemically possibly correct.

7. Addressing Constraint 1:
A Different Tack—Knowledge-Theoretic Semantics

So what now? Well, a suspension, perhaps temporary, of confidence in the basic revisionary argument in this context need not surrender all prospect of a strong motivation for an intuitionistic approach to the logic of vague discourse. The basic revisionary argument attempts to garner the desired result without any particular assumptions about semantics. Let us therefore now instead consider directly what might be the most desirable shape for a semantics to take that is to be adequate for a language—a *minimally sufficient soritical language for F*—that has just enough resources to run instances for a particular vague predicate F of both the Tolerance and No Sharp Boundaries paradoxes. Such a language thus contains the predicate F, a finite repertoire of names, one for each member of a suitable soritical series, brackets subject to the normal conventions, and the standard connectives and quantifiers of first-order logic.

Let L be such a language. Since we wish to avoid any commitment to the idea that when F is applied to an object that is surveyable for it, the result can take a truth-value beyond our ken, we have no interest in any semantic theory for L which works with an evidentially unconstrained notion of truth. But nor, since we are now (even if temporarily) agnostic about EC (and therefore also about VE), should such a semantics work instead with a verificationist notion of truth. It follows that we should not choose a truth-theoretic semantics at all.[157] But then what? Well, what

---

[157] I am not here assuming that merely to give a truth-theoretic semantics for some region of discourse must involve explicit commitment to one horn or the other of this alternative. But the question may

any competent practitioner of L has to master are the conditions under which its statements may be regarded as known or not. We may therefore pursue a semantic theory that targets such conditions directly, in a spirit of aiming at a correct description of what we are in position to regard as knowledgeable linguistic practice. It will be for the critic to make the case, if there is a case to make, that we thereby misdescribe the practice we actually have.

How to make a start? We don't have much to go on. What is solid to begin with is only that there is a range of polar cases where there is no doubt that Fx may be known, a range of polar case where there is no doubt that $\sim Fx$ may be known, and a range of cases that manifest the uncertainty and variability of judgement that our governing deflationism regards as constitutive of vagueness. But consider the following controversial principle (CP):

All the *knowable* statements in L are knowable by means of knowing the truth-values of atomic predications—(which we are assured of being able to do only in polar cases)

According to CP, any of the molecular statements of L can be known, if it can be known at all, by knowing some of L's atomic statements. So the semantic clauses for the connectives and quantifiers by means of which any particular molecular statement is constructed ought—if that statement is to be reckoned knowable—to reflect an upwards path, as it were, whereby the acquisition of such knowledge might proceed. If we accept CP, we will be looking therefore for a semantics which recursively explains conditions of knowledge for the molecular statements of L in terms of those of their constituents.[158]

Presumably, we are not going to want to accept CP. "Controversial" somewhat flatters the principle. We will surely want to admit a range of exceptions, cases where a molecular statement plausibly holds good even when its constituents are borderline. Some, for instance, will be general statements that are arguably analytic of the specific vague predicate concerned, like "Everything red is coloured"; others may be nomologically grounded in the property concerned, like maybe "All heaps are broadest in the base." A more interesting class of exceptions are what Kit Fine once characterized in terms of the notion of *penumbral connection*.[159] They will concern vague predicates in general. Epistemicists will regard some instances of the law of excluded middle as coming into this category. We will not follow them in that, but we should want to allow, for example, that no matter what F may be, all instances the law of non-contradiction are knowable as, with respect to the kind of series we are concerned with, are all *monotonicity conditionals*, that is, statements of the form

$$Fx' \to Fx,$$

legitimately be pressed, and the point I am making in the text is that we cannot answer unless at the cost of surrender of agnosticism about EC. Better, therefore, not to invite the question. However, there is more to say about the motivation for the style of semantics about to be proposed. I'll return to the matter at the end below.

[158] For ease of formulation, I here count the instances of a quantified statement as among its 'constituents.'

[159] Fine (1975).

—notwithstanding whether x is borderline for F. The same will hold for the corresponding generalizations:

$$(\forall x) \sim(Fx \ \& \ \sim Fx), \quad (\forall x)((Fx' \rightarrow Fx), \quad (\forall x)((\sim Fx \rightarrow \sim Fx'), ...$$

To be sure, that such claims are knowable is not uncontroversial. It is a familiar feature of many-valued treatments of vagueness that principles like these are sometimes parsed as indeterminate—when, for instance, indeterminacy in a conjunct is treated as depriving a conjunction of determinate truth, or a conditional with an indeterminate antecedent and consequent is regarded as thereby indeterminate. We are not here taking a stand on the question whether such treatments are appropriate when one accepts their governing assumption, viz. that being a borderline case is a kind of *alethic* status, contrasted with both truth and falsity. But we are, recall, rejecting that governing assumption. And when instead borderline-case status is viewed as a cognitive status, as on *our* governing assumption, there is no evident reason to demur at the suggestion that principles of penumbral connection can be known. We can know of structural constraints that knowledge, were it but attainable, of the truth-values of a range of statements would have to satisfy without having any guarantee that we can get to know those truth-values.

These considerations suggest we pursue a theory of knowledge for L that has CP as a motivating base but includes a range of permitted exceptions to it. The theory will incorporate a knowledge-conditional semantics for L and a logic based upon it, but may also contain additional, primitive axioms of penumbral connection and perhaps other axioms analytic of or otherwise somehow guaranteed for a particular choice for F. The semantics will comprise recursive clauses that determine, for each of the quantifiers and connectives of L, the conditions that are necessary and sufficient for knowledge of any L-statement—outside the permitted exceptions—in which that operator is the principal operator on the basis of the knowledge-conditions of its constituents.

The natural approach will be something in the spirit of the Brouwer–Heyting–Kolmogorov (BHK) interpretation of intuitionist logic,[160] which, as is familiar, proceeds in proof-theoretic rather than truth-theoretic terms. There is, however, an important point about the BHK interpretation that we need to flag before moving to propose specific clauses for the theory for L that we seek. In logic and mathematics, or so one might plausibly hold, all knowledge (other than of axioms) is conferred by, and only by, proof. So it can look as though BHK-style semantics is already nothing other than a local version of knowledge-conditional semantics. So it is, but expressing matters that way may encourage an oversight. While proofs in logic and mathematics confer knowledge of what they prove, that is not all they do. They also vouchsafe knowledge of what is proved *as* knowledge. Someone who comprehendingly works through a mathematical proof that P learns not merely that P is true but also— assuming their grasp of the concept of knowledge, etc.—that P may now be taken to be part of their knowledge. Logical and mathematical proofs establish a right to include P as part of what one may legitimately claim to know. Say that knowledge is

---

[160] See, for example, Troelstra (1991: §5.2).

certified when accomplished in a fashion that legitimizes that claim: accomplished in such a way that a fully epistemically responsible, sufficiently conceptually savvy epistemic agent will be aware that they have added to their knowledge. The clauses to follow are to be understood in terms of knowledge that is certifiable— *c-knowledge*—in this sense.[161]

Adapting BHK-style clauses in a natural way, we may accordingly propose:

'A&B' is knowable just if it's knowable that 'A' is knowable and that 'B' is knowable

'A∨B' is knowable just if it's knowable that either 'A' is knowable or 'B' is knowable

'(∀x)Ax' is knowable just if it's knowable that for any object in the soritical domain and term, '*a*,' known to denote that object, 'A*a*' is knowable[162]

'(∃x)Ax' is knowable just if it's knowable that, for some object in the soritical domain and term, '*a*,' known to denote that object, 'A*a*' is knowable

What about the conditional? Actually, we don't strictly need a treatment of the conditional for the present purposes.[163] And this is fortunate, since the natural proposal:

'A → B' is knowable just if it's knowable that if 'A' is knowable, 'B' isknowable,

raises an awkwardness which I will explain below.[164]

We can now assert the following <u>Thesis</u> (verification is left to the reader):[165]

> Where validity is taken as c-knowability-preservation, and c-knowledge is taken to be factive and closed over c-knowable logical consequence, the clauses above justify rules of deduction for the listed operators coinciding with the common ground for those operators—the standard rules for conjunction introduction and elimination, disjunction introduction and elimination, universal generalization and instantiation, and existential generalization and instantiation—recognized by both classical and intuitionist first-order logic.

---

[161] It is a consequence of some kinds of knowledge-externalism that not all knowledge need be c-knowledge. But the externalist will presumably grant that knowledge *often* is c-knowledge, since it is not supposed to be a consequence of externalism that our claims to knowledge are mostly imponderable without further investigation. The crucial assumption I am making in what follows is that knowledge achieved by canonical means—typically, casual observation—of clear cases of the 'usual suspects' will be c-knowledge.

[162] Recall that L will contain a known name for every element of the soritical domain.

[163] When the major premise for the Tolerance paradox is formulated, as standardly, as a universally quantified conditional (rather than, e.g., as involving a binary universal quantifier) then the paradox does of course depend on the unrestricted use of *modus ponens*. But the intuitionist resolution of the paradox to be proposed will pick no quarrel with that, and is thus neutral on the semantics of the conditional to that extent.

[164] See n. 42.

[165] "And how," the Dear Reader may ask, "am I supposed to do that when you have nominated no specific logic for the metalanguage—here English!—in which I am supposed to run through the relevant reflections?" *Touché*. But the meta-reasoning concerned will require, besides the noted properties of c-knowledge, no more than the rules of inference for 'and,' 'or,' 'any,' 'some,' and 'if' which constitute common ground between classical and intuitionist first-order logic.

What about negation? An adaptation of the BHK-style clause along the above lines would run:

'~A' is knowable just if it is knowable that 'A' is not knowable

But that, obviously, will introduce calamity into any account that accepts Verdict Exclusion.[166] Our official stance at this point is one of agnosticism towards VE, but it would be good to have the resource of a treatment of the paradoxes that would be robust under the finding that VE was after all philosophically mandated. In any case, and perhaps more telling, BHK-style negation has always been open to the intuitive complaint that it provides a licence to convert grounds for thinking we are doomed to ignorance on some matter into grounds for denial and thus distorts negation as intuitively understood.

There is, however, a natural and much more intuitive replacement:

'~A' is knowable just if some 'B' is knowable that is knowably incompatible with 'A,'[167]

or more generally

'~A' is knowable just if some one or more propositions are knowable that are conjointly knowably incompatible with 'A.'

There is no space here to undertake a proper exploration of the philosophical credentials of this proposal. Still the reader may find it intuitively plausible that mastery of negation, at least at the level of atomic statements, is preceded in the order of understanding by mastery of which of them exclude which others: being not yellow, for example, is, among coloured things, initially understood as the having of

---

[166] Or, as Rumfitt observes in Chapter 6, this volume, into any account that accepts, for Williamsonian reasons or otherwise, that a witness to a sharp cut-off in a soritical series could not knowledgeably be identified.

[167] The reader should note that there is a question, drawn to my attention by Tim Williamson, whether we may stably combine this proposed knowledge-theoretic clause for negation with the knowledge-theoretic clause for the conditional flagged earlier:

'A → B' is knowable just if it's knowable that if A is knowable, B is knowable.

For suppose Verdict Exclusion E is accepted and A is such that

  (i) It is knowable that A is borderline
  (ii) Then it is knowable that A is not knowable (by VE)
  (iii) So it's knowable that if A is knowable, then B is knowable (by substitution in ~A ⇒ (A→ B) (*ex falso quodlibet*) and closure of knowledge across knowable entailment)
  (iv) So it's knowable that if A, then [take some arbitrary contradiction for B] (from iii, by the knowledge-theoretic clause for the conditional)
  (v) So it's knowable that ~A (by the proposed clause for negation, letting B be: if A, then [contradiction], and presuming that to be knowably incompatible with A)

So the proposed clause degrades after all into the BHK-style knowledge-theoretic clause for negation:

'~A' is knowable just if it is knowable that 'A' is not knowable,

which is what we were trying to improve on. True, the argument as presented depends on VE, which we have not endorsed. But it will run for any 'A' that is knowably unknowable. Maybe there are no such statements formulable in a minimally sufficient soritical language. Maybe one should look askance at *ex falso quodlibet*. Still a concern is raised that will need disinfection in a fully satisfactory general treatment. I cannot pursue the matter here.

some colour that rules out being yellow. The above proposal reflects the suggestion that we may take incompatibility among atomic statements as epistemically primitive. Matters change, of course, once molecular statements enter the mix. For molecular statements, incompatibility will, conversely, sometimes be recognizable only by recognizing that one or more of them entail the negation of something entailed by others. That is,

> A pair of (sets of) propositions are knowably mutually incompatible if there is some proposition 'A' such that the one knowably entails 'A' and the other knowably entails '~A'

If we now, for convenience, avail ourselves of a dedicated constant, '⊥,' to express the situation when a set of propositions, X, incorporates each of some pair of knowably incompatible propositions, thus:

$$X \Rightarrow \perp,$$

then the first of the displayed clauses above mandates the following negation introduction rule

$$\sim \text{Intro.} \quad \frac{X \cup \{A\} \Rightarrow \perp}{X \Rightarrow \sim A}$$

while the second displayed clause mandates the following negation elimination rule:

$$\sim \text{Elim.} \quad \frac{X \Rightarrow \sim A, \; Y \Rightarrow A}{X \cup Y \Rightarrow \perp}$$

That is, intuitively, if a set of propositions entails the negation of some proposition, then adding to it any set of propositions that entail that proposition will result in incompatibility.

Whatever deep justification these proposals may be open to, it will be enough for present purposes if they seem plausibly knowability-preservative in the light of the reader's intuitive understanding of negation. Their most immediately significant consequence is that they allow us to justify intuitionist *reductio* as a derived rule.[168] Given the Thesis flagged above, we thus have all the rules (&I, ∃I, *reductio*) needed to run both the No Sharp Boundaries paradox and—assuming no question is raised about *modus ponens*—the Tolerance paradox as well. The upshot, in the presence of assumed knowledge of the polar assumptions, is the following important corollary:

---

[168] At least they do so if we may assume the Cut rule. Intuitionist *reductio* may be represented as the pattern:

$$\frac{X \cup \{A\} \Rightarrow B; \; Y \Rightarrow \sim B}{\{X \cup Y\} \Rightarrow \sim A}$$

Suppose we have an instance of right-hand premise. From that and B⇒B we may obtain by ~Elim:

$$Y \cup \{B\} \Rightarrow \perp$$

From that and the lh premise, we have, by Cuts

$$\{X \cup Y\} \cup \{A\} \Rightarrow \perp$$

So by ~Intro, we have {X∪Y} ⇒ ~A

Corollary:
When the quantifiers and connectives are understood as above, there is no option but to regard the negation of the major premise of an NSB Sorites as known.

8. Addressing Constraint 1 (*cont.*): The Payoff

Constraint 1 requires that we explain how and why the *reductio* of the major premises accomplished by the paradoxical reasoning fails to justify the unpalatable existential. This is now straightforward. By the clause for '∃,' the knowability of the unpalatable existential requires that for some object in the soritical domain and term, '*a*,' known to denote that object, '*Fa* & ~*Fa'*' is knowable—requires, in short, the knowability of a witness to a sharp cut-off. We neither have nor have any reason to think we can obtain that knowledge: no '*a*' denoting any clear case furnishes such a knowable witness. And we have absolutely no reason, either given by the sorites reasoning itself or otherwise, to think that such a witness may be knowledgeably identified in the borderline area. Since, by the Corollary emphasized at the conclusion of the preceding section, we do know the negations of each of NSB and TP, the respective classical inferences from the negations of NSB and TP to the unpalatable existential fail to guarantee knowability and are thus are invalid in the present knowledge-theoretic setting.

So there is the needed daylight. Constraint 1 is met and the discomfort involved in regarding the soritical reasoning simply as a *reductio* of its major premise is thus relieved.

9. Addressing Constraint 2

At least, it is relieved if, as required by constraint 2, we can neutralize the persistent temptation to regard the major premises for the paradoxes as true. The epistemicist— and indeed almost all theorists of this topic[169]—also share this obligation, of course, so here we, most of us, can march in step. There are a number of sources for the temptation. I'll touch on five.

9.1 Projective Error

The core attraction of NSB is, naturally, simply the other face of the unpalatability of the Unpalatable Existential. And that in turn springs from our inclination to accept that NSB is simply a statement of what it is for *F* to be vague in the series in question. On the present overarching conception of what vagueness is, this is a tragic mistake. It is, indeed, the pivotal mistake, "the decisive step in the conjuring trick" that our intuitive thinking plays on us here. For *F* to be vague is for it to have borderline cases, but its possession of borderline cases is, according to the overarching deflationary conception of vagueness here proposed, a matter of our propensity to certain

---

[169] The exceptions are those theorists who prefer to look askance at the underlying logic of the paradox; for instance, at the assumption of the transitivity of logical consequence in this setting (Zardini 2015) or at intuitionist *reductio* (Fine 2017).

dysfunctional patterns of classification outside the polar regions.[170] F's being vague is thus a fact *about us*, not about the patterns that may or may not be exhibited by the Fs and the non-Fs in a sorites series. Nothing follows from F's vagueness about thresholds, or the lack of them, in a sorites series, or indeed about the details of its extension at all.

The diagnosis of projective error chimes nicely with constraint 3: the overarching conception of borderline case vagueness we are working under is invoked to undergird the proposed means of satisfying constraint 2.

9.2 Inflated Normativity

However there are other kinds of seductive untruth at work in conjuring the attraction that the major premises exert. One such involves an implicit inflation of the legitimate sense in which competent practice with the usual suspects is constrained by *rule*, and is a crucial factor in the allure of tolerance premises. An example is the general thought that the rules for the use of any of the usual suspects which can be justifiably applied or denied purely on the basis of (varying degrees of casual) *observation*, must mandate that elements in the soritical domain between which there is no relevant (casually) observable difference should be described alike. (For how otherwise could mere observation enable us to follow the rules?) In fact, none of the expressions with which we are concerned is governed by rules that mandate any such thing. But the illusion that they are—indeed must be—so governed has deep sources. As announced earlier, I must forbear to go further into these matters here.[171]

9.3 An Operator Shift?

Both the foregoing, though ultimately misguided, are nevertheless respectable reasons for our inclination to accept the major premises, involving subtle philosophical mistakes. I am not completely confident that some of us, over the last four decades of debate of these paradoxes, may not have fallen prey to a less respectable reason. (I am sure no present reader would be guilty of this.) There is a fallacious transition available in this context of a kind that we know it is easy to slip into: an operator shift fallacy. The transition concerned is that from

Nothing in the meaning of F (the way we understand it) mandates a discrimination between adjacent elements in the soritical series,

to

The meaning of F (the way we understand it) mandates that there is to be nodiscrimination between adjacent elements in the soritical series, so that 'Fx & $\sim Fx'$' is everywhere false.

---

[170] This much is common ground with epistemicism as I understand it. The difference is that we reject the further step of postulating a sharply bounded property our inability to keep track of whose extension explains the dysfunctionality.

[171] Wright (1975) rehearses what is still the best case known to me for thinking otherwise and points an accusatory finger at the implicit inflation of normative constraint; the inflation is further explained and debunked in Wright (2007).

9.4 Irrelevant Truths

There are a number of truths in the vicinity that may tempt one to accept NSB but which actually simply have no bearing on it. Each of the following, for example, is true:

- that no clear cases bear witness to the unpalatable existential;
- that nobody could justify claiming to have identified a witness in the borderline area;
- that we (normal speakers) have no conception of what it would be like even to have the impression that we had identified a witness.

But these all merely reinforce the impression that the unpalatable existential is nothing we can justify. The mere possibility of a coherent epistemicism should teach us, if nothing else, at least that such considerations do not parlay into good reasons for its *denial*.

9.5 Oversight of the Implications of Seamless Transition

The temptation to think otherwise, though, dies hard. "Granted," it may be said, "that if we are epistemicists, considerations like the above provide no good reason for denial. But what if we are *not* epistemicists? What if our attitude is, as the governing conception of vagueness that you yourself are proposing involves, that there are here no facts behind the scenes: that our best practice exhausts the relevant facts—that 'nothing is hidden'? If there are no truth-makers for predications of F and not-F save aspects of our best practice, then—given that our best practice determines no sharp cut-off for F in a suitable series—must we not conclude that there is none? And then is not some version of NSB, perhaps employing some suitably 'wide' notion of negation, going to be forced on us?"

Well, no. To acknowledge that there is neither a transcendent nor a practice-constituted truth-maker for any instance of 'Fx & $\sim Fx$' in the series ought not to be allowed to service anything stronger than the thought that the transition between F cases and not-F cases in the series is *seamless* in both directions. Since it is a datum of the problem that there *is* nevertheless a transition, that should not and must not be allowed to admit of conversion into any thought to the effect that F-relevant status is unrestrictedly *preserved* across the individual steps, for that is inconsistent with any transition taking place at all. But NSB is exactly such a thought, as the paradoxical reasoning itself shows. Ultimately, then, the plausibility of NSB derives from our failure to understand seamless transition, for taking the phenomenon seriously requires that we uphold neither cut-offs nor tolerance. Intuitionistic distinctions provide a way of stably maintaining the needed neutrality.

10. Addressing Constraint 3

The third constraint we imposed on an intuitionistic treatment of the sorites was the requirement that the first two constraints—explaining how and why there can be a deductive gap between the negation of the major premise and the unpalatable existential, and explaining the spurious plausibility of the different forms of major premises—be met in a way that is informed by an overarching conception of what vagueness consists in. Have we accomplished this?

It is arguable that the second constraint is not really motivated in the case of the Tolerance paradox. To be sure, the vagueness of a predicate, deflationarily conceived, is nothing that should suggest that it be tolerant. However, the principal motivations to regard, for example, the usual suspects as tolerant have, as remarked, little to do with their vagueness per se and need a separate treatment, not embarked on here. On the other hand, the diagnosis of projective error as responsible for the thought that an NSB premise just states what it is for *F* to be vague in a relevant soritical series draws heavily and specifically on the deflationary conception of vagueness that I have represented as the heartbeat of an intuitionistic approach.

What about the first constraint—explaining the deductive gap? Assume that the knowledge-theoretic semantics offered performs as advertised. We have to acknowledge that a semantic theory of this kind might be proposed, for certain purposes, for almost any factual discourse. So the question becomes: what, if anything, is it about vagueness as deflationarily conceived that makes such a semantics specifically appropriate for vague discourse?

Recall that it is essential to our deflationism not merely to regard certain judgemental patterns among competent judges as constitutive of an expression's vagueness but to reject the demand for explanation of these patterns in terms of underlying semantic phenomena—for instance, sharply bounded but imperfectly understood semantic values, incomplete (or conflicting) semantic rules, or the worldly side of things being such as to confer truth-statuses other than truth and falsity. That precludes any semantic theory that works with a bivalent notion of truth, truth-value gaps, or postulates any kind of third truth-status. Admittedly, the possibility is left open of working with a verificationist truth-conditional theory, as would be mandated by EC. However, no reason is evident why the knowledge-theoretic style of semantics proposed could not amount to one way of implementing the semantic import of EC, nor hence why all the crucial parts of the treatment of the paradoxes proposed could not survive were we to quash any reservations about EC. So we have not closed that particular road by going about things the way I have here. But nor have we committed to travelling it.

11. Rumfitt's Semantics

Ian Rumfitt outlines a semantics which, he suggests, has some prospect of underwriting an intuitionist treatment of the Sorites reasoning, although he canvasses reasons for misgivings. At the risk of seeming ungrateful, I have to say that I too have misgivings—or perhaps just questions—about his proposal.

As I understand it, everything turns on the intuitive thought that we may model the vagueness of a predicate by thinking of its extension as an *open set* on some appropriately ordered field, F, of elements, where a set S is open in F just in case to any of its members corresponds a 'neighbourhood' of elements of F each of which is likewise a member of S. We may, to fix ideas, assume we are dealing with a simple linear order. So the suggestion is that the vagueness of a relevant predicate, *F*, shows in the circumstance that each F element in F is, as it were, surrounded in both directions by (possible) F-elements.

I have two problems with this intuitive idea. First, it will itself spawn a Sorites unless we can think of the neighbourhood of an F-element as becoming smaller and smaller the further we move in the direction of non-F-ness. This will require that the field is at least densely ordered, and the availability of such an ordered field is not a characteristic of all sorites-prone predicates. Wang's Paradox itself—the Sorites paradox for "intelligibly representable in decimal notation" over the series 0, 1, 2, 3, ... etc.—is a counterexample. So are the original paradoxes for *heap* and *bald*.

Second, the proposal is too generous as a characterization of what vagueness *is*. Precise predicates may determine open sets. For instance "x<1," for example, determines an open set on the real line, but is of course perfectly precise.[172] These points matter if we are operating—as Rumfitt must conceive he is not—under the third of the constraints I imposed on any satisfactory treatment of the Sorites, namely, that any restrictions imposed on the logic should flow from a satisfactory characterization of the nature of vagueness. That characterization needs to be inclusive of all vague predicates if a fully satisfying treatment is to be in prospect. But it also needs to be exclusive of precise predicates since, for them, the Sorites reasoning, when run on as a *reductio* of the major premise, does not culminate in an *unpalatable* existential but in a perfectly acceptable one, and should accordingly not be blocked.

Finally, notwithstanding my own sortie into the knowledge-theoretic semantics outlined above, this is perhaps the point to sound a note of reservation about the need for any such underpinning for logical restrictions in the presence of paradox, and hence about the proper understanding of constraint 1. There are delicate questions in the vicinity here concerning what should as a *solution* to a paradox—concerning how much, and what kind of, explaining of what is going wrong one is required to accomplish. One is reminded of Schiffer's distinction between "happy face" and "unhappy face" cases.[173] With paradoxes of the latter kind, there may be little to offer by way of explanation other than to say that concepts or practices that are in some way inherently incoherent have become entrenched, and that the only solution is to modify them in ways which, perhaps because of their entrenchment, may seem to lack independent accountability.

Suppose someone robustly takes the following stand:

- that the major premise of a Sorites *has* to be false because inconsistent, by absolutely elementary reasoning, with truths;
- that we clearly do not know that the unpalatable existential has a witness; and hence
- that there has to be something wrong with the reasoning—classical reasoning—that forces us to deny the latter ignorance if we think we have the former knowledge.

---

[172] Indeed the same point engages Rumfitt's earlier suggestion (Chapter 6, this volume, p. 151) that the extension of a vague predicate is *regular open*—that is, is the interior of its own closure, where the closure of X is the complement of the interior of the complement of X. For the interior of the complement of the interior of the complement of {x:x < 1} is {x:x < 1}. Rumfitt does disclaim any intention to suggest that possession of a regular open extension is a feature of all (including "non-polar") vague predicates; but the critical point is that such an extension is possessed by some *precise* predicates.

[173] Schiffer (1996, 2003).

Why do we need a semantic theory to underwrite and explain the gap that classical reasoning here evidently illicitly crosses?

I can imagine two answers. One would be that if classical reasoning is here unjustified, it must be recognizably so in the light of a proper account of how we already implicitly understand the logical operators involved. There is therefore a theoretical obligation to give a correct account of that prior implicit understanding. This response can be appropriate, however, only if we take the view that the sorites-paradoxical reasoning involves a *mistake* that is in principle recognizable as such by the lights of the understanding of the key notions involved that we already have. We have to be, in other words, in the territory of a possible "happy face" solution. And if we are confident that that is so, there will now be a constraint on any semantic theory to be offered that it present a plausible account both of the understanding of ordinary thinkers and of why they are here inclined to mistake its requirements. Highly abstract topological proposals don't promise well for that constraint.

The other answer adopts a reformist stance. The semantic project is not to recover an account of extant distinctions which, when we are seduced by the paradoxical reasoning, we overlook but, in the wake of *already* well-motivated revisions of classical logic, to propose a framework in which those revisions have an independent sanction, so that we can restore a sense of knowing what we are doing in inferential practice and why the suspect transitions fail. It is in this spirit that I offer the knowledge-theoretic clauses proposed. However, if it is only for this purpose that it is needed, then semantic theory is precisely not needed to *justify* the relevant logical revisions but instead, rather like Hegel's Owl of Minerva, "spreads its wings only with the coming of the dusk."

Coda: The 'Forced March' Paradox

As noted in my introductory remarks earlier, "the" Sorites paradox is a misnomer. Sorites paradoxes come in various stripes. All purport to elicit some form of aporia from the assumptions—surely incontestable—concerning any predicate F among the 'usual suspects' that it may, without qualification, truly be applied at one pole of a suitable soritical series for it and that its application at the opposite pole is, without qualification, false. So in every case there are these *polar* assumptions. But what else the paradox-merchant adds to the brew in order to foment aporia varies. In the kind of case on which we have been focused, she adds a plausible-seeming extra premise— the *major premise*—and deduces a contradiction by seemingly unexceptionable deductive moves from its combination with the polar assumptions. But what of the quite different routine involved in the so-called Forced March sorites? Here there is no deductive path to a formal contradiction. Rather, we are invited to envisage a hapless subject who is, as it were, marched through the successive elements in a suitable soritical series and required to return a verdict in point of F-ness about each element. If the subject is competent, she must return the appropriate verdicts concerning the clear cases at the poles. Hence—the paradox-merchant continues— since the verdicts required at the poles are different, she must at some point give some kind of discriminatory responses to an adjacent pair of elements between which—for so the series is constructed—she can discern no relevant difference.

Since it is a form of incompetence to purport to discriminate cases between which one discerns no relevant difference, it appears to follow that, when a sufficient range of cases has to be considered, consistently competent use of any of the usual suspects is metaphysically impossible.

That's an uncomfortable conclusion rather than a contradictory one. We might just accept it. Indeed, it may seem that that the epistemicist, for one, is committed to some such conclusion in any case since it goes with the view, at least as developed by Williamson, that no set of verdicts about F's application throughout a sorites series can be both comprehensive and everywhere *knowledgeable*. Still, that is not quite the same point. What the Forced March sorites seems to foist on us is the conclusion that no such comprehensive set of verdicts can both respect the clear cases of F and not-F and be everywhere *principled*—that is, reflective only of distinctions that the subject can actually draw.

Whether that conclusion is something we could live with or not, there is a natural response to block it. It is true, of course, that as the march progresses, the hapless subject must *respond* differently to some adjacent pairs of elements if she is not to misdescribe polar cases. At some stage, she must stop returning the verdict, 'F'; and at some stage, she must start returning the verdict, 'not-F'. But whenever she does something different, this does not have to amount to the returning of a, perhaps subtly,[174] differing *verdict*. She may, for instance, simply fail to come to a view. And if she does, she is not forced to, as it were, project this change in her response onto the items concerned and pretend to have discerned a relevant difference *in them*. The reasoning of the Forced March paradox misses the distinction between differential responses that purportedly mark a relevant distinction between adjacent pairs of elements in the series and differential responses that are merely different. It is only the occurrence of the latter that is imposed by the requirement of differential verdicts at the poles.

If what is troublesome about the Forced March routine is as suggested, then this point, it seems to me, defuses any attendant paradox. And the point carries no commitment to any particular theory of vagueness but may be made by all hands. Still, there is a respect in which the deflationary conception of vagueness with which we have been working may here claim an additional advantage. Both epistemicism and semantic indeterminism are committed to countenancing further facts of the matter to which the subject's differential responses, when they occur, may or may not correspond. Suppose a is the last case to which she returns a clear verdict, 'F.' Confronted with its successor, a', she does something different. For the epistemicist, her latter response may actually serendipitously (albeit unknowledgeably and no doubt improbably) have aligned with the actual cut-off of F's extension in the series. Likewise for the indeterminist, the subject's response to a' may mark the transition to a point at which the semantic rules for 'F' have ceased to mandate any determinate verdict. Even without any problematical suggestion that full competence should involve sensitivity to such differences, both views must still admit that there is a

---

[174] "Perhaps subtly" differing because the most challenging development of the paradox will impose no limitations on the repertoire of qualifiers—including intonation, facial expression, linguistic hedges ("sort of F," "F-ish," etc.)—with which the subject may be permitted to soften or qualify a verdict.

question of *alignment* between the subject's response and a further fact about how matters really stand with *a* and *a′* in point of *F*-ness; hence they must allow that there is a question of the extent of the *felicity* in general of the Forced March subject's responses, understood in terms of the extent to which, outside the polar regions, they actually align with whatever *F*-relevant changes are taking place as she progresses along the series. I propose that it is desirable for a would-be theorist of vagueness to avoid any such admission. The intuitionistic approach, with its integral repudiation of any idea of vagueness as constituted in semantic facts that somehow underlie and explain the distinctive patterns of use of vague expressions, and its consequent commitment to liberalism about verdicts in the borderline region, does exactly that.

References

Berker, Selim (2008). "Luminosity Regained," *Philosophers' Imprint* 8, 1–22.

Cargile, James (1969). "The Sorites Paradox," *The British Journal for the Philosophy of Science* 20 (3), 193–202.

Chambers, Timothy (1998). "On Vagueness, Sorites, and Putnam's "Intuitionistic Strategy,'" *The Monist* 81 (2), 343–8.

Dummett, Michael (1975). "Wang's Paradox," *Synthese* 30 (3–4), 301–24.

Dummett, Michael (2007). "Reply to Crispin Wright," in Randall E. Auxier and Lewis Edwin Hahn (eds.), *The Philosophy of Michael Dummett*, Chicago, IL: Open Court, 445–54.

Fine, Kit (1975). "Vagueness, Truth and Logic," *Synthese* 30 (3–4), 265–300.

Fine, Kit (2017). "The Possibility of Vagueness," *Synthese* 194 (10), 3699–3725.

Horwich, Paul (1990). *Truth*, New York: Oxford University Press.

Mott, Peter (1994). "On the Intuitionistic Solution of the Sorites Paradox," *Pacific Philosophical Quarterly* 75 (2), 133–50.

Putnam, Hilary (1983). "Vagueness and Alternative Logic," *Erkenntnis* 19 (1–3), 297–314.

Putnam, Hilary (1985). "A Quick Read is a Wrong Wright," *Analysis* 45 (4), 203.

Putnam, Hilary (1991). "Replies and Comments," *Erkenntnis* 34 (3), 401–24.

Rea, George (1989). "Degrees of Truth versus Intuitionism," *Analysis* 49 (1), 31–2.

Read, Stephen, and Crispin Wright (1985). "Hairier than Putnam Thought," *Analysis* 45 (1), 56–8.

Rosenkranz, Sven (2003). "Wright on Vagueness and Agnosticism," *Mind* 112 (447), 449–63.

Schiffer, Stephen (1996). "Contextualist Solutions to Scepticism," *Proceedings of the Aristotelian Society* 96, 317–33.

Schiffer, Stephen (2003). *The Things We Mean*, Oxford: Oxford University Press.

Schiffer, Stephen (2016). "Vagueness and Indeterminacy: Responses to Dorothy Edgington, Hartry Field, and Crispin Wright," in Gary Ostertag (ed.), *Meanings and Other Things*, Oxford: Oxford University Press, 458–81.

Schwartz, Stephen P. (1987). "Intuitionism and Sorites," *Analysis* 47 (4), 179–83.

Schwartz, Stephen P., and William Throop (1991). "Intuitionism and Vagueness," *Erkenntnis* 34 (3), 347–56.

Sorensen, Roy A. (1988). *Blindspots*, Oxford: Oxford University Press.

Srinivasan, Amia (2015). "Are We Luminous?," *Philosophy and Phenomenological Research* 90 (2), 294–319.

Troelstra, Anne S. (1991). "History of Constructivism in the Twentieth Century," ITLI Prepublication Series ML–1991–05. Amsterdam. Final version in Juliette Kennedy and

Roman Kossak (eds.), *Set Theory, Arithmetic, and Foundations of Mathematics: Theorems, Philosophies*, Lecture Notes in Logic 36, Cambridge: Cambridge University Press, 150–79.

Williamson, Timothy (1994). *Vagueness*, London: Routledge.

Williamson, Timothy (1996). "Putnam on the Sorites Paradox," *Philosophical Papers* 25 (1), 47–56.

Williamson, Timothy (2000). *Knowledge and its Limits*, Oxford: Oxford University Press.

Wright, Crispin (1975). "On the Coherence of Vague Predicates," *Synthese* 30 (3–4), 325–65.

Wright, Crispin (1976). "Language-Mastery and the Sorites Paradox," in Gareth Evans and John McDowell (eds.), *Truth and Meaning*, Oxford: Clarendon Press, 223–47.

Wright, Crispin (1982). "Strict Finitism," *Synthese* 51 (2), 203–82.

Wright, Crispin (1984). "Second Thoughts about Criteria," *Synthese* 58 (3), 383–405.

Wright, Crispin (1987). "Further Reflections on the Sorites Paradox," *Philosophical Topics* 15 (1), 227–90.

Wright, Crispin (1992). *Truth and Objectivity*, Cambridge, MA: Harvard University Press.

Wright, Crispin (2001). "On Being in a Quandary: Relativism Vagueness Logical Revisionism," *Mind* 110 (437), 45–98. Reprinted in Patrick Grim, Peter Ludlow, and Gary Mar (eds.), *The Philosopher's Annual*, vol. 24 (from the literature of 2001). New York: Center for the Study of Language and Information, 2003, 273–325.

Wright, Crispin (2003a). "Rosenkranz on Quandary, Vagueness and Intuitionism," *Mind* 112 (447), 465–74.

Wright, Crispin (2003b). "Vagueness: A Fifth Column Approach," in J. C. Beall (ed.), *Liars and Heaps*, Oxford: Oxford University Press, 84–105.

Wright, Crispin (2007). "'Wang's Paradox,'" in Randall E. Auxier and Lewis Edwin Hahn (eds.), *Library of Living Philosophers volume XXI, The Philosophy of Michael Dummett*, Chicago, IL: Open Court, 415–44.

Wright, Crispin (2010). "The Illusion of Higher-Order Vagueness," in Richard Dietz and Sebastiano Moruzzi (eds.), *Cuts and Clouds: Vaguenesss, its Nature and its Logic*, Oxford: Oxford University Press, 523–49.

Zardini, Elia (2015). "Breaking the Chains: Following-From and Transitivity," in C. Caret and O. Hjortland (eds.), *Foundations of Logical Consequence*, Oxford: Oxford University Press, 221–75.

Replies to Part III
Logical Revisionism

Replies to Tennant and Shieh

In the course of his career Michael Dummett ran at least three *prima facie* independent lines of argument against the semantic assumptions that have traditionally underwritten classical logic. One, prominent in his meta-semantical essays in the 1960s and 1970s,[175] was the contention that the orthodox coupling of the principle of Bivalence with the assumption that a semantic theory should be uniformly truth conditional, makes a mystery of how we could either acquire or manifest a proper understanding of any of the huge sweep of potentially unverifiable statements—all that concern, for instance, distant regions of space and time, or others' mental states, or that involve unrestricted generalization, or counterfactual conditional constructions—that we plainly do perfectly well understand.[176] This general species of Dummettian argument, with its attendant promise of obtaining significant metaphysical conclusions from considerations in the philosophy of language and mind, captivated much of my attention in the first part of my career.[177] Indeed, I still feel its draw. It will be the focus of attention below in my comment on Sanford Shieh's Chapter 8 (this volume).

Another, independent argument of Michael's, local to the philosophy of mathematics and surfacing most explicitly in writings published in the early 1990s,[178] claims that the classical conception of quantification is unsustainable in areas of pure mathematics that concern *indefinitely extensible* populations of objects.[179] I will take the opportunity to present what I take to be the essence of the indefinite extensibility argument in §5 below. Sandwiched between these, however, came the ostensibly quite different line of revisionary reasoning that draws on central strands in the work of Gentzen[180] and Prawitz[181] and constitutes the heart of the philosophy

---

[175] See the essays collected in Dummett (1978), especially "Truth," "The Reality of the Past," "Realism," and "The Philosophical Basis of Intuitionistic Logic."

[176] Someone who finds this Dummettian line of argument persuasive has accordingly two options: rejection of the unrestricted validity of the principle of Bivalence and rejection of truth-conditional semantics. It is striking that Dummett's own response was to embrace both.

[177] See Wright (1987) and extended second edition (1993).   [178] Dummett (1991: 317–19).

[179] Dummett's argument from indefinite extensibility to intuitionistic revisions of classical logic has generally been received with scepticism (see Boolos 1993; Clark 1993, 1998; Oliver 1998; Rumfitt 2015). More sympathetic interpretations are proposed in Heck (1993) and Wright (2018); the core argument of the latter is given in §4 below.

[180] The *locus classicus* is Gentzen (1935).   [181] Prawitz (1965, 1974).

of logic in *The Logical Basis of Metaphysics*, to the effect that that the good standing of a set of principles of inference for a logical operation depends upon their possessing a certain kind of mutual *harmony*, and that while the standard intuitionistic rules for negation pass muster by this criterion, the classical rules, incorporating the principle of double negation elimination, conspicuously fail to do so. I shall refer to this as the *harmonic* argument.

Below, I'll begin by offering some reflections about the distinctive, rigorous characterization of harmony developed by Neil Tennant in earlier work and further refined in his contribution to the present volume, and then pass to some more general thoughts about the importance assigned to the notion in recent philosophy of logic and its alleged serviceability for revisionary argument. The latter is something I have gradually changed my mind about—indeed, the harmonic argument for intuitionistic revisions of classical logic has come to seem to me the least convincing of the three Dummettian lines distinguished. The principal reasons for my misgivings are independent of Tennant's specific proposals and will foreseeably apply to anything that is recognizable as an explication of the notion of harmony as intuitively understood.

1. Tennant

Neil Tennant has been a staunch supporter of the harmonic argument throughout his career. His characteristically interesting and trenchant Chapter 9 (this volume), besides responding to a technical glitch[182] in the characterization of harmony offered in his *Anti-Realism and Logic*, provides a useful insight into the kind of defence of the requirement that he regards as cogent. I hope that my responding by expressing scepticism both about his attempt to fix the glitch and about the importance that he attaches to harmony will not seem too ungrateful a reception of his contribution.

Two preliminary remarks may be apt. First, the requirement of harmony, however exactly it proves best to characterize it, pertains in the first instance to the relationship between proposed introduction and elimination rules for a logical operator. Systems of logic may accordingly be characterized as harmonious or not only insofar as they allow of formalization within the framework of a natural deductive or sequent calculus, in which each primitive logical operator is associated with its own such rules. Moreover, although the discussion in the literature has been for the most part content to focus on cases where each operator is associated with a proprietary pair of such rules, the question of harmony—provided good sense can be made of it at all—should still arise in cases where an operator, λ, is controlled by a number of independent rules for its introduction, or for its elimination; in such a case, the issue will be whether the collective proof theoretic requirements of λ's introduction rules are properly in kilter with the deductive power assigned to sentences in which it is the principal operator by the collective import of its elimination rules.

It is also worth stressing, second, that there is—at least as far as I can see, and contrary, I think, to Tennant's perspective on the matter—no essential connection

---

[182] As he reports, I sent him notification of this in an email back in 2005, playing Russell to his Frege. "No worse fate can befall a scientific writer than . . . ."

between a requirement of harmony and the general programme of inferential role semantics. One who rejects the former is still at liberty to argue for the meta-semantical view that the meaning of expressions generally, or of logical operators in particular, is fixed by their contribution to the inferential powers of sentences containing them. Conversely, one who rejects inferential role semantics in general may still endorse reasons for thinking that harmony within its inferential rules is a prerequisite for the good standing of a logical operator. Indeed below we will briefly review two well-known types of consideration that have seemed to some to motivate a requirement of harmony but entrain no obvious commitment to inferential role semantics in general.

Deferring the question why harmony might be important, what, to a first approxi-mation, is it? On the most general and intuitive understanding of the idea, I take it to be the proposal that the introduction and elimination rules governing a logical operator, λ, should make *mutual* sense: a requirement that it should be possible to explain, and justify, the consequences assigned by λ's elimination rule(s) to any statement, S, in which it is the principal operator by reference to the conditions cited in λ's introduction rule(s), and conversely, possible to explain why exactly those conditions are cited in the introduction rules by reference to the consequences that λ's elimination rules permit us to derive from any such statement, S.

One tradition, originating with Gentzen, views the introduction rules for an operator as the source of its meaning, and the elimination rules as merely unpacking the implications of its having the meaning it does. This direction has sometimes been reversed; it can seem more plausible, for example, to think of *modus ponens* as a meaning-fixer for the conditional, and (some form) of conditional proof as articu-lating the consequentially required justification-conditions for a conditional state-ment if it is to underwrite inferences by *modus ponens* from it. However, I do not think endorsement of a requirement of harmony carries per se any commitment to the semantic primacy either of introduction or elimination rules. It is their mutual support and mutual explicability that is the crucial aspect.

This conception of the matter would seem nicely to accord with Tennant's account,[183] which exactly proposes that the introduction and elimination rules for a logical operator should be mutually justifying. However, his account offers a particular, distinctive interpretation of what such mutual justification should consist in. Harmony, Tennant originally proposed, should consist in satisfaction of the following principle:

Introduction and elimination rules for a logical operator λ must be formulated so that a sentence with λ dominant expresses the *strongest* proposition which can be inferred from the stated premises when the conditions for λ introduction are satisfied; while it expresses the *weakest* proposition that can feature in the way required for λ elimination.[184]

Here, Tennant explains, the reader is to understand the strongest proposition with a certain feature *F* to be a proposition that entails any proposition with feature *F*; while

---

[183]   First proposed in his (1978: 74) and following. The account is further developed in chapters 9 and 10 of Tennant (1987), but the basics remain the same. See also Tennant (1997: ch. 10).

[184]   Tennant (1978: 74).

the weakest proposition with feature F should be entailed by any proposition with feature F.[185]

Why these requirements of strength and weakness? Tennant is not explicit, but the underlying thought is, presumably, that in order to maximize the potentialities for valid inference provided for by a system in which λ is to feature, we should want a proposition with λ-dominant to be the logically most powerful proposition that is justifiable in the manner codified by the λ-introduction rule. And if, correspondingly, we want the λ-elimination rule to enable us fully to exploit the logical power thereby built into such a proposition, and hence to be as productive as is justifiably possible, then—since the strength of such a rule will vary inversely in proportion to the strength of its principal premise—we also need the premise for a λ-elimination step to be the weakest that can validly sustain that step.

However, the question next arises: what means are to be admissible in establishing that these requirements of strength and weakness are satisfied in a particular case? It is here that Tennant makes his distinctive move. On the face of it, the requirements are hostage to a certain relativity. Let λ be a binary connective. Then you might expect that whether $A\lambda B$ satisfies the strength and weakness requirements may depend on the background proof-theory in terms of which the question whether a given range of statements entail or are entailed respectively by $A\lambda B$ is to be answered. Tennant, however, thinks—or so I conjecture—that we can finesse any threat of relativity by calling on the very proof-theoretic resources provided by the λ-introduction and λ-elimination rules themselves. According to the *Anti-Realism and Logic* characterization,[186] the rules harmonize just in case two conditions, of strength and weakness, respectively, are met:

S: Any binary connective @ for which the pattern of λ-introduction is valid may be shown *by λ-elimination* to be such that $A\lambda B \Rightarrow A@B$—so that $A\lambda B$ is, by the lights of λ-elimination, the strongest statement justified by the premises for λ-introduction;

W: Any binary connective @ for which the pattern of λ-elimination is valid may be shown *by λ-introduction* to be such that $A@B \Rightarrow A\lambda B$—so that $A\lambda B$ is, by the lights of λ-introduction, the weakest statement that can validly play the role assigned to it by λ-elimination.

"To prove S," Tennant summarizes, "one appeals to the workings of [λ-elimination]; and to prove W one appeals to the workings of [λ-introduction]. The rules are thus caused to interanimate to meet the requirement of harmony."[187]

I have always felt nervous about this tactic of "interanimation." The question which a demonstration of harmony is meant to address concerns whether the rules proposed for an operator are properly balanced and so in good standing. A demonstration that they are so that involves deployment of those very rules only has whatever credibility we may antecedently assign to them in advance of the

---

[185] Naturally there need be no strongest and weakest propositions so associated with an arbitrary feature F.

[186] Tennant (1978: 96–8). I have, I trust harmlessly, paraphrased Tennant's original formulations.

[187] Chapter 9, this volume, p. 239.

demonstration. Why is Tennant's procedure not merely epistemically circular, requiring that we trust in the good standing of the rules before we have shown that they deserve it?

Worse: may it not be possible to devise variously defective pairs of I- and E-rules that can nevertheless be made out to support each other by their underwriting 'proofs' of the strength and weakness requirements in the manner outlined? Well, indeed—the glitch I communicated to Tennant and that he responds to in his Chapter 9, this volume, is exactly an example where that is so. Here it is again:

> Let λ-introduction have the pattern of Vel-introduction, and let λ-elimination pattern with conjunction-elimination. So, to establish strength, we need to show by the use of λ-elimination that any connective, @, that has the pattern of Vel-introduction, (so that we have both $A \Rightarrow A@B$, and $B \Rightarrow A@B$), is such that $A\lambda B \Rightarrow A@B$.
>
> *Proof:* Assume $A\lambda B$. Use λ-elimination to infer one of A, or B—as you choose. Either will then suffice for $A@B$, by the assumption that @ has the pattern of Vel-introduction.
>
> As for weakness, we need to show by the use of λ-introduction that any connective, @, that has the pattern of conjunction-elimination (so that we have both $A@B \Rightarrow A$, and $A@B \Rightarrow B$) is such that $A@B \Rightarrow A\lambda B$.
>
> Proof: Assume $A@B$. By hypothesis, the pattern of conjunction-elimination is valid for @, so we can infer both A and B. Either will then suffice via λ-introduction for the proof of $A\lambda B$.

So *tonk*, thinly disguised as λ, flies below Tennant's harmonic radar in *Anti-Realism and Logic*.

Tennant's response in Chapter 9 (this volume) is to place additional constraints on the *manner* in which the proofs of strength and weakness may allowably proceed. Basically, he now requires that when an operator λ is associated with more than one allowable pattern of introduction, or more than one allowable pattern of elimination, the proofs of strength and weakness must make use of *all* the allowable λ-elimination and λ-introduction patterns respectively. Notably, however, in the above 'proof' of strength for λ, although proceeding via λ-elimination as required, we needed to use only one arm of the λ-elimination rule. And in the 'proof' of weakness for λ, while proceeding via λ-introduction as required, we needed to use only one arm of the λ-introduction rule. So these proofs don't count.

The reader may be tempted to dismiss this manoeuvre as *ad hoc* monster-barring, but I think that would be unfair. Tennant conceives his project, after all, as that of determining under what conditions an operator has been assigned a coherent sense. It is in the spirit of that project that we should regard all aspects of its assigned introduction and elimination rules as entering into the determination of λ's sense, and in that case it is perfectly reasonable to require that a proof that it is harmonious should need to exploit *all* those resources.

Uneasiness remains, though. We assigned both patterns of Vel-introduction and both patterns of conjunction-elimination to our tonkish λ, but what if we had artfully assigned just one half of each? Thus:

$$(\lambda\text{-}I) \quad \frac{\Gamma \Rightarrow A}{\Gamma \Rightarrow A\lambda B}$$

$$(\lambda\text{-}E) \quad \frac{\Gamma \Rightarrow A\lambda B}{\Gamma \Rightarrow B}$$

Can we not now very easily still establish strength and weakness as follows? To establish strength, we now need to show by what is now the *single allowable* pattern of λ-elimination that any connective, @, that has the pattern of λ-introduction, is such that $A\lambda B \Rightarrow A@B$.

Proof: Assume $A\lambda B$. Use λ-elimination to infer A. Then use λ-introduction to infer $A\lambda(A@B)$. Then use λ-elimination again to infer A@B.

To establish weakness, we now need to show by what is now the *single allowable* pattern of λ-introduction that any connective, @, that has the pattern of λ-elimination, is such that $A@B \Rightarrow A\lambda B$.

Proof: Assume A@B. Use λ-introduction to infer $(A@B)\lambda A$. Use λ-elimination to infer A. Then use λ-introduction again to infer $A\lambda B$.

We have complied with Tennant's additional constraints. What now?

In Chapter 9, Tennant anticipates this wrinkle. He responds by adding two extra conditions on legitimate proofs of strength and weakness—respectively that a legitimate proof that λ meets the strength condition must not only make full use of all admissible forms of λ-elimination but *must make no use* of λ-introduction; and that a legitimate proof that λ meets the weakness condition must not only make full use of all admissible forms of λ-introduction but *must make no use* of λ-elimination. These conditions are violated by the respective middle steps in each of the above proofs.

Once again, the reader may be inclined to regard this manoeuvre as *ad hoc*. And this time, I am not sure how to defend Tennant against the charge. For while it may indeed compromise the project of demonstrating that a coherent sense has been assigned to an operator if the purported demonstration does not avail itself of all the putatively sense-determining aspects of that operator's use, Tennant's new restriction is going in the opposite direction: that of *banning* appeal to certain aspects of an operator's proposed proof-theoretic profile in any legitimate such demonstration. One salient underlying rationale for such a ban could be that a convincing demonstration that the introduction rule, or the elimination rule, for an operator is in good standing, should appeal only to independent resources. But if that were the rationale, then the thought ought to reactivate the worry about epistemic circularity lurking in the strategy of "interanimation."

These are, it seems to me, significant and damaging concerns about the motivation for Tennant's proposal. But there are also doubts about its results. I'll return to the latter below. But first let us ask:

2. Why Care about Harmony?

At first blush, a requirement that the introduction and elimination rules for an operator should be in harmony can seem like the merest epistemological common sense. We want logical inference, after all, to provide a means for valid extensions of our knowledge. So, for each logical operator, λ, in the language, surely we need an assurance that whatever conditions are assigned as validating the inference *to* a statement, S, in which λ is the principal operator do indeed provide a guarantee of the consequences derivable *from* such a statement by λ's elimination rules. Otherwise logic will over reach itself, risking nemeses of error, inconsistency, or explosion. Conversely, we should also require that the elimination rules permit us fully to mine the consequences we implicitly build into S by assigning to λ the introduction rules that we have. Otherwise, logic will under-reach, passing up the opportunity to give us knowledge of things that actually lie within sound inferential range.

Call that the simple epistemological argument for harmony. Initially impressive, it is not at all clear that it really carries much clout. For one thing, it handsomely begs the question against an extremely important anti-realist trend in twentieth-century philosophy of logic, represented by the conventionalism of the positivists and the work of the later Wittgenstein, that disputes the implicit assumption that logic is accountable to any *independently determined* alethic or epistemic, status of the statements that it allows us to link as premises and conclusions in arguments. What the relevant status will be taken to be depends on what it is that one takes logical consequence to be in the business of conserving—truth, assertibility, provability, or knowability more generally, are some of the notable options. But whatever is proposed about that, there is nothing that harmony is needed to vouchsafe unless we assume that the distribution of *status-values*, as we may call them, proceeds, metaphysically speaking, in advance of and independently of our logical practices.[188] Dummett, of course, was well aware of this, regarding the justifiability of a harmony requirement as resting on a rejection of the radical semantic holism that he reckoned was implicit in any view that repudiates the relevant kind of status-independence. Such a holism regards our canons of inference as contributing towards the further determination of the meanings of the statements that they allow us to connect as premises and conclusions in argument, rather than as answerable in any straightforward way to antecedent such meanings. Whether there are compelling arguments against any such view is another question.[189]

There is, though, a less controversial shortcoming in the simple epistemological argument for harmony. Even if it be granted that it is a requirement on logic to conserve some form of independently determined positive alethic, or epistemic, status across inference, and at least a desideratum that logic maximize the opportunities for the authentic recognition of such status by inference, it is a further, so far unjustified step to conclude that individual rules of inference need to be so selected that they each meet the requirement, and the desideratum, *in isolation* and independently of the behaviour of the others. The requirement motivated by the simple

---

[188] Wittgenstein (1956/1978: I, 155).   [189] Dummett (1993b: ch. 10).

epistemological argument, if any, is only that a *system* of inference conserve the targeted status and maximize the opportunities for its legitimate inferential recognition. The former is just the requirement of conservative extension; the latter, the desideratum of a certain kind of completeness. But unless it can somehow be shown that a system can attain neither benefit unless its rules are individually harmonious, no argument for harmony can be forthcoming purely from the reflection that logic should neither over- nor under-reach.

Whether or not in recognition of this, the dominant motivation, I think it fair to say, for those from Gentzen onwards who have supported a requirement of harmony has been not epistemological but *semantic*. The central thought has been that the justification of a principle of inference for a certain operator must derive from the meaning attached to that operator; and that while we may, in a basic class of cases, simply fix such meanings by stipulation of rules of inference, we are constrained thereafter to accept only rules that cohere with the meanings so fixed. Harmony is simply the requirement of such coherence. Where it is missing, the supporter of harmony contends, no satisfactory understanding is obtainable of the operator concerned.

It certainly seems that there is something right about this line of thought. That there is *some* kind of connection between harmony and 'making sense' is vividly illustrated by any of a familiar range of intuitively disharmonious examples. Prior's *tonk* is explosive, to be sure, permitting the derivation of anything from anything. But a more basic, though less spectacular, difficulty with it is that there is, arguably, simply no such thing as the proposition, *A tonk B*. For there is in general, for arbitrary *A* and *B*, no proposition for whose truth both that of *A* and that of *B* are logically necessary—so that each can be validly inferred from it—but for which *either is individually logically sufficient*. This is not the situation of a simple contradiction— a proposition whose truth-condition places inconsistent demands upon the world but can itself be perfectly consistently stated in second intension, as the content, say, of reported speech. There is a distinction between having logically unsatisfiable truth-conditions—that is, truth-conditions which place contradictory demands upon the world but allow of consistent characterization—and having a 'truth-condition' whose very description is incoherent. The predicament of *tonk* is of the latter kind.

This distinction is reinforced if we reflect on *tonk*'s consistent, non-explosive dual, *tunk*, characterized by these introduction and elimination rules:

$$(Tunk\text{-}I) \quad \frac{\Gamma \Rightarrow A; \Delta \Rightarrow B}{\Gamma \cup \Delta \Rightarrow A \, tunk \, B}$$

$$(Tunk\text{-}E) \quad \frac{\Gamma \Rightarrow A \, tunk \, B; \Delta \cup \{A\} \Rightarrow C; \Sigma \cup \{B\} \Rightarrow C}{\Gamma \cup \{\Delta \cup \Sigma\} \Rightarrow C}$$

This time, we do at least have the possibility of a coherent *interpretation* of the operator, since *tunk*'s rules are validated by conjunction. But so to interpret it will immediately enforce cogent criticism of the elimination rule. If *tunk* is conjunction, then *tunk*-E is too demanding; we don't need *both* the second *and* the third premises.[190] What we cannot do, however, is stipulate that *tunk* is not to be conjunction, for there is then no other proposition for A *tunk* B to be. There is no proposition for whose truth A and B do not individually but only conjunctively suffice but whose consequences are restricted—*modulo* the contribution of side-premises—to those in common between A and B. If someone insists that *tunk* is not conjunction, and that the contrast between its E-rule and that for conjunction above exactly captures their distinction, then the right description of the situation is not that the truth of A *tunk* B has contradictory truth-conditions but that it has no truth-conditions.

Tunk illustrates how a determinate (non-explosive) inferential practice governed by certain rules may—when the rules are disharmonious—yet lack the wherewithal for the logical operator concerned to take on an intelligible meaning. The practice with *tunk* is perfectly definitely characterized. We can follow its rules with confidence that they will not lead us into trouble. But intuitively we will not understand the practice—will not have any sense that, in following them, we know what we are doing. Use, then, does not suffice for meaning *tout court*. Harmony, it may now be suggested, is prerequisite for a well-defined use for a logical operator to guarantee that it has meaning.

One more illustration. *Tonk* has, of course, a quantificational cousin: a rogue quantifier harnessed to these rules:

$$(Quonk\text{-}I) \qquad \frac{\Gamma \Rightarrow At,}{\Gamma \Rightarrow (Qx)Ax} \qquad \text{where '} t \text{' is a term}$$

$$(Quonk\text{-}E) \qquad \frac{\Gamma \Rightarrow (Qx)Ax}{\Gamma \Rightarrow At,} \qquad \text{where '} t \text{' is any term}$$

These rules are explosive among predications of 'A' on terms—they allow us to infer any such predication from any other. But their more basic shortcoming is that there simply is no quantified proposition P such that for an arbitrary domain of objects and predicate, Ax, it *suffices* for P to be true that some object in the domain is A and is *necessary* for P to be true that any object of the domain is A. Quantified propositions specify, intuitively, the extent of the range of application enjoyed in a domain of entities

---

[190] Rather we should replace *tunk*-E with a pair of elimination rules:

$$\frac{\Gamma \Rightarrow A\,tunk\,B; \Delta \cup \{A\} \Rightarrow C}{\Gamma \cup \Delta \Rightarrow C}$$

and

$$\frac{\Gamma \Rightarrow A\,tunk\,B; \Sigma \cup \{B\} \Rightarrow C}{\Gamma \cup \Sigma \Rightarrow C}$$

by the predicate into which they quantify. There simply is, in general, no possible such extent that, while guaranteed by a single instance, requires exceptionlessness.

And likewise for *Quonk*'s dual:

(*Junk-I*) $$\frac{\Gamma \Rightarrow At,}{\Gamma \Rightarrow (Jx)Ax}$$ where all members of Γ are 't'-free;

(*Junk-E*) $$\frac{\Gamma \Rightarrow (Jx)Ax; \Delta \cup \{At\} \Rightarrow C,}{\Gamma \cup \Delta \Rightarrow C}$$ where C and all members of Δ are 't'-free.

As with *tunk*, so with *junk*. '$(Jx)\ldots x$' is interpretable as a universal quantifier, but only at the cost of making a mystery of the restrictions concerning occurrences of 't' imposed in the elimination rule. These restrictions are designed to ensure a certain generality in the second premise, $\Delta \cup \{At\} \Rightarrow C$. Their intended effect is to ensure that C would be validly derivable from the combination of Δ with *any* statement of the form, At. If '$(Jx)\ldots x$' is a universal quantifier, there is no point in such a precaution. The elimination rule would be valid in the unrestricted form

$$\frac{\Gamma \Rightarrow (Jx)Ax; \Delta \cup \{At\} \Rightarrow C,}{\Gamma \cup \Delta \Rightarrow C}$$

But what can "$(Jx)Ax$" mean if not universality—what kind of distribution of A-elements in the domain is to be necessary and sufficient for "$(Jx)Ax$" to be true? For if any less than universal distribution is to suffice, then it is now *Junk-I* that makes no sense since the restriction imposed therein precisely ensures that *any* statement of the form "At" will hold on the assumptions, Γ. What sense is there in that requirement if the truth of "$(Jx)Ax$" requires no more than partial instantiation of "Ax" throughout the domain?

I have presented the examples of *tunk* and *junk* as illustrating how a logical practice can be operationally perfectly clear, and even consistent, and yet make no intuitive sense. But "making no sense" may seem tendentious as a description of these examples. Is the possibility illustrated not rather that a putative logical operator may be annexed to a determinate, conservative inferential practice and yet have no *reference*—that no possible truth function, or quantifier, exactly fits the requirements imposed by the introduction and elimination rules for *tunk* and *junk*, respectively? Well, the supporter of harmony will not be handicapped if that account of the matter is preferred. A convincing case that a logical operator can have a determinate reference only if its proprietary inferential rules are in harmony would for provide for no less effective a critique of disharmonious systems.

The question is whether the examples really do make for a convincing such case. They are convincing, no doubt, that some putative logical operators that are charac-terized by disharmonious rules are semantically defective in one way or another. It is, however, a different, stronger claim that their semantic defects are *owing* to their disharmony. Might there be some further factor, present in these cases but missing in some of the intended target cases—including, crucially, the case of classical negation—which contributes to the manifest deficiencies of the examples we have reviewed?

The defective operators so far reviewed fail to illustrate a further kind of harmony-compromised case. Let L be a language for first-order logic with identity as standardly formalized and consider an enrichment of it, L+, achieved by adding the countably infinite suite of numerically definite existential quantifiers, ("There are at least n A's"), that are standardly characterized recursively by the following clauses:

$$(\mathcal{E}_0 x)Ax \quad \leftrightarrow \quad \sim (\exists x)Ax$$

$$(\mathcal{E}_1 x)Ax \quad \leftrightarrow \quad (\exists x)Ax$$

and thereafter

$$(\mathcal{E}_{n+1} x)Ax \quad \leftrightarrow \quad (\exists x)[Ax \,\&\, (\mathcal{E}_n y)(Ay \,\&\, y \neq x)]$$

These clauses provide for the translation each sentence of L+ involving one or more occurrences of the new quantifiers into the base language, L. Let us call the final translation thereby obtainable of any L+ sentence, free of all occurrences of the new quantifiers, its L-translation. Suppose that we elect to secure the inferential effect of the translations by associating each of the new quantifiers with an introduction rule that allows us to infer any sentence in which it is the principal operator from its L-translation, and an elimination rule that simply reverses that transition.[191] All is therefore perfectly harmonious, so far. But suppose we were now to add a further infinite suite of elimination rules, each of the following form:

---

[191] Thus we have, for example:

$(\mathcal{E}0 - I)$
$$\dfrac{\Gamma \Rightarrow \sim(\exists x)Ax}{\Gamma \Rightarrow (\mathcal{E}_0 x)Ax}$$

$(\mathcal{E}0 - \mathcal{E})$
$$\dfrac{\Gamma \Rightarrow (\mathcal{E}_0 x)Ax}{\Gamma \Rightarrow \sim(\exists x)Ax}$$

$(\mathcal{E}1 - I)$
$$\dfrac{\Gamma \Rightarrow (\exists x)Ax}{\Gamma \Rightarrow (\mathcal{E}_1 x)Ax}$$

$(\mathcal{E}1 - \mathcal{E})$
$$\dfrac{\Gamma \Rightarrow (\mathcal{E}_1 x)Ax}{\Gamma \Rightarrow (\exists x)Ax}$$

$(\mathcal{E}2 - I)$
$$\dfrac{\Gamma \Rightarrow (\exists x)(Ax \,\&\, (\exists y)(Ay \,\&\, y \neq x))}{\Gamma \Rightarrow (\mathcal{E}_2 x)Ax}$$

$(\mathcal{E}2 - \mathcal{E})$
$$\dfrac{\Gamma \Rightarrow (\mathcal{E}_2 x)Ax}{\Gamma \Rightarrow (\exists x)(Ax \,\&\, (\exists y)(Ay \,\&\, y \neq x))}$$

...and so on.

$$\frac{\Gamma \Rightarrow At, \text{ (where all members of } \Gamma \text{ are '}t\text{'-free)}; \Delta \Rightarrow (\mathcal{E}_n x)Ax}{\Gamma \cup \Delta \Rightarrow C,}$$

where "C" is to be replaced by L-translation of '$(\mathcal{E}_{n+1}x)Ax$.' The effect of the addition is that whenever we have the wherewithal to establish a universally quantified conclusion on certain assumptions, and a specific "at least n many" conclusion involving the same predicate on certain (other) assumptions, the pool of the assumptions concerned will now furnish a proof that there are at least $n+1$ elements satisfying the predicate in question, no matter what one chooses for n. Unless therefore we have an independent proof of the countable infinity of the domain, the additional $\mathcal{E}_n$-elimination rules will make for a non-conservative extension of the underlying proof-theory and introduce disharmony into the rules both for the numerically definite quantifiers and for the regular existential quantifier in terms of which the former are ultimately defined.

The example is highly contrived, of course. But I want it to illustrate a crucial point. It would be a possible but an exceedingly strange response to regard the stipulation of the additional rules as compromising the *intelligibility* of the existential quantifier, or calling into question its possession of reference to any genuine logical operation. The salient issues raised by the added rules concern not meaning but justification. Specifically, are they *valid*? Are we entitled to take them to be so? And the answer to that might be: Well, they simply amount to one way of stipulating an axiom of infinity. Such as stipulation might, of course, raise epistemological issues on its own account. But they will not be issues concerning whether, after the stipulation, we retain any coherent understanding of '∃,' or any right to suppose that it refers to an operation of logic in good standing.

Not all examples of disharmony, then, immediately raise questions of intelligibility, or coherence of understanding. What explains the difference between the $\mathcal{E}_n x$-case and those of *tonk, tunk, quonk,* and *junk*? The obvious answer is that while the augmented rules for the $\mathcal{E}$-quantifiers non-conservatively extend the consequence relation in the base language, they do so in a way that *is* conservative of consequence, as *intuitively* understood, *modulo* a certain hypothesis: to wit, that the domain of the quantifiers of L is at least countably infinite.

Call such a hypothesis a *content-saving* hypothesis. And let us grant that disharmony in the proprietary inferential rules of a logical operator, λ, always at least *threatens* to undermine the possibility of a coherent understanding of it. Then the lesson to draw from our contrived example is, I propose, that this threat may always be deflected, at least in cases where the disharmony resides in an excessive strength of λ's elimination rule(s), provided that there is a saving hypothesis to hand to underwrite the additional inferential potential—exceeding that conferred on it by its introduction rule(s)—of any statement in which λ is the principal operator. It is when there is no salient such hypothesis that we may tend to feel that "we do not know what we are doing." Otherwise the question will be, rather: what justification have we, in the context in question, for the relevant hypothesis?

Notably, even an explosive operator like *tonk*, may be saved in this way in certain specialized contexts—for example, if the class of sentences to which the *tonk*-rules are to be applied contains nothing but classical logical truths and if the notion of logical consequence is interpreted as classical.[192] Interestingly, however, there can be no saving the assignment of *under*-extensive elimination rules for an operator in this kind of way. It may be possible to explain why the legitimate consequences of a statement in which λ is the principal operator exceed those implicitly assigned to it by the introduction rules for λ—by, for instance, as illustrated, appeal to special features of an intended domain of application—but it makes no sense to suppose that special circumstances might be such that *less* can be attributed to such a statement by way of consequences than is justified by the relevant introduction rules. To suggest that is so would be to allege, rather, that the relevant introduction rules are invalid in the circumstances in question. There is thus no saving the ilk of *tunk* and *junk*.

We asked whether there might be some further factor, present in cases where disharmony unquestionably seems to obstruct coherent understanding but potentially absent in other cases, perhaps including some where disharmony has prompted some theorists to argue for logical revision. If the foregoing is accepted, though, the answer is the other way round. It is the *absence* of saving hypotheses that allows disharmony, when it does, to obstruct coherent understanding.

Where do these reflections leave the harmonic critique of classical logic? We may take it, without loss of generality, that the crux of the critique will be the alleged disharmony between the classical introduction and elimination rules for negation in a formalization where the former comprises an intuitionistically acceptable version of *reductio ad absurdum* and the latter is the rule of double negation elimination. Intuitionist *reductio* is the principle that if the members of a set of premises are collectively inconsistent, then the negation of any of them may be inferred from a pool of assumptions consisting in the remainder. We may represent it like this:

$$(RAA) \quad \frac{\Gamma \cup \{A\} \Rightarrow \bot,}{\Gamma \Rightarrow \sim A}$$

where '$\bot$' is, as usual, a constant expressing absurdity. Double negation elimination is, naturally, the rule

$$(DNE) \quad \frac{\Gamma \Rightarrow \sim\sim A,}{\Gamma \Rightarrow A}$$

The latter harmonizes with the former only if any proof of its premise yielded by *RAA* must already provide the materials for a proof of its conclusion without the detour, as it were, through successive steps of *RAA* and *DNE*, respectively. A proof of the premise, $\Gamma \Rightarrow \sim\sim A$, achieved by a step of *RAA*, will have consisted in establishing the sequent, $\Gamma \cup \{\sim A\} \Rightarrow \bot$. We have harmony, therefore, only if any such proof must already supply the materials for a proof of $\Gamma \Rightarrow A$ without recourse to *DNE*. To assess the question, we must first determine what it takes to establish a sequent whose succedent is '$\bot$.' The intuitively correct answer is: a demonstration that its antecedent

---

[192]   As many have noted—see, for example, Cook (2005) and Warren (2015).

entails a contradiction. There are thus two ways that $\Gamma \cup \{\sim A\} \Rightarrow \perp$ might have been obtainable. First, we might have had an independent proof of $\Gamma \Rightarrow \perp$, obtaining $\Gamma \cup \{\sim A\} \Rightarrow \perp$ by augmentation of the antecedent. Second we might have had an independent roof that $\Gamma \Rightarrow \sim\sim A$, whence we could have obtained $\Gamma \cup \{\sim A\} \Rightarrow \sim A \,\&$ $\sim\sim A$ by conjunction introduction. But in neither case is there any guarantee of prior materials sufficient to establish $\Gamma \Rightarrow A$. True, we *might* have had an independent proof that $\Gamma \Rightarrow A$, and thereby been in position to establish $\Gamma \cup \{\sim A\} \Rightarrow A \,\& \sim A$. But there is no general reason why that should have had to be so. Double negation elimination thus over-extends the warrant for negated statements conferred on them by *RAA* in the stated form.

To be sure, the immediate conclusion should only be that one, albeit standard formulation of classical sentential logic is disharmonious. It is a further step to conclude that classical logic is *essentially* disharmonious. And indeed one well-explored direction of response by its supporters has been to argue, in various ways, that it also permits harmonious formulation. Multiple conclusion sequent-calculus formulations provide one implementation of that project.[193] Another, more interesting in my view, is Ian Rumfitt's argument[194] that the classical rules governing negation may be harmonized in a framework which employs a primitive notion of rejection, as well as a notion of assertion. The salient issue, however, is whether such manoeuvres are really called for, whatever independent interest they may have. If the philosophical interest of disharmony is its *prima facie* threat to intelligibility, the crucial question is whether, even if classical logic is essentially disharmonious, there is any serious threat to intelligibility in its particular case. And since what we have to deal with is an over-extensive elimination rule in a system that remains nonetheless perfectly consistent, the question is whether some suitable saving hypothesis is available, to rationalize the additional strength of the rule.

Once the question is put like that, the impression that the harmonic argument provides an independent line of revisionary criticism of classical logic immediately evaporates. For there is—*of course!*—an eminently suitable saving hypothesis for the additional strength in the classical negation elimination rules. It is exactly the hypothesis incorporated in classical bivalent truth-conditional semantics, that each statement in a proposed domain of application for classical logic is associated with truth-conditions that determinately either obtain or do not. It is only if that hypothesis is *already* rejected that the harmonic argument can carry any bite. The argument therefore stands in need of the support of the kind of meta-semantical considerations that were the hallmark of the first of the Dummettian lines of argument distinguished above, and carries no independent thrust.

3. Another Puzzle

There is another puzzle about the harmonic argument that arises even if it is granted (if only for the sake of argument) that it does have that background meta-semantical support, so that the claim that harmony is a constraint on the acceptability of rules of

---

[193] See Read (2000).   [194] Rumfitt (2000).

inference is now advanced in a setting in which the meaning of a statement is already thought of determined not by bivalent truth-conditions but by its verification conditions or, more generally, by some privileged selection of the grounds for accepting it. In brief: it appears that an insistence on harmony, so far from being necessary if logical operators are to possess coherent sense, may now actually threaten the very utility of deduction itself.

The difficulty is brought into focus if we think of harmony in terms of Michael Dummett's image of the 'levelling of local peaks.' That image comes to this: that any application of an introduction rule for an operator, λ, should be such that, if we then immediately apply its elimination rule, we will be taken only to statements (or sequents) which we have either already proved or else anyway have the materials to prove 'above the line' as it were, that is, in advance of the λ-introduction step, without any application of the λ-rules. But to demand that the λ-rules be harmonious in this sense is dangerously close to requiring that they be useless for the purpose of extending our knowledge by deduction. More carefully, it is to insist that they be useless for the purpose of deductive extension of our knowledge of statements in which λ is not the principal operator. For the requirement of harmony is that the only such statements that we may admissibly derive from a statement in which λ is the principal operator have to be such that we could already have known them independently, without the detour through the proof, or other form of acquisition of knowledge, of that statement. Generalized to every logical operator, the point thus becomes that deduction is epistemically pointless unless one's interest is merely in the acquisition of knowledge of statements in which logical operators occur. But it is hard to see what might motivate so restricted an interest. Why have the specialized vocabulary of logic at all unless it falls within the competence of logic to increase our knowledge, in the right circumstances, not merely, as it were incestuously, of logical compounds but, for example, of atomic statements or others formed by the application of non-logical (e.g., modal or attitudinal) operators?

Now, this reasoning tacitly assumes that the only grounds on which knowledge of a λ-compound—a statement in which λ is the principal operator—might be based are those given by λ's introduction rules. But of course, it may be objected, we have to allow other kinds of ground too. I may, for example, remember, or learn by testimony, that a disjunction is true without recalling, or being told, which of its disjuncts is; or I may deduce that it is true because all other possibilities besides the truth of one of its disjuncts have somehow been excluded, even though I have so far no case for thinking that either of its disjuncts in particular is true. In this case, reasoning by cases on the disjuncts may well lead me to new atomic knowledge, as it were, which I could not otherwise have obtained. The problem for the harmonist, though, is that an admission that the general utility of deduction depends upon our willingness to allow other kinds of ground for λ-compounds besides those articulated by λ's introduction rules would appear simply to surrender the requirement of harmony. For consider a disjunction—say, "The canary escaped by the kitchen window or by the back door"—based not on grounds for one of its disjuncts in particular but on, as it were, non-distributive information (suppose the canary is gone and no other possible avenue of escape is open) and suppose we think it likely that the neighbour's cat will have caught the bird in either case. Still there need be no justifiable route open to the conclusion that the bird has indeed suffered that demise *unless* we go through the disjunction; in particular the pool of information consisting of the facts that the bird cage is open, that the bird has gone, that the back door and window are also both open, that no other exit is open, and that the neighbour's cat habitually prowls the garden area just outside, offer no prospect of justifying it by any other route.

To be sure, this kind of example may yet be reconciled with the requirement of harmony if it is insisted that the only admissible additional grounds for an λ-compound besides those cited in λ's introduction rules—call those, as is usual in the literature, λ's *canonical* grounds—must be grounds for thinking that information could *in principle* be (or have been) obtained that would *canonically* justify the λ-compound.[195] The above grounds for the disjunction about the canary, for example, are indeed grounds for thinking that a suitably placed observer would have seen the bird fly through the open door or the window. That, though, is a feature of the particular example. The reader will freely be able to think of cases where a disjunction impresses itself on us simply because all possibilities besides those detailed by its disjuncts are excluded by our evidence, but where there is no prospect that we can advance, or could have advanced, to evidence selectively for any particular one of them.[196]

4. Tennant's Rogues' Gallery

Let us return to Tennant. His present proposal, exactly stated for the case of a binary connective, is this (I elaborate slightly):

A connective λ is now proved to be harmonious if and only if the putative demonstration, D, of the strength condition—that AλB is the strongest conclusion sustained by the conditions described by λ-introduction—meets the following conditions:

(i)   D exploits *all* the conditions described by λ-introduction;
(ii) D makes *full use* of λ-elimination; but
(iii) D *does not make any use* of λ-introduction,

while the putative demonstration, D′, of the weakness condition—that AλB is the weakest possible major premise that sustains the consequences described by λ-elimination—meets these conditions:

(iv) D′ exploits *all* the conditions described by λ-elimination;
(v)   D′ makes *full use* of λ-introduction; but
(vi) D′ *does not make any use* of λ-elimination.[197]

There is, in my mind at least, some unclarity about how exactly in these clauses Tennant intends the notions of "exploiting all the conditions" described by

---

[195] This is, in effect, what in *The Logical Basis of Metaphysics* Dummett famously styles the "Fundamental Assumption." Notably, he concludes (p. 277) that it is "very shaky."

[196] For cogent further elaboration of this point, see Rumfitt (2017).

[197] For Tennant's own formulation, see Chapter 9, this volume, at pp. 241–2.

[a standard statement of] an operator's introduction rule(s), and of "making full use" of its elimination rule, to be understood.[198] One also wonders what motivation the new restrictions might be given other than to block extant 'bad guys'; for if that is all they have going for them, such tactics offer little prospect of insight into how harmony should be understood or its putative value. However, it is noticeable that Tennant's discussion is preoccupied with the case of connectives. The reader may accordingly be interested to ponder whether the following roguish examples do not in any case all pass muster by Tennant's reformulated tests.

(1) Consider again the quantifier *Quonk*, given by the rules:

$$(Quonk\text{-}I) \quad \frac{\Gamma \Rightarrow At,}{\Gamma \Rightarrow (Qx)Ax} \qquad \text{where 't' is a term}$$

$$(Quonk\text{-}E) \quad \frac{\Gamma \Rightarrow (Qx)Ax}{\Gamma \Rightarrow At,} \qquad \text{where 't' is any term}$$

Quonk differs from the universal quantifier only in that its introduction rule imposes no restrictions on the kind of part played by 't' in the premises in the pool, Γ. And by lacking any such restrictions, its introduction and elimination rules are rendered explosive among atomic predications of 'A.' But we can nevertheless very easily provide this defective quantifier with proofs of the strength and weakness conditions of the kind that Tennant's clauses above, naturally understood, demand. Thus:

Strength: we need to show by the use of Quonk-elimination that any quantifier, Q^*, that shares the Quonk-introduction rule is such that $(Qx)Ax \Rightarrow (Q^*x)Ax$; and we have to do this in such a way that (i) we exploit *all* the conditions described by Quonk-introduction; (ii) make *full use* of Quonk-elimination; and (iii) do not make any use of Quonk-introduction.

Proof: trivial. Suppose Q^* is subject to the Quonk-introduction rule:

$$\frac{\Gamma \Rightarrow At,}{\Gamma \Rightarrow (Q^*x)Ax} \qquad \text{where 't' is a term}$$

Since   (1) $(Qx)Ax \Rightarrow At$ by Quonk-E,

we have   (2) $(Qx)Ax \Rightarrow (Q^*x)Ax$ by our supposition and transitivity of '$\Rightarrow$' QED.

---

[198] Indeed, when I first read the passage, I took clauses (i) and (iii), to be contradictory—likewise (iv) and (vi)—and assumed there must be some kind of typo in Tennant's text. On reflection, though, it occurred to me that what he must have in mind is that the two clauses, (i) and (iv), relate to the handling of the 'dummy'—parametric—operator in candidate proofs of strength/weakness for a targeted operator, λ. This reading is applied in the examples to follow.

Moreover (condition i) we assigned to Q* exactly the *unique* pattern of introduction possessed by Quonk and appealed to exactly that in the second line of the proof; and we used *the only type* of elimination step permitted by Quonk-elimination at line 1 (condition ii); and we made no use of Quonk-introduction (condition iii).

Weakness: we need to show by the use of Quonk-intoduction that any quantifier, Q*, that shares the Quonk-elimination rule is such that $(Q^*x)Ax \Rightarrow (Qx)Ax$; and we have to do this in such a way that (i) we exploit *all* the conditions described by Quonk-elimination; (ii) make *full use* of Quonk-introduction; and (iii) do not make any use of Quonk-elimination.

Proof, trivial: suppose Q* is subject to the Quonk-elimination rule:

$$\frac{\Gamma \Rightarrow (Q^*x)Ax,}{\Gamma \Rightarrow At} \qquad \text{where '}t\text{' is a term}$$

Since	(1)	$(Q^*x)Ax \Rightarrow At$	by our supposition
and	(2)	$At \Rightarrow (Qx)Ax$	by Quonk-introduction,
we have	(3)	$(Q^*x)Ax \Rightarrow (Qx)Ax$	by transitivity of '$\Rightarrow$.' QED.

Moreover (condition i) we assigned to Q* exactly the unique pattern of elimination possessed by Quonk and appealed to exactly that in the first line of the proof; and we used the only type of introduction step permitted by Quonk-introduction at line 2 (condition ii); and we made no use of Quonk-elimination (condition iii).[199]

I will leave it to the reader to verify that it is no more difficult to meet Tennant's revised conditions, thus naturally understood, in giving similar proofs of Strength and Weakness for the combination of conditional proof and *modus ponens*, and for a truth-operator characterized by the standard disquotational and enquotational schemata. Indeed it is straightforward to do the same for the course-of-values operator of Frege's *Grundgesetze* characterized by this introduction rule:

$$\frac{\Gamma \Rightarrow (\forall x)(Fx \leftrightarrow Gx)}{\Gamma \Rightarrow \{x{:}Fx\}{=}\{x{:}Gx\}}$$

---

[199] That Tennant's account runs into difficulties with 'de-restricted' quantifiers has been noted by Steinberger (2009). Tennant (2010) has replied that the proofs of strength and weakness that the objection exploits depend, at least in a sequent calculus setting, on applications of transitivity (Cut) which has no independent sanction in this context. This response initially impressed me as egregiously *ad hoc*. However, Ian Rumfitt has reminded me that Tennant has long argued that Cut should be jettisoned in any case—at least, if one aspires to a logic whose consequence relation reflects relations of relevance between premises and conclusion. However that may be, the point remains that Tennant's move marks a surrender of the most distinctive feature of his original proposal, the idea, gestured at by the metaphor of "interanimation" noted earlier, that harmony should be an *absolute* feature of introduction and elimination rules, invariant under their incorporation within systems differing in their structural rules of inference. Steinberger (2011) offers further technical criticisms of Tennant's revised account.

and this elimination rule:

$$\frac{\Gamma \Rightarrow \{x{:}Fx\}{=}\{x{:}Gx\}}{\Gamma \Rightarrow (\forall x)(Fx \leftrightarrow Gx).}$$

But the latter pair are, of course, inconsistent in full classical second-order logic, while the standard rules for the conditional, augmented by the cited rules for a truth operator, suffice in the presence of the Curry sentence for the ipsonymous paradox.

I argued above that harmony, intuitively understood, is not in general a necessary condition for a logical operator to possess a coherent sense. It now appears that the harmony of the rules for an operation, at least when understood specifically as by Tennant, is also no sufficient condition for the coherence of its proof-theory.[200] Nor, I am bound to say, do I think that the latter point necessarily reflects badly on Tennant's account, since it does seem that some of the roguish cases are indeed harmonious by the lights of the notion as intuitively understood. But then if harmony is neither necessary for coherence of understanding, nor sufficient to guarantee inferential safety, what is the point of it?[201]

5. Dummett on Indefinite Extensibility

Dummett countenanced a generous range of application for the notion of indefinite extensibility. Each of *set, cardinal number, ordinal number, real number, arithmetical proof, arithmetical truth,* and even *natural number* are in various places in his writings suggested to be indefinitely extensible.[202] Briefly characterized, an indefinitely

---

[200] Of course, other explications of harmony than Tennant's may do better with some of the roguish cases—for instance, in enabling the classification of the de-restricted quantifier rules as disharmonious. At the risk of displaying my ignorance, however, I confess that I do not know of any that would diagnose disharmony in the rules for the operators involved in the Curry and Basic Law V paradoxes. My own inclination is to say, rather, that it is not in terms of disharmony that the problems with these examples should be diagnosed. Julien Murzi has reminded me that in *The Logical Basis of Metaphysics* (p. 258) Dummett seems to have agreed with this, seeking instead to block paradox by imposing his notorious "complexity condition," to the effect that the conclusion of any legitimate introduction rule must be logically more complex than any of its premises. The question of motivation aside, that particular proposal promises serious trouble for perfectly legitimate-seeming premise-discharging rules like intuitionist *reductio* and conditional proof.

[201] For all I have said, there may yet, of course, be some other kind of virtue in harmony, at least when the notion is best explicated, and thus it might still be a strike against classical logic were it to prove essentially disharmonious. One suggestion in this direction (Julien Murzi, personal communication) is that harmony in its proprietary rules may be an essential characteristic of the *logicality* of an operation, and correspondingly that only theorems that can be justified by the application of harmonious principles of inference can properly be accredited as *logical* truths.

[202] As is familiar, Dummett derived the notion from Russell, who regarded it as the kernel of the correct diagnosis of the paradoxes. Here is an illustrative quotation from the latter's 1906 paper, "On some difficulties in the theory of transfinite numbers and order types": . . . the contradictions result from the fact that . . . there are what we may call *self-reproductive* processes and classes. That is, there are some properties such that, given any class of terms all having such a property, we can always define a new term also having the property in question. Hence we can never collect *all* of the terms having the said property into a whole; because whenever we hope we have them all, the collection which we have immediately proceeds to generate a new term also having the said property.

extensible concept, according to Dummett, is one that is essentially associated with a "principle of extension"—a function—that takes as argument any definite totality, t, of objects each of which falls under the concept and produces as value an object that also falls under the concept, but is not in t. In Dummett's view, when we are dealing with a domain comprising the instances of such a concept, the general validity of the principle of Bivalence for quantifications over it must be forfeit.

In order to understand why Dummett may have held this view, we must first surmount a salient problem with his various characterizations of indefinite extensibility—a problem of (what one hopes is an avoidable) *circularity*. For let P be any concept falling within the scope of Dummett's intent and $\mathcal{F}$ the relevant principle of extension. Evidently not any old totality of items falling under P is meant to be an admissible argument for $\mathcal{F}$. In particular, *indefinitely extensible* sub-totalities, including of course P itself, are excluded. Rather we are supposed to restrict $\mathcal{F}$'s domain of arguments to *definite* sub-totalities of P. But 'definite' here is just the complement of 'indefinitely extensible.' So someone who has not yet grasped that notion won't understand the implicit restriction either, and so will not be able to acquire any specific understanding from Dummett's characterizations.

There is at least one extant solution to this problem.[203] Here I must give only the briskest statement of it. We can begin with an explicitly relativized notion. Let P be any concept and Π a higher-order property of concepts of the type of P. Then we say that P is *indefinitely extensible with respect to* Π if and only if there is a function $\mathcal{F}$ from items of the same type as P to items of the type of the instances of P such that if Q is any sub-concept of P such that ΠQ, then

(1) $\mathcal{F}Q$ falls under the concept P,
(2) It is not that case that $\mathcal{F}Q$ falls under the concept Q, and
(3) $\Pi Q'$, where Q' is the concept instantiated just by $\mathcal{F}Q$ and by every item which instantiates Q (in set-theoretic terms, i.e., Q' is $(Q \cup \{\mathcal{F}Q\})$).

The idea is that the sub-concepts of P of which Π holds have no maximal member. For any sub-concept Q of P such that ΠQ, there is a proper extension Q' of Q such that Q' is likewise a sub-concept of P and such that $\Pi Q'$.'

This relativized notion is straightforward to illustrate. By its lights, *natural number* is indefinitely extensible with respect to *finite* and the operation of selecting the successor of the greatest member of a finite set of natural numbers, *real number* is indefinitely extensible with respect to *countable* and Cantor's diagonal construction, and *arithmetical truth* is indefinitely extensible with respect to

---

Compare this formulation from Dummett (1993a: 441):

> [An] *indefinitely extensible* concept is one such that, if we can form a definite conception of a totality all of whose members fall under the concept, we can, by reference to that totality, characterize a larger totality all of whose members fall under it.

See also Dummett (1991: 316–19) where he cites the above passage from Russell.

[203] The following solution is proposed in Shapiro and Wright (2006). For further exploration of the connection of indefinite extensibility, so characterized, with the paradoxes, see Wright (forthcoming).

recursively enumerable and Gödel's key construction in the proof of incompleteness of arithmetic.

Now reflect that in some cases, the process of extension is not actually *indefinitely* possible but stabilizes at some ordinal point. These are cases where—helping ourselves to the classical ordinals—we can say that some ordinal λ places a lowest limit on the length of the series of Π-preserving applications of F to any Q such that ΠQ. In such cases, any series of extensions whose length is less than λ results in a collection of P's that is still Π, but once the series of iterations extends as far as λ, the resulting collection of P's is no longer Π, and so the "process" stabilizes. Thus ω provides such a limit for the case of *natural number, finite* and *successor of the greatest member*, and ω1 provides such a limit for *real number, countable* and the diagonal construction.[204]

Say in such a case that P is *up-to-λ-extensible* with respect to Π. And now let us say that P is *properly indefinitely extensible* with respect to Π just if P, Π, and some 𝓕 meet the conditions for the relativized notion as originally defined but *there is* no λ such that P is merely up-to-λ-extensible with respect to Π. Finally, say that P is *absolutely indefinitely extensible* just in case there is some Π such that P is properly indefinitely extensible with respect to Π.

So that finesses the circularity. It is salient, however, that our revamped characterization embeds an unrestricted quantification over the ordinals. This is, plausibly, no artefact of our characterization. Something of the sort must be a feature of any satisfactory account of the intent of "indefinite extensibility" which by the very phrase adverts to the idea of serial but limitless iteration of some or other process of expansion. Assuming so, I think the point may shed some light on Dummett's generosity with the notion, remarked on above. We observed that *natural number*, for instance, is merely up-to-ω extensible with respect to *finite* and *successor of the greatest member*. But that claim requires that we countenance ω, the first infinite ordinal. Suppose a finitist about the ordinals. From the point of view of such a theorist, *natural number* will qualify as absolutely indefinitely extensible. And while there are not many contemporary sceptics about the simple infinite, sceptics about the Cantorian uncountable are not so scarce. These theorists will acknowledge only countable ordinals, so for them *real number* will, by our characterization, likewise rank as absolutely indefinitely extensible. In short, which are the absolutely indefinitely extensible concepts will, on our characterization, depend upon what one takes to be the extent of the ordinals, the measures of all possible series. We may conjecture that Dummett's 'generosity' reflected a sense of this, coupled with the reflection that for any but the most theological of platonists, the extent of the ordinals is a matter that is—metaphysically—*open*.

It is, however, the latter thought that is, I think, the key to the argument against the assumption of Bivalence. Why should a concept's possession of an indefinitely extensible domain, characterized as we have, put an obstacle in the way of bivalent semantics for statements concerning its instances—in particular, for statements

---

[204] I do not know what if anything definite may be said by way of specifying a least ordinal bound on the sequence of iterated 'Gödelizations' of sets of arithmetical truths that conserve recursive enumerability. The smallest non-recursive ordinal, i.e., the so-called Church–Kleene ordinal, usually written: ω_1^{CK}, sets a bound on it.

involving quantification over the entire domain? The crucial point would appear to be that Dummett is thinking of indefinite extensibility as a distinctive genre of *vagueness*—a kind of essential haziness of extension.[205] And this, he is taking it, impacts distinctively on the meaning we may legitimately attach to the quantifiers. For where we generalize over the instances of such an extensionally hazy concept, we may not legitimately suppose, as classical bivalent semantics does, that truth-values will invariably be conferred on a generalization 'upwards' as it were: that whenever '$(\forall x)Ax$' is true, it will be made so by the truth of each of a determinate range of instances, 'Aa.' Rather the truth of such generalizations, when they are true, must somehow be grounded intensionally, in the concepts in play in the open sentence, Ax.[206]

How, though, are we to understand this putative genre of extensional vagueness? Not straightforwardly on the model of borderline case vagueness of the domestic or garden variety exhibited by "red," "bald," etc. *Ordinal number* is itself the paradigm (both intuitively and by the above characterization) of an absolutely indefinitely extensible concept, but if there is vagueness in it, it cannot be cashed out by reference to the idea that there are or could be items which were borderline cases for it—neither clearly ordinals nor clearly not.

Still, we can approach a partial understanding of the Dummettian thought and its impact on the meaning of quantification by focusing on a garden-variety example. Consider a double sorites. Suppose we learn that a linear spatial array of coloured two-dimensional figures has been constructed that is simultaneously a sorites series both for *red* and for *round*. We understand the instructions to have been that a bright red, circular figure was to be placed at one edge and a pale orange, elliptical figure at the other, and that the series of figures was to be arranged in such a way that each element in the array was, by eye, indistinguishable in both hue and shape from its immediate neighbours. Knowing only this, what should we think about this generalization over the series:

All the red figures are round?

Assuming for the sake of the example that garden-variety vagueness already poses obstacles to Bivalence, there is, in our present sate of information, no evident reason why the generalization has to be either determinately true or determinately false. If, as there may be, there are figures in the series of which it is indeterminate both whether they are red or not and whether they are round or not, then—since there is no internal relation between the two characteristics—it will be indeterminate of any such figure whether it is red and not round and so indeterminate whether it is a counter-example to the generalization. And now we are free to suppose that the array has been so constructed that the worst cases for the generalization are all cases like that. If so, it will then be indeterminate whether it has any counter-examples, so indeterminate whether it is true or false.

---

[205] In his (1963), Dummett is repeatedly explicit that the indefinite extensibility of a concept should be conceived as a "variety of inherent vagueness."

[206] The distinction between the two models—I have heard them termed *instantial* and *generic* respectively—of truth-making for generalizations is very illuminatingly explored in Linnebo (2018).

But of course the situation may change if we learn that there were additional instructions—for example, an instruction to ensure that changes in hue took place more rapidly than changes in shape, so that the first figures to be placed that are borderline in hue were still to be definitely round. Or conversely.

For our purposes, the crucial points to take from the example are two. First, the reds and the rounds are of *indeterminate extent* in the imagined array. Second, generalizations over them are accordingly at risk of indeterminacy unless it is determined otherwise, by appropriate additional rules of construction for the series. These are exactly the essential elements in the way that Dummett is thinking of generalizations over the instances of any indefinitely extensible concept: they are concepts whose instances, under some canonical order, are likewise indeterminate in extent, so that generalizations over them can be presumed to be determinate in truth-value only in so far as determined as true or false by essential aspects of the very concept concerned or by the 'rules of construction' for the relevant ordering. A flow of truth-value fixation upwards, from the truth-values of the instances, cannot be relied upon.

So, the relevant species of vagueness for Dummett's purpose is essential *indeterminacy of extent*. Still, the analogy lets us down at a crucial point. The reds and the rounds *peter out* in the array we imagined. There is a smooth slide towards oranges and ellipses. But the ordinals, for example, do not 'peter out'—we do not, if we run the series of ordinals on and on, gradually slide into a region where we no longer deal with ordinals but something else. The analogy uses the reds and the rounds to illustrate indeterminacy in extent—that is how Dummett is thinking of the instances of an indefinitely extensible concept. But it does not, so far, convey any understanding of how a concept that is not borderline-case vague can nevertheless *be* indeterminate in extent in something relevantly similar to the way in which, in the imagined array, *red* and *round* are. The reason for the indeterminacy of the extent of the reds is the borderline case-vagueness of *red*. The reason for indeterminacy in extent for the case of *ordinal* cannot be that. So what is it?

According to the proposed characterization of absolute indefinite extensibility, the extension of any absolutely indefinitely extensible concept spreads out in tandem with that of *ordinal*. But how far do the ordinals go? Let an *unrestrictionist* be anyone who allows, as for example a finitist does not, that every well-ordered series has an ordinal number, and one moreover greater than that of any of its initial segments. Then the crucial point is that the extent of the ordinals is something that we—unrestrictionists—have not merely not fully determined but *cannot* determine. The ordinals, unrestrictedly conceived, are extended by *any possible* number of iterations of successor and of limit. But how many such iterations *are* possible? The rub is that, for the unrestrictionist, the extent of the ordinals themselves is a parameter in the relevant notion of possibility. If we enquire what determines how far the possibilities run, there is no answer for the unrestrictionist to give than: well, for any ordinal number λ, a series of iterated operations of successor and limit of length λ is always possible. It follows that there *is* no explaining the extent of the ordinals. Sure, they are to run on "without bound." But we have no grip on what a *bound* is here except: some specific ordinal limit. We would thus need an antecedent grasp of the extent of the ordinals to determine the extent of the possibilities. To say that the series of ordinals goes on as far as is possible is to say nothing that we can non-circularly explain. Nor, therefore—since anything we can understand must somehow have been explained to us without presupposition that we understood it already—is it anything of which we possess a clear concept.

This is of no consequence for a theological platonist. All that follows for that theorist is that in ordinal arithmetic we investigate a domain of whose overall structure we have no clear concept. But if we hold that the determinacy of a mathematical domain depends on *our achieving* a determinate concept of it— something which, as Ian Rumfitt has rightly emphasized, is an essential part of Dummett's outlook[207]—then we must acknowledge that the extent of the ordinals is indeed indeterminate. With that acknowledgement, we surrender the right to think of the truth-values of generalizations over the ordinals as settled, after the fashion of classical bivalent semantics, by those of their instances. And crucially the same will then go, in tandem, for absolutely indefinitely extensible concepts as a class. Rather, as illustrated with the reds and the rounds, the only possible ground for the truth of such generalizations must reside in the 'rules of construction.'

That, I take it, is the essential nerve of Dummett's argument. If my exegesis is right, however, then a shadow lingers over it: the unrestrictionist conception of the ordinals has—of course—to deal with the Burali–Forti paradox. Indefinite extensibility, as characterized above, is not inextricably tied to paradox. A concept can be absolutely indefinitely extensible and yet paradox free. What spawns paradox is what I have elsewhere called *reflexive* indefinite extensibility: the predicament where a concept is not merely absolutely indefinitely extensible by the lights of the characterization offered but where it is itself an instance of the, or some, relevant higher-order concept P with respect to which it is indefinitely extensible. Ordinal, intuitively understood, provides a paradigm instance of that predicament: the key trigger of Burali–Forti is exactly the thought that the ordinals are themselves well-ordered.[208]

This is too huge an issue to try to treat adequately in the present context, even if I had a clear idea how an adequate treatment might proceed. Still, the shape of the proposal that a defender of Dummett's argument must make is clear at least in outline. What the revisionary argument as here developed needs if it is to be successful is some way of conserving the basis for the extensional indeterminacy of the ordinals while avoiding paradox, for a revisionary argument based on an inco- herent conception can hardly be persuasive. In effect then, a proponent of the argument needs to re-engineer the concept *ordinal* in such a way that it remains indefinitely extensible in the light of something close to the characterization above but ceases to be *reflexively* so. I know of no reason to despair of that project. However, I can offer no evidence that Dummett ever saw matters in exactly this light, let alone that he made any concrete suggestion about how best the trick might be pulled.[209]

---

[207] See the extended passage from *The Seas of Language* (Dummett 1993a; quoted at Rumfitt 2015: 265).

[208] For further discussion of reflexive indefinite extensibility, see Wright (forthcoming).

[209] For a somewhat hand-wringing discussion of the challenges here, see Shapiro and Wright (2006: §9).

6. Shieh

The misgivings about classical logic that belong with the first phase of Michael Dummett's writings on these matters—predating the 'harmonic' and the indefinite extensibility revisionary arguments on which we were just now focused—draw on a putative tension between (potential) undecidability and our right to regard the Law of Excluded Middle as valid. But there are many different ways of developing such misgivings. Sanford Shieh's intricate and interesting Chapter 8 (this volume) submits to critical scrutiny no less than *five* distinguishable such lines of argument that may be interpreted as present either in Dummett's writing or the secondary literature it has generated. He finds difficulties with all of them and, although he does conclude his discussion with a suggestion of an alternative, leaves the reader on a note of scepticism whether any forceful revisionary argument of this broad kind has yet been developed at all. I shall here attempt to undermine this scepticism.

First, though, a caveat. Introducing the kind of revisionary argument on which he has set his sights, Shieh writes of Dummett that he

like Brouwer, insists on our cognitive limitations—there are undecidable statements whose truth or falsity we might never discover. While for Brouwer these limitations constrain mathematical reality, for Dummett they constrain the truth-values of statements. Specifically, *undecidable statements are neither true nor false, and hence are counter-examples to the principle of bivalence.*

—my emphasis. Shieh continues:

Finally, as in the case of Brouwer, for Dummett the law of excluded middle founders on our epistemic limitations. A statement has to be false in order for its negation to be true; so, since an undecidable statement is not false, its negation is not true. Thus, for any undecidable statement, neither it nor its negation is true, which yields a counter-example to the law of excluded middle.

This is a serious misrepresentation of the general gist of Dummett's—and indeed the historical intuitionists'—thinking. I think the point is now pretty widely understood that the intuitionists' reservations about the law of excluded middle (henceforward EM), and/or about Bivalence as its underlying semantic principle, having nothing to do with the spectre of truth-value gaps. Intuitionists hold neither that there are undecidable statements that are neither true nor false, nor that there are any kind of counter-examples to Bivalence (henceforward BV). They do indeed hold that our right to assume the validity of excluded middle is compromised by our epistemic limitations, but not for that kind of reason. Any truth-value gap interpretation of the intuitionists' position is confounded by their acceptance of the validity of the double negation of EM. No cases are recognized where each of its disjuncts fails.

Fortunately, I don't think the most important substance of Shieh's discussion is compromised by this initial misstep, though it does impact on it throughout, as I shall indicate below. He begins by isolating two 'Naïve Revisionary Arguments.' The first is what he calls the *Two-Step Naïve Revisionary Argument*. It works by a *reductio* on three initial assumptions: Bivalence, Undecidability, and Epistemic Constraint for truth. Letting '$R(p)$' express that we can, in this world, feasibly recognize that it is the case that p, Shieh formulates them thus:

$$(\forall p)(True(p) \vee False(p)) \hfill \text{BV}$$

$$(\exists p)(\neg R(True(p)) \ \& \ \neg R(False(p))) \hfill \text{Und}_T$$

$$(\forall p)((True(p) \rightarrow R(True(p))) \ \& \ (False(p) \rightarrow R(False(p)))) \hfill \text{EC}_T$$

The argument then simply notes (step 1) that these compose an inconsistent triad, and hence that acceptance of epistemic constraint alongside the existence of undecidable statements forces rejection of Bivalence. That in turn forces rejection of EM if (step 2) we add the usual principle defining falsity in terms of truth and negation:

$$(\forall p)(T(\neg p) \leftrightarrow F(p)), \hfill \text{FN}$$

together with the Equivalence Schema for truth:

$$(\forall p)(p \leftrightarrow T(p)) \hfill \text{EP}$$

The latter principles provide for the interderivability of LEM and BV. (They are also presumably required for a realist defence of LEM on the basis of BV.)

The *One-Step Naïve Revisionary Argument*, by contrast, differs from its Two-Step cousin by dropping any play with the truth-predicate. It accordingly works with modified versions of Undecidability and Epistemic Constraint and finesses any need for a second step involving FN and EP, simply observing that the following three assumptions once again compose an inconsistent triad:

$$(\forall p)(p \vee \neg p) \hfill \text{EM}$$

$$(\exists p)(\neg R(p) \ \& \ \neg R(\neg p)) \hfill \text{Und}_P$$

$$(\forall p)(p \rightarrow R(p)) \hfill \text{EC}_P$$

There is, however, as Shieh immediately notes, an apparently fatal flaw with the One-Step argument: Und_P and EC_P are already *pairwise* incompatible, since EC_P entails of any witness to Und_P that both its negation and its double negation hold. There is therefore no pressure exerted on EM by the inconsistency of $\{\text{EM}, \text{Und}_P, \text{EC}_P\}$.

There is no such problem with the Two-Step argument: Und_T and EC_T are not pairwise incompatible, and their joint acceptance does accordingly enforce rejection of BV. Specifically, their joint acceptance enforces the admission that any witness to Und_T is neither true nor false. However, in order to convert the admission of counter-examples to BV into a strike against EM, we now need to call on the resources for the elimination of 'true' and false' provided by EP and FN. But, avers Shieh, EP *fails in the presence of truth-value gaps*, since instances of it may now have gappy antecedents but false consequents. So the premises of the first step of the Two-Step argument undermine its second step.[210]

A proponent of the Naïve Two-Step argument might wonder about the force of this objection. They might wonder, for example, with what justification Shieh assumes an account of the conditional that, when truth-value gaps are a possibility,

---

[210] There are similarities here with Ian Rumfitt's discussion of what he calls the "Simple Argument" in chapter 10 of Rumfitt (2015).

counts a conditional statement as failing in any other than the clear case when it has a true antecedent and an other-then-true consequent. If p is indeterminate in truth-value, should we not say the same about $True(p)$?

I think that question may well have merit on some understandings of *indeterminacy*. Gappiness, however, is not indeterminacy of truth-value but a *determinate lack* of truth-value. If we were to try to say that EP should hold, even in the presence of truth-valueless statements, then on the assumption of the biconditional equivalence of the negations of biconditional equivalents, it will directly follow that negation and *True* will commute as prefixes. But one who believes in truth-value gaps must precisely deny that every instance where it is not true that p is a case where the negation of p is true.

So I reckon Shieh's respective objections to the two Naïve Revisionary arguments hit their respective marks. A good thing, then, that, for the reason I gave at the start, neither argument has anything to do with intuitionism! Intuitionists do not believe in truth-value gaps, nor accordingly—since they do of course accept (one or another form of) epistemic constraint—do they endorse either of the Undecidability theses in the forms given above. Their reservations about classical logic arise, rather, from their conviction that there is a very wide class of regions of discourse where, while we have no assurance that every statement is decidable, there are nevertheless good reasons for believing that truth cannot outrun decidability. Since, in the presence of epistemic constraint, classical logic—and EM in particular—provides us with the means to 'prove' that every statement is decidable, we should conclude that we have no assurance of the soundness of classical logic for the regions of discourse concerned.

This general strategy of argument involves what Shieh terms an "epistemic ascent." We are no longer concerned with arguments that purport to show that classical logic is unsound but rather with arguments to the effect that we have no grounds for confidence that it is sound. The shift gives rise to the third and fourth revisionary arguments that Shieh considers, amounting respectively to 'epistemically ascended' versions of the two Naïve arguments latterly reviewed.

Shieh's third revisionary argument—the *Revised One-Step Revisionary Argument*—is exactly what I have elsewhere termed the "Basic Revisionary Argument." It develops a contradiction from the joint assumptions of knowledge of the soundness of classical logic (specifically, knowledge of the validity of EM), knowledge of epistemic constraint, and ignorance of decidability. Shieh formulates these assumptions as follows:

$$K[(\forall p)(p \vee \neg p)] \qquad \text{KEM}$$

$$\neg K[(\forall p)(R(p) \vee R(\neg p))] \qquad \text{NKD}_P$$

$$K[(\forall p)(p \rightarrow R(p))] \qquad \text{KEC}_P$$

Assuming closure of knowledge over known entailment, it is easily seen that knowledge of epistemic constraint—KEC$_P$—forecloses, in the presence of NKD$_P$, on knowledge of the validity of excluded middle. Someone who considers that they know, of some range of statements, that any of them that are true are decidably so, but also recognizes that they lack any guarantee that all are decidable, should accordingly eschew excluded middle.

This one-step argument, like its Naïve predecessor, rests on no challenge to Bivalence but bears directly on classical logic. However, Shieh's fourth argument, the *Revised Two-step Revisionary Argument*, once again targets Bivalence first. Noting that

$$(\forall p)((T(p) \rightarrow R(T(p)) \,\&\, (F(p) \rightarrow R(F(p)))) \qquad \text{EC}_T$$

and

$$(\forall p)(T(p) \vee F(p)) \qquad \text{BV}$$

together entail

$$(\forall p)(R(T(p)) \vee R(F(p))) \qquad \text{DEC}_T$$

we can infer, appealing again to closure of knowledge over known entailment, that knowledge of EC_T and ignorance of DEC_T require ignorance of BV.

We can parlay that ignorance into ignorance of EM if we have knowledge of FN and EP. But Shieh now objects that we do not have knowledge of the latter unless *we know* that there are no truth-value gaps—for if there are, then EP fails for the reason we reviewed above, Shieh writes,

If we don't know that every statement is either true or false, then it is consistent with what we know that not every statement is either true or false. Now, that not every statement is either true or false implies, by classical reasoning, that there are statements that are neither true nor false. So, if we allow classical logic, then we can conclude from our ignorance of bivalence that it is consistent with what we know that there are truth-value gaps, so we don't know that there are no gaps. But then, *classically, ignorance of bivalence implies ignorance of the Equivalence Principle.*[211]

In short, since its assumptions are not co-tenable in a classical logical context, the fourth argument is, he alleges, "question-begging."

The reader should ponder that last claim. What are the rules of the dialectic here? Shieh is assuming a principle that in Wright (2001) I dubbed *AG*; roughly, that in order justifiably to profess ignorance of p, a subject must be in an information state that underwrites their regarding it as an epistemic possibility that not-p. That is what allows him to convert an assumption of ignorance of Bivalence into a commitment to the (epistemic) possibility of truth-value gaps. Let me not demur at that for the moment. But he also seems to be assuming that, in order not to 'beg the question,' an argument purporting to make a case against a logical principle must use only premises which one who regards that principle as sound could simultaneously rationally accept. Yet how, in that case, could the argument succeed in putting any pressure on the principle in question? Wouldn't you expect that the premises for any *successful* argument against (our knowledge of) the validity of a logical principle

---

[211] Chapter 8, this volume, p. 196.

would have to be such as to generate contradiction or some or other form of aporia when reasoning exploiting that principle is applied to them? I'll come back to this.

Shieh develops three objections to the Revised One-step—alias the Basic Revisionary—Argument. Two of them have a significantly *ad hominem* flavour, relating to aspects of my earlier work on this topic.[212] For reasons of space, and in deference to the likely limits of the reader's patience, I forbear to try to respond to them here. In any case, I think his most important objection is the third. He summarizes it as that

> Wright's revisionary arguments do not, in the end, escape an epistemic version of the original tension, in one Naïve Argument, between anti-realism's epistemic conception of truth and the existence of undecidable statements.[213]

The focus of Shieh's objection to the Basic Revisionary Argument is thus on its second premise, NKD_P. The Naïve One-step Revisionary Argument foundered on a tension between Epistemic Constraint and the assumption of Undecidability. Shieh thinks there a similar tension between *knowledge* of Epistemic Constraint and *Ignorance* of Decidability.

We here confront the 'awkward wrinkle' that I grappled with in Wright (2001):[214] if we accept AG—if justified profession of ignorance of p requires having no sufficient grounds to discount not-p—then justified profession of ignorance of Decidability requires having no sufficient grounds to discount the truth of its negation,

$$\neg[(\forall p)(R(p) \lor R(\neg p))].$$

But we know that in the presence of epistemic constraint, the conjunction, $\neg R(p)$ & $\neg R(\neg p)$, leads to contradiction. So we have

$$\neg(\exists p)(\neg R(p) \,\&\, \neg R(\neg p))$$

And in classical logic, that entails $(\forall p)(R(p) \lor R(\neg p))$. So from a classical standpoint, and given AG, knowledge of epistemic constraint *does* preclude ignorance of decidability. Not so from an intuitionist standpoint, of course. But it is an intuitionistic standpoint that the Basic Revisionary Argument is supposed to be motivating—so, as I then thought and as Shieh thinks, we cannot just presuppose it.

In Wright (2001) my response was to reject AG and therefore to attempt to explain how someone can be knowingly ignorant of a proposition while knowingly in position to discount its negation. Successfully to explain that would, of course, amount to an explanation of a principled rejection of the rule of double negation elimination. My suggestion exploited the following replacement for AG. Consider any compound statement, A, whose truth requires that there be a specific pattern of distribution of truth-values across its constituents. And let the constituents in question be subject to evidential constraint. Then the principle the revisionist needs, I contended, is this:

(AG^+) A is known only if there is an assurance that a suitably matching pattern of evidence for (or against) its (relevant) constituents may feasibly be acquired.

---

[212] In Wright (1981, 1992).   [213] Chapter 8, this volume, p. 180.   [214] Wright (2001: 67–9).

My idea was to bring AG^+ to bear alongside the notion of a *quandary* that took centre stage in the (2001) discussion. A quandary is there characterized as any statement with respect to which we labour under four levels of ignorance. For any such statement, we

1 do not know whether it is true or false,
2 do not know how we might come to know whether it is true or false,
3 can produce no reason for thinking that there is any way of coming to know whether they are true or false, and,
4 have no basis to think that anything amounting to knowledge of whether it is true or false "is metaphysically provided for."[215]

To illustrate how the resulting package works, it follows, as the reader will speedily appreciate, that if p is both subject to evidential constraint and a quandary, then an acceptance of AG^+ will allow us to regard $p \lor \neg p$ is not known without incurring any commitment to the epistemic possibility of its negation.

Shieh objects that this train of thought doesn't help the revisionist to explain the target case, viz. how a justified profession of ignorance specifically of Decidability in the form:

$$(\forall p)(R(p) \lor R(\neg p))$$

can fail to involve commitment to the epistemic possibility of its negation. But I am not sure what he has in mind. My idea, to the contrary, was that if some p in the relevant range is both subject to evidential constraint and a quandary, then $R(p)$ and $R(\neg p)$ will be likewise; so by AG^+, $R(p) \lor R(\neg p)$ will not be known; so its generalization will not be known either. And nothing follows about the epistemic possibility of the negation.

Be all that as it may. Shieh next points out that to allow that there are statements which are subject to evidential constraint but are also quandaries provides in any case for a new revisionary argument, distinct from the Basic Revisionary Argument. This is his fifth revisionary argument, the *Quandary Argument*. It runs as follows:

$\neg\neg(p \lor \neg p)$ is a theorem of minimal logic, so known for any p.

Where p is a quandary but subject to EC, $p \lor \neg p$ is not known (by AG^+).

So double negation elimination (DNE) is not known to be valid.

So no classical principle whose proof essentially depends on DNE is known to be valid.

Shieh's objection to this, and presumably to any anti-realism inspired argument for the revision of classical logic based on the notion of quandary, concerns condition 4 in its characterization. "[A]ssuming anti-realism," he writes, "and so assuming knowledge of the Epistemic Constraint, can we profess ignorance of the metaphysical possibility of deciding any given statement?" If we think we know evidential

---

[215] Wright (2001: 71, 75).

constraint holds for a range of statements, can we also know that any of them are quandaries, so meet condition 4?

The reader will by now have anticipated the difficulty Shieh has in mind. We can take it that if we have "no basis to think that anything amounting to knowledge" of whether p is true or false "is metaphysically provided for," then we do not know that p is decidable, that is we do not know that

(i) $R(p) \lor R(\neg p)$.

But we do know that EC entails:

(ii) $\neg(\neg R(p) \& \neg R(\neg p))$

And classically, (ii) entails (i). So in a classical framework, the assumption that we can know that EC is good for a known quandary is inconsistent. Hence we can only run the quandary argument in a setting where we have *already* dropped classical logic.

Shieh's assumption, as noted, is that these considerations show that the Quandary Argument—and, analogously, the Basic Revisionary Argument—are question-begging. But at this point we should step back and ask, what would it be for an argument against (our knowledge of) the validity of a system of logic L *not* to be question-begging by such lights. It would have to be an argument whose premises we could consistently accept along with acceptance of the inferential framework supplied by L but which, within that framework, entailed that we should suspend commitment to some principle of L. That is an incoherent description.

An argument for the revision of a system of logic ought ideally to provide cogent considerations even for those antecedently committed to it. I say "ideally" because such an argument might still have dialectical value even if its force were appreciable only by the uncommitted. But what we cannot coherently require of such an argument is that acceptance of its premises be consistent with acceptance of the very inferential principle(s) whose good standing it is intended to call into question. That, in effect, is ¬ to require that the argument be ineffective!

What is the correct conception of the dialectical rules that should govern argument for the revision of principles of logic? What conditions does a successful revisionary argument have to meet? Familiarly, it's a complex and difficult question. But clearly such an argument must draw only on well-motivated assumptions. And the reasoning involved must be common ground to all parties to the debate. However, there is and can be no requirement that all parties to the debate could accept its assumptions consistently with their varying background inferential commitments, since it is precisely one in particular of those background sets of commitments that the argument is designed to challenge. It therefore cannot, without further ado, be an objection to such an argument that, by reasoning in exactly the ways that it targets, we can make trouble for the co-tenability, or mutual coherence, of its premises.

Really, the point is no more subtle than this: it is no sort of reply to an argument that purports to show you are wrong about something that, if you are not wrong, its premises cannot all be acceptable! The kind of revisionary arguments here reviewed, whatever their special individual strengths and weaknesses, all call for the exercise of *philosophical judgement*. If the application of the very classical principles of inference that such an argument challenges calls the co-tenability of its premises into question, the next thing to do is to ask; but *ought* they, intuitively, to be co-tenable? If the judgement is "Yes," then that's a point at which classical logic is doing us a disservice and where the revisionary argument gets its revisionary bite.

References

Boolos, George (1993). "Whence the Contradiction?," *Proceedings of the Aristotelian Society, Supplementary Volumes* 67, 213–33.

Clark, Peter (1993). "Sets and Indefinitely Extensible Concepts and Classes," *Proceedings of the Aristotelian Society Supplementary Volumes* 67, 235–49.

Clark, Peter (1998). "Dummett's Argument for the Indefinite Extensibility of Set and Real Number," *Grazer Philosophische Studien* 55, 51–63.

Cook, Roy T. (2005). "What's Wrong with Tonk(?)," *Journal of Philosophical Logic* 34 (2), 217–26.

Dummett, Michael (1963). "The Philosophical Significance of Gödel's Theorem," reprinted in *Truth and Other Enigmas*, London: Duckworth, 1978, 186–201.

Dummett, Michael (1978). *Truth and Other Enigmas*. London: Duckworth.

Dummett, Michael (1981). *Frege: Philosophy of Language*, second edition, London: Duckworth.

Dummett, Michael (1991). *Frege: Philosophy of Mathematics*, Cambridge, MA: Harvard University Press.

Dummett, Michael (1993a). "What is Mathematics About?," in *The Seas of Language*, Oxford: Clarendon Press, 429–45.

Dummett, Michael (1993b). *The Logical Basis of Metaphysics*, Cambridge, MA: Harvard University Press.

Dummett, Michael (1994) "Chairman's Address: Basic Law V," in *Proceedings of the Aristotelian Society* 94, 243–52.

Gentzen, Gerhard (1935). "Untersuchungen über das logische Schließen" ("Investigations into Logical Deduction"), *Mathematische Zeitschrift* 39, 176–210.

Heck, Richard Kimberly (1993). "Critical Notice of Review of Frege: Philosophy of Mathematics, by Michael Dummett," *The Philosophical Quarterly* 43 (171), 223–33.

Linnebo, Øystein (2018). "Impredicativity and Two Conceptions of Collection and Generality," in Ivette Fred and Jessica Leech (eds.), *Being Necessary: Themes of Ontology and Modality from the Work of Bob Hale*, Oxford: Oxford University Press.

Oliver, Alex (1998). "Hazy Totalities and Indefinitely Extensible Concepts: An Exercise in the Interpretation of Dummett's Philosophy of Mathematics," *Grazer Philosophische Studien* 55, 25–50.

Prawitz, Dag (1965). *Natural Deduction: A Proof-Theoretical Study*, London: Courier Dover Publications.

Prawitz, Dag (1974). "On the Idea of a General Proof Theory," *Synthese* 27 (1/2), 63–77.

Read, Stephen (2000). "Harmony and Autonomy in Classical Logic," *Journal of Philosophical Logic* 29 (2), 123–54.

Rumfitt, Ian (2000). "'Yes' and 'No,'" *Mind* 109 (436), 781–823.

Rumfitt, Ian (2015). *The Boundary Stones of Thought: An Essay in the Philosophy of Logic*, Oxford: Oxford University Press.

Rumfitt, Ian (2017). "Against Harmony," in Bob Hale, Crispin Wright, and Alexander Miller (eds.), *A Companion to the Philosophy of Language*, second edition, Chichester: John Wiley & Sons, Ltd, 225–49.

Russell, Bertrand (1906). "On Some Difficulties in the Theory of Transfinite Numbers and Order Types," *Proceedings of the London Mathematical Society* 4 (14), 29–53.

Shapiro, Stewart, and Crispin Wright (2006). "All Things Indefinitely Extensible," in A. Rayo and G. Uzquiano (eds.), *Absolute Generality*, Oxford: Oxford University Press, 255–304.

Steinberger, Florian (2009). "Not So Stable," *Analysis* 69 (4), 655–61.

Steinberger, Florian (2011). "Harmony in a Sequent Setting: A Reply to Tennant," *Analysis* 71 (2), 273–80.

Tennant, Neil (1978). *Natural Logic*, Edinburgh: Edinburgh University Press.

Tennant, Neil (1987). *Anti-Realism and Logic: Truth as Eternal*, Oxford: Oxford University Press.

Tennant, Neil (1997). *The Taming of the True*, Oxford: Clarendon Press.

Tennant, Neil (2010). "Harmony in a Sequent Setting," *Analysis* 70 (3), 462–68.

Warren, Jared (2015). "Talking with Tonkers," *Philosopher's Imprint* 15 (24).

Wittgenstein, Ludwig (1956). *Remarks on the Foundations of Mathematics*, G. H. von Wright, Rush Rhees, and G. E. M. Anscombe (trans.), Oxford: Blackwell.

Wright, Crispin (1981). "Dummett and Revisionism," *The Philosophical Quarterly* 31 (122), 47–67. Reprinted as "Anti-realism and Revisionism," in *Realism, Meaning and Truth*, Oxford: Basil Blackwell, 1987, 433–57.

Wright, Crispin (1987). *Realism, Meaning and Truth*, Oxford: Basil Blackwell.

Wright, Crispin (1992). *Truth and Objectivity*, Cambridge, MA: Harvard University Press.

Wright, Crispin (1993). *Realism, Meaning and Truth*, second edition, Oxford: Basil Blackwell.

Wright, Crispin (1999). "Is Hume's Principle Analytic?," *Notre Dame Journal of Formal Logic* 40 (1), 6–30.

Wright, Crispin (2001). "On Being in a Quandary: Relativism Vagueness Logical Revisionism," *Mind* 110 (437), 45–98.

Wright, Crispin (2018). "How High the Sky? Rumfitt on the Putative Indeterminacy of the Set-Theoretic Universe," *Philosophical Studies* 175, 2067–78.

Wright, Crispin (forthcoming). "How did the Serpent of Inconsistency Enter Frege's Paradise?," in Philip Ebert and Marcus Rossberg (eds.), *Essays on Frege's Basic Laws of Arithmetic*, Oxford: Oxford University Press.

Reply to Part IV

The Epistemology of Metaphysical Possibility

Reply to Hale

Bob Hale's Chapter 10 (this volume) is a characteristically searching critique of an assessment I wrote some years ago[216] of the famous challenge to physicalism about the mental that is outlined in the concluding pages of Saul Kripke's *Naming and Necessity*. Hale and I are agreed that Kripke's argument comes short. The interesting question is exactly *why* it comes short. It's a question that takes us to the heart of one central issue concerning the epistemology of the metaphysical modalities.[217]

1. Kripke's Challenge

Let us speedily review the background. *Naming and Necessity* was a veritable cluster-bomb of original, interrelated, deep-reaching ideas. Among the marquee notions it introduced were included:

- The idea of rigid designation and the thesis that the proper names of natural language are, characteristically, rigid designators;
- The contention that the same is true of many common nouns including a large class of terms standing for natural kinds of things, states and stuffs—'water,' 'heat,' 'tiger,' 'elm,' 'diamond,' etc.;
- The idea of natural kinds as identified by essential characteristics of which an ordinary competent understanding of words standing for them may be entirely innocent;

---

[216] Wright (2002).

[217] I should advise the reader that while I shall eventually offer some further reflections about how physicalism may defend against Kripke's argument, it will be with the epistemology of metaphysical modality that I will be mainly concerned here. I will not, for the most part, be further exploring other issues to which Kripke's argument gives rise—for instance whether his argument, even if successful as given, can engage token-token forms of physicalism, whether natural kind terms are indeed, characteristically, rigid designators, whether, if so, their reference has to be construed at a deep, rather than a superficial level, what indeed rigidity comes to once singular reference and singular terms are left behind. These are all questions which a full appraisal of the argument would demand we look at, and Wright (2002) does have things to say about them. Here, though, our primary focus will be on the epistemology of possibility. And while I shall be drawing on some of the details of my earlier discussion, I have done my best to presuppose no acquaintance with that.

- The modern conception of the metaphysical modalities as grounded not in the ways we talk and think but in the essential characteristics of what we talk and think about; and
- The consequent separation of the necessary from the *a priori* and resultant hospitality to a raft of *a posteriori* necessities.

It is a consequence of these proposals that any statement purporting to identify the essence of a natural kind will be, if true at all, necessarily so.[218] The contention that 'water,' for example, rigidly designates a natural kind requires that the use of that English word (with its present meaning) in speaking of any scenario, actual or counterfactual, always denotes the same kind of stuff, just as the use of "Richard Milhous Nixon" in speaking of any scenario, actual or counterfactual, always denotes the man who was actually president of the United States in 1970. Hence, as Putnam famously argued, (assuming that water is indeed a chemical kind) the proper description of a hypothetical situation where some colourless, tasteless liquid, naturally occurring in rainfall, lakes and rivers, etc., etc., but of a differing chemical constitution, displayed all the characteristic surface properties of water, would be not that in that situation some water would have a different chemical constitution to what water actually characteristically has, but that the lakes and rivers, etc., would not be filled with water. If 'water' rigidly designates a chemical kind and it is true that water is H_2O, then it is necessarily true.

It immediately follows that any evidence against the necessity of such an identification is *eo ipso* evidence against its truth. But what should count as evidence against the necessity of such an identification in the first place? As I interpreted Kripke, his discussion implicitly rests on a major assumption on which we should pause: that *all* purportedly metaphysically necessary statements, even those grounded in essences knowable only *a posteriori*, are hostage to what we can, to borrow Descartes's happy phrase, "clearly and distinctly" conceive; that—short of its actuality—a clear and distinct conception of a situation is the best possible evidence of its possibility. This principle—what in Wright (2002) I called the *Counter-Conceivability Principle*[219]—invites us, of course, to provide an account of when a conception should rank as relevantly clear and distinct. But without taking that issue on directly, we can cash the principle's operational content as being this: that if one has what at least *appears to be* a lucid, detailed conception of how it might be that not *P*, then that should count as a good, albeit defeasible ground for its being possible that not *P*, and hence its not being necessary that *P*, whatever the subject matter of *P*.[220]

By the Counter-Conceivability Principle, then, all putative metaphysical necessities, even *a posteriori* ones, have to run the gauntlet of what we can, as we think,

---

[218] At least, it is a consequence provided the characterization of the essence appealed to (namely, in the stock example, "H_2O") is itself a rigid designator.

[219] "*Counter*-Conceivability" as a reflection of the envisaged use of the principle against claims of metaphysical necessity.

[220] The Counter-Conceivability Principle is something which might really have *deserved* the title of "Hume's Principle," now of course purloined by the neo-Fregean programme for the foundations of arithmetic. Recall *Treatise* Bk. I, pt. II, §II: "*whatever the mind clearly conceives, includes the idea of possible existence, or in other words, that nothing we imagine is absolutely impossible.*"

clearly and distinctly conceive. If, to all appearances, we can indeed clearly and distinctly conceive of a scenario in which the proposition that P would be false, then that will be to disclose, *a priori*, a (defeasible) defeater of the claim of P to record a metaphysical necessity, even if the sole possible form of justification for P would have to incorporate empirical information.

Now, the Counter-Conceivability Principle (henceforward CCP) is apt to impress as a controversial assumption at best—and at worst, in the present context, as a crude mistake. For conceivings, even at their clearest and most distinct, are constrained only by the requirements of the *concepts* that they configure. The metaphysical-modal status of a proposition, by contrast, at least in a range of central cases where it is recognized *a posteriori*, is grounded in requirements imposed by the essential natures of non-conceptual *things*—objects, properties, and kinds, etc.—which its ingredient concepts are possibly inadequate, possibly incomplete, concepts *of.* Conceivings, at their best, may tell us what is and is not permitted by the concepts they work with; may tell us, in short, what is *conceptually* possible, impossible or necessary. But Kripke's own work is exactly the *locus classicus* for the existence and importance of the distinction between the conceptual and the metaphysical modalities. The CCP, it may be charged, when harnessed to an attempt to engage the latter, implicitly surrenders that distinction. So any argument based on the CCP is dialectically unavailable to anyone who endorses the distinction.

That's an impressive protest. There is surely something right about it. On the other hand, if claims of (unactualized) metaphysical possibility cannot be justified, even if defeasibly, by considerations of what is lucidly conceivable, how *can* they be justified? What is the epistemology of metaphysical possibility to be?

For the time being I am going to leave this issue as an Elephant in the Room. We'll come back to it below. First, let's review Kripke's argument, my original response, and Hale's alternative proposals.

So: assume that water, the stuff, is indeed a natural kind, correctly identified at the level of Daltonian chemical theory as H_2O, and that "water," the noun, is a rigid designator of that kind. Then

(i)   Water is H_2O

is not merely true but necessarily true. Kripke anticipated[221] the natural objection to this upshot: surely it is readily conceivable that water might have turned out *not* to be H_2O. We can easily imagine the science having worked out quite differently, even that water might have proved not to be any particular chemical kind at all but, as it were, a syndrome of surface properties with a multiplicity of variable underlying causes. But if the CCP is in force, then—since what might have turned out to be the case might thereby *have been* the case—that is a *prima facie* case that (i) is not necessary. And in that case we will have to deny on purely philosophical grounds that it is even true—which is absurd.

---

[221] The discussion of this point in Lecture III of *Naming and Necessity* actually proceeds in terms of other examples—Hesperus and Phosphorus, heat and molecular motion. So there is an element of interpretation in the assumption that Kripke would have endorsed the account to follow.

The objection is heavily invested in the CCP. One response for a defender of Kripke's modal views would be simply to drop the principle. Famously, though, that is not what Kripke did. Rather, he allowed that the objector may be lucidly conceiving *something* perfectly coherent but contended that what is so conceived is nothing to the purpose. What *is* conceivable is that a stuff that presents on the surface as water does should not be H_2O. But to conceive of a stuff that presents on the surface as water not being H_2O is not to conceive of *water's* not being H_2O.

Again, since both "Hesperus" and "Phosphorus" are, plausibly, rigid designators, the statement,

(ii)   Hesperus is Phosphorus,

must, if true at all, be necessarily true. But it took quite painstaking astronomical observation and calculation to figure out that (ii) is true and it might seem readily conceivable that the enquiry might have turned out differently: that Hesperus, the brightest star-like body seen in the evening sky, might have turned out to be a distinct entity from Phosphorus, the bright 'star' characteristically visible low in the eastern sky just before sunrise. The Kripkean reply will then be, similarly: No, what is indeed conceivable is that a heavenly body that presented in the evening sky exactly as Hesperus does should have turned out not to be Phosphorus. But to conceive of a heavenly body that presented itself exactly as Hesperus actually does turning out not to be Phosphorus is not to conceive of *Hesperus'* turning out not to be Phosphorus.

In sum: Kripke's response to protect the necessity of true identifications of the relevant kind from the depredations of apparently lucid counter-conceivings is to insist on a distinction between conceiving of X not being F and conceiving of an *epistemic counterpart* of X not being F. Something which presents as water in all (surface) respects covered by our pre-theoretic conception of water—the indicators that, prior to the scientific investigation, we would use to classify a sample as one of water—need be no more than an epistemic counterpart of water. Something that presents as Hesperus in all respects covered by the characteristics incorporated in our folk-identifications of Hesperus need be no more than an epistemic counterpart of Hesperus. If, in accordance with the CCP, an episode of imagination, however lucid, is to support a judgement of genuine possibility, it must be shown to engage with X itself, rather than a mere epistemic counterpart of X. Otherwise the claimed necessity of [X is F] cannot be brought thereby into question. The apparent counter-conceivability of [Water is H_2O], and of [Hesperus is Phosphorus], are *modal illusions*, spawned by this conflation.

The crux of Kripke's anti-physicalist argument is now, famously, there is no such corresponding distinction, or available conflation, for pain. Pain has no epistemic counterpart that is not pain: whatever presents on the surface as pain (i.e., whatever hurts!) is pain. Using 'C-fibre-excitation' as parametric for any putative identification of pain in neurophysical terms, we thus have both that

(iii)   [Pain is C-fibre excitation] is, if true, necessary,

—since both ingredient terms are, plausibly, rigid designators—and that

(iv)   [Pain is C-fibre excitation] is counter-conceivable,

since it does seem perfectly possible fully lucidly to conceive of oneself, or others, being in pain in circumstances where their, or one's own, C-fibres prove, by appropriate scientific tests, to be inactive. And now there is no defending the purported identity by the charge that the imagination here deals merely in an epistemic counterpart of pain. Again, there *are* no mere epistemic counterparts of pain, no states that present in all respects as pains do but are not pains. Any epistemic counterpart of pain *is* pain; the imaginative episode *does* engage its target. So by the CCP, a case, so far undefeated, is made that pain is not necessarily C-fibre excitation. It follows within Kripke's assumptions that pain is not C-fibre excitation at all and hence, since the play with the latter was merely parametric, that pain is *no* type of physical state.

2. One Proposed Response

How should a physicalist respond to this? The most salient direction is to pursue the thought that, granting that the apparent possibility of pain's not being C-fibre excitation cannot be a modal illusion in *that* kind of way, still maybe that is not the only possible template that an illusion of possibility may instance. Perhaps there are other ways whereby an apparently lucid conception of how P might be true can miss its intended mark—can fall short of an indication of P's genuine possibility—other than by trading on a conflation between events, states, or stuffs, etc., and mere epistemic counterparts of them. Perhaps there are other ways whereby an apparently lucid conception of P's obtaining can indeed actually be lucid and adequately detailed but be *of* something else, something that masquerades, as it were, as the obtaining of P.

In the discussion of mine to which Hale is responding, I offered two putative examples to suggest one such alternative model. The first was the following fantasy:

> Suppose—Kripke would agree—that I am essentially a human being, and that it is an essential characteristic of human beings to have their actual biological origins. So it is an essential characteristic of mine to be the child of my actual parents. Still I can, it seems, lucidly conceive of my not having had those parents but others, or even—like Superman—of my originating in a different world, of a different race, and having been visited on Earth from afar and brought up as their own by the people whom I take to be my biological father and mother. I can, it seems, lucidly imagine my finding all this out tomorrow.[222]

Now, this example, I claimed, cannot be dismissed

> on the ground that it fails to be sensitive to the distinction between myself and a mere epistemic counterpart, a mere "fool's self," as it were, sharing the surface features by which I identify myself but differing in essence. It cannot be so dismissed because I don't, in the relevant fashion, identify myself by features, surface or otherwise, at all. The point is of a piece with Hume's observation that, in awareness of a psychological state as one's own, one is not presented as an object to

---

[222] Wright (2002: 435).

oneself. When I conceive some simple counterfactual contingency—say, my being right at this moment in the Grand Canyon—I do not imagine someone's being there who presents themselves, on the surface, as being me. Rather I simply imagine *my* having relevant kinds of experience—imagine, that is to say, the relevant kinds of experience from my first personal point of view. No mode of presentation of the self need feature in the exercise before it can count as presenting a scenario in which *I* am in the Grand Canyon; *a fortiori*, no *superficial* mode of presentation, open to instantiation by someone other than myself.[223]

The suggestion is that a sweeping range of lucid and detailed flights of fancy running counter to presumed metaphysical necessities are open to the First Person—fancies of different biological origins, alternative modes of embodiment (Kafka's Beetle), even disembodied existence and conscious survival of the dissolution of one's body. And this is so precisely because first-personal thought is unconstrained by any mode of presentation of the self that stands to it as the surface-indicators of water stand to the natural kind. Nothing in my concept of myself—the concept, that is to say, which I exercise just by the intelligent use of "I"—obstructs the coherence of these fantasies. They can be as lucid and detailed as you like, yet are not open to diagnosis as missing their mark by virtue of working with mere epistemic counterparts. Still, the scenarios they construct appear to stand in flat conflict with widely accepted necessities of origin and constitution.

A different but broadly analogous challenge to the CCP, I contended, is provided by certain examples of apparent conceivability in mathematics. I wrote,

We can rest assured, I suppose, that Andrew Wiles really has proved Fermat's "Last Theorem," which therefore holds good as a matter of conceptual necessity. But we can imagine a sceptic about the result who flatters himself that he can still conceive of finding counterexamples to the theorem, and of finding mistakes in Wiles' proof. Of course there will be limits on the detail of these or the sceptic would be thought-experimentally finding *real* counterexamples and mistakes. Still, we should not deny that he could be conceiving *something*, and doing so moreover—subject only to the preceding point—in as vivid and detailed a way as could be wished.

About this case, however,

the last diagnosis we should propose . . . is that his conceivings are insensitive to the distinction between finding counterexamples to Fermat's theorem and finding counterexamples to an *epistemic counterpart* of it!—or to the distinction between finding a mistake in Wiles' reasoning and finding a mistake in an epistemic counterpart of that.[224]

To be sure, the issues pointed to by this example go beyond those raised by the question whether it is indeed conceivable that any particular identification of pain with a neurophysical kind should fail. For this time it is, plausibly, as suggested in the quoted passage, a *conceptual* impossibility which is presented as conceivable. The bearing of the example is directly on the tenability of the CCP even where it is conceptual possibility that is at stake. It accordingly raises very vividly the question, just what should we be asking of a relevantly "lucid and detailed" conception.

---

[223] Wright (2002: 435–6).   [224] Wright (2002: 436–7).

A natural objection to the case—tendered, indeed, by Hale—is that, while we may indeed grant that the sceptical mathematician is coherently conceiving something, the acknowledged unavoidable limitations to the detail of his conceivings rob the example of its intended force. The CCP, after all, speaks of clear and distinct—lucid and detailed—conceptions: a defender of the principle has no reason to deny that impossibilities may be *hazily* conceived. Well, no doubt. But must the conceivings of the sceptic in the example be *relevantly* hazy? There is no reason not to grant that whatever ideas our mathematician had, at a time before Wiles's discovery, both of what a proof of Fermat would have to be, and of what a counter-example would have to be, were definite enough to put him in position to recognize any instantiation of either, provided it could be surveyably written down—definite enough, indeed, to leave him with no space for discretion were he to be presented with such an instance of either. And presumably they are no less definite after the announcement of Wiles's result. But are not concepts that are definite enough to enforce recognition of any instance of them, with no foreseeable room left for discretion, thereby definite enough to provide the materials for conceivings that are relevantly lucid and detailed for the purposes of the CCP? If not, what is the standard to be?

These two cases, then, are *prima facie* relevantly lucid and detailed conceivings of impossibilities. The choices they present to a friend of the CCP are, accordingly, two: either—*the radical option*—to admit them and then to identify some suitable quali-fication of the Principle that allows it still to be of some modal-epistemological use; or—*the conservative option*—to follow Kripke's example in trying to make a case that they do not, appearances notwithstanding, involve genuine conceivings of what they purport to be conceivings *of*.

My suggestion in my earlier discussion was of the second, conservative kind. The examples indicate, I suggested, that there

has to be a category of conceivings that fall short of being counter-conceptions to a given proposition, not because their detail fails to be sensitive to the distinction between items that the proposition is about and 'fool's' equivalents of them

—epistemic counterparts—

but because it is insensitive to another distinction: that between genuinely conceiving of a scenario in which P fails to obtain and conceiving, rather, of what it would be like if, *per impossibile*, P were (found to be) false.[225]

On this proposal, there are examples not only of metaphysical but also of conceptual impossibility, P, such that we should not deny that one can perfectly lucidly and in detail conceive of scenarios that deserve the description: this is how things would/ could be if P were to be false. What we deny is that such conceivings are tantamount to conceiving of P's being false. The CCP is therefore spared. But at the same time we still provide the physicalist with the resources to write off the apparent counter-conceivability of [Pain is C-fibre activity] as a modal illusion. She may maintain that it is a modal illusion of this second kind: that is, she can allow that while it is apparently possible clearly and distinctly to conceive of pain in the absence of C-fibre

---

excitation and, conversely, of C-fibre-excitation in the absence of felt pain, it is not really possible. Rather, analogously to Kripke's explanation of the illusion of conceivability of water's not being H_2O, the physicalist can contend that while *something* is indeed lucidly and distinctly conceived in such flights of fancy, what is conceived is not the intended target. One conceives, it may be suggested, not of pain's not being C-fibre excitation, but—what is not the same thing—of what things would be like if, *per impossibile*, it were something else.

3. Hale's Objections and Countersuggestions

This response to Kripke's argument is not to Hale's liking. We should distinguish, he urges, between, on the one hand,

(a) The question whether there are indeed other relevant species of illusion of possibility in addition to the species, highlighted by Kripke, involving play with mere epistemic counterparts of a targeted kind, or stuff, etc.,

and, on the other,

(b) The question whether the putative *per impossibile* species invoked in my treatment of the 'SuperWright' and 'Sceptical Mathematician' examples is either (i) needed to account for either of those cases, or indeed (ii) is anyway in good standing.

Hale thinks the answer to (a) is 'yes,' and that the other relevant modes that he goes on to describe will sustain no less a negative appraisal of the anti-physicalist argument. But he has doubts about both (b)(i) and (b)(ii).

To focus first on (b)(ii). Hale reacts to the formulation in the passage from my (2002) most recently quoted above—the putative contrast between "genuinely conceiving of a scenario in which *P* fails to obtain and conceiving, rather, of what it would be like if, *per impossibile*, *P* were (found to be) false." I rather regret that formulation insofar as it may encourage a reader to wonder—as does Hale—whether the impossibility marked by "*per impossibile*" is to be an ingredient in the second kind of conceiving. The answer of course is that it is not: the second kind of conceiving is to be the source, after all, of a kind of *illusion* of possibility—so naturally there needs to be scope for someone to be taken in by it; the impossibility cannot be part of the explicit content of the conceiving itself.

Hale's principal doubt, however, about the bona fides of *per impossibile* conceivings as characterized, focuses on the subjunctive conditional: "what it would be like if...*P* were (found to be) false." Against this he urges the obvious question: Where a proposition is impossible, would *anything* distinctively be the case if it were to hold?

Standard—Lewis–Stalnaker-style—semantic treatments of counterfactuals say "No," of course. In such treatments, a counterfactual conditional holds just if its consequent holds in all relevant worlds in which its antecedent does. Since, when a counterfactual has an impossible antecedent, there are no worlds—*a fortiori*, no relevant worlds—in which the antecedent does hold, there are no worlds in which the counterfactual fails to hold. So standard semantics declares such conditionals necessarily true, irrespective of the content of their consequents. But if all such conditionals are true, then there is nothing distinctive that the situation would be like if, say, the square root of 2 were rational or if I had been born of different parents. So there is nothing distinctive to be the content of the appropriate conceiving: you can conceive what it would be like if, *per impossibile*, P were to obtain, by conceiving anything at all.

Although I have no space to pursue the matter here, my own view is that, its theoretical naturalness notwithstanding, this consequence reflects badly on the standard semantical treatment of counterfactuals—that it indicates more about a degradation of content implicit in such treatments than about the putatively illusory character of any worthwhile distinction between, say, "If the square root of 2 were rational, it would be equivalent to *m/n* for some pair of integers, *m* and *n*" and "If the square root of 2 were rational, it would be equivalent to π." There is an interesting growing literature in sympathy with this reaction,[226] though how to develop a semantics for counterfactuals to do better than the standard treatment in this regard remains difficult and controversial.[227] I shall not pursue that matter further here. As things stand, it seems fair to concede to Hale that the prospects for a successful response to the anti-physicalist argument by deployment of the *per impossibile* template for modal illusion have not yet been clearly made out.

May we do better by enlisting a further template of the conservative kind—one that stops short of the idea that examples like those considered involve genuine conceivings of the impossible? Hale suggests that we may and offers two further potential sources of illusion of possibility in particular. First on the Sceptical Mathematician:

The sceptic starts with the hypothesis—in fact necessarily false, although he does not, of course, take it to be so—that there are counter-examples to FLT [Fermat's Last Theorem]. He then imagines someone finding one—*without, however, attempting to imagine it in any detail*—and goes on to imagine the impact of this 'discovery' on the mathematical community: the initial reactions of shock and even disbelief which, however, quickly disappear (since simple and apparently flawless calculations show the putative counter-example to be the genuine article), the embarrassment of those (virtually everyone) who had been 'taken in' by Wiles's proof, Wiles's own chagrin and reluctant admission that there must be some flaw in his work, the ensuing hunt for a mistake—perhaps a hitherto unnoticed but false assumption—in that 'proof,' and so on. There is, it seems, no doubt that this is a perfectly conceivable flight of fancy, and that it could be elaborated in as much vivid detail as you like, subject to the proviso that it remains silent on the mathematical detail of the supposed counter-example.... What is right is that this is a perfectly intelligible piece of conceiving. What is wrong—or at least, in my view, very much open to question—is that the flaw in it, viewed as an attempt to counter-conceive FLT, should be diagnosed as Wright proposes, that is, as a matter of insensitivity to a distinction between conceiving of FLT's being false and conceiving of what it would be like *if, per impossibile*, FLT *were (found to be) false*. Nothing in what the sceptical mathematician actually *imagines*—as distinct from the background doubt as to the truth of FLT which *fuels* his

[226] Nolan (1997, 2013); Kment (2006a,b); Brogaard and Salerno (2013); Berto (2014); Berto and Schoonen (2017).
[227] See Berto (2014).

imaginative flight—*demands* its description in those terms, *rather than* in these quite different terms: conceiving what it would be like *if, as is perfectly possible, FLT were... thought* to be counter-exemplified.[228]

Now, I already argued that the level of necessarily missing mathematical detail in the Sceptical Mathematician's imaginings should not immediately disqualify them from the status of adequately lucid and detailed conceivings in the spirit of the CCP. But I won't pursue that further now. Hale's proposal is that the example is better described—or anyway as well described—in terms of his *Doxastic Consequences* model. According to this proposal, we are apt to confuse conceiving of *P*'s being (found to be) false with conceiving of what it would be like if, as is perfectly possible, *P were thought to be* false. And, Hale contends, nothing in the description of the case requires that we think of the Sceptical Mathematician as doing anything beyond what can be accommodated by the latter description.

We will consider that claim in a moment. But first here is Hale on SuperWright:

> We should now come back to the question suggested just now, whether one should describe this kind of conceiving in terms of imagining (our) finding out or discovering that such-and-such, or rather in more guarded terms as imagining (our) having compelling evidence or good reason to believe that such-and-such. If Hume's Principle is true—that is, if conceivability implies possibility—then we should insist upon the more guarded description. For if we really could conceive of finding out that *P*, then it would be possible that we should find out that *P*. And since 'find out' is factive—we can only find out that *P*, if *P*—it would follow (by the sound modal principle that ◊*A* entails ◊*B*, when *A* entails *B*) that it is possible that *P*. It is true enough, or at least widely accepted, that in many—perhaps the vast majority of—cases, it is metaphysically, even if not epistemically, possible that the causes of events and states of affairs should have been other than they actually were. In such cases, we may agree that we can conceive of finding out that the causes were otherwise. But if some of the facts about causation are metaphysically necessary—as they must be, if the doctrine of essentiality of origin is true—then whatever it is that we can conceive in regard to causes being other than they actually were, it cannot be that we can conceive of *discovering* that they were other than they actually were. The closest we can come to that is (something like) conceiving of our having compelling evidence, or good reason to believe, that the causes were otherwise.[229]

This reasoning seems to me compelling, as far as it goes. That is, if—as Hale is assuming—genuine conceivability really does imply possibility, and *P* is necessarily false, then of course, given the factivity of 'finding out,' we cannot genuinely conceive of finding out that *P*. In that case, we have no option but to redescribe in lesser terms whatever it is that is apparently conceived lucidly and in detail in the SuperWright example. Hale's suggestion is that the example exhibits what he calls *The Misleading Evidence model*—we think it possible that *P* because we can conceive of our having compelling evidence or good reason to believe that *P*, and we fail to distinguish this from conceiving of its being the case that *P*.

Now, I certainly don't want to deny that misjudgements of possibility *might* be spawned by either of the kinds of conflation that Hale draws attention to—by confusing how things would be if *P* with how things would be if we (all) believed *P*

---

or if we had misleading evidence that *P*. But there is an immediate concern: it is not clear that either the Doxastic Consequences model or the Misleading Evidence model cuts any mustard when it comes to trying to explain away the appearance of the lucid and detailed conceivability of pain without C-fibre excitation or its converse. Someone who thinks—as most of us may be presumed to do—that one can vividly imagine what it would be like to be in pain absent any particular type of putative characteristic physical basis for it is most unlikely to be prepared to accept the watered down suggestion: "No: you are really just imagining what it would be like if you believed you were in pain while, for example, none of your C-fibres were firing," or "No: you are really just imagining what it would be like if you had misleading evidence that you were in pain while, for example, none of your C-fibres were firing." There is no clear sense to the idea of misleading evidence that one is oneself in pain. And there is no plausibility whatever in the suggestion that, in the imagined scenario, one is merely imagining consequences and surroundings of the belief that one is in pain. Not a bit of it: one imagines *pain*—the very phenomenal state itself, not a doxastic state and its surroundings and consequences. For one out to put a roadblock in the way of Kripke's anti-physicalist argument, Hale's two suggested additional models of misbegotten conceivings seem, so far, not to promise any leverage at all.

Hale, expectably, thinks otherwise. He writes,

In fact, Kripke's argument is in trouble anyway ... so long as my *Doxastic Consequences* model is workable. For if that is so, then—whether or not it is agreed to be the best model for explaining the Sceptical Mathematician's modal illusion(s)—it could, clearly, be invoked by the identity-theorist to explain away the apparent conceivability of painless C-fibre [excitation] and C-fibreless pain. We can easily imagine how things might be, if pain were believed to be other than C-fibre [excitation]. We imagine, for example, media reports of neurophysiologists discovering that as well as C-fibres, normal human beings have D-fibres. Since these are distributed about our bodies much as C-fibres are, it is readily understandable how it came to be thought that pain is C-fibre [excitation]. Analysis reveals, however, that whilst most pain-killing drugs reduce *both* C- and D-fibre activity, some newly developed ones reduce only the latter and are everywhere as effective as the usual drugs in reducing pain. Perhaps we go on to imagine reading newspaper articles reporting massive investment by the big pharmaceutical companies in research programmes aimed at developing newer and even better D-fibre 'dullers,' and less well-funded investigations into the (now obscure) function of C-fibres, etc.[230]

But this train of thought would seem to be point-missing. The question is not whether we can in fact easily imagine what things might be like if pain were believed to be other than C-fibre excitation (even if it is essentially C-fibre excitation) but whether the characterization of what we are doing when we do that, outlined as by Hale, does justice to what we actually, as it seems to us lucidly and in detail, imagine when we imagine *pain without C-fibre excitation*. The fact is that it surely does not.

I envisage a protest. Surely, it may be suggested, this is to take Hale's proposals the wrong way round: they will fare better for the physicalist's purpose if we focus not on pain but on the second term in the putative identity, viz. C-fibre excitation (or whatever, parametrically, the physical basis of pain is taken to be). Let us grant the

---

[230] Chapter 10, this volume, p. 267.

contrasts between having misleading evidence that one is in pain and really being in pain, or between merely believing that one is pain and really being in pain, struggle for any genuine phenomenological content. Still, the ideas of having misleading evidence of the absence of C-fibre excitation, as opposed to C-fibre excitation genuinely being absent, or of merely believing that there is no occurrent C-fibre excitation, as opposed to there really being none, do not, it is plausible, present the same difficulties. And if they do not, cannot Hale's models be harnessed after all to help the physicalist account for an illusion of possibility in the relevant case? What, the physicalist may argue, is conceivable, is not pain without C-fibre excitation but pain in the presence of misleading evidence that there is no concurrent C-fibre excitation, or pain in circumstances where it is generally believed that there is no concurrent C-fibre excitation. Are not these 'alibis' as good as, though different in content to, the play with the idea of an epistemic counterpart marshalled in Kripke's original defence of the necessary identity of water with H_2O?

I don't think this improves matters. There is a dilemma. Is it or is it not to be granted that there is a difference in general between conceiving of an absence of C-fibre excitation in one's occurrent neuro-physical make-up and conceiving of a situation where all the evidence misleadingly suggests such an absence or where it is at any rate accepted by all concerned that there is such an absence? If these differences are conceivable, then—however exactly the difference is accounted for—surely there can be no obstacle to one's conceiving of the former state, rather than one of Hale's surrogates. And then, since such a conception will presumably be entirely non-committal as far as one's occurrent phenomenological state is concerned, there should be no obstacle to adjoining to it in imagination the felt experience of pain. Just that was exactly the nub of Kripke's original argument. But if we say instead that the distinctions in question *cannot* feature in the content of an episode of conceiving—that while, of course, perfectly *intelligible*, they are, strictly, *inconceivable* distinctions—then, sure, pain without C-fibre excitation will now be inconceivable but so will be pain *with* C-fibre excitation and pain, or anything else, with, or without, any of the range of circumstances which have now implicitly been pushed beyond conceivability. I won't here speculate by what principle that range might be determined, but one would suppose that once the distinction between C-fibre excitation and misleading evidence of C-fibre excitation is held to be ineligible to feature within the content of any relevant conceiving, the extent of the parallel casualties thereby entrained will be very wide indeed—wide enough to put a huge sweep of metaphysical possibility claims beyond the reach of corroboration by lucid and detailed conceiving. The cost of protecting the necessity of the putative identity of pain with C-fibre excitation in this way is thus utterly to emasculate the CCP.[231]

There is one more move to consider. We may continue to allow that the contrast between conceiving of an absence of occurrent C-fibre excitation and conceiving of a situation in which we merely have misleading evidence of such an absence, for example, can still be active in the description of the content of a subject's conceivings,

---

[231] See Wright (2002: 413–17).

but that the distinction is *phenomenologically* underdetermined: that which charac-terization is to be preferred will turn, rather, on the modal status of the proposition whose possibility, when the episode of conceiving is described in in the former way, it would, via the CCP, support. So if pain is necessarily C-fibre excitation, then the characterization of the relevant episode of conceiving as being of pain without C-fibre excitation will be illicit. But if pain is not necessarily C-fibre excitation, then that characterization can stand—even though subjectively the episode in question may impress as exactly the same.

I won't need to dwell on the obvious methodological difficulty with this. The CCP can have no part to play in an interesting epistemology of the modal if whether anyone counts as entertaining an adequately lucid and detailed conception of circumstances in which P would hold is a question that cannot in general be answered positively without an independent assessment of P's modal status; if the distinction, that is, between a genuine and a merely apparent counter-conception cannot in general be drawn purely by reference to the actual detail of the conceivings concerned. In general, if the CCP is to hold as a *basis for judgement*, then warrant for a conceivability claim has to be available independently of collateral assessment of the actual modality of the proposition concerned. The distinction between genuinely conceiving [not-P] and merely conceiving [P^*], somehow confused with [not-P], has to be available, and operational, prior to any verdict on the modal status of P.

4. A Different Response

So let's backtrack. It is high time to pay attention to our neglected Elephant in the Room. Let us note, to begin with, that the force of the SuperWright and Sceptical Mathematician examples does not depend on their being presented under the *per impossibile* template—that is, as cases where a lucid and relevantly detailed concep-tion succeeds in being a conception not of the truth of P but—what need not amount to that—only instead of how things would be in certain respects if P, *per impossibile*, were true. The issues about the status of counterfactual conditionals with impossible antecedents raised by so presenting the examples may accordingly be sidestepped. The challenge to Kripke's argument that they pose will hardly be diminished if they may be taken to illustrate how it can be *prima facie* possible to conceive in a relevantly lucid and detailed manner of the *actual* obtaining of what are, in meta-physical fact, metaphysical impossibilities. Both the SuperWright and Sceptical Mathematician examples can be easily addressed to this purpose: *re* the former, it doesn't detract from the credentials of the relevant play in the imagination if I am deluded enough to suppose that what I imagine about my origins and early history is *actually* true; *re* the latter, it was already a feature of the example that the Sceptical Mathematician does exactly that. Provided such *prima facie* conceivings resist defla-tion by the Kripke-style of explanation in terms of epistemic counterparts, they will present an unanswered challenge both to the CCP and to its annexure to the anti-physicalist argument.

To receive the examples in this way is, of course, to take the *radical* option distinguished earlier:[232] it is to regard them as so far undefeated counter-examples to the CCP, and hence to be prepared to allow that conceivability itself can be a source of modal illusion in a fashion contrasting with any model that postulates that in illusory cases our conceivings miss their mark—that we conceive of something *other than* the truth of the targeted proposition. I am suggesting, therefore, that we should abandon the CCP-conservative approach to modal illusion shared by the epistemic counterpart, *per impossibile*, Misleading Evidence and Doxastic Consequences models and take seriously the salient doubts, voiced earlier, about the standing of the CCP itself. Conservative approaches attempt to maintain that lucid and detailed conceivability—*real* lucid and detailed conceivability—suffices for meta-physical possibility. So where *P* is impossible yet we have what appears to be a lucid and detailed conception of *P*'s obtaining, that conception must either be defective in some way or be a perfectly good conception of a surrogate scenario which we somehow mistake for the obtaining of *P*. Yet if conceivability is to supply a sure criterion for metaphysical possibility, this requires that in any case where an appar-ently lucid and detailed conception seems to be available of the obtaining of what is in fact an impossibility, we are guaranteed to be able *either* to find relevant fault with its credentials to be lucid and relevantly detailed *or* to make a good case that it deals with a surrogate. I suggest that, even over the short space of the preceding discussion, a serious question has emerged whether we can in general redeem any such guarantee. No: when it is metaphysical possibility that is at issue, why not just recognize that some metaphysical impossibilities may be perfectly lucidly conceivable—precisely because the impossibilities concerned are not grounded in the first place (purely) in our *concepts* of the events, states, or stuffs, etc., concerned?[233]

We noted the outstanding objection to this move: that to abandon (or in any way significantly qualify) the CCP is to risk leaving ourselves with no effective epistem-ology of metaphysical possibility—no means for properly controlling judgements of metaphysical possibility save when they are grounded in actuality. This, though, should impress as an overreaction. For one thing, provided we retain some form of effective control on ascriptions of metaphysical *necessity* (and thereby on impossi-bility = the necessity of a contradictory), it is not clear that we anyway also need an effective epistemology of possibility. Rather, the metaphysical possibility of *P* can be treated as a default status, defeasible by grounds for the necessity of some *Q* that is inconsistent with *P*.[234] But, for another thing, it is in any case still open to us to treat lucid and detailed conceivings as providing positive though defeasible grounds for possibility claims. What is required is only that we acknowledge that they can misfire in either of two general ways:

(i)   First, as grounds for *conceptual possibility* claims, they can misfire if they suggest the possibility of something that is indeed conceptually impossible but for reasons sufficiently subtle to lie well beyond anything whose recognition is

---

[232] Chapter 10, this volume, p. 423.
[233] Kung (2010 and 2014) is another who is supportive of this direction.
[234] Divers (2002); Hale (2003, 2013).

required by an ordinary competent, reflective understanding of the concepts concerned. (The Sceptical Mathematician.)

(ii) As grounds for *metaphysical possibility* claims, lucid and detailed conceivings can misfire if they work with concepts that misrepresent,[235] or are silent on, aspects of the essential nature of the objects they concern. Misrepresentation may result in the apparent exclusion of genuine possibilities; silence in the recognition of spurious possibilities. The latter case gets us something close in spirit to Kripke's diagnosis of the water = H_2O case, even though our outlook now on the CCP is radical rather than conservative; for we are enabled to diagnose the case not indeed as involving imaginative play with an epistemic counterpart of water but as working with an *incomplete* concept of water. Simply, the ordinary concept of water is silent on the essential nature of water. So it allows for the detailed and lucid imaginability and thereby the conceptual possibility of scenarios in which water is other than H_2O. The conclusion that so much is metaphysically possible would indeed be licensed if it could be assumed that the concepts exercised in the imaginative scenario were fully indicative of the essential nature of the stuff. However, they aren't. There is no need to say that what is conceived is merely that an epistemic counterpart of water is not H_2O. What is clearly and lucidly conceived is, rather, the conceptual possibility of a metaphysical impossibility.

Crucially, however, unlike the diagnosis in terms of imaginative play with mere epistemic counterparts, the same form of debunking explanation may be applied smoothly to SuperWright. One who subscribes to the (widely accepted) essential animal nature of human selves, and the necessity of our actual biological origins, should insist that the concept of the self expressed by "I" that is active in the flights of fancy of the kind concerned is simply silent on these essential matters. It therefore does indeed generously allow of conceptual possibilities that transcend the metaphysical possibilities. The fancy of being born of other parents, etc., is such a possibility, lucidly imaginable in detail without conceptual incoherence, but metaphysically impossible nonetheless.

Coda

So, what of Kripke's anti-physicalist argument? It should now be clear what a physicalist must say in response. She must counter that any purely phenomenal concept of pain is similarly inadequate to the essential nature of pain. Pain, for physicalism, must have a *dual essence*, with ineliminable phenomenal and physical components.[236] The apparent conceptual possibilities opened up by concentration on just one of these are or can be genuine conceptual possibilities, imaginable lucidly and in detail, and are properly described as realizations in imagination of conceptual

---

[235] I am assuming that our concept of something may misrepresent it in certain details yet still succeed in referring to it, that is, succeed in being *of* the item concerned.

[236] This can be a stance available for type–type physicalism, though the physicalist may do well to leave room for a variety of essential physical types of pain, some perhaps species-wide (but maybe no wider), others tied to varieties in our phenomenal feels.

possibilities for pain. But the conceptual possibilities for pain are a broader class than the metaphysical possibilities, and such flights of fancy bear on the latter only given the—for the physicalist—erroneous assumption that the phenomenal concept of pain is adequate to its essence.

"But how can pain have a *dual essence* of this kind?" Well, that was a conundrum implicit in physicalism from the start. In effect, it is just the form in which the so-called Hard Problem of Consciousness presents itself to physicalism. It is, in one shape or another, a problem for everyone.

References

Berto, Francesco (2014). "On Conceiving the Inconsistent," *Proceedings of the Aristotelian Society* 114 (1), 103–21.

Berto, Francesco, and Tom Schoonen (2017). "Conceivability and Possibility: Some Dilemmas for Humeans," *Synthese* 195 (6), 1–19.

Brogaard, Berit, and Joe Salerno (2013). "Remarks on Counterpossibles," *Synthese* 190 (4), 639–60.

Divers, John (2002). *Possible Worlds*, London: Routledge.

Hale, Bob (2003). "Knowledge of Possibility and of Necessity," *Proceedings of the Aristotelian Society* 103 (1), 1–20.

Hale, Bob (2013). *Necessary Beings: An Essay on Ontology, Modality, and the Relations between Them*, Oxford: Oxford University Press.

Hume, David (2003). *A Treatise of Human Nature*, Mineola, NY: Dover Publications.

Kment, Boris (2006a). "Counterfactuals and Explanation," *Mind* 115 (458), 261–310.

Kment, Boris (2006b). "Counterfactuals and the Analysis of Necessity*," *Philosophical Perspectives* 20 (1), 237–302.

Kripke, Saul (1980). *Naming and Necessity*, Cambridge, MA: Harvard University Press.

Kung, Peter (2010). "Imagining as a Guide to Possibility," *Philosophy and Phenomenological Research* 81 (3), 620–63.

Kung, Peter (2014). "You Really Do Imagine It: Against Error Theories of Imagination," *Noûs* 50 (1), 90–120.

Nolan, Daniel (1997). "Impossible Worlds: A Modest Approach," *Notre Dame Journal of Formal Logic* 38 (4), 535–72.

Nolan, Daniel (2013). "Impossible Worlds," *Philosophy Compass* 8 (4), 360–72.

Wright, Crispin (2002). "The Conceivability of Naturalism," in Tamar Szabo Gendler and John Hawthorne (eds.), *Conceivability and Possibility*, Oxford: Oxford University Press, 401–39.

Index Nominum

Index